AF616457

ADVANCES IN RAPID THERMAL PROCESSING

Proceedings of the Symposium

Editors

Fred Roozeboom
Philips Research
Eindhoven, The Netherlands

Jeffrey C. Gelpey
STEAG RTP Systems
Peabody, Massachusetts, USA

Mehmet C. Öztürk
North Carolina State University
Raleigh, North Carolina, USA

Jim Nakos
IBM Microelectronics
Essex Junction, Vermont, USA

ELECTRONICS, DIELECTRIC SCIENCE AND TECHNOLOGY, AND HIGH TEMPERATURE MATERIALS DIVISIONS

Proceedings Volume 99-10

THE ELECTROCHEMICAL SOCIETY, INC.,
10 South Main St., Pennington, NJ 08534-2896, USA

Published by:

The Electrochemical Society, Inc.
10 South Main Street
Pennington, New Jersey 08534-2896, USA

Telephone 609.737.1902
Fax 609.737.2743
e-mail: ecs@electrochem.org
Web: http://www.electrochem.org

Library of Congress Catalogue Number: 99-66173

ISBN 1-56677-232-X

Printed in the United States of America

PREFACE

The international symposium on Advances in Rapid Thermal Processing (RTP) was held from May 3 through May 6, 1999 in Seattle, Washington as part of the 195th Meeting of the Electrochemical Society, Inc. For the first time this annual symposium was held under the auspices of the ECS, with 48 papers presented and session attendance reaching 150.
The purpose of this symposium, as that of its preceding events was to provide an international forum for those working in the field of Rapid Thermal Processing to discuss all aspects of RTP. The program addressed recent innovations in RTP equipment issues as well as RTP processes and their applications in the fabrication of advanced semiconductor microelectronics and other devices.

This book contains all refereed papers presented at the symposium. From the contents one can conclude that the focus in RTP has shifted over the past years from mainly instrumental issues (problems of temperature control, uniformity and reproducibility) to issues concerning pushing the limits of ultrathin gate dielectrics and ultrashallow junctions in MOSFET gate stack engineering and MOSFET source/drain engineering. This is clearly caused by the fact that modern RTP equipment has far more reliable temperature measurement and control to produce today's devices with 0.18 μm channel length, and below.
Of course the improvement of equipment and the development of new equipment still continue. For example a new, interesting and revolutionary new floating wafer reactor concept was presented, where by means of heating in the hot-wall reactor wafers are gas-levitated.
The symposium illustrated that the range of applications is still growing, but also that new challenges appear, such as Transient Enhanced Diffusion which requires higher temperature ramp rates than currently used and possible, to manufacture ultrashallow junctions in future devices.

A panel discussion was held on the key barriers emerging from this symposium, and this discussion drew a large audience. Our thanks are due to the panellists G. Bai (Intel), D.-L. Kwong (University of Texas at Austin), T.P. Ma (Yale University), G. Lucovsky (North Carolina State University) and H.Huff (Sematech). Panel moderator was R. Thakur (STEAG RTP Systems). The main issue here was the perspective of advanced dielectric and metallization materials, and the (short-time) processing technologies needed in the future for sub-0.1 μm channel length.

The work presented here includes comprehensive reviews as well as invited and contributed papers. The book contains seven major sections: 1) ultrathin gate dielectrics, 2) ultrashallow junctions, 3) metals and silicides, 4) contacts, 5) novel applications, 6) RTCVD and Epitaxy of Si and SiGe and 7) equipment & temperature issues and modeling
Section 5 also includes the Dielectric Science and Technology Callinan Award Address, given by A. Rohatgi of Georgia Institute of Technology.

We anticipate that this volume will provide scientists and technologists working in the field of RTP with up-to-date knowledge, as well as understanding of the directions in which the key issues involved are evolving.

The editors are grateful to all authors for the in-depth treatment of their fields of expertise, and for sharing their recent findings. Our sincere appreciation goes to those individuals who co-chaired sessions, and reviewed the papers. We would like to acknowledge the financial support of our sponsors, being three ECS Divisions (Electronics Division, Dielectric Science and Technology Division and High Temperature Materials Division) and AG Associates, Applied Materials, ASM America, CVC Products, Eaton Thermal Processing Systems, J.I.P.ELEC, Luxtron, Mattson Technology, SensArray Corporation, STEAG RTP Systems.

Fred Roozeboom
Jeff C. Gelpey
Mehmet C. Öztürk
Jim Nakos

June 1999

SYMPOSIUM ON
ADVANCES IN RAPID THERMAL PROCESSING

Symposium Co-Chairmen:
F. Roozeboom, Philips Research, Eindhoven, The Netherlands
J.C. Gelpey, STEAG RTP Systems, Peabody, MA,USA
M.C. Öztürk, North Carolina State University, Raleigh, NC, USA
J. Nakos, IBM Microelectronics, Essex Junction, VT, USA

Monday, May 3, 1999

MOSFET GATE STACK ENGINEERING: ULTRATHIN GATE DIELECTRICS
Session Co-Chairman: J. Nakos, IBM Microelectronics, Essex Junction, VT, USA
Session Co-Chairman: B. Froeschle, STEAG RTP Systems, Dornstadt, Germany

MOSFET GATE STACK ENGINEERING:
ADVANCED GATE DIELECTRICS FOR NANOSCALE MOSFETs
Session Co-Chairman: M.C. Öztürk, North Carolina State University, Raleigh, NC, USA
Session Co-Chairman: R.P.S. Thakur, STEAG RTP Systems, San Jose, CA, USA

PANEL DISCUSSION
Session Co-Chairman: R.P.S. Thakur, STEAG RTP Systems, San Jose, CA, USA

Tuesday, May 4, 1999

MOSFET Gate Stack Engineering: ULTRATHIN GATE DIELECTRICS
Session Co-Chairman: M.C. Öztürk, North Carolina State University, Raleigh, NC, USA
Session Co-Chairman: D.-L. Kwong, University of Texas at Austin, TX, USA

MOSFET SOURCE/DRAIN ENGINEERING: ULTRASHALLOW JUNCTIONS
Session Co-Chairman: W. Lerch, STEAG RTP Systems, Dornstadt, Germany
Session Co-Chairman: S. Felch, Varian Semiconductor Equipment, Gloucester, MA, USA

Session Co-Chairman: A.Fiory, Lucent / Bell Laboratories, Murray Hill, NJ, USA
Session Co-Chairman: S. Shishigushi, NEC Corporation, Sagamihara, Japan

MOSFET SOURCE/DRAIN ENGINEERING: METALS AND SILICIDES
Session Co-Chairman: P. Agnello, IBM Microelectronics, Hopewell Junction, NY, USA
Session Co-Chairman: J. Nakos, IBM Microelectronics, Essex Junction, VT, USA

Wednesday, May 5, 1999

MOSFET SOURCE/DRAIN ENGINEERING: CONTACTS
Session Co-Chairman: M.C. Öztürk, North Carolina State University, Raleigh, NC, USA
Session Co-Chairman: J. Nakos, IBM Microelectronics, Essex Junction, VT, USA

NOVEL APPLICATIONS
Session Co-Chairman: F. Roozeboom, Philips Research, Eindhoven, The Netherlands
Session Co-Chairman: R.P.S. Thakur, STEAG RTP Systems, San Jose, CA, USA

RTCVD AND EPITAXY OF Si AND SiGe
Session Co-Chairman: I. Raaijmakers, ASM America, Phoenix, AZ, USA
Session Co-Chairman: F. Roozeboom, Philips Research, Eindhoven, The Netherlands

Thursday, May 6, 1999

EQUIPMENT & TEMPERATURE ISSUES AND MODELING
Session Co-Chairman: J. Hebb, Eaton Thermal Processing Systems, Peabody, MA, USA
Session Co-Chairman: J.C. Gelpey, STEAG RTP Systems, Peabody, MA, USA

TABLE OF CONTENTS

Preface ix
Conference Organization xi

Section I 1
MOSFET Gate Stack Engineering: Ultrathin Gate Dielectrics

1.* Enabling Single-Wafer Process Technologies for Reliable Ultra-Thin Gate Dielectrics
G. Miner, G. Xing, H.S. Joo, E. Sanchez, Y.Yokota, C. Chen, D. Lopes, and A. Balakrishna 3

2. Evaluation of Ultra-Thin Gate Oxides using Different Ambients in a Rapid Thermal Processing System
Y.B. Jia, J.Y. Choi, J. Schuur, J.H. Das, R. Sharangpani, and R.P.S. Thakur 15

3. Dilute Steam Rapid Thermal Oxidation for 30 Å Gate Oxides
K.G. Reid, H. Tseng, R. Hegde, G. Miner, and G. Xing 23

4. Preparation of Ultra-Thin Gate Oxides with annealing in Nitric Oxide
B. Froeschle, N. Sacher, and F. Glowacki 31

5.* High-k Gate Stack for sub-0.1 μm CMOS Technology
G. Bai 39

6.* Recent Developments in Ultrathin Nitride Gate Stack Prepared by In-Situ RTP Multiprocessing for CMOS ULSI
S.C. Song, B.Y. Kim, H.F. Luan, D.-L. Kwong, M. Gardner, J. Fulford, D. Wristers, J. Gelpey, and S. Marcus 45

7.* Advanced Gate Dielectrics Synthesized by JVD
T.P. Ma 57

8.* Integration of Alternative High-k Gate Dielectrics into Aggressively Scaled CMOS Si Devices: Chemical Bonding Constraints at Si-Dielectric Interfaces
G. Lucovsky 69

9. Evaluation of Ultra-Thin Gate Evaluation of Ultra-Thin Gate Stack Dielectrics for 0.1 μm PMOSFETs
A. Srivastava, C.M. Osburn, K.F. Yee, H.H. Heinisch, E.M. Vogel, K.Z. Ahmed, Z. Wang, K. Min, B. Timberlake, C. Parker, J.J. Wortman, and J.R. Hauser 81

10. Growth Kinetics and Modeling of Direct Oxynitride Growth with NO-O2 Gas Mixtures
R. Sharangpani, S.P. Tay, R. Thakur, S. Everist, J. Nelson, and P.M. Smith 89

11. Interfacial Properties of Si-Si_3N_4 Formed by Remote Plasma and Rapid Thermal Processing
H. Lazar, V. Misra, Z. Wang, M. Mulkarni, W. Li, M. Mahler, and J.R. Hauser 95

* *Invited paper*

Section II 103

MOSFET Source/Drain Engineering: Ultrashallow Junctions

12.* Shallow Junction Formation by Low Energy Implant and High Ramp-Up Rate RTA Process
S. Shishiguchi, A. Mineji, T.Y. Matsuda and H. Kitajima 105

13. Spike Anneals in RTP : Kinetic Analysis
E.G. Seebauer 117

14. Transient Enhanced Diffusion and Ostwald Ripening of Ion-Implantation Generated Defects in Silicon
N.E.B. Cowern, G. Mannino, F. Roozeboom, P.A. Stolk, H.G.A. Huizing, J.G.M. van Berkum, N.N. Toan, P.H. Woerlee, F. Cristiano, and A. Claverie 125

15. Electrical Measurements of Annealed Boron Implants for Shallow Junctions
A.T. Fiory, K.K. Bourdelle, M.E. Lefrancois, D.M. Camm, and A. Agarwal 133

16. Influence of Thermal Nitridation on the Diffusion of Arsenic during Rapid Thermal Annealing
W. Lerch, N.A. Stolwijk, S.D. Marcus, D.F. Downey, and M. Schäfer 141

17.* Doping and Annealing Requirements to Satisfy the 100 nm Technology Node
D.F. Downey, S.B. Felch, and S.W. Falk 151

18. Direct Correlation Between Defects and Thermal Stress in Rapid Thermal Processing
V. Parihar, S. Venkataraman, R. Singh, K.F. Poole, and R.P.S. Thakur 163

19. Application of Excimer Laser Annealing in the Formation of Implanted Shallow Junctions
L.K. Nanver, E.J.G. Goudena, Q.W. Ren, M. van de Berg, R. Mallee, and J. Slabbekoorn 171

20. Ultra-Shallow P+ -N Junctions for 50-70 nm CMOS Using Selectively Grown In-Situ Boron-Doped Silicon Films
I. Ban and M.C. Őztűrk 179

21. Shallow Junction Fabrication by Rapid Thermal Outdiffusion from Implanted Oxide
J. Schmitz, M. van Gestel, P.A. Stolk, Y.V. Ponomarev, F. Roozeboom, F.N. Cubaynes, J.G.M. van Berkum, W.M. van de Wijgert, P.C. Zalm, and P.H. Woerlee 187

Section III 195

MOSFET Source/Drain Engineering: Metals and Silicides

22.* RTP for Advanced Device Fabrication using Shallow, Elevated, and Silicided Junctions
C.M. Osburn 197

23. Pre-Deposition Treatments for Selective Rapid Thermal Chemical Vapor Deposition of $TiSi_2$ on Arsenic-Implanted Silicon Substrates
H. Fang, M.C. Őztűrk, P.A. O'Neil, and E. Seebauer 207

* *Invited paper*

Section IV 215
MOSFET Source/Drain Engineering: Contacts

24.* Junction Perimeter Leakage Considerations for the Integration of $CoSi_2$ and Damascene W Local Interconnect in Dynamic Logic Compatible, Sub-0.25 μm CMOS Technologies
P.D. Agnello 217

25. Influence of Rapid Thermal Ramp Rate on Phase Transformation of Titanium Silicides
Y.Z. Hu, S.P. Tay, J. Yang, R. Thakur, P.M. Smith, and G.Bailey 229

26.* Low Resistivity Contacts to Ultra-Shallow Junctions in ULSI Devices
L.J. Chen, S.L. Cheng, and L.W. Cheng 237

27. Attainment of Low Resistivity Polycide Films Using Rapid Thermal Annealing
H.A. Yoon, C. Chen, A. Singhal, D. Lopes, G. Miner, S. Hong, M. Yamazaki, and Y. Maeda 249

28. Thermal Stability Improvement of Cobalt Disilicide Thin Films on (001)Si by High Temperature Sputtering Deposition
H.Y. Huang, L.J. Chen, W.F. Wu, and R.P. Yang 257

Section V 263
Novel Applications

29.[+] Rapid Thermal Processing of High Performance Dielectrics and Silicon Solar Cells
A. Rohatgi 265

30. Poly-Si_{1-x} Ge_x Process Integration for Low Resistance Gate CMOS Technology
H. Takeuchi and T.-J. King 277

31. Scanning Rapid Thermal Annealing Process for Low Temperature Poly-Silicon Thin Film Transistors
T.-K. Kim, G.-B. Kim, Y.-G. Yoon, C.-H. Kim, B.-I. Lee, and S.-K. Joo 285

32. Enhanced Mobility in Buried SiGe Channel PMOS Fabricated using Rapid Thermal Processing
D.J. Tweet and S.T. Hsu 291

33. Strain Relaxation of Si/ $Si_{1-x-y}Ge_xC_y$/Si Quantum Wells Grown by RTCVD
M.H. Lee, Y.D. Tseng, C.W. Liu and M.Y. Chern 299

Section VI 307
RTCVD and Epitaxy of Si and SiGe

34.* Emissivity Effects in Low-Temperature Epitaxial Growth of Si and SiGe
W.B. de Boer and D. Terpstra 309

35. Suppressed Phosphorus Autodoping in Silicon Epitaxy for Ultrasharp Phosphorus Profiles by Low Temperature Rapid Thermal Chemical Vapor Deposition
M. Carroll, M. Yang, and J.C. Sturm 319

[+] *Dielectric Science and Technology Callinan Award Address; * Invited paper*

36. Comparative Study of Crystallinity and Surface Roughness of RTCVD versus LPCVD Deposited Polysilicon
J.W.H. Maes, C. Pomarede, M. Mansoori, C.W. Werkhoven, and I.J. Raaijmakers 327

37. As Peaks in Si (100) Films Fabricated with Rapid Thermal Epitaxy
W.D. van Noort, L.K. Nanver, C.C.G. Visser, A. van de Boogaard, and J.W. Slotboom 335

Section VII 343
Equipment & Temperature Issues and Modeling

38.* Critical Considerations and Integration Issues in the Design of an RTP System
A. Gat, Z. Koren, P.J. Timans, and R.P.S. Thakur 345

39.* Temperature Calibration in Microelectronic Manufacturing
P. Vandenabeele and W. Renken 359

40. Passive and Active Pyrometry in RTP and RTCVD Systems
E.D. Glazman, A.E. Glazman, Z. Atzmon, H. Gilboa, E. Iskevitch, and A. Thon 371

41. Temperature Measurement, Uniformity, and Control in a Furnace-Based Rapid Thermal Processing System
J. Hebb and A. Shajii 375

42. Emissivity Compensated Wafer Temperature Measurement Using Intensity-Modulated Lamp Light
M. Hauf, H. Balthasar, C. Merkl, S. M□ller, and C. Striebel 383

43. Floating Wafer Reactor: RTP Based on Thermal Conductive Heat Transfer
V.I. Kuznetsov, S. Radelaar, and E.A.H. Granneman 391

44. Dynamic Uniformity Control in a Rapid Thermal Processing System
K.S. Balakrishnan, S. Shooshtarian, N. Acharya, P.J. Timans, and R.P.S. Thakur 399

45.* A Novel Full-Quartz Open Cluster Platform for Advanced Rapid Thermal Processing

R. Bremensdorfer, H. Walk, E. Merz, and S. Paul 407

46. Modeling Chamber Radiation Effects on Radiometric Temperature Measurement in Rapid Thermal Processing
F. Rosa, Y.H. Zhou, Z.M. Zhang, D.P. DeWitt, and B.K. Tsai 419

47. Emissivity of Bare and Coated Si Wafers: Theoretical Studies
B. Sopori, W. Chen, Y. Zhang, J. Madjdpour, and N.M. Ravindra 427

48. An Advanced Radiation Model for Thermal Processing of Wafers
S. Mazumder and A. Kersch 435

* *Invited paper*

Section VIII 443
Author Index and Key Word Index

Author Index 445

Key Word Index 449

FACTS ABOUT THE ELECTROCHEMICAL SOCIETY, INC.

The Electrochemical Society, Inc., is an international, nonprofit, scientific, educational organization founded for the advancement of the theory and practice of electrochemistry, electrothermics, electronics, and allied subjects. The Society was founded in Philadelphia in 1902 and incorporated in 1930. There are currently over 7,000 scientists and engineers from more than 70 countries who hold individual membership; the Society is also supported by more than 100 corporations through Contributing Memberships.

The Technical activities of the Society are carried on by Divisions and Groups. Local Sections of the Society have been organized in a number of cities and regions. Major international meetings of the Society are held in the Spring and Fall of each year. At these meetings, the Divisions and Groups hold general sessions and sponsor symposia on specialized subjects.

The Society has an active publications program which includes the following:

Journal of The Electrochemical Society - The *Journal* is a monthly publication containing technical papers covering basic research and technology of interest in the areas of concern to the Society. Papers submitted for publication are subjected to careful evaluation and review by authorities in the field before acceptance, and high standards are maintained for the technical content of the *Journal*.

Electrochemical and Solid-State Letters - *Letters* is the Society's rapid-publication, electronic journal. Papers are published as available at http://www3.electrochem.org/letters.html. This peer-reviewed journal covers the leading edge in research and development in all fields of interest to ECS. It is a joint publication of the ECS and the IEEE Electron Devices Society.

Interface - *Interface* is a quarterly publication containing news, reviews, advertisements, and articles on technical matters of interest to Society Members in a lively, casual format. Also featured in each issue are special pages dedicated to serving the interests of the Society and allowing better communication among Divisions, Groups, and Local Sections.

Meeting Abstracts *(formerly Extended Abstracts)* - Meeting Abstracts of the technical papers presented at the Spring and Fall Meetings of the Society are published in serialized softbound volumes.

Proceedings Series - Papers presented in symposia at Society and Topical Meetings are published as serialized Proceedings Volumes. These provide up-to-date views of specialized topics and frequently offer comprehensive treatment of rapidly developing areas.

Monograph Volumes - The Society sponsors the publication of hardbound Monograph Volumes, which provide authoritative accounts of specific topics in electrochemistry, solid-state science, and related disciplines.

For more information on these and other Society activities, visit the ECS Web site:

http://www.electrochem.org

Section I

MOSFET Gate Stack Engineering: Ultrathin Gate Dielectrics

ENABLING SINGLE-WAFER PROCESS TECHNOLOGIES FOR RELIABLE ULTRA-THIN GATE DIELECTRICS

Gary Miner, Guangcai Xing, Hyun Sung Joo, Errol Sanchez,
Yoshitaka Yokota, Chiliang Chen, Dave Lopes and Ajit Balakrishna
Applied Materials, Santa Clara, CA 95054, USA

Continued scaling of MOS gate dielectrics challenges the limits of current process and equipment technologies. The extendibility of SiO_2 dielectrics is limited by the exponential increase in direct tunneling gate current and the decreased reliability as well as the process control of ultra-thin films. Higher dielectric constant materials will soon be required to address these challenges. In this paper we present data suggesting single-wafer process technologies can both extend the life of SiO_2-based dielectrics and address the inevitable change to higher dielectric constant materials.

INTRODUCTION

The advancement of device capability and speed requires the continued aggressive scaling of the gate dielectric. Table 1 shows the 1997 Semiconductor Industry Association (SIA) targets for gate dielectric equivalent oxide thickness ($t_{ox,eq}$) as a function of time and device technology generation. Also shown are the Advanced $t_{ox,eq}$ requirements driven by the most advanced device manufacturers, which tend to lead this SIA roadmap by a full generation.

Table 1. SIA Roadmap for Gate Dielectric

Calendar year	1997	1999	2001	2003	2006
Technology (μm)	0.25	0.18	0.15	0.13	0.10
Equivalent oxide thickness (nm)	4.0-5.0	3.0-4.0	2.0-3.0	2.0-3.0	1.5-2.0
Advanced equiv. ox. thickness (nm)	3.0-4.0	2.5-3.5	1.5-2.5	1.0-1.5	< 1.0

This roadmap presents difficult challenges to gate dielectric performance, most significantly for reliability and tunneling leakage. Recent research has shown fundamental reliability may limit silicon dioxide (SiO_2)-based dielectrics to 2.6 ± 0.15 nm, as shown in Fig. 1 [1]. This indicates a new or fundamentally more reliable dielectric will be required, otherwise the scaling targets for gate thickness and operating voltage may not be met. This projection was based on oxides grown in a conventional batch furnace, and was independent of oxidation ambient for oxygen (O_2), nitrous oxide (N_2O) and nitric oxide (NO) gases. However, we will show that certain growth conditions in a single-wafer processing chamber differ from batch furnaces and consistently produce more reliable oxides for the same thickness. This may enable the extended use of SiO_2 below the 2.6 nm limit.

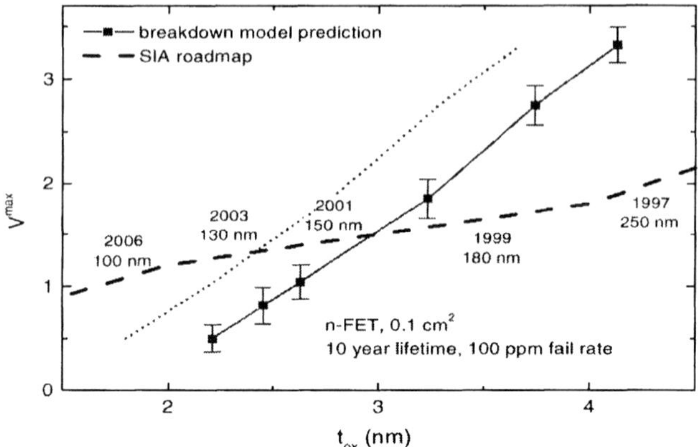

Figure 1. Reliability limit projection for SiO_2 [1]

Leakage is also a limiting factor in gate dielectric scaling. Figure 2 shows gate leakage as a function of thickness [1]. Based on a maximum tolerable leakage current of 1 A/cm^2, the minimum thickness is 2.1 nm, again indicating scaling limits will be reached at the 0.13 μm technology node.

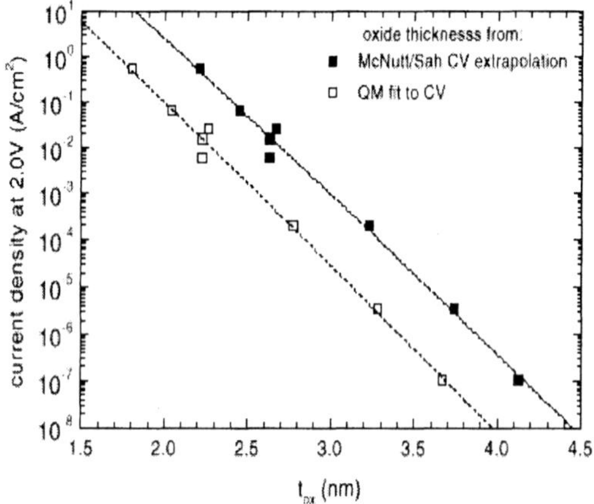

Figure 2. Gate leakage versus thickness [1]

Single-wafer technologies also offer improvements in ultra-thin film process control and flexibility for integration of multiple process steps to address more advanced dielectric solutions. Multi-step gate processes will become prevalent with stacked or graded dielectrics utilizing higher dielectric constant materials such as silicon nitride.

This paper will focus on 0.18 and 0.13 μm gate dielectric solutions in the 1.5-3.0 nm range processed in the Applied Materials Centura™ cluster. The performance improvements shown will bring about the rapid adoption of single-wafer cluster tools as the preferred production technology for 0.13 μm and beyond.

In-Situ Steam Generation (ISSG)

In-Situ Steam Generation (ISSG) is a unique process for performing steam oxidation. Typically batch furnaces use an external pyrogenic torch to produce steam. In the external torch configuration, hydrogen (H_2) and oxygen (O_2) are combusted at atmospheric pressure in proximity to a hot element or in a hot-wall reaction chamber which ignites the reaction producing steam (H_2O). The resulting steam is then introduced into the hot-wall furnace tube to oxidize the wafers.

ISSG is a low pressure process whereby H_2 and O_2 are introduced to the process chamber directly, without precombustion. The process is performed in the RTP Centura™, a cold wall, rapid thermal processor. Process gas flows across a rotating wafer heated by tungsten-halogen lamps. The hot wafer is the ignition source and the reaction between H_2 and O_2 occurs at the wafer surface. The process is kept at low pressures, near 10 Torr, to ensure safety and to enable the unique reaction kinetics of this process.

The primary gas reactions in this low pressure process are [3]:

$$H_2 + O_2 \rightarrow 2OH \quad (1)$$
$$H_2 + OH \rightarrow H_2O + H \quad (2)$$
$$O_2 + H \rightarrow OH + O \quad (3)$$
$$H_2 + O \rightarrow OH + H \quad (4)$$

Using a model of the RTP Centura including temperature and gas flow dynamics, the primary species at the wafer are found to be O_2, H_2, H_2O and O. Figure 3 shows the H_2O and O components as a function of position as the gas moves across the chamber and is rapidly heated by the wafer. The reaction rates of each of these species with the silicon substrate are not known. However, we would expect the atomic oxygen to be highly reactive and we believe it to be responsible for the unique reaction kinetics observed.

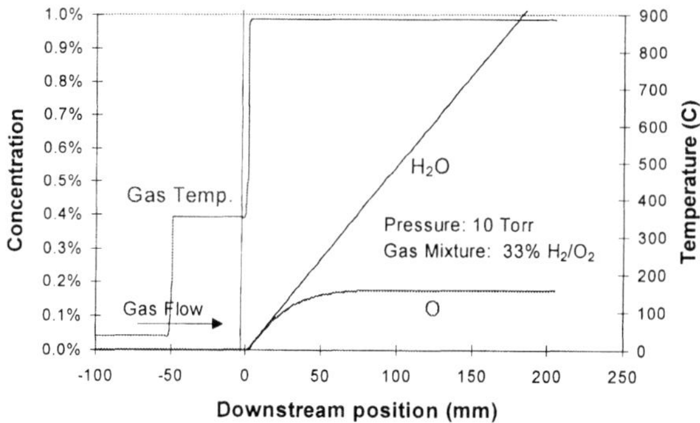

Figure 3. Model of $H_2 + O_2$ reaction across a wafer

Figure 4 shows the oxide thickness grown using ISSG for a fixed thermal cycle of 1050 °C, 60 seconds. Oxidation rates typically show a square root dependence on pressure. This is true for pressures from atmospheric pressure down to 50 Torr (data not shown). However, near 10 Torr the growth rate suddenly rises to nearly twice the atmospheric dry growth rate. We believe that this anomalous high growth rate is caused by the atomic oxygen. At pressures above 20 Torr the mean free path is reduced, leading to more rapid recombination from O to O_2, and the high growth rate disappears.

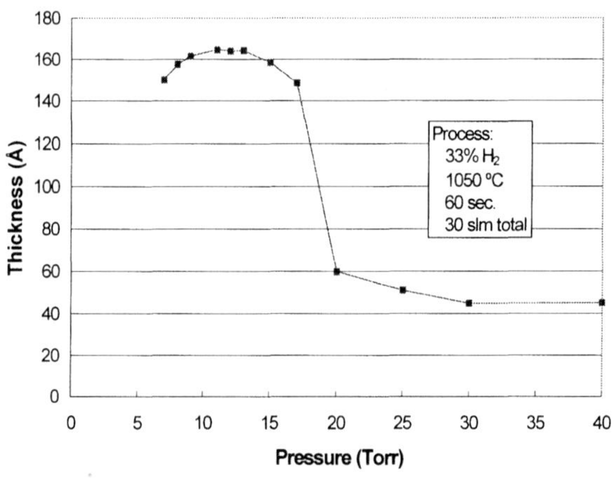

Figure 4. Pressure dependence of ISSG growth

In addition to the differences in reaction chemistry and growth kinetics, ISSG produces fundamentally different film properties compared to conventional dry and wet oxides. As shown in Fig. 5, when etched in a highly dilute HF solution (H_2O:HF > 5,000:1), an ISSG oxide grown with 5% H_2 in O_2 has 4% lower etch rate versus a dry oxide grown at the same temperature. Since this etch rate difference is seen at the top surface of the oxide, it indicates a difference in physical or chemical structure of the bulk oxide (i.e. away from the interface).

This difference in bulk material or bonding structure also improves electrical performance. Thin ISSG gate oxides have consistently yielded significantly enhanced reliability, as well as reduced gate leakage, versus dry oxides and batch furnace oxides [2]. Figure 6 shows improvement of more than seven times in Time-to-Breakdown measurements with constant field stress, and Fig. 7 shows a ten-fold improvement in gate leakage compared to 3.0 nm, dry rapid thermal oxides (which were comparable to batch furnace oxides). Figure 8 shows the consistent reliability improvement seen compared to batch furnace oxides, both wet and dry, from several independent experiments.

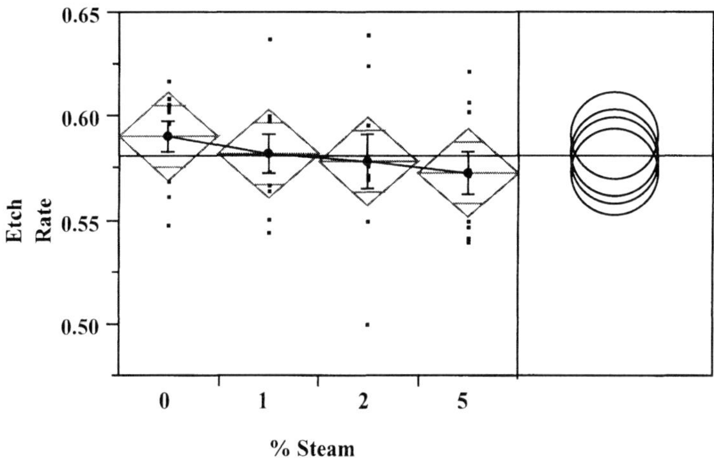

Figure 5. Dilute HF etch rate of ISSG oxides

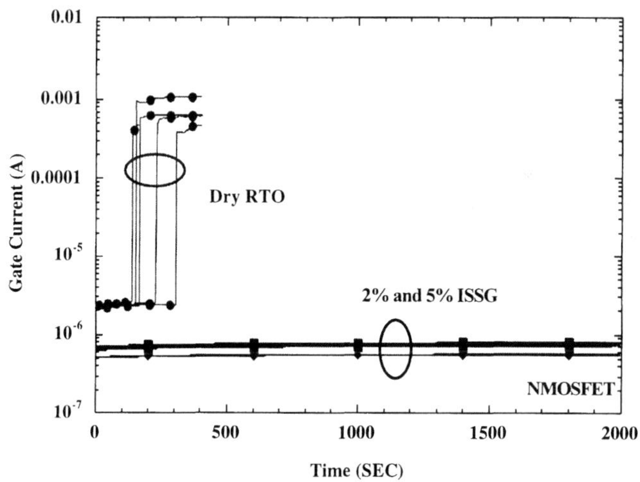

Figure 6. Time-to-Breakdown comparison between dry and ISSG oxides [2]

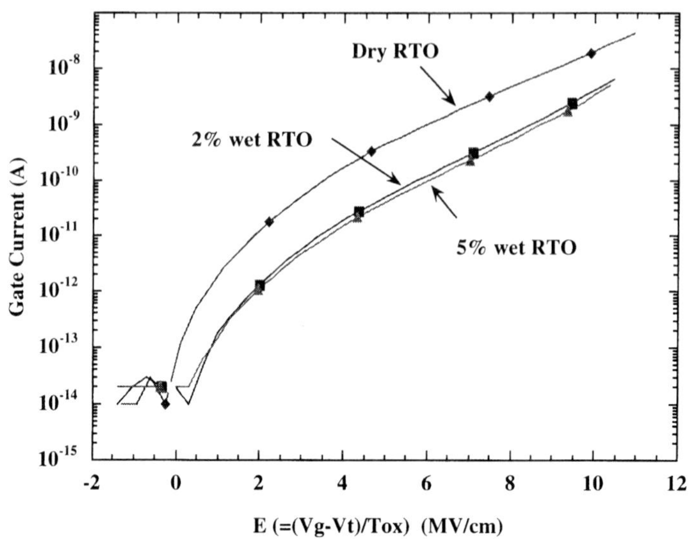

Figure 7. Gate leakage comparison between dry and ISSG oxides [2]

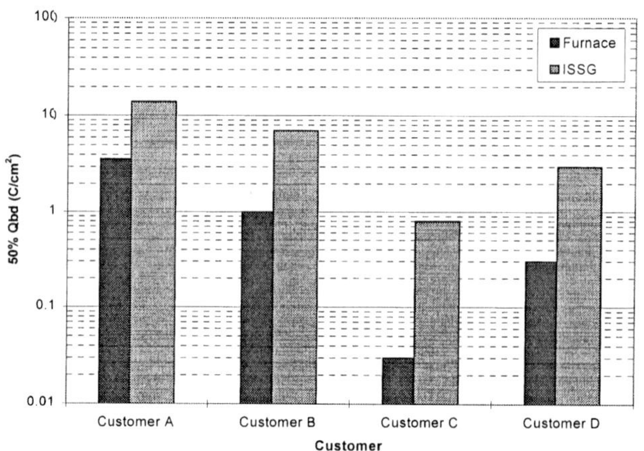

Figure 8. Q_{bd} comparison between furnace and ISSG oxides

Consistent improvement in reliability of roughly an order of magnitude has been demonstrated in multiple tests comparing ISSG with furnace dry and wet oxides as well as RTP dry oxides. These improvements using ISSG can extend SiO_2.

N_2O Oxidation

Oxidation using N_2O has been performed in both batch and single-wafer systems, and again the single-wafer process is fundamentally different.

At high temperatures, N_2O decomposes. The primary reactions are: [6]

$$N_2O \rightarrow N_2 + O \quad (5)$$
$$N_2O + O \rightarrow 2NO \quad (6)$$
$$O + O \rightarrow O_2 \quad (7)$$
$$N_2O + O \rightarrow N_2 + O_2 \quad (8)$$

At oxidation temperatures the N_2O decomposition occurs quickly such that in a batch system by the time the gas has reached the wafers, the ambient is composed of primarily N_2, O_2, and NO.

In a single-wafer chamber, such as the RTP Centura, the decomposition occurs at or near the wafer surface, and therefore the atomic oxygen created in reaction 1 plays an important role (4-6). The resulting differences include improved reliability, reduced surface nitrogen incorporation, and improved wafer-to-wafer repeatability.

Figure 9 shows a distinct improvement in charge-to-breakdown of oxides grown in N_2O when compared to both RTO and furnace oxides [7]. This result has been verified in several experiments.

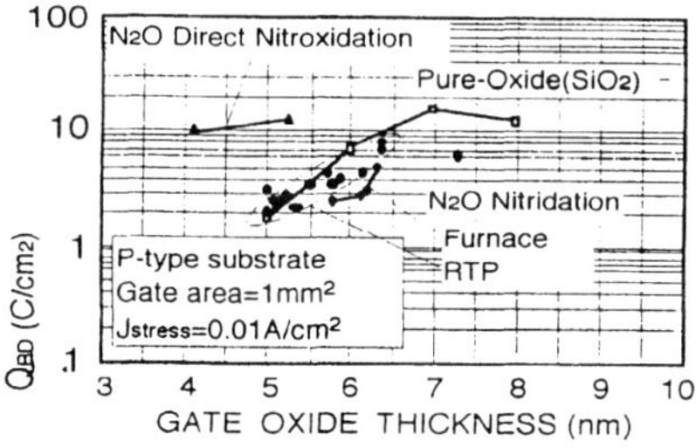

Figure 9. Charge-to-Breakdown of N_2O-grown oxides [7]

N_2O oxides have been shown to inhibit boron penetration from p+ polysilicon gates. In a recent demonstration N_2O was found to be a more effective barrier to boron penetration than an NO annealed oxide despite having a lower peak nitrogen concentration [7]. The nitrogen profiles in Fig. 10a show a typical broad profile for the N_2O-grown sample with a peak concentration of 1.5 atomic percent versus the NO-annealed sample with a peak concentration of 5 atomic percent. Figure 10b shows that the flatband voltage change resulting from boron penetration is smallest for the N_2O-grown oxide.

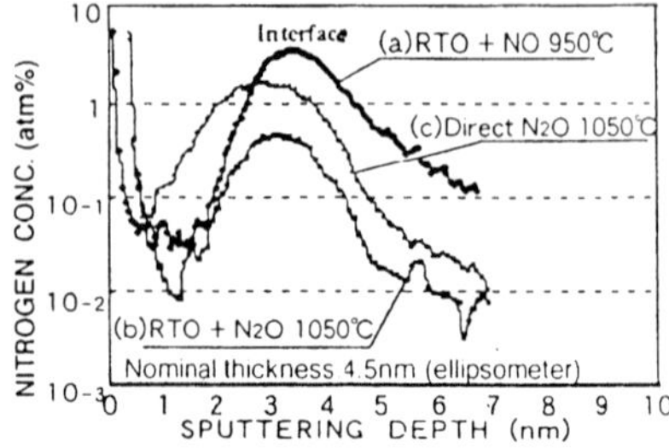

Figure 10a. SIMS nitrogen profiles of N_2O-grown and NO-nitrided oxides [7]

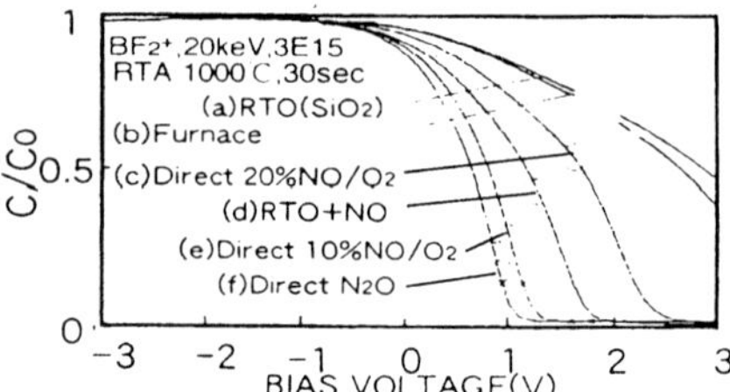

Figure 10b. V_{fb}-shift caused by boron penetration [7]

Because of these demonstrated advantages, single wafer N_2O oxidation is being adopted for production use at 0.18 μm. This is the first production use of RTO for gate dielectrics. This milestone is important not only because it recognizes the superior electrical properties of a single-wafer technology, but because it attests to the production worthiness of the RTO technology. The transition to single wafer has begun at 0.18 μm and by 0.13 μm should be the dominant technology.

0.13 μm GATE DIELECTRIC TECHNOLOGIES

If improved SiO_2-based technologies such as ISSG and N_2O-grown oxides are unable to meet the device requirements at the 0.13 μm technology node, it is expected that films with higher dielectric constant will be adopted, such as silicon nitride or silicon oxynitride. Several candidate process technologies, which are summarized in Table 2, have demonstrated improvements in gate leakage, reliability, and general transistor performance over SiO_2.

Table 2. 0.13 μm Gate Dielectric Technologies:		References:
1.	Integrated Oxynitride (ION) RT-N_2O / CVD SiON / N_2O Post Anneal	[8]
2.	In-Situ Rapid Thermal CVD Si_3N_4 RT-NO / CVD Si_3N_4 / NH_3 and N_2O Post Anneals	[9]
3.	Remote Plasma Nitrided Oxide (RPNO) RTO / RPN / N_2 Post Anneal	[10]
4.	Remote Plasma Enhanced CVD Si_3N_4 RPO / RPCVD Si_3N_4 / N_2 Post Anneal	[11,12]
5.	Jet Vapor Deposited (JVD) Si_3N_4 JVD Si_3N_4 / Post Anneals	[13,14]

Each of these technologies relies on an integration of multiple single-wafer processes to form the gate dielectric. However, only technologies 1 and 2 can be performed using currently available production equipment, such as the Applied Materials Centura cluster. Modules for RTO, CVD Si_3N_4 or SiON, as well as CVD polysilicon can be integrated to produce these integrated stacks.

Figure 11 shows a cross-section Tunneling Electron Micrograph (TEM) of an integrated stack deposited in the Applied Materials Centura, including a 1.4 nm base RTO oxide plus a 1.5 nm CVD Si_3N_4 deposition followed by a 150 nm undoped CVD polysilicon gate. The TEM shows atomically smooth interfaces between each layer.

A similar tool configuration was used to produce the Integrated Oxynitride (ION) dielectrics in [8]. These ION stacks showed improved properties versus standard thermal oxides. Gate leakage was reduced for both PMOS and NMOS devices by 2 orders of magnitude, as shown in Fig. 12. The ION stack also had equivalent or better interface state densities and was equally resistant to current stress as measured by charge pumping current (I_{cp}), as is also shown in Fig. 12. Boron penetration was completely eliminated with the ION stack.

The RTCVD Si_3N_4 stacks in [9] can also be produced with a cluster tool combining RTO using NO with CVD Si_3N_4 deposition and RTA using NH_3 and N_2O. These stacks have shown excellent improvement over standard SiO_2 gates, particularly in terms of gate leakage, reliability, and stress-induced leakage current.

Similarly, integrated CVD and RTP processes are expected to address the requirements of the 100 nm technology node. Here a CVD high-K dielectric will likely be combined with RTP used for interface preparation and for post CVD annealing. Polysilicon or metal gate electrodes may also be included in this gate stack.

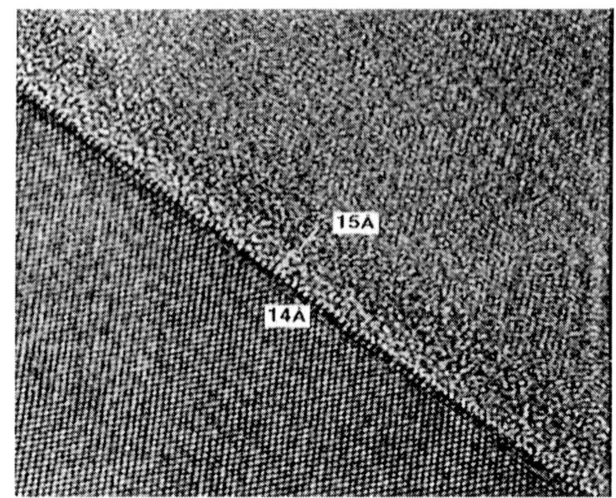

Figure 11. Cross-Section TEM of RTO / CVD Si_3N_4 / CVD polysilicon stack

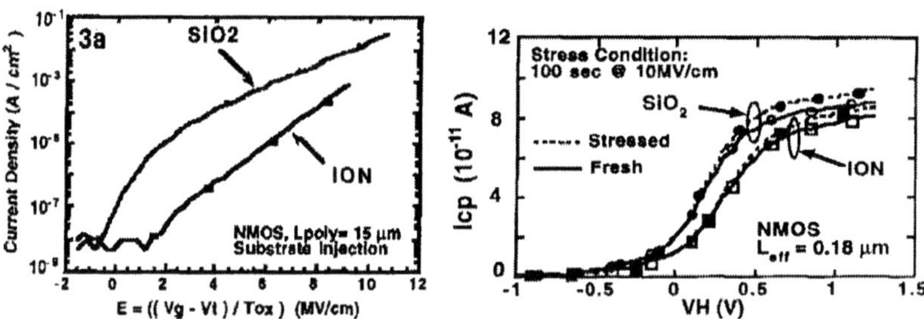

Figure 12. V_{fb}-shift caused by boron penetration [7]

CONCLUSIONS

The unique thermal environment of a single-wafer chamber enables new process technologies for gate dielectrics. ISSG and N_2O oxidation both generate atomic oxygen during the reaction/decomposition which significantly enhances gate dielectric integrity and reduces gate leakage.

These capabilities will result in the first production use of single-wafer for gate dielectrics in 0.18 μm devices. The same technologies will provide a reliable base oxide for stacks including nitride and oxynitride which are the most promising candidates for the 0.13 μm device node. Integrated single wafer processes are predicted to be the dominant technology for the 0.13 μm node and beyond.

REFERENCES

1. J.H. Stathis and D.J. DiMaria, "Reliability Projection for Ultra-Thin Oxides at Low Voltage", *Int. Electron Device Meeting*, San Francisco, CA (1998).
2. K. Reid, H. Tseng, R. Hegde, G. Miner and G. Xing, *Electrochem. Soc. Symp. Proc.* **99**-10 (1999) 23 (*these proceedings*).
3. D.G. Vlachos, "Reduction of Detailed Kinetic Mechanisms for Ignition and Extinction of Premixed Hydrogen/Air Flames," *Chem. Engineering Science,* **51** (1996) 3979.
4. P.J. Tobin, Y. Okada, S.A. Ajuria, V. Lakhotia, W.A. Feil and R.I. Hegde, "Furnace formation of silicon oxynitride thin dielectrics in nitrous oxide (N_2O): the role of nitric oxide (NO)", *J. Appl. Phys.*, **75** (1994) 1811.
5. E.C. Carr, K.A. Ellis and R.A. Buhrman, "N depth profiles in thin SiO_2 grown or processed in N_20: the role of atomic oxygen", *Appl. Phys. Lett.,* **66** (1995) 1492.
6. K.A. Ellis and R.A. Buhrman, "Furnace gas-phase chemistry of silicon oxynitridation in N_2O", *Appl. Phys. Lett.,* **68** (1996) 1696.
7. Y. Yoneda and A. Ishinaga, "The Dielectric Breakdown and Interface Characteristics of Nitrided Oxide Films Formed by N_2O Direct Nitroxidation and NO Nitridation," *International Symposium on Advanced ULSI Technology - Challenge and Breakthroughs,* Tokyo University, Tokyo, (1998).
8. H.H. Tseng, D. O'Meara, P. Tobin, V. Wang, X. Guo, R. Hegde, I. Yang, P. Gilbert, R. Cotton, and L. Hebert, "Reduced Gate Leakage Current and Boron Penetration of 0.18 μm 1.5 V MOSFETS Using Integrated RTCVD Oxynitride Gate Dielectric," *Int. Electron Device Meeting*, San Francisco, CA (1998).
9. S. Song, H. Luan, Y. Chen, M Gardner, J. Fulford, M. Allen, and D.-L. Kwong, "Ultra-Thin (<20 Å) CVD Si_3N_4 Gate Dielectric for Deep-Sub-Micron CMOS Devices," *Int. Electron Device Meeting*, San Francisco, CA (1998).
10. S. Hattangady, D. Grider, R. Kraft, W-T. Shiau, M. Douglas, P. Nicollian, M. Rodder, G. A.Brown, A. Chatterjee, J. Hu, S. Aur, H.-L. Tsai, R.A. Chapman, R.H. Eklund, I-C. Chen, and M.F. Pas, "Remote Plasma Nitrided Oxides for Ultrathin Gate Dielectric Applications," *SPIE 1998 Symposium on Microelectronic Manufacturing*, Santa Clara, CA (1998).
11. C.G. Parker, G. Lucovsky, and J.R. Hauser, "Ultrathin Oxide-Nitride Gate Dielectric MOSFET's" *IEEE Electron Device Letters*, **19** (1998) 106.
12. G. Lucovsky, *Electrochem. Soc. Symp. Proc.*, **99**-10 (1999) 69 (*these proceedings*).
13. G. Cui, T. Tamagawa, B.L. Halpern, J.J. Schmitt, X. Wang, M. Khare, Y. Shi and T.P. Ma, "Low Temperature Deposition of gate dielectrics by the Jet Vapor Deposition^{TM} Process" *4th Int. Conf. On Advanced Thermal Processing of Semiconductors*, RTP '96, p. 219.

14. T.P. Ma, *Electrochem. Soc. Symp. Proc.*, **99**-10 (1999) 57 (*these proceedings*).

EVALUATION OF ULTRA-THIN GATE OXIDES USING DIFFERENT AMBIENTS IN A RAPID THERMAL PROCESSING SYSTEM

Y. B. Jia, J. Y. Choi, and J. Schuur
Integrated Device Technology, Inc., 2975 Stender Way
Santa Clara, CA 95054, USA

J. H. Das, R. Sharangpani, and R. P. S. Thakur
STEAG RTP Systems, 4425 Fortran Drive, San Jose, CA 95134-2300, USA

Ultra-thin gate oxides grown by Rapid Thermal Processing in different ambience, including O_2, steam, and N_2O, were evaluated and compared with furnace grown gate oxides. Gate oxide integrity and uniformity were improved by using RTP. Oxides grown in N_2O and steam showed the best electrical data.

INTRODUCTION

As the feature size of CMOS devices scales down, the quality of ultra-thin gate oxides are becoming more and more critical. Rapid thermal processing (RTP) is an attractive candidate for forming ultra-thin gate oxides due to tighter process control, fast ramp rate, and fast gas switching capability. This allows short oxidation time and high process temperature, resulting in better oxide quality and a smoother Si/SiO_2 interface. Another advantage of RTP is the lower thermal budget compared to conventional furnace processes.

Rapid thermal oxidation in dry O_2 and nitrogen containing environment has been studied extensively[1,2]. Higher charge to breakdown (Q_{BD}) and better hot carrier immunity were reported for nitrided oxides. Furthermore, nitrided oxides are found to be a better barrier for boron penetration from p^+ polysilicon gate into the channel region of a p-channel MOSFET. Recently, rapid thermal oxidation in wet O_2 (steam mixed with O_2) was reported[3]. The integrity of oxide grown in wet O_2 ambient was found to be better than that of dry O_2 oxide, especially for oxides grown at low temperatures. The thermal budget can be further lowered by rapid thermal oxidation in wet O_2 without degrading oxide integrity due to the high oxidation rate.

In this paper, electrical data of ultra-thin oxides grown by RTP in different oxidation ambient are presented. The data are compared to those grown by furnace processes using the same ambient. Based on our experimental results RTP is demonstrated to be a superior process for gate dielectric formation in deep sub-micron CMOS devices.

EXPERIMENTAL DETAILS

In this work, O_2, steam, and N_2O were used in RTP oxidation. For comparison, the same ambients were also used in furnace oxidation processes. Table I lists the growth methods used for this study, including one-step, two-step, and three-step processes, some using gas mixtures: (1) RTO, grown in O_2 (one-step process). (2) RTON, grown in O_2 followed by an anneal in N_2O/O_2 gas mixtures (two-step process, equivalent to RTO + RTN). (3) RTONO, grown in O_2 followed by an anneal in N_2O/O_2 gas mixtures and re-oxidation in O_2 (three-step process). (4) RTN, grown in N_2O/O_2 gas mixtures (one-step process). (5) WRTO, grown in steam mixed with O_2/N_2 (one-step process). (6) WRTON, grown in steam mixed with O_2/N_2 followed by an anneal in N_2O/O_2 gas mixtures (two step process). In this study, oxide thicknesses were 50Å, 40Å, 30Å, and 22Å. For two-step processes, 30Å or 20Å was grown in the first step, and 10Å or 20Å was grown in the second step, taking oxide with 40Å final thickness as an example. For three-step processes, 20Å was grown in the first step, and each 10Å was grown in each of the following two steps, also taking oxide with 40Å final thickness as an example. RTP oxidations were done in AG's 8800 and steampulse systems.

MOS capacitors were fabricated for electrical measurement. The substrates were n-type Si wafers with n-well implant. P^+ polysilicon was used as gate material. Charge to breakdown was measured at constant current of 5 $nA/\mu m^2$. C-V, field to breakdown, and leakage current measurements were also performed.

Table I. Summary of growth methods and oxidation ambients

Method	Oxidation ambient		
	STEP - I	STEP - II	STEP - III
RTO	O_2	-	-
RTON	O_2	Mixture of N_2O/O_2	-
RTONO	O_2	Mixture of N_2O/O_2	O_2
RTN	Mixture of N_2O/O_2	-	-
WRTO	Steam mixed with O_2/N_2	-	-
WRTON	Steam mixed with O_2/N_2	Mixture of N_2O/O_2	-

EXPERIMENTAL RESULTS

The oxide uniformity of RTP oxides was found to be better than furnace oxide. Figure 1 shows a typical uniformity data of RTP oxides. The standard deviation is less than 1%.

Similar field to breakdown and gate leakage were found among RTP and furnace oxides. However different Q_{BD} data were observed for oxides grown by different processes.

Figures 2-5 show the Q_{BD} data for the growth methods listed in Table 1. In Figure 2, Q_{BD} data of oxides grown in O_2 (RTO) at three different temperatures (950 °C, 1000 °C, and 1050 °C) are presented. Figure 3 shows the same data of oxides grown in steam mixed with O_2 (WRTO) at 900 °C, in steam mixed with O_2 and N_2 (WRTO) at 1000 °C, and in steam mixed with O_2 at 900 °C followed by N_2O/O_2 gas mixture anneal (WRTON). Figure 4 compares oxides grown in O_2 (RTO) at 1000 °C, in O_2 followed by N_2O/O_2 gas mixture anneal (RTON), and in O_2 followed by an anneal in N_2O/O_2 gas mixtures and re-oxidation in O_2 (RTONO). Figure 5 compares RTON and RTN (grown in N_2O/O_2 gas mixtures) gate oxidation processes. The oxide thickness formed in each step of RTON is 30Å/10Å for RTON -L and 20Å/20Å for RTON -H in Figure 5, respectively.

The Q_{BD} data can be summarized as following. (1) For oxides grown in O_2, high temperature results in higher oxide integrity (Figure 2); For oxides grown in steam mixed with O_2/N_2, similar correspondence between Q_{BD} and temperature was observed (Figure 3). (2) Oxide grown in steam mixed with O_2/N_2 shows improved oxide integrity compared with O_2 oxide grown at the same temperature (Figures 2 and 3). (3) N_2O/O_2 anneal of O_2 and steam oxides improves Q_{BD} significantly (Figures 3 and 4). (4) No Q_{BD} difference was observed for re-oxidation of oxide grown by RTON (Figure 4). (5) For an oxide layer of fixed thickness grown in RTON process, Q_{BD} increases with increasing the oxide thickness grown in the second step (i.e. RTN step) (Figure 5). Table II lists the Q_{BD} values of 50% cumulative probability for oxides grown by different methods.

Table II. Summary of Q_{BD} values of 50% cumulative probability

	RTO 950 °C	RTO 1000 °C	RTO 1050 °C	WRTO 900 °C	WRTO 1000 °C	RTON -L	RTON -H	RTN	RTONO	WRTON
Q_{BD}	0.5	1.0	1.8	1.0	6.5	2.8	6.2	12	2.8	6.5

Figure 6 shows nitrogen profiles in RTON and RTN oxides determined by SIMS analysis. The maximum nitrogen content is 0.7% for RTON and 1.2% for RTN. The peak of nitrogen content is closer to the interface in RTON oxide compared with that in RTN oxide.

DISCUSSION

(1) The data showed that high temperature oxidation results in better oxide integrity. This is due to good flow of oxide at high temperatures, so that the weak points of the oxide layer are densified. The data also suggest that oxide flows better in steam ambient than dry O_2. (2) Nitridation of O_2 and steam oxide in N_2O/O_2 gas mixtures improves oxide integrity. It is believed that nitrogen incorporation hardens the oxide which is harder to break. (3) One-step process using N_2O/O_2 mixture gas results in the best electrical data. In dual-gate CMOS technology, p+ polysilicon is used as the gate electrode for p-channel MOSFET. Boron penetration through the gate oxide from polysilicon to the channel region results in Vt shift when gate oxide scales down. RTN is proposed to be most suitable for thin gate oxide formation in dual-gate CMOS process due to higher nitrogen incorporation and good breakdown data. For thicker oxide, two-step process (RTON) is proposed. This is based on the fact that nitrogen incorporation in RTN process slows down oxide growth rate, thus, the thermal budget may be too high for thick oxide grown by RTN process. (4) Steam oxide grown at high temperatures shows similar breakdown data as RTN, thus, is also a promising process for thin gate oxide. (5) RTP oxide is found to be better than or equivalent to furnace oxide in the same ambient. Furthermore RTP offers better process control and lower thermal budget. Thus RTP is proposed for ultra-thin gate oxide formation.

CONCLUSION

In conclusion, RTP has been investigated for ultra-thin gate oxide formation using different ambients. Oxides grown in N_2O/O_2 gas mixtures and in steam mixed with O_2/N_2 at high temperatures showed the best integrity. These two processes indicate to be the most promising processes for ultra-thin gate oxide formation down to 20Å.

REFERENCES

1. H. Fukuda, T. Arakawa, and S. Ohno, IEEE Trans. Electron Dev. **39**, 127 (1992)
2. Y. Okada, P. J. Tobin, P. Rushbrook, and W. L. DeHart, IEEE Trans. Electron Dev. **41**, 191 (1994)
3. R. Sharangpani, R. P. S. Thakur, N. Shah, and S. P. Tay, Solid State Technology **41**, 171 (1998)

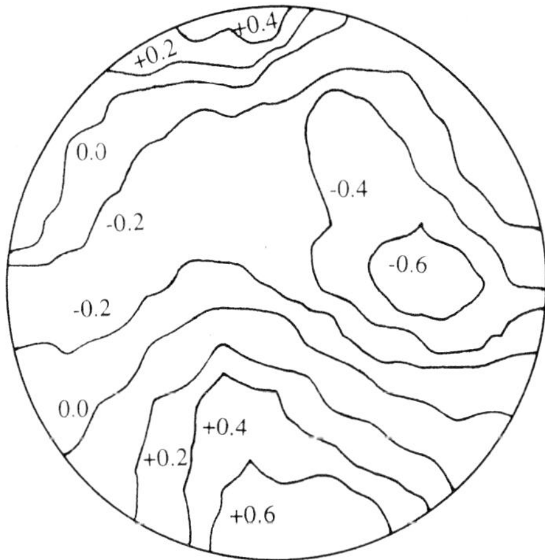

Figure 1. Oxide thickness map. Growth method: RTON (grown in O_2 followed by an anneal in N_2O/O_2 gas mixtures). Mean: 40.63Å; Standard deviation: 0.98%; Contour interval: 0.2 Å.

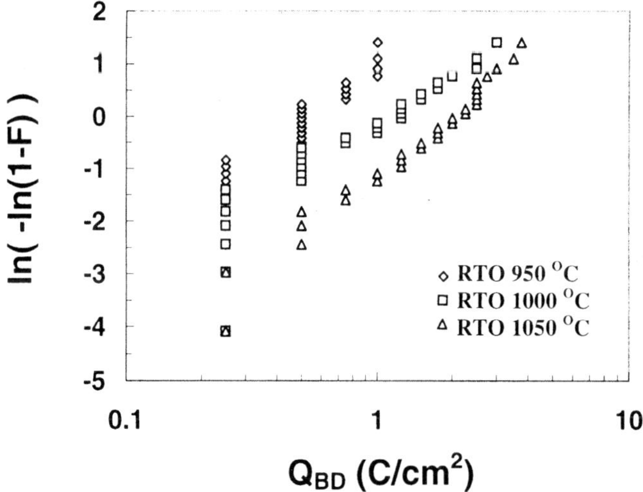

Figure 2. Charge to breakdown (Q_{BD}) data measured at constant current: 5 nA/μm^2. Growth method: RTO (grown in O_2 at 950 °C, 1000 °C, and 1050 °C). Oxide thickness is around 40Å.

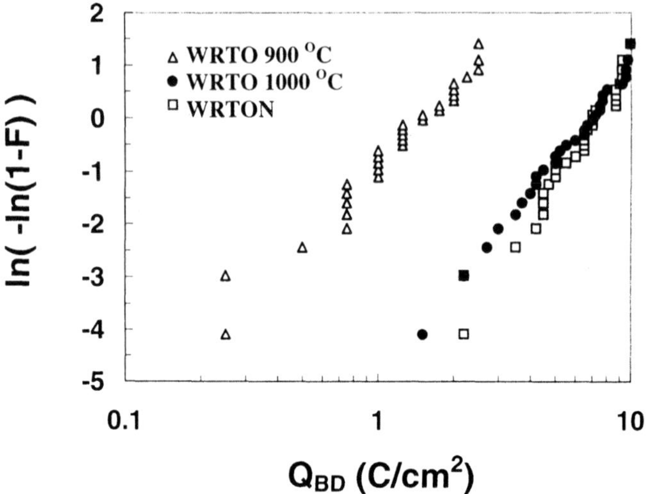

Figure 3. Charge to breakdown (Q_{BD}) data measured at constant current: 5 nA/μm^2. Growth method: (1) WRTO (grown in steam mixed with O_2 at 900 °C); (2) WRTO grown in steam mixed with O_2/N_2 at 1000 °C: (3) WRTON (grown in steam mixed with O_2 at 900 °C followed by an anneal in N_2O/O_2 gas mixtures). Oxide thickness is around 40 Å.

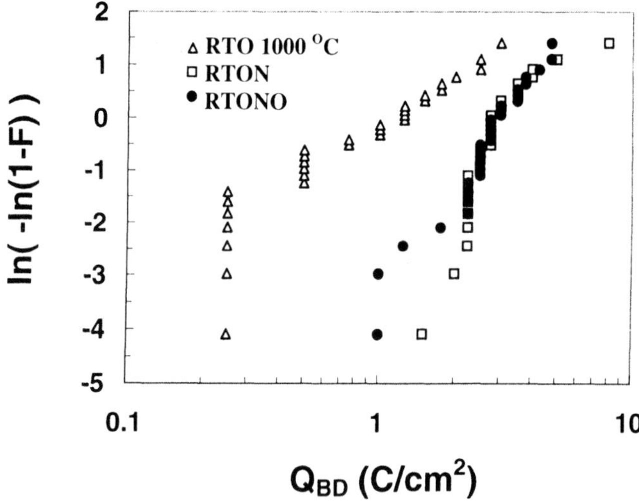

Figure 4. Charge to breakdown (Q_{BD}) data measured at constant current: 5 nA/μm^2. Growth method: (1) RTO (growth in O_2 at 1000 °C); (2) RTON (grown in O_2 followed by an anneal in N_2O/O_2 gas mixtures); (3) RTONO (grown in O_2 followed by anneal in N_2O/O_2 gas mixtures and re-oxidation in O_2). Oxide thickness is around 40Å.

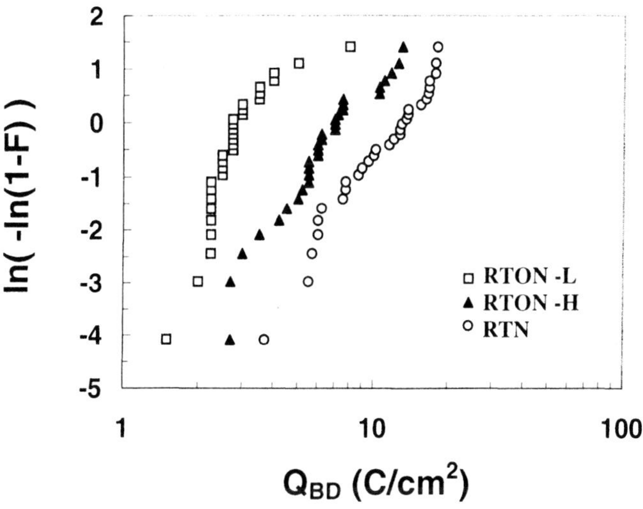

Figure 5. Charge to breakdown (Q_{BD}) data measured at constant current: 5 nA/μm^2. Growth method: (1) RTON -L (grown in O_2 (grow 30Å) followed by anneal in N_2O/O_2 mixture gases (grow another 10Å)); (2) RTON -H (grown in O_2 (grow 20Å) followed by anneal in N_2O/O_2 mixture gases (grow another 20Å)); (3) RTN (grown in N_2O/O_2 mixture gases). Oxide thickness is around 40Å

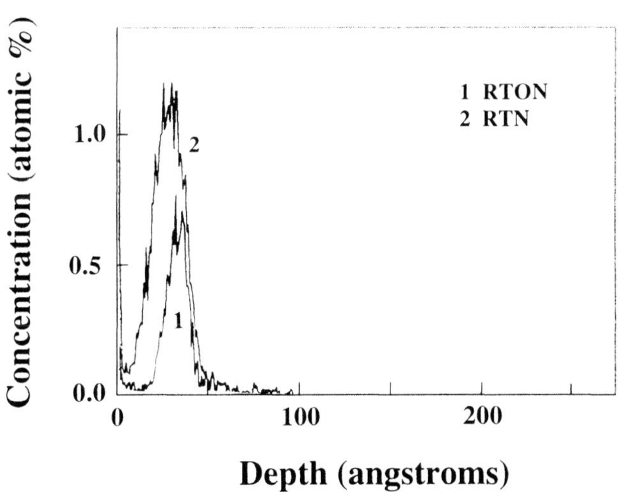

Figure 6. SIMS profiles of nitrogen distribution in oxides grown by: (1) RTON (growth in O_2 followed by anneal in N_2O/O_2 mixture gases); (2) RTN (growth in N_2O/O_2 mixture gases). Oxide thickness is around 50Å.

DILUTE STEAM RAPID THERMAL OXIDATION FOR 30Å GATE OXIDES

Kimberly G. Reid*, Hsing Tseng*, Rama Hegde*, Gary Miner**, Guangcai Xing**
*Advanced Products Research & Development Laboratory, Motorola, Austin, TX 78721
** Applied Materials, Santa Clara, California 95054

Gate dielectrics are scaling rapidly, with an industry preference to maintain SiO_2 as the gate dielectric for as long as possible. As the physical thickness is reduced to meet transistor speed requirements, the ultimate scaling limit of SiO_2 may be caused by the direct tunneling through the SiO_2, reliability of sub 30Å SiO_2, and issues associated with uniform growth of such thin SiO_2 films in manufacturable equipment. This paper investigates dilute steam oxidations compared to dry O_2 oxidations in a single wafer rapid thermal processor (RTP). The advantages of the RTP include more precise control of the gas ambient that allow careful study of the process effects on the gate dielectric as well as uniform thickness control for the very thin gate oxides in use today for CMOS technology. It is found that the reliability of the dilute steam processes is significantly better than the dry process, and the leakage current is reduced compared to dry RTO gate oxides, such improvement in the quality of the gate oxide can extend use of SiO_2 as the gate dielectric.

INTRODUCTION

Gate dielectric for MOS transistors at the 0.15μm and below technology node has become the leading front end integration issue as the dielectric scales to sub 35Å regime for improved current drive and transconductance (1). Direct tunneling through the gate dielectric at operating voltages has been identified as a key issue along with oxide reliability (2). Several approaches to extend the use of silicon dioxide or silicon oxynitrides have been investigated. The use of nitric oxide and nitrous oxide during thermal oxidations or as post oxidation anneals has been extensively explored (3).

Recently Si_3N_4, Ta_2O_5, TiO_2, and other higher dielectric constant materials are being proposed as the next generation gate dielectrics. Although the transition to a new material is inevitable for CMOS gate dielectric, SiO_2 will be used as long as possible. Therefore, processes which can improve the reliability and reduce the leakage of SiO_2 films are needed. In this paper the use of dilute steam oxidation for 30Å gate oxides is investigated with respect to these key metrics.

EXPERIMENTAL

N+ and P+ gated MOSFETs were fabricated using twin well CMOS technology on 200 mm N type silicon wafers (4). A trench isolation module was followed by a 100Å sacrificial oxidation process. The sacrificial oxide was stripped with 100:1 HF. The wafers were then split into groups for the gate oxidation process. The polysilicon gates were deposited in a standard furnace process and all subsequent processing was done with the wafers combined into one group.

The 30Å gate oxidation processes were all done by RTP using a two step process of an oxidation followed by a post oxidation anneal (POA). The RTP is capable of low pressure oxidation, wafer rotation, and has twelve zones of temperature control for across wafer uniformity optimization (5). The processes were all done at 1050°C for both the oxidation and the POA. The splits included a dry O_2 only oxidation at 80 Torr with an oxidation time of 43 seconds followed by a post oxidation anneal in N_2 for 10 seconds, a 2% H_2 in O_2 dilute steam oxidation at 10 Torr for 15 seconds followed by a 10 seconds N_2 post oxidation anneal, and finally a 5% H_2 in O_2 dilute steam oxidation at 7.2 Torr for 8 seconds again with the 10 seconds N_2 POA. The post oxidation anneal was done at the oxidation pressure in all cases and at 1050°C.

The dilute steam process uses a novel means of generating steam within the RTP chamber, also called In-Situ Steam Generation (ISSG), whereby a mixture of H_2 in O_2 is reacted at the hot wafer surface (6). The reaction produces H_2O and additional products, including atomic oxygen and OH, which may be responsible for the unique properties of the resulting SiO_2.

The physical oxide thicknesses were measured using a single wavelength ellipsometer assuming a refractive index of 1.465. Nine points were measured on test wafers run with each split and the data is shown in Table 1. The physical thicknesses were very similar.

Table 1 Physical thicknesses of test wafers run with each split

Split	t_{ox} (Å)	1 σ (Å)
Dry RTO	29.34	0.09
2% Dilute Steam RTO	30.31	0.44
5% Dilute Steam RTO	30.97	0.28

RESULTS AND DISCUSSION

The oxide thicknesses were measured from transistors in inversion and were about 30Å. Figure 1 shows that the electrical thicknesses of all three splits were very similar, within 3%. Time-To-Breakdown under constant E-field stressing at 15 MV/cm for dilute steam RTO was more than 7X longer than dry RTO as shown in Figure 2. The dry RTO Time-to-Breakdown is comparable to similar thickness oxides grown in a vertical batch furnace. Figure 3 shows the gate leakage current for dilute steam RTO was about 10X lower than dry RTO at 5 MV/cm E-field. Charge pumping results shown in Figure 5 indicate that the interface state density for three splits were about the same which suggested the advantages observed may be due to a bulk property difference.

Several physical analyses were done on either the 30Å samples or 200Å samples grown under identical oxidation conditions by extending the oxidation times. Composition measurements were carried out on 30Å films using secondary ion mass spectrometry (SIMS) to determine the hydrogen profile and interfacial hydrogen content if any, as described elsewhere (7). The SIMS data, shown in Figure 5, was acquired using Cs+ bombardment while monitoring clustered secondary species CsH and CsSi. The SIMS analysis showed that hydrogen content is the same in all three cases with hydrogen detected through the film, even in the dry RTO case. The dry RTO case was grown in the same chamber as the dilute steam, perhaps resulting in hydrogen "contamination" of the film to the same levels as detected in the post annealed dilute steam samples. It has been demonstrated elsewhere that post annealing can remove hydrogen from SiO_2 grown in a steam ambient.

Cross sectional TEM was done to examine the interface of the silicon substrate and the SiO_2. Figures 6, 7 and 8 show the dry RTO, 2% dilute steam, and 5% dilute steam interfaces respectively. There was no significant difference observed in the physical interface by TEM.

Figure 9 shows the etch rate data using a dilute H_2O:HF mixture of greater than 1000:1. The wafers were etched at 2, 10, 20, and 30 minutes time points. The etch rate was determined from nine points per wafer. Thicknesses were measured using single wavelength ellipsometry. The etch rate plotted is for each point oxide removed divided by the delta time. The first 2 minutes was discounted as contaminant removal. The figure shows the means for each oxidizing ambient as well as all the data. The statistical analysis using 95% confidence t-test showed only significant difference between the dry and the 5% dilute steam. However, it is worth noting that the trend in the mean etch rate decreased with increasing dilute steam.

Obviously it is very difficult to use physical analysis to determine the role steam is playing in these 30Å oxides. The etch rate data along with the TEM of the interface, may provide support to the theory that the reliability enhancement and leakage current reduction are due to the "bulk" properties of these thin oxides. Of course, bulk in this case means not the interface as the entire film is extremely thin.

CONCLUSIONS

Ultra-thin dilute steam RTO with 30 Å thickness shows very encouraging gate oxide reliability. The Time-To-Breakdown under constant E-field stressing for dilute steam RTO was significantly longer than dry RTO. In addition, the gate leakage current for dilute steam RTO was about 10X lower than dry RTO which is very attractive for gate oxide scaling. Physical analysis and electrical results indicate that the interface quality is similar and that perhaps the "bulk" oxide quality plays the dominant role in providing a more reliable and less leaky SiO_2 film using dilute steam ambient.

ACKNOWLEDGEMENTS

The authors would like to acknowledge the management support of Victor Wang, Phil Tobin, Joe Mogab, Betsy Weitzman, and Fabio Pintchovski. The analysis support from Ted Neil, Jim Conner, and Lata Prabhu was greatly appreciated.

REFERENCES

1. Iwai and H. Momose, IEDM Tech. Digest, p. 163, (1998).
2. M. Depas, T. Nigam, and M. Heyns, IEEE Trans. Electron Devices, **43**, p. 1499, (1996).
3. B. Maiti, P. Tobin, V. Misra, R. Hegde, K. Reid, and C. Gelatos, IEDM Tech. Digest, p. 651, (1997)
4. M. Luo, P. Tsui, W.M. Chen, P. Gilbert, B. Maiti, A.R. Sitaram, and S.W. Sun, IEDM Tech. Digest, p. 691, (1995).
5. B. Peuse, G. Miner, M. Yam and C. Elia, in *Rapid Thermal and Integrated Processing VII*, M. Öztürk, F. Roozeboom, P.J. Timans and S.H. Pas, Editors, PV 525, p.71, Materials Research Society Proceedings Series, Warrendale, PA (1998).
6. G. Miner, G. Xing, H.S. Joo, E. Sanchez, Y. Yokota, C. Chen, and D. Lopes, Electrochem. Soc. Proc. (1999).
7. R. Hegde, B. Maiti, R. Rai, K. Reid, and P. Tobin, J. Electrochem. Soc. 145, L13-L15, (1998).

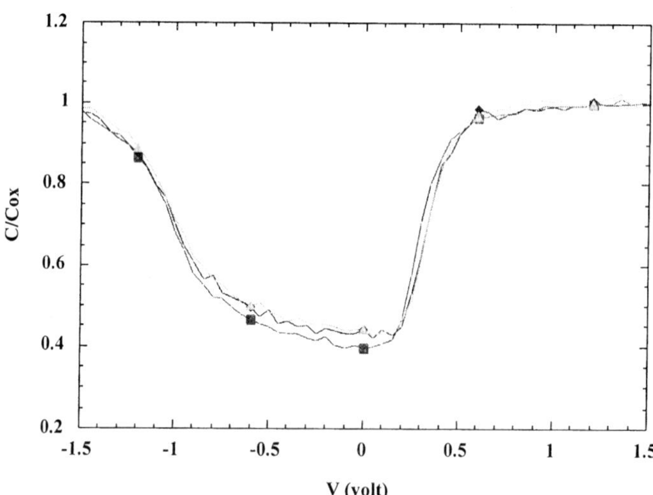

Figure 1 CV Plot demonstrating the electrical thickness to be very similar for the 3 processes

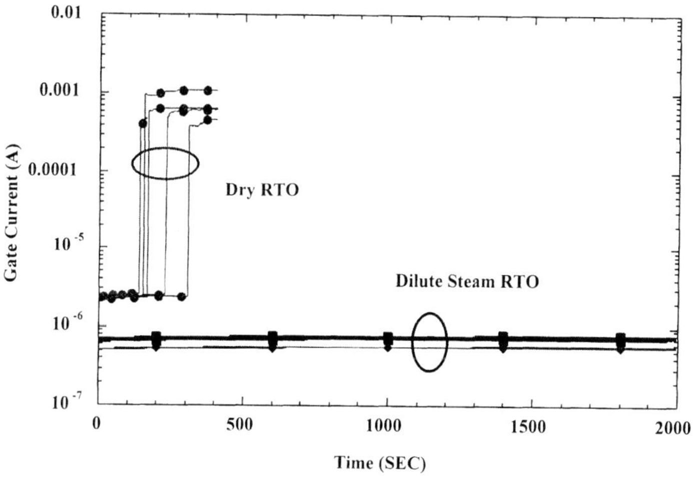

Figure 2 Time-To-Breakdown under constant E-field stressing at 15 MV/cm.

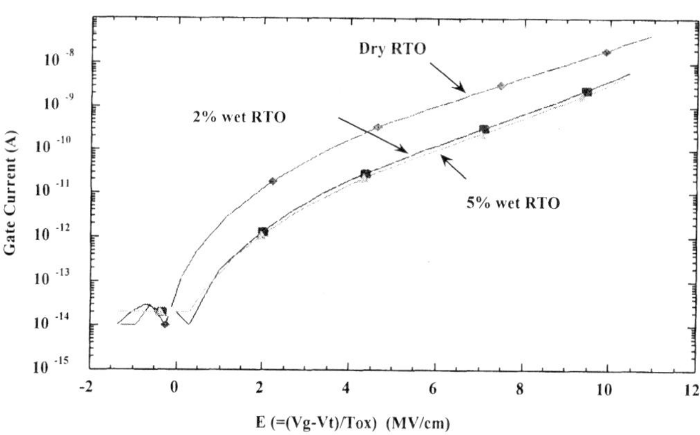

Figure 3 Gate leakage current comparing Dry RTO to 2% and 5% dilute steam RTO.

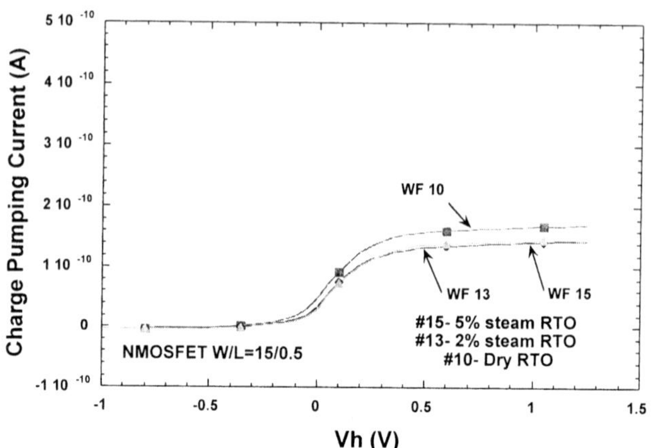

Figure 4 Charge pumping current for each oxidation process.

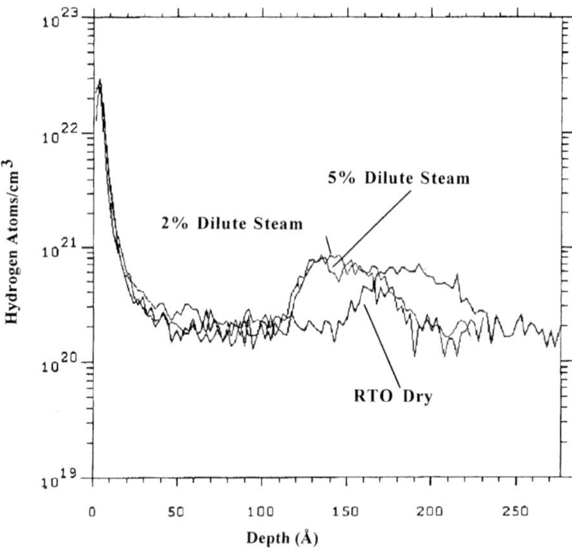

Figure 5. Hydrogen SIMS depth profiles for 5% and 2% dilute steam and dry RTO for 30Å samples.

Figure 6. TEM for the dry RTO 30Å sample

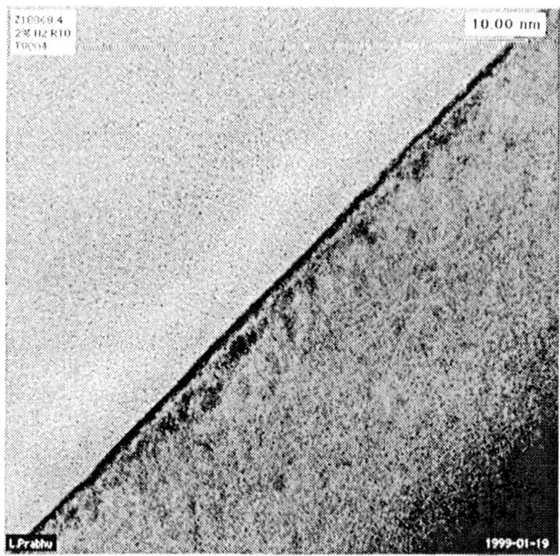

Figure 7. TEM for the 2% dilute steam RTO.

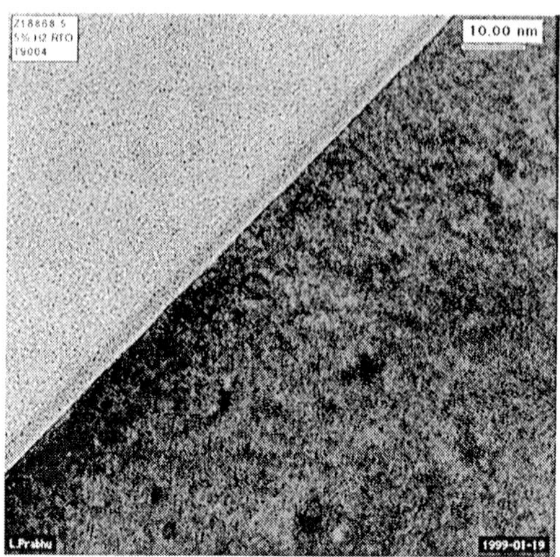

Figure 8. TEM for the 5% dilute steam RTO.

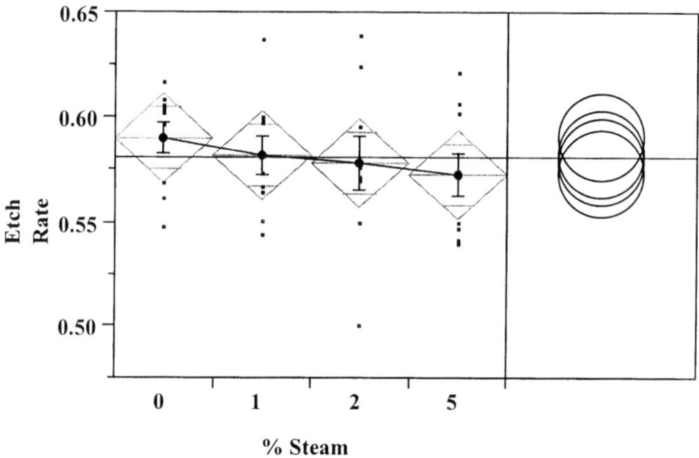

Figure 9 Etch rate shown from 200Å samples using ≥1000:1 dilute HF.

PREPARATION OF ULTRA-THIN GATE OXIDES WITH ANNEALING IN NITRIC OXIDE

Barbara Froeschle, Nicole Sacher, Frédérique Glowacki*
STEAG RTP Systems,
Daimlerstrasse 10,
D-89160 Dornstadt, Germany

Thin oxides and nitrided oxides, which are prepared under different process conditions, are characterized using Secondary Ion Mass Spectroscopy (SIMS) and Time Dependent Dielectric Breakdown measurements. The results show the positive effect of annealing in pure NO gas on the Charge-to-Breakdown (Q_{BD}) value. The improvement seems to be dependent on the amount of nitrogen incorporated at the silicon/silicon dioxide interface and not on the annealing temperature. The first results using an integrated Vapor Phase Cleaning before oxidation show comparable results to standard cleaned wafers.

INTRODUCTION

As gate oxide thickness decreases, the properties of the interfaces become more and more important. Therefore, cluster tools, which can perform the entire gate sequence without breaking vacuum, are expected to be important for future device technologies. [1]. Vapor Phase Cleaning (VPC) using Anhydrous HF (AHF) / methanol for native oxide removal then becomes more important because standard wet cleaning cannot be easily integrated in a vacuum cluster tool.

Due to their superior electrical characteristics and their efficient boron diffusion barrier properties thin nitrided gate oxides have gained a lot of interest [2]. One possibility of nitridation is the annealing of the oxides in pure NO gas or the nitridation in nitrous oxide gas (N_2O). The advantage of annealing in NO gas is the high amount of nitrogen, which can be incorporated into the oxides without additional oxide growth. Therefore, this study will focus on gate stacks prepared in a cluster environment which includes Vapor Phase Cleaning (VPC) and NO annealing.

EXPERIMENTAL

Cluster Process

The experiments are performed in an Advance 800 Polygon Hot Cluster, which was developed within the framework of a European ESPRIT-SEA technology development project by ASM International and STEAG RTP Systems. This single-wafer 200 mm

* Present address: Applied Materials, 11B Chemin de la Dhuy, F-38246 Meylan Cédex, France

cluster tool integrates Vapor Phase Cleaning (VPC), Rapid Thermal Oxidation/ Nitridation (RTO/RTO-N), nitride deposition (CVD) and in-situ doped polysilicon deposition (CVD) [3].

Vapor Phase Cleaning (VPC)

The vapor phase cleaning uses Anhydrous HF (AHF)/Methanol for the removal of native oxide prior to the oxidation or nitridation. The silicon wafers are standard wet cleaned before transferring to the cluster tools. The remaining thin chemical oxide is etched in the VPC module using AHF/methanol. The etching step was performed at a temperature of 40 °C and a pressure of 100 mbar using AHF, diluted with nitrogen.

Rapid Thermal Oxidation and Nitridation (RTO/N)

Directly after the cleaning the oxidation (RTO) is carried out in the RTO module in pure oxygen (O_2) at a pressure of 500 mbar and at a temperature of either 1050 °C or 1150 °C. Nitrided oxides (RTO-N) are produced by annealing the oxides in pure nitric oxide gas (NO) at a temperature of 900 °C and 1050 °C and at the same pressure than the oxidation step.

RESULTS

Analysis of Nitrogen Content by Secondary Ion Mass Spectroscopy (SIMS)

The nitrogen incorporation and distribution in the oxide are analyzed for different NO annealing times and temperatures by Secondary Ion Mass Spectroscopy (SIMS) using Cs^+ sputtering and Cs-^{14}N, Cs-^{28}Si and Cs-^{16}O detection.

To analyze the 5-nm thin oxide it is necessary to cover the thin oxide with a thicker CVD oxide, to overcome the influence of the first nanometer non-equilibrium sputtering of the samples. The profile of a nitrided oxide sample is shown in Fig. 1 for an anneal of 1 s in NO at 1050 °C. The oxidation temperature was 1050 °C. To calculate the amount of nitrogen incorporated in the oxide the peak concentration of nitrogen is set into relation to the total amount of atoms in the SiO_2 network, as follows:

$$D_A(SiO_2) = D\,(SiO_2) * A = 6.81 \times 10^{22} \text{ atoms / cm}^3 \qquad (1)$$

where $D_A(SiO_2)$ = density of the atoms in SiO_2,
$D\,(SiO_2)$ = (molecule) density of SiO_2 (2.27×10^{22} molecules / cm^3), and
A = number of atoms per SiO_2 molecule (being 3)

$$C(N_{interface})\ [\text{at. }\%] = C(N_{interface})\ [\text{atoms/cm}^2] * 100/D_A(SiO_2) \qquad (2)$$

where $C(N_{interface})$ [at%] = nitrogen concentration at the interface measured in at. %, and

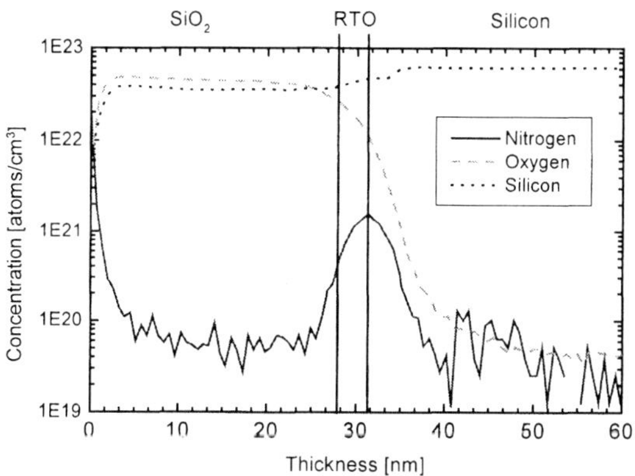

Fig. 1: SIMS profile of a nitrided oxide sample with the capping LTO layer on top. Process conditions: oxidation at 1050 °C, subsequent annealing in pure NO 1050 °C for 1 s. (LTO: Low Temperature Oxide by CVD of TEOS)

$C(N_{interface})$ [atoms/cm^2] = nitrogen concentration at the interface measured in atoms / cm^2.

The nitrogen is located at the Si/SiO_2 interface as demonstrated by a SIMS analysis of a 14-nm thick oxides in Fig. 2. The "pile-up" of the nitrogen at the interface is located where the oxygen concentration drops to 33 atomic %, which is defined as the oxide-substrate interface [4]. The oxidation was done at 1050 °C at a pressure of 1000 mbar. The annealing in NO gas was performed at the same temperature and pressure for 120 s.

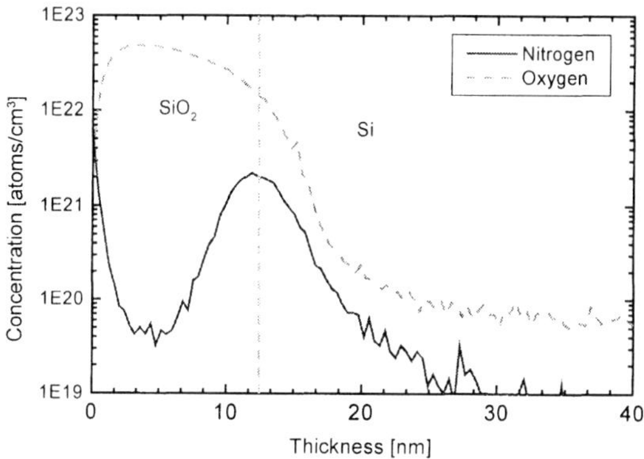

Fig. 2: SIMS profile of a 13 nm thick oxide, prepared by oxidation at 1050°C and 1000 mbar, followed by an NO anneal for 120 s at the same temperature and pressure.

For the oxides annealed at 900 °C and 1050 °C the nitrogen concentration at the Si/SiO_2 interface varies from 0.5 at. % for a 5 s NO anneal at 900 °C to 3.2 at. % for 5 s NO anneal at 1050 °C as shown in Fig. 3.

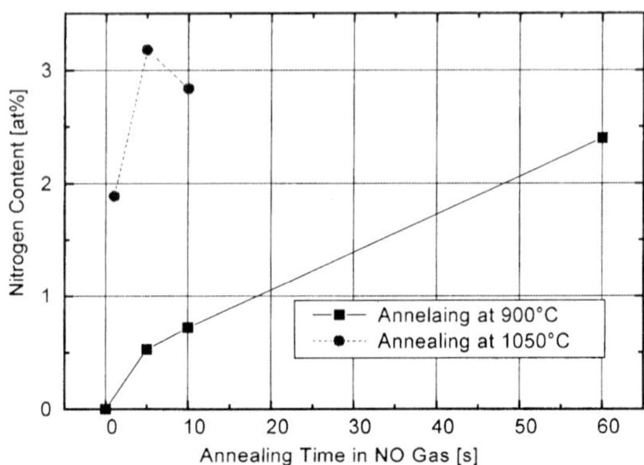

Fig. 3. Peak nitrogen concentration vs. annealing time in NO gas for annealing at 900 °C and 1050 °C.

Comparison of Annealing in Nitric Oxide (NO) and Nitrous Oxide (N_2O)

As shown in Fig. 4, the nitrogen content of a 5 nm thin oxide is more than 5 times higher after annealing in NO gas than after annealing at the same temperature in N_2O gas.

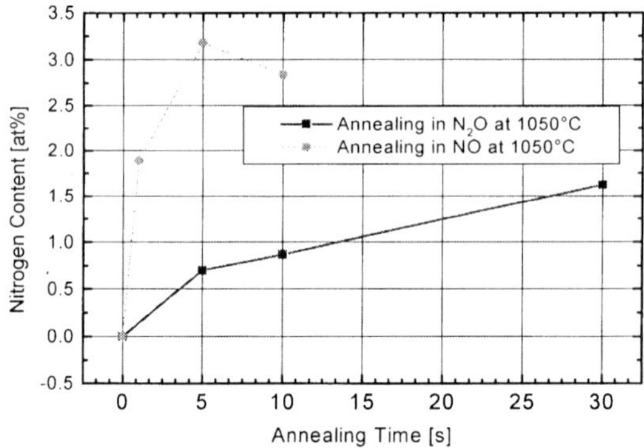

Fig. 4: Dependence of nitrogen incorporation on annealing in N_2O and NO at the same temperature

Electrical Characterization

MOS capacitors with gate oxide thickness of 5 nm have been fabricated to investigate the electrical properties obtained for different cleaning methods and different types of oxidation. For all types of samples a 300 nm thick in situ doped polysilicon is deposited after the oxidation in the same cluster. The gate contacts of area 0.1 mm^2 are defined by photolithography and wet etching. The reliability of the dielectric films is evaluated from time-dependent dielectric breakdown (TDDB) characteristics. The TDDB measurements are performed at negative gate bias on 100 capacitors by applying a constant current density of 100 mA/cm^2.

Influence of Nitrogen Content on Q_{BD} Values

Time-dependent dielectric breakdown results are shown in Fig. 5 for samples annealed in pure NO gas at 900 °C, and in Fig. 6 for samples annealed at 1050 °C.

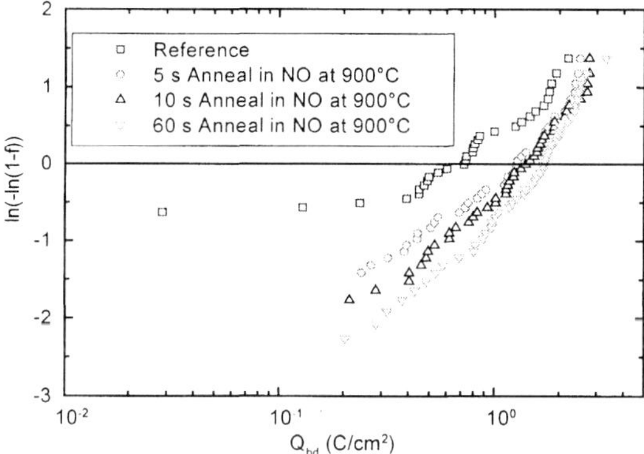

Fig. 5: Charge to breakdown characteristics of oxides with different annealing in NO gas at 900°C

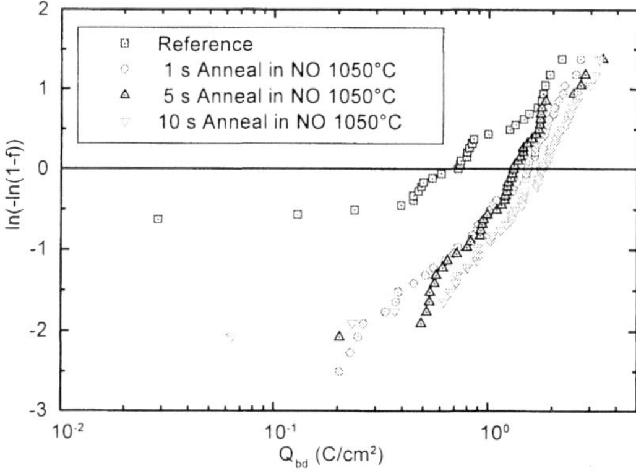

Fig. 6: Charge to breakdown characteristics of oxides with different annealing in NO gas at 1050°C

For both annealing temperatures of 900 °C and 1050 °C the Q_{BD} values increase with increasing annealing time.

The dependence of the Q_{BD} (63%) value on the nitrogen content for different annealing times is shown in Fig. 7. The increase in the Q_{BD} value is mainly dependent on the nitrogen concentration at the interface and not on the annealing conditions.

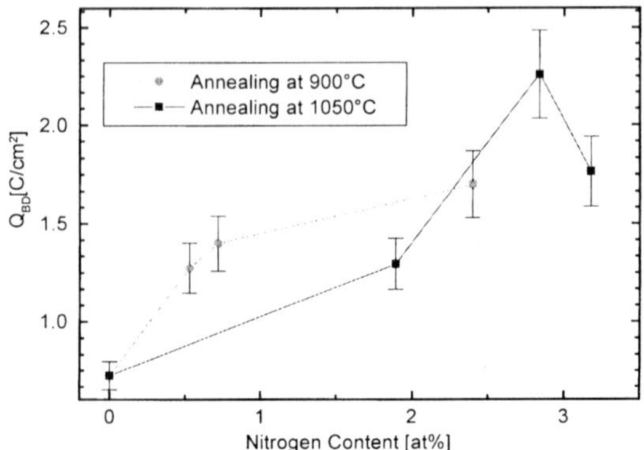

Fig. 7: Dependence of Q_{BD} on nitrogen at the interface for different annealing temperatures.

Influence of Nitrogen Content on Q_{BD} Values for different Oxidation Temperatures

Fig. 8 shows the dependence of the Q_{BD} values on the oxidation temperature. The samples annealed at 900 °C and 1050 °C for 5 s after oxidation at 1050 °C show a much

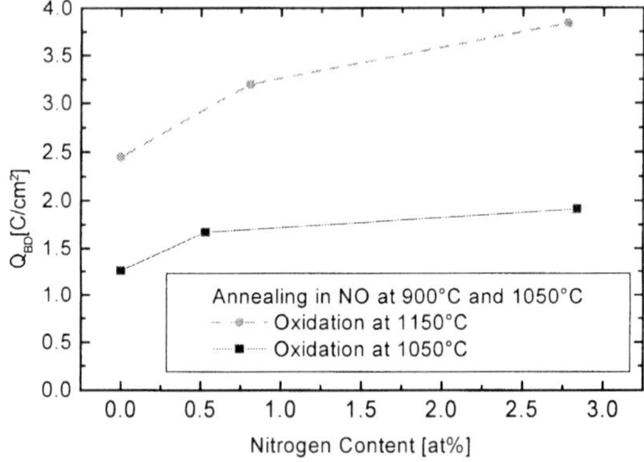

Fig. 8: Dependence of Q_{BD} on nitrogen at the interface for different oxidation conditions

higher Q_{BD} value than samples annealed under the same conditions but oxidized at 1050 °C. These results show the influence of the higher oxidation temperature on the Q_{BD} value.

Influence of Cleaning on the Q_{BD} Values

As shown in Fig. 9 the Weibull plot for standard cleaned samples is comparable that for Vapor Phase Cleaning type cleaned samples.

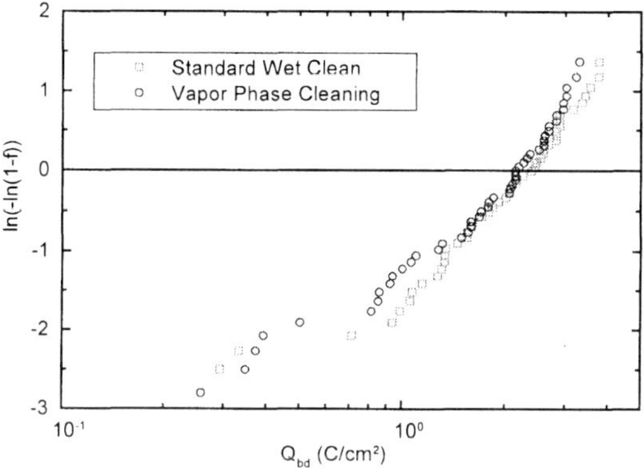

Fig. 9: Q_{BD} distribution after different cleaning techniques

CONCLUSION

Ultra-thin gate oxide and nitrided oxides have been prepared using different pre-cleaning methods. The results show the influence of different nitridation conditions on the nitrogen incorporation and on the electrical properties of the samples. Further investigations are now underway to investigate the influence of different vapor phase cleaning techniques on the annealing in NO gas and on the resulting electrical properties.

ACKNOWLEDGMENTS

The authors would like to thank A. Gschwandtner, G. Innertsberger and A. Grassl from Siemens AG in Munich for the work together within the ESPRIT / SEA CICDIP project. The authors are also grateful for the support and the analytical measurements of A. Bauer, M. Beichele, G. Rattmann, M. Lucassen from the Fraunhofer Institute for Integrated Circuits in Erlangen.

REFERENCES

[1] The National Technology Roadmap for Semiconductors, Technology Needs, Semiconductors Industry Association, 1997 edition

[2] M. Bhat, L.K. Han, D. Wristers, D.-L. Kwong and J. Fulford, Appl. Phys. Lett. **66** (1995) 1225

[3] B. Froeschle, N. Sacher, F. Glowacki, T. Pompl, G. Innertsberger, and A. Gschwandtner, "*In-situ cleaning and thin oxide processing capabilities for sub-quarter micron technology in integrated process modules*", Proceedings of the 6th RTP Conference, Kyoto, Sept. 9-11, 1998, p. 90

[4] J. M. Grant, Mat. Res. Soc. Symp. Proc. **387** (1995) 175

HIGH K GATE STACK FOR SUB-0.1 UM CMOS TECHNOLOGY

Gang Bai
Intel, 3065 Bowers Ave., Santa Clara, CA 95052

To scale transistors beyond 0.1 um, it is imperative to find a high k dielectric to replace current SiO2 as gate insulator. Realizing the promise of high k in an integrated CMOS process is an enormously challenging task and requires effective collaboration among researchers from academics and industry at this pre-competitive phase. Required solutions include (1) formation of high k sandwiched between Si without any "low" k interfacial layer so that an equivalent SiO2 thickness of < 1 nm can be achieved; (2) atomic layer deposition control and manipulation so that k value is maintained down to thickness < 10 nm; (3) stability upon post thermal processing such as dopant activation up to 1000C/10s. Furthermore, interface between high k gate insulator and Si channel must have minimum dangling bonds with Dit < 1e11/cm2ev so that charge inversion and mobility is not degraded. Based on above mentioned constraints and framework, potential candidates of high k materials such as ZrO2 and Al2O3, and deposition methods such as MBE and ALCVD are identified. Preliminary results are discussed in reference to above established criteria. None of high k gate dielectric investigated to date meets all requirements. Possible solution path and research opportunities are proposed.

INTRODUCTION

Continued thickness scaling of SiO2 gate insulator in past decade enables technologists to build deep sub-micron transistors with decreased supply voltage while maintaining sufficient drive current. Unfortunately, current SiO2 gate thickness is quickly approaching atomic scale of ~ 1 nm where quantum tunneling current exceeds acceptable level of gate leakage of ~ 1A/cm2. In principle, a high k dielectric can deliver an equivalent SiO2 thickness of 1 nm with a greater physical thickness (10 nm if k=40) and hence a lower leakage current. Enabling Moore's law beyond 0.1 um technology requires a high k gate stack solution with unit gate capacitance of > 30 fF/um^2 to enable continuing transistor scaling of 0.7x voltage and gate length reduction per technology generation without reducing drive current. Many groups have started research project to explore high k gate stack solution. A multi-university center with significant effort on high k research was funded jointly by SRC and Sematech in 1998.

TECHNOLOGY OBJECTIVE AND SOLUTION PATH

The driving force of the high k is to build high performance CMOS transistors at lower operating voltage, which dictates high gate capacitance requirement (Table I). Today's state of the art 0.18 um CMOS technology has a gate capacitance of ~ 15 fF/um2 and a gate voltage of 1.3-1.5 V (1). Transistor scaling requires voltage reduction to <1V in order to have manageable power consumption for chips with > 10M transistors. Maintaining drive current with low voltage requires gate capacitance scaling to > 30 fF/um2. Furthermore, with gate length scaled to sub-0.1 um, high gate capacitance is needed to maintain gate control and reduce short-channel effect. A gate leakage of < 1 A/cm2 is required to have an acceptable standby power. One primary reason of Si as the semiconductor of choice is the superior interface between Si and its oxide, SiO2, with interface trap density, Dit < 5x1010/cm2. A low Dit is required for effective channel charge inversion and for minimizing charge scattering centers that can degrade mobility. With high k gate dielectric, channel mobility can not be degraded in order to have a performance transistor. To implement high k gate stack solution in manufacturing CMOS chips, the solution must be compatible with CMOS process flow and can be effectively integrated. Finally, the high k gate stack must be reliable with > 10 year lifetime for the products to be competitive in the market place.

Table I High k gate stack technical metric and success criteria.

Metric	Success Criteria	Comments
Gate Capacitance (C)	> 30 fF/um^2	(1) C is stuck at ~20 fF/um^2 because of ~1nm interfacial SiO_2;
Gate Leakage (J)	< 1 A/cm^2	
Interface Trap (D_{it})	< $10^{11}/cm^2$	
Channel Mobility	No Degradation	(2) Mobility is degraded because of "high D_{it}."
Process Integration	Compatible with CMOS	
Reliability	> 10 years	

Today's start-of the art 0.18 um technology uses ultra-thin SiO2 and complementary poly-Si electrodes as gate stack (1). Figure 1 shows one possible gate stack solution path. By replacing SiO2 with Si-friendly high k gate dielectric, one may achieve a gate capacitance of ~ 30 fF/um2 with poly-Si/high k/Si structure. This gate capacitance can support 0.1 um technology. By further replacing poly-Si electrode with complementary silicides, one may achieve a gate capacitance of ~ 45 fF/um2 by eliminating poly-depletion. This gate capacitance is needed beyond 0.1 um technology.

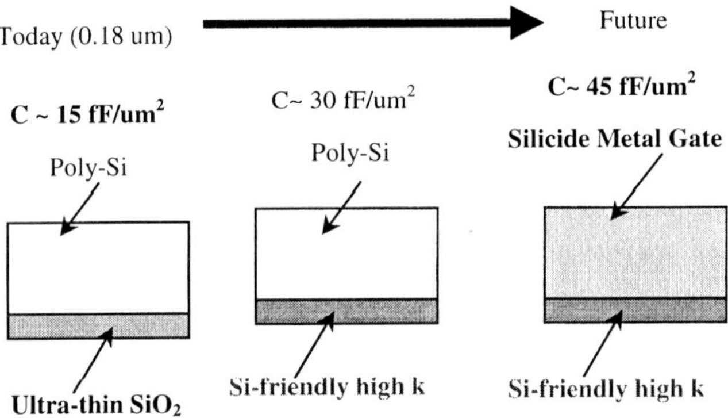

Fig. 1 Schematic showing one possible solution path to achieve high gate capacitance to enable transistor scaling beyond 0.1 um.

There are several classes of dielectric, such as oxides and fluorides. Among oxides, one can categorize them into two types according to their thermodynamic stability when in contact with Si: (1) Si-friendly oxides which are thermodynamically stable on Si and (2) other oxides which react with Si upon thermal annealing (2). For type (2) oxides, a dielectric barrier layer such as an ultra-thin Si3N4 is needed to prevent reaction between Si and these oxides. The presence of such a barrier layer complicates process and limits equivalent oxide thickness (EOT) to be > 5 A. Therefore we want to focus on Si-friendly oxides as gate dielectric to replace SiO2. Table II lists Si-friendly oxides with a dielectric constant > 12 (3x of SiO2 dielectric contact). The simplest such oxides have one metallic element, which are rare-earth metals except Al. Their dielectric constants range from 12 to 40. To achieve a EOT of 5A, the physical thickness of such oxides varies from 15 to 50 A. One issue with these rare-earth oxides is that they are good oxygen conductors. This implies that an interfacial SiO2 can be formed between Si and high k if there is sufficient oxygen supply and these oxygen diffuse through rare-earth oxides to Si interface. Al2O3 is advantageous in this respect because it is a good oxygen barrier. More complex such oxides have two different metallic elements. They tend to have higher dielectric constants. One such oxide, SrTiO3, deserves special notice. In bulk form, SrTiO3 reacts with Si to form SrSiO3 and TiSi2, with a small driving force of –19 kcal/mole. However, preliminary results indicate that epitaxial SrTiO3 can be stable on Si up to ~ 800C. This stability could be induced by strain and reduced interfacial energy of epi-SrTiO2/Si.

Table II Selected list of Si-friendly oxides with a dielectric constant, k, > 12.

Silicon-Compatible Oxides

Material	***K***
La_2O_3	21
HfO_2	18 – 40
ZrO_2	12 – 20
Y_2O_3	14
Al_2O_3	11 -13
Y_2O_3—ZrO_2	30
$LaAlO_3$	25
$HfSiO_4$	13
$ZrSiO_4$	13
(*)$SrTiO_3$	**60-200**

Thin film deposition is a mature technology, with many deposition methods, from physical deposition (such as sputter, MBE) to chemical deposition (such as RTCVD, PECVD, ALCVD, UHV-CVD). For gate dielectric deposition on Si, two features are critical: (1) one must be able to produce a pristine Si surface without native SiO2; and (2) one must be able to manipulate deposition at atomic level and control low oxygen partial pressure to ensure no SiO2 will form during high k deposition. MBE or its variant, UHV-CVD, is best suited for this application (Table III).

Table III Characteristics of MBE (or UHV-CVD) deposition platform.

Deposition	Features	Comments
MBE **UHV-CVD**	(1) UHV environment to achieve fresh Si surface (2x1 reconstruction) (2) Controlled oxygen ambient to manipulate oxide growth without forming SiO_2	All other depositions (such as PVD, ALCVD, RTCVD, PECVD) show presence of interfacial SiO_2 from either native oxide prior to deposition or oxide grown during high k deposition

As dielectric thickness is scaled to < 10 A regime, poly-Si depletion becomes a significant portion of the total electrical thickness of the gate capacitor. To increase gate capacitance, one must eliminate this poly-depletion by using metal electrodes. To build high performing, low threshold voltage CMOS transistors, metal electrodes must have work functions close to that of n+ and p+ poly-Si. In other words, two types of metals must be used, one with a work function close to ~ 4 eV for nMOS and another with a work function close to ~ 5 eV for pMOS. There are several classes of metals, such as elemental metals, silicides, and conductive nitrides and oxides. Table IV shows metals with work function close to either 4 or 5 eV.

Table IV Work functions of selected metals.

Work function (eV)	Al	Ti	TiSi2	ZrSi2	Pt	Pd	MoSi2	WSi2
N+-poly Si like	4.1	4.0	4.0	4.1				
P+-poly-Si like					*5.3*	*5.0*	*4.7-6.0*	*4.6-5.0*

Integrating dual metals into a CMOS process flow is a major roadblock to implement metal electrodes. One scheme is to build transistor in standard process flow, and then remove "dummy" poly-Si and SiO2 gate stack, and deposit a new high k and metal gate stack (3). Another scheme is to build transistors first, and then convert poly-Si to silicides by depositing metal on top of poly-Si and inducing reaction at high temperature (Fig. 2). This scheme is much simpler, and can be effectively integrated in existing CMOS process flow. The key is that the high k gate dielectric must be able to stop silicide reaction at silicide/high k interface. This scheme also requires that the high k dielectric is stable up to ~ 1000C/10s for dopant activation.

Fig. 2 Schematic drawing of complementary silicides gate electrode with n+ and p+ poly-Si like work functions.

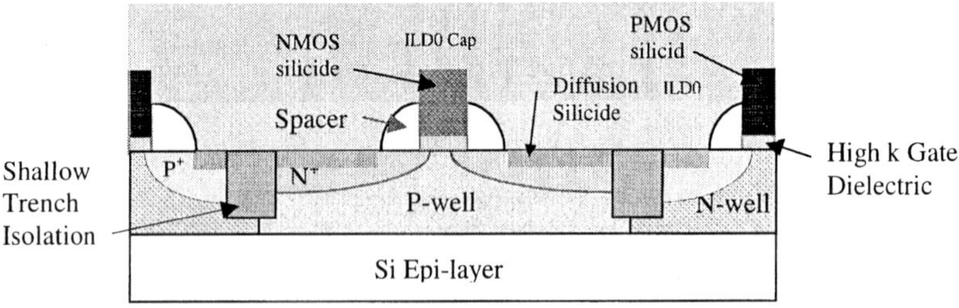

EXPERIMENTAL RESULTS AND DISCUSSION

Many groups from university, industry, and national labs have started research projects on high k gate stack in past year (4-9). Most groups started by directly applying high k oxides (such as Ta2O5, BST) extensively studied for DRAM application to gate. One immediately finds that such oxides react with Si upon annealing and a Si3N4 barrier layer is required to prevent such a reaction. As a result, the EOT is stuck at > 10 A (Table V).

Table V Summary of high k results to date from different research groups.

High k Dielectric	**Deposition**	**Results**
$SrTiO_3$/Si (ORNL, Motorola)	**MBE**	+ up to 36 fF/um^2; + no interfacial SiO_2; - many unknowns -> require more research
ZrO_2, HfO_2, Al_2O_3	ALCVD	+ up to 20 fF/um^2 - interfacial SiO_2 > 1 nm -> must limit this SiO_2
$ZrSiO_4$, $HfSiO_4$ (TI, NCSU)	PVD, CVD	+ up to 20 fF/um^2 - little data -> require more research to explore the limit
TiO_2/Si_3N_4, Ta_2O_5/Si_3N_4 (U.Minnesota,Yale, UT/Austin, Berkeley	CVD	- need of a Si_3N_4 barrier layer limits EOT to > 1 nm

We started with Si-friendly oxides such as HfO2, ZrO2, Al2O3, Y2O3, and used ALCVD to deposit such oxides. We also found that there is an interfacial SiO2 (> 10 A) between high k and Si (Fig. 3). This SiO2 comes from two sources: (1) native oxide before deposition starts and/or (2) growing oxide during high k deposition when oxygen diffuses through high k to Si interface. As a result, the gate capacitance is stuck at < 20 fF/um2 (Fig. 4). To achieve gate capacitance of 30 fF/um2, one must eliminate this interfacial SiO2. This implies that (1) one must have a fresh pristine Si surface prior to high k deposition and (2) one must be able to manipulate deposition at atomic layer to prevent SiO2 formation. MBE system becomes the tool of choice for such application. Using a MBE, McKee has demonstrated that SrTiO3 can be formed on Si without any interfacial SiO2 and achieved a gate capacitance of ~36 fF/um2 (4).

99Å ZrO_2 on Si(100)

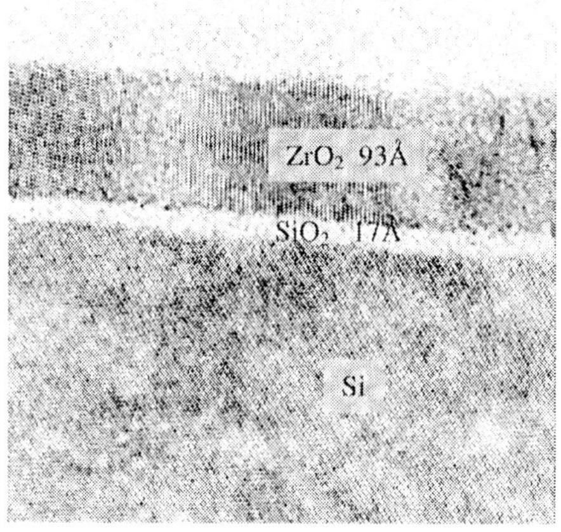

Fig. 3 XTEM of ALCVD ZrO2.

Fig. 4 Leakage current density of MOS capacitors with ALCVD high k dielectric.

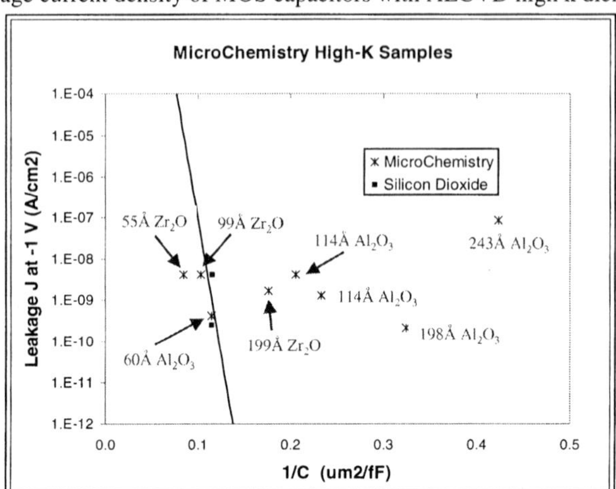

CONCLUSIONS

To scale transistors beyond 0.1 um requires a high k gate dielectric to replace current SiO2. None of high k gate dielectric investigated to date meets all requirements. SrTiO3 is promising because epitaxial SrTiO3 has been grown on Si without any interfacial SiO2 and has a gate capacitance up to 36 fF/um2. Realizing the promise of high k in an integrated CMOS process is an enormously challenging task and requires effective collaboration among researchers from academics and industry.

ACKNOWLEDKEMENTS

I thank C. Perkins, B. Roberds, J. Shaw for analysis and discussion, J. Carruthers, B. Doyle, B. Triplett, R. Chau and L. Yau for discussion, P. Coon, J. Mardinly, C. Matos, J. Duan for physical characterizations. I would also like to thank J. Skarp and S. Haukka for ALCVD samples, D. Schlom, A. Kingon for sharing their insights of high k oxides.

REFERENCES

1. S. Yang, et al, IEDM 98, 197.
2. K.J. Hubbard and D.G. Schlom, J. Mater. Res. 11, 2757 (1996).
3. A. Chatterjee, et al. IEDM 98, 777.
4. R.A. McKee, F.J. Walker, and M.F. Chisholm, Phys. Rev. Lett. (1998)
5. B.He, T.Ma, S. Campbell and W. Galdfelter, IEDM 98, 1038.
6. B. Wu, H. Luan, L.Kang, B.Kim, R.Vrtis, D.Roberts, and D. Kwong, IEDM 98, 609.
7. L. Manchanda, IEDM 98, 605
8. X.Guo, T.Ma, T.Tamagawa, and B.Halpern, IEDM 98, 377.
9. D.Park, Q.Liu, T.King, C.Hu, A.Kalnitsky, S.Tay, and C.Cheng, IEDM 98, 381.

Recent Developments in Ultrathin Nitride Gate Stack Prepared by In-situ RTP Multiprocessing for CMOS ULSI

S. C. Song, B. Y. Kim*, H. F. Luan, and D. L. Kwong
Microelectronics Research Center
Department of Electrical and Computer Engineering
The University of Texas at Austin
Austin, TX

M. Gardner, J. Fulford, and D. Wristers
Advanced Micro Devices
Austin, TX

J. Gelpey and S. Marcus
Steag RTP
Tempe, AZ

ABSTRACT

This paper reviews recent advancements in ultra thin nitride gate stack fabricated by in-situ RTP multiprocessing for advanced dual-gate CMOS devices. Compare to control SiO_2 devices of identical equivalent oxide thickness (Teq), devices with RT-CVD Si_3N_4 gate dielectric show significant reduction of tunneling leakage current, higher drive current and peak transconductance, enhanced immunity to hot carrier stress, and complete suppression of boron penetration.

INTRODUCTION

As MOS devices are scaled down, gate oxide thickness has been reduced aggressively. Aside from reproducibility and Manufacturability issues, tunneling currents are a major barrier to be overcome. In addition, boron penetration in P+-poly PMOSFETs with such thin gate oxides increases significantly. Therefore, ultra thin alternative gate dielectrics with lower leakage and enhanced boron diffusion barrier properties are needed. Si_3N_4 has received considerable attention for such application [1]. Previously, it was shown that MOSFETs with relatively thick CVD Si_3N_4 gate dielectric showed high trap density and inferior hot carrier reliability [2]. Recently, we have demonstrated that ultra thin CVD Si_3N_4 gate stack fabricated by in-situ rapid thermal CVD (RTCVD) technique shows significant lower leakage current, lower electron trapping, superior interface properties and excellent boron diffusion barrier properties compared to SiO_2 of identical Teq. In this paper, we report high quality RT-CVD Si_3N_4 films and their applications to advanced N and PMOSFETs. Results will be presented to demonstrate superior performance and reliability of RT-CVD Si_3N_4 gate stack.

B. Y. Kim*. Currently with IBM, Hopewell Junction, NY

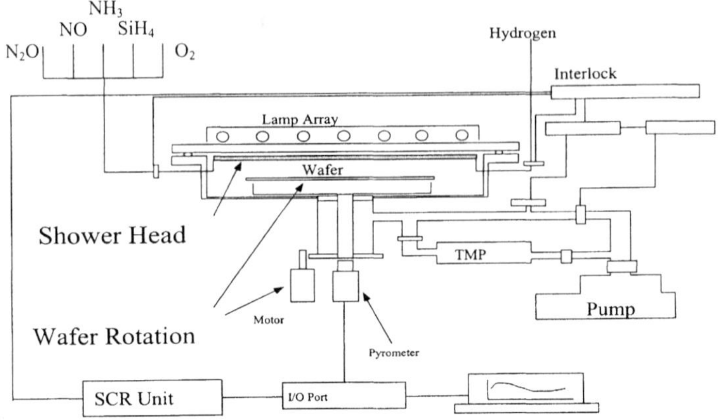

Fig. 1 Schematic of RTCVD module. Wafer rotation and shower head are used to achieve uniform film. Pyrometer is for temperature measurement of substrate

EXPERIMENT

Both MOS capacitors and MOSFETs are fabricated. For MOSFET fabrication, conventional CMOS process is used. In our experiment, both n+poly PMOSFET and p+poly PMOSFET are fabricated. A standard RCA cleaning is used in the sequence of SC1, HF dip, SC2, and final HF dip. Entire gate dielectric films are formed by RTCVD process as described in [3]. The schematic diagram of in-situ RTP CVD system is shown in Fig. 1. In order to obtain uniform film, wafers are rotated (4~5 rpm) during process. Turbo pump is used to reduce base pressure down to 10^{-5} Torr before gas flow. Wafer temperature is measured by pyrometer with wavelength ~3 μm. The temperature-time profile and schematic procedure of gate dielectric formation are illustrated in Fig. 2. RTCVD Si_3N_4 gate dielectric process comprises four consecutive steps, such as interface passivation in NO, Si_3N_4 deposition, further nitridation in NH_3, and final N_2O anneal. The whole steps are carried out sequentially in one chamber without breaking vacuum during process (i.e., in-situ). Control SiO_2 films are also grown in the same RTCVD system in O_2 ambient. Equivalent oxide thickness is calculated based on high frequency C-V measurement in strong accumulation without considering quantum-mechanical effects.

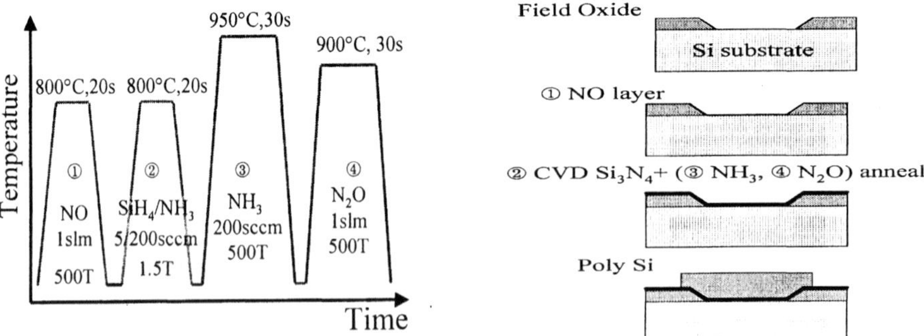

Fig. 2 Time-Temperature profile and schematic procedure of CVD Si_3N_4 gate dielectric formation

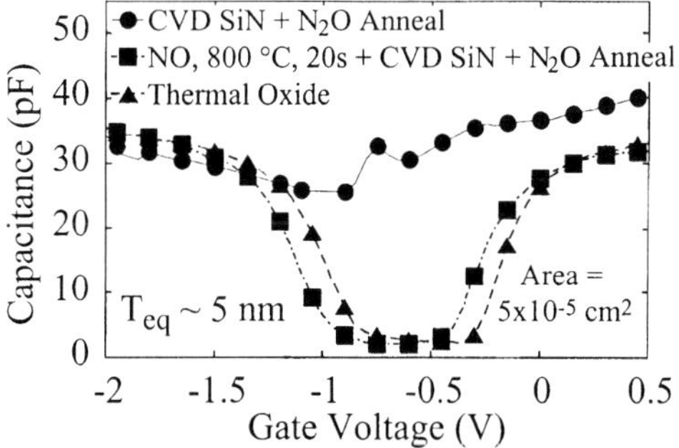

Fig.3 The quasi-static C-V curves of SiO_2 and CVD Si_3N_4 with and without NO layer. Distorted C-V curve indicates poor interface properties. NO layer provides excellent passivation of interface properties

Fig. 4 TEM cross section of nitride/oxide stack gate structure. Extremely smooth interface is observed with NO grown interface passivation layer

Results and Discussion

Fig. 3 shows quasi-static C-V curves of thermal oxide and CVD Si_3N_4 with and without NO grown interfacial layer. The quasi C-V curve of Si3N4 device without the interfacial layer is severely distorted due to large amount of interface state density. Si_3N_4 with interface layer, however, exhibits smooth transition from inversion to accumulation, suggesting improved interface properties.

TEM cross section of the Si_3N_4/ SiO_2 gate stack structure with poly gate and silicon substrate is shown in Fig. 4. The total physical thickness of the stack layer is 28Å, and the equivalent oxide thickness is 20 Å from C-V measurement. As shown in TEM picture, interface passivation layer grown in NO before CVD Si_3N_4 deposition provides extremely smooth interface between CVD Si_3N_4 and silicon substrate. Since direct CVD Si_3N_4 deposition on silicon substrate generates lots of interface states which degrade MOSFET performance, NO grown interfacial passivation layer is indispensable in order to achieve high performance MOSFET with benefits of Si_3N_4.

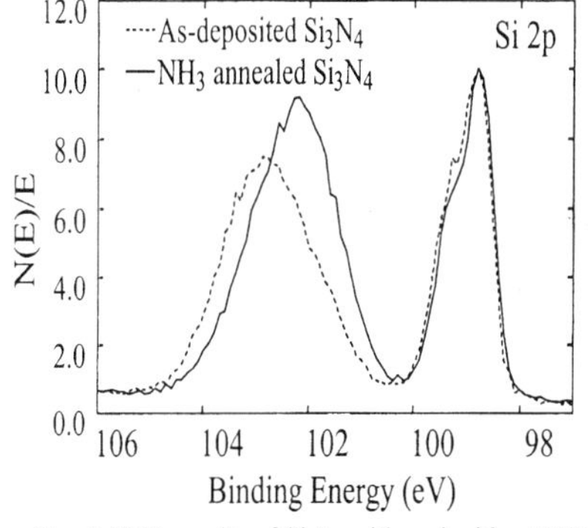

Fig. 5 XPS results of Si 2p with and without NH_3 nitridation. NH_3 nitridation shifts Si 2p peak negatively due to more stoichiometric composition.

The effects of NH_3 anneal after Si_3N_4 deposition is clear from XPS analysis. Fig. 5 shows XPS results of as deposited CVD Si_3N_4 and NH_3 nitrided Si_3N_4 film in which electron binding energy at Si 2p state shifts negatively after NH_3 nitridation. Even though exact binding energy of Si_3N_4 layer is rather difficult due to interference with bottom oxide layer, NH_3 nitridation is clearly shown to convert as deposited Si rich CVD Si_3N_4 to more stoichiometric film.

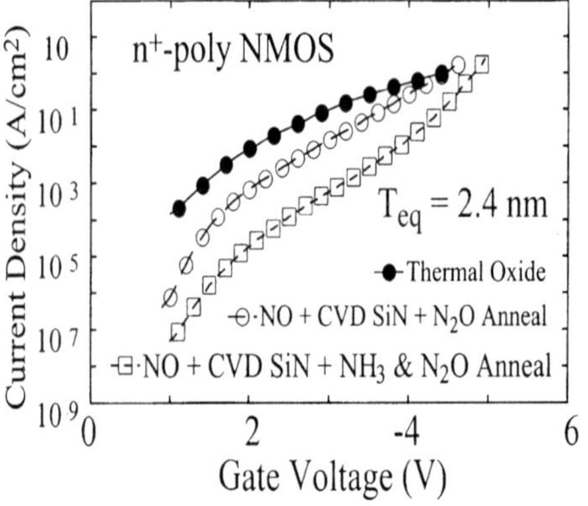

Fig. 6 NH_3 nitridation reduces tunneling current significantly by incorporating more N and eliminating traps

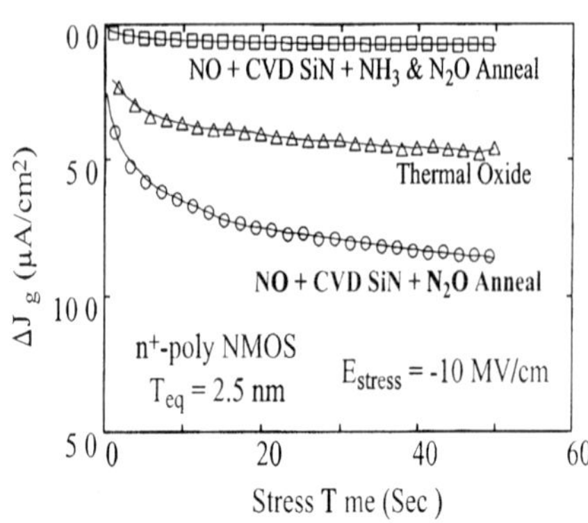

Fig. 7 NH_3 nitridation fills nitrogen vacancies that act as hole traps, resulting in negligible leakage current increase during stress.

NH_3 nitridation reduces hole traps related to excess Si in as-deposited Si_3N_4 layer, and effective to reduce trap-assisted tunneling leakage current as shown in Fig. 6. Since excess silicon, or nitrogen vacancy (similar to oxygen vacancy in SiO_2), act as a hole trap [4], leakage current increases significantly during high field stress due to positive charge build up in the film as shown in Fig. 7. However, NH_3 annealed CVD Si_3N_4 shows negligible leakage increase, meaning that nitrogen vacancies are filled by NH_3 nitridaiton.

Typical high frequency C-V and I-V curves of CVD Si_3N_4 and control SiO_2 are compared in Fig. 8 (a), (b). Flat band voltage difference between these two films is as small as 50mV, which is much smaller compared to Si_3N_4 film prepared by other method [5]. Due to thicker physical thickness, CVD Si_3N_4 film exhibits ~100X lower leakage current density in entire gate bias range compared to SiO_2 of identical equivalent oxide thickness.

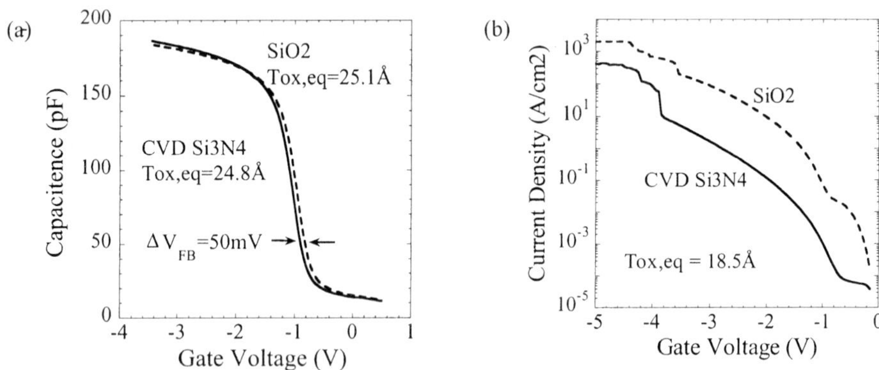

Fig. 8 (a) Typical high frequency CV curves of CVD Si_3N_4 and SiO_2. V_{FB} difference is as small as 50mV. (b) Gate leakage current density of CVD Si_3N_4 and SiO_2 of identical Tox,eq. ~100X lower gate leakage current is observed from CVD Si_3N_4 compared to SiO_2

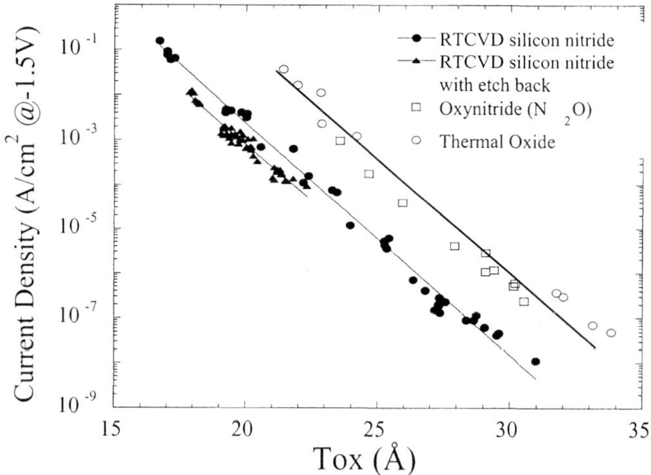

Fig. 9 Current density @Vg=-1.5V of SiO_2, N_2O oxynitride and CVD Si_3N_4 devices with respect to Tox,eq. CVD Si_3N_4 films show more than 10^2X lower leakage current compared to SiO_2 or N_2O oxynitride. Removing top oxide layer grown during N2O anneal using HF solution helps to reduce thickness and leakage current.

In Fig. 9, leakage current density (@Vg=-1.5V) is plotted as a function of equivalent oxide thickness for CVD Si_3N_4, N_2O grown oxynitride, and SiO_2. It's worth to note that HF dip of CVD Si_3N_4 film, which is called "Etch-back process", is effective to reduce thickness and leakage current. Since N_2O anneal step in CVD Si_3N_4 process grows oxide layer on both top and bottom of Si_3N_4 film, removing top oxide layer using dilute HF solution may increase dielectric constant of CVD Si_3N_4 film. Thanks to extremely slow etch rate of Si_3N_4 layer in HF solution, thickness uniformity after etch-back is excellent.

Contrary to conduction in thick Si_3N_4 film which is known as Frenkel-Pool emission related to defect sites inside of nitride layer [6], ultra thin CVD Si_3N_4 film with appropriate post

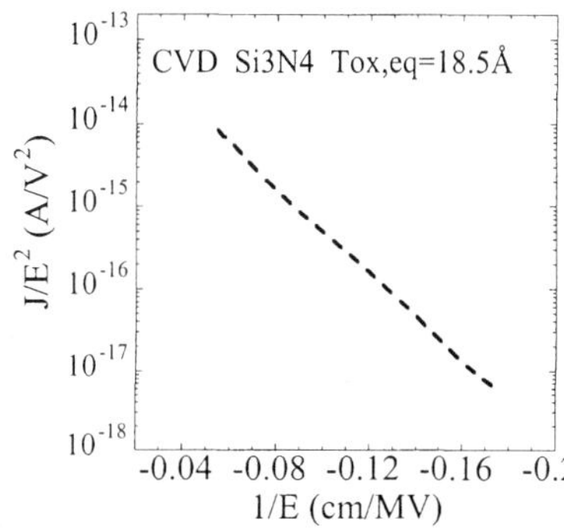

Fig. 10 F-N plot of ultra thin CVD Si_3N_4 film. Linear behavior in F-N plot indicates conduction in ultra thin CVD Si_3N_4 film is tunneling

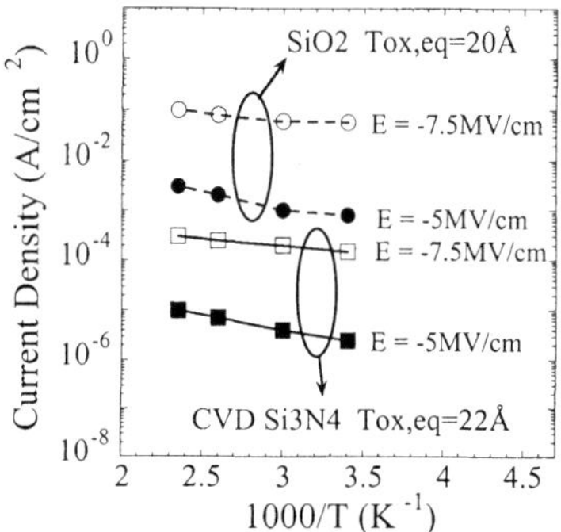

Fig. 11. Negligible dependence of leakage current on temperature suggests conduction mechanism of ultra thin CVD Si_3N_4 film is FN tunneling rather than PF emission.

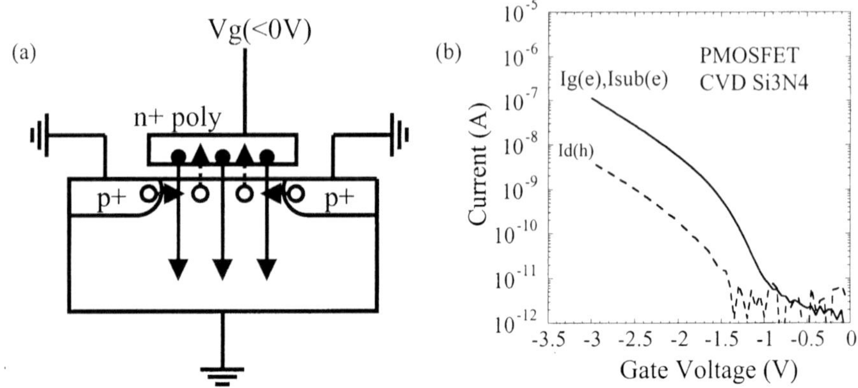

Fig. 12 (a) Carrier separation measurement configuration using n+ poly PMOSFET. Negative bias is applied on n+ gate and souce/drain/substrate are grounded. (b) The results of carrier separation of PMOSFET with CVD Si_3N_4. Most of gate current is composed of substrate current in which electrons flow from n+ gate to substrate.

annealing process exhibits Fowler-Nordheim tunneling as shown by the F-N plot in Fig. 10. This is further confirmed by the temperature independence of leakage current shown in Fig. 11.

It is found that electron is the main carrier composing gate leakage current of ultra thin CVD Si_3N_4 based on carrier separation method using PMOSFET [7] contrary to thick Si_3N_4 in which hole dominates [8]. Fig. 12 (a) illustrates configuration of carrier separation measurement using n+ poly PMOSFET. When negative bias is applied to gate, electrons in poly gate flow to substrate, and holes in source and drain flow into gate. The result of this measurement on CVD Si_3N_4 is shown in Fig. 12 (b). Gate current is almost composed of substrate current which comprises electron flow from gate to substrate, while portion of drain current which means hole flow from source and drain to gate electrode in gate current is less than 5%. Since hole conduction in conventionally thick CVD Si_3N_4 is due to trap levels located close to valence band [9], electron conduction in ultra thin CVD Si_3N_4 film suggests that those hole traps are effectively eliminated by post deposition anneal.

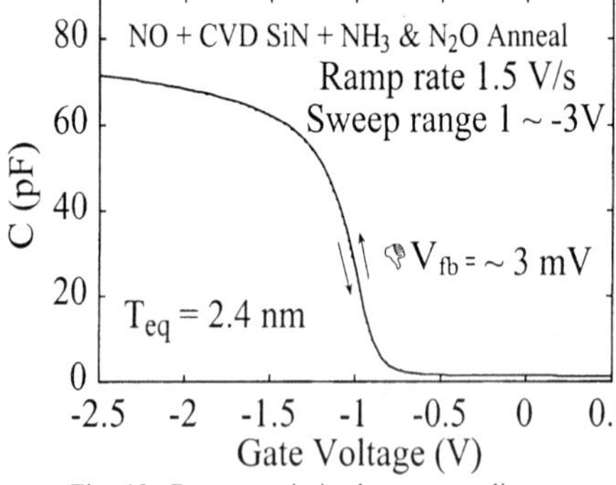

Fig. 13 Due to optimized post annealing process, CVD Si_3N_4 film shows negligible hysteresis.

Hysteresis of the C-V curve of CVD Si_3N_4 film is evaluated by the amount of flatband voltage shift between two different sweep directions (i.e., Vg = 1 to –3V, and –3 to 1V) with gate voltage ramp rate of 1.5V/s in high frequency C-V curve. As can be seen in Fig. 13, virtually no hysteresis characteristics are observed, meaning complete anneal away of shallow bulk traps due to optimized annealing combination of NH_3 and N_2O as well as thinning effect of Si_3N_4 film.

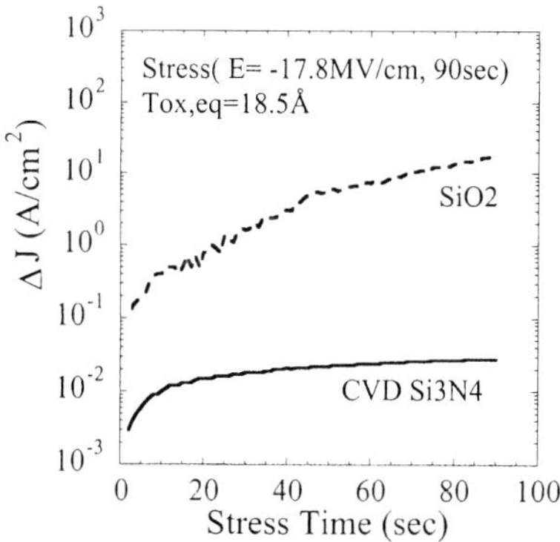

Fig. 14 Current evolution with stress time of SiO_2 and CVD Si_3N_4. Due to large amount of hole trapping, SiO_2 exhibits higher leakage current increase

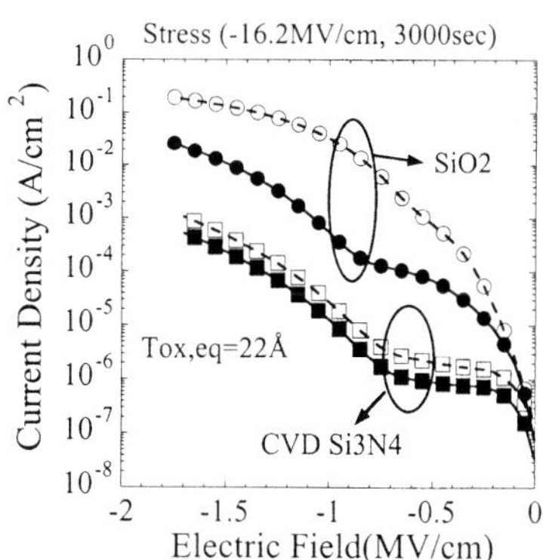

Fig. 15 I-V curves of SiO_2 and CVD Si_3N_4 before and after stress. For SiO_2, leakage current at low field region increases significantly after stress

The gate leakage current at low electric field is increased by high field stress, which is known as SILC (Stress Induced Leakage Current). The reason for increasing leakage current is believed to be due to the build-up of holes originated by high field stress. Fig. 14 shows the leakage current evolution with respect to stress time both from CVD Si_3N_4 and SiO_2. As can be seen, leakage current increases in order to maintain constant field (17.8 MV/cm) across the SiO_2, which means considerable hole trapping. On the other hand, CVD Si_3N_4 shows negligible leakage current change, suggesting far less hole trapping during stress. Fig. 15 depicts the I-V curves of CVD Si_3N_4 and SiO_2 of identical thickness before and after stress. The leakage current increases significantly in SiO_2 after stress especially at low field region, while slight leakage current increase is observed in CVD Si_3N_4. This results indicate that charge trapping, especially hole trapping is considerably suppressed in ultra thin CVD Si_3N_4 film by employing appropriate annealing condition.

TDDB (Time Dependent Dielectric Breakdown) is measured by stressing SiO_2 and CVD

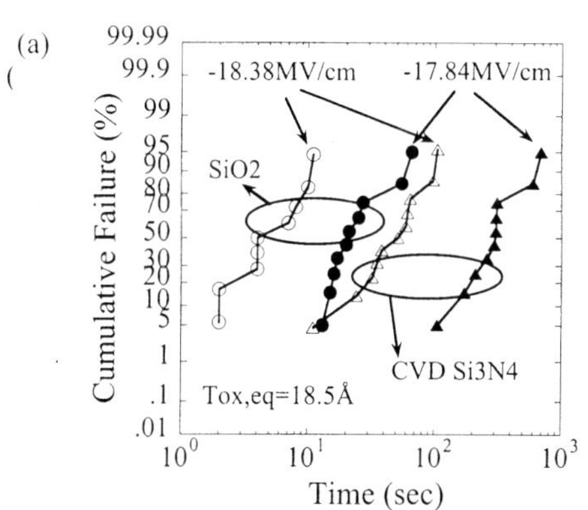

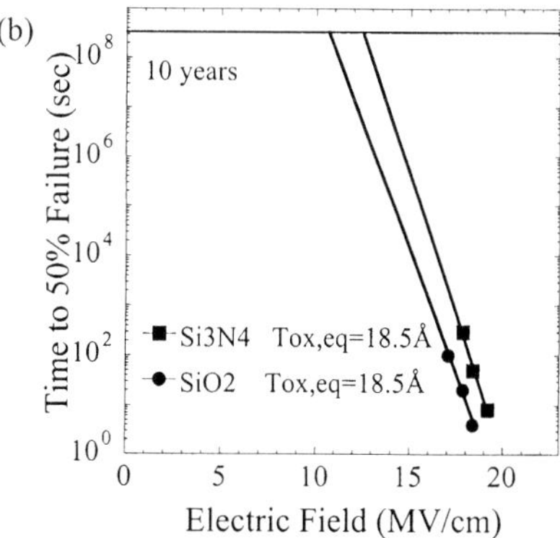

Fig. 16 (a) TDDB characteristics of SiO_2 and CVD Si_3N_4. CVD Si_3N_4 shows almost 10X longer time to breakdown. (b) 10 years lifetime extraction based on TDDB data. –10.5MV/cm and –12.2 MV/cm are the electric field at 10 years for SiO_2 and CVD Si_3N_4 respectively.

Si_3N_4 capacitors with 5X10-5cm2 area at a constant field. Fig. 16 (a) shows the measured cumulative percent failure as a function of stress time at different electric fields. A 10X longer time-to-breakdown is observed in CVD Si_3N_4 film at a given electric field as shown in Fig. 16 (b). Recently, gate oxide failure is reported to be limiting factor for scaling of oxide thickness, since time-to-breakdown decreases exponentially as gate oxide becomes thinner due to increasing tunneling current [10]. Due to significantly lower leakage current, CVD Si_3N_4 can extend scaling limit of SiO_2 in terms of dielectric reliability as well as stand-by power consumption.

The initial drain current of Si_3N_4 device is compared with that of SiO_2 device in Fig. 17 (a), (b). Si_3N_4 devices show distinct higher drain current than SiO_2 devices for both N and PMOSFETs. NO annealed oxynitride has been reported to produce smoother interface and less coulombic scattering compared to N_2O oxynitride or SiO_2 [11], which is speculated to be the reason of improved drivability. Log Id vs. Vg curves are plotted in Fig. 17 (c). It is shown that subthreshold swing values of both Si_3N_4 and SiO_2 devices are excellent and comparable which suggests that CVD Si_3N_4 with NO passivation layer has comparable or even less interface states density compared to control SiO_2.

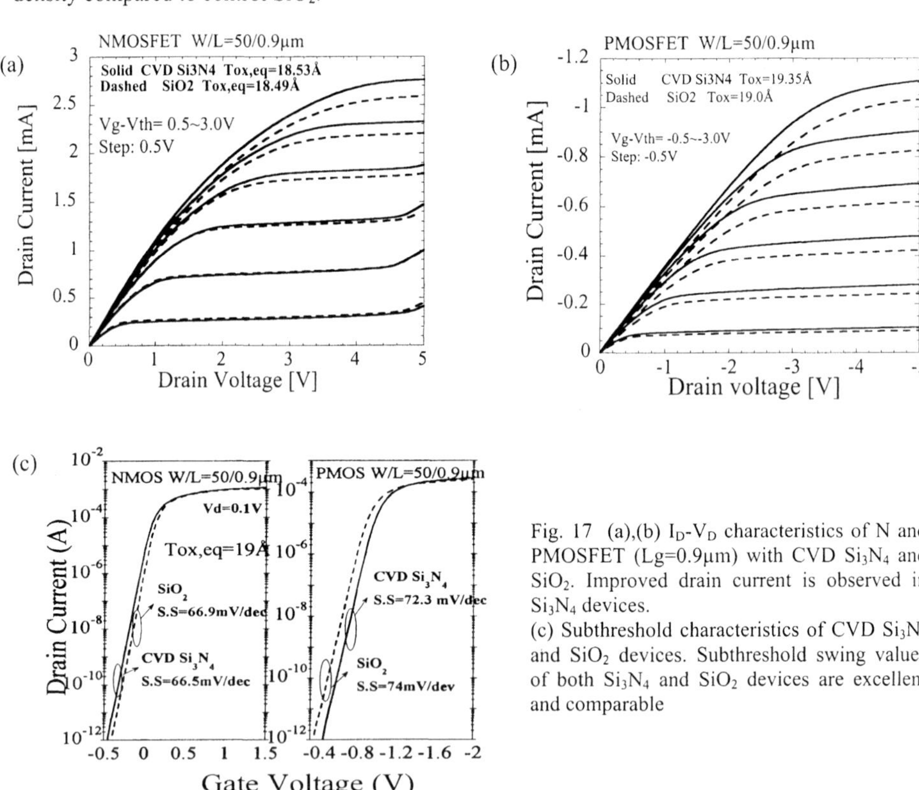

Fig. 17 (a),(b) I_D-V_D characteristics of N and PMOSFET (Lg=0.9μm) with CVD Si_3N_4 and SiO_2. Improved drain current is observed in Si_3N_4 devices.
(c) Subthreshold characteristics of CVD Si_3N_4 and SiO_2 devices. Subthreshold swing values of both Si_3N_4 and SiO_2 devices are excellent and comparable

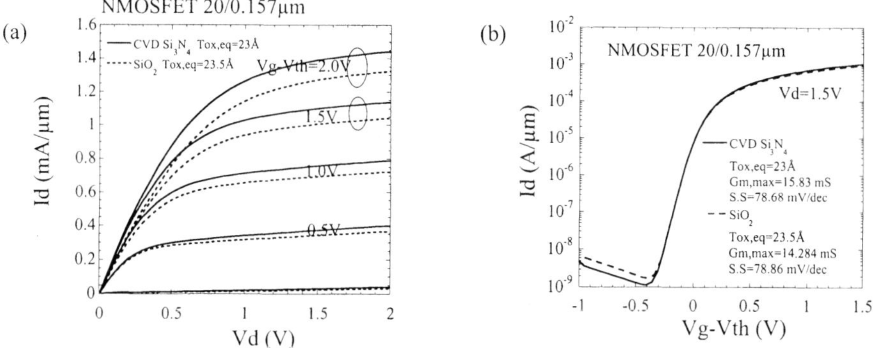

Fig. 18 (a) Id-Vd characteristics of sub-micron NMOSFETs with CVD Si_3N_4 and SiO_2. (b) Log Id-Vg characteristics. CVD Si_3N_4 device exhibits improved characteristics both in "on" and "off" state.

Improved drivability is also observed in deep-sub-micron NMOSFET (Lg=0.157μm) with CVD Si_3N_4 as shown in Fig 18 (a). In addition, CVD Si_3N_4 NMOSFET shows comparable subthreshold characteristics to SiO_2 device (Fig. 18 (b)). NMOSFET (Lg=0.157μm) with CVD Si_3N_4 exhibits Idsat (@Vg=Vd=1.5V)=0.9mA/μm with Idoff (@Vd=1.5V, Vg=0V)=39nA/μm.

Fig. 19 illustrates the normalized transconductance of Si_3N_4 and SiO_2 for both NMOSFET and PMOSFET at linear region (|Vd|=0.1V). Si_3N_4 device shows comparable peak Gm for NMOSFET, whereas PMOSFET with Si_3N_4 shows higher peak Gm value compared to control device. It is known that coulomb scattering is responsible for peak Gm value [12]. Therefore, comparable peak Gm value suggests that CVD Si_3N_4 and SiO_2 have similar interface state and fixed charge that cause coulomb scattering. In the high field region, the Gm value of PMOSFET with Si_3N_4 is lower than control device, while it is still higher in the NMOSFET with Si_3N_4. This asymmetric behavior of high field Gm in Si_3N_4 devices may be due to different type of interface states between N and PMOSFET as described in [13]. However, complete mechanism is not yet clear.

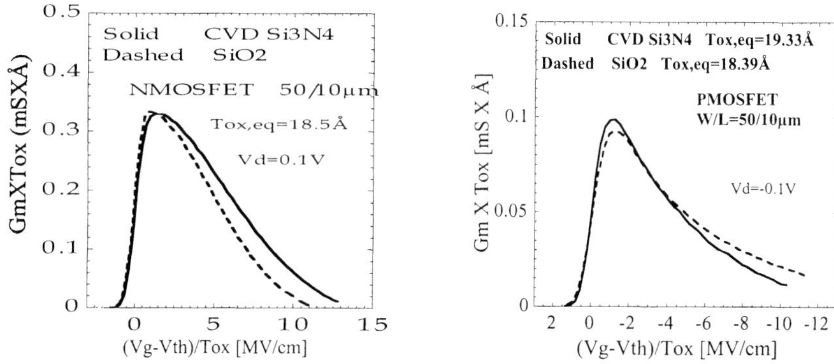

Fig. 19 Normalized transconductance of CVD Si_3N_4 and SiO_2. CVD Si_3N_4 devices exhibit comparable peak G_m values both in N and PMOSFET. Asymmetric G_m behavior between N and PMOSFET is observed at high field region.

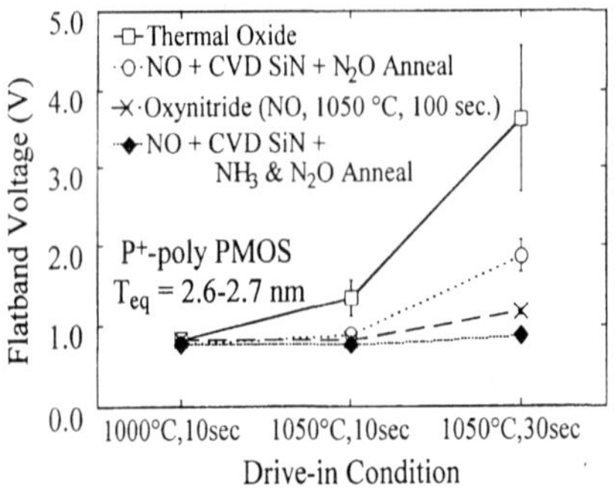

Fig. 20 CVD Si_3N_4 with NH_3 nitridaiton shows negligible V_{FB} shift due to strong resistance to boron diffusion

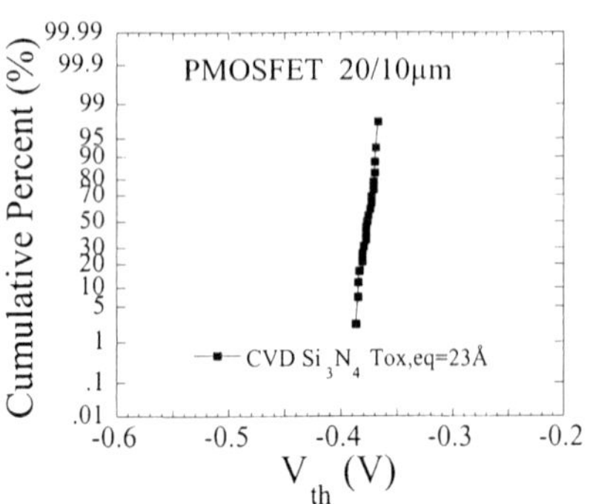

Fig. 21 Vth distribution of p+ poly PMOSFETs with CVD Si_3N_4. Tight Vth distribution indicates complete elimination of boron penetration

One of the major problems associated with ultra thin SiO_2 film in dual gate CMOS process is boron penetration through gate oxide in p+ poly PMOSFET, causing Vth instability and degradation of device performance. Fig. 20 shows boron penetration characteristics of various gate dielectric films. CVD Si_3N_4 shows almost no flat band voltage shift even up to 1050C, 30sec BF_2 drive-in condition, while significant Vfb shift is observed in other films. It's worth to note that CVD Si_3N_4 without NH_3 nitridation exhibits relatively large Vfb shift compared to CVD Si_3N_4 with NH_3 nitridaion, which implies that NH_3 nitridation densifies the as-deposited Si_3N_4 films. Since boron penetration becomes harder as N concentration of gate dielectric increases [14], complete suppression of boron penetration in CVD Si_3N_4 is attributed to high N concentration. Fig. 21 shows Vth distribution of PMOSFETs with CVD Si_3N_4 in 200mm wafer. Since boron penetration causes Vth instability, extremely tight Vth distribution indicates complete elimination of boron penetration in CVD Si_3N_4 film.

Scaling of critical dimensions to sub-micron ranges without proportional scaling of the power supply voltage results in a significant increase of the horizontal and vertical electric fields

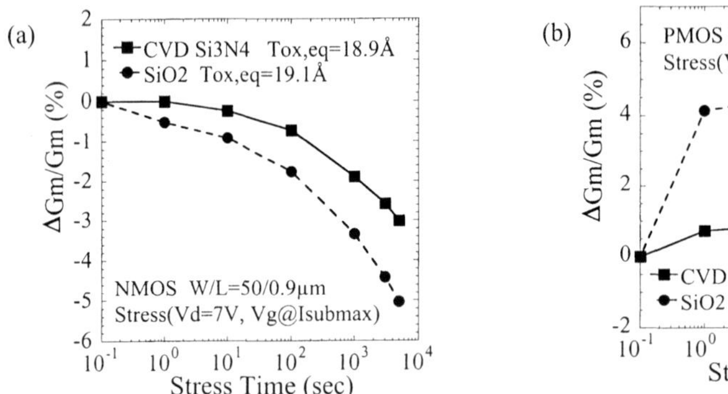

Fig. 22 (a) Less Gm degradation in NMOSFET with CVD Si_3N_4 is attributed to stronger Si-N bonds than strained Si-O bonds at the interface. (b) PMOSFET with SiO_2 exhibits turn-around phenomenon due to electron trapping initially and interface state generation later.

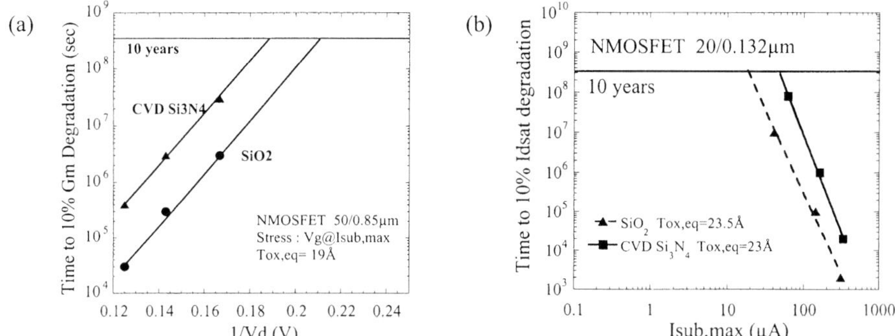

Fig. 23 Hot carrier reliability comparison of NMOSFET ((a) Lg=0.85μm (b) Lg=0.132μm) between CVD Si_3N_4 and SiO_2 under DAHC condition. Due to stronger Si-N bonds at the interface, CVD Si_3N_4 has 10X longer lifetime.

in the channel region, which increases likelihood of hot carrier induced degradation in aggressively scaled MOSFETs. Under DAHC (Drain Avalanche Hot Carrier injection) condition for NMOSFETs, CVD Si_3N_4 device shows superior hot carrier reliability compared to control SiO_2 device as shown in Fig. 22 (a). Less Gm degradation is observed in CVD Si_3N_4 NMOSFETs, and this is attributed to stronger Si-N bonds than strained Si-O bonds at the interface, which suppresses interface state generation during hot carrier stress. Significant electron trapping and interface state generation are observed for PMOSFETs with SiO_2 under DAHC condition. As shown in Fig. 22 (b), electron trapping dominates the degradation initially, which results in channel shortening, thus Gm increases. However, the interface state generation becomes more important later, which reduces Gm, resulting in turn-around phenomena in Gm degradation. CVD Si_3N_4 PMOSFET, however, shows strong immunity to hot carrier stress due to low electron trap density as well as stronger interface properties.

Gm degradation of NMOSFETs with CVD Si_3N_4 and SiO_2 is monitored under different Vd, and device lifetime against hot carrier stress, defined as time to 10% Gm degradation, is extracted. As shown in Fig. 23 (a), 10X longer lifetime is observed in CVD Si_3N_4 compared to SiO_2 devices. Deep-sub-micron (Lg=0.132μm) NMOSFETs are also evaluated for hot carrier reliability, and the extracted lifetime is shown in Fig. 23 (b). Under identical maximum substrate current, CVD Si_3N_4 device shows ~10X longer time to 10% Idsat degradation compared to SiO_2.

CONCLUSION

Proposed fabrication process of CVD Si_3N_4 film comprises 4 steps (1) growth of interface passivation layer using nitric oxide (NO), (2) deposition of Si_3N_4 through reaction of SiH_4 and NH_3. (3) further nitridation in pure NH_3 ambient, and (4) post nitridation annealing with nitrous oxide (N_2O). CVD Si_3N_4 with NO interface layer has comparable or even lower interface state density compared to SiO_2. In addition, stronger Si-N bonds at the interface inhibit interface state generation during hot carrier stress, which results in enhanced immunity of CVD

SiO_2 devices to hot carrier stress. Higher dielectric constant of Si_3N_4 allows thicker physical thickness without sacrificing equivalent oxide thickness, resulting in significant lower leakage current. Further nitridation in pure NH_3 ambient is very effective to densify as-deposited Si_3N_4 layer due to more stoichiometric composition and reduced charge trap density. Hydrogen-containing species, acting as electron traps originated from NH_3 and SiH_4 in Si_3N_4 film, are effectively eliminated due to stronger oxidizing effect of N_2O anneal. Due to higher N concentration, CVD Si_3N_4 film eliminates boron penetration completely. MOSFETs with CVD Si_3N_4 show improved drivability, and no indication of channel mobility degradation.

CVD Si_3N_4 with appropriate anneal processes can be used as advanced alternate gate dielectric material for future ULSI technology due to lower leakage current, improved MOSFET performance, higher reliability to electrical stress, and easier implementation into real process.

REFERENCES

[1] H. S. Momose, T. Moromoto, Y. Ozarwa, K. Yamabe, H. Iwai, *Trans. Electron Devices*, 1995, v. 42, no.4, p. 704

[2] H. Iwai, H. S. Momose, T. Morimoto, S. Takagi, and K. Yamabe, *ESSDERC* Sept. 1990, p. 287

[3] B. Y. Kim, H. F. Luan, and D. L. Kwong, *IEDM*, 1997, p.163

[4] M. H. Woods, R. Williams, *J. Appl. Phys*. 47, 1997, p. 1082

[5] H. H. Tseng, P. Tsui, P. Tobin, M. Khare, *IEDM*, 1997, p. 647

[6] S. M. Sze, *J. Appl. Phys*. V. 38, 1967, p. 2951

[7] M. K. Mazumder, K. Kobayashi, J. Mitsuhashi, H. Koyama, *Trans. Electron Devices*, 1995, v. 41, no.12, p. 2417

[8] F. T. Liou, S. O. Chen, *Trans. Electron Devices*, 1984, v. 31, no.12, p. 1736

[9] V. J. Kapoor, S. B. Bibyk, *The Physices of MOS Insulator*, New york: Pergamon, 1980, p. 117

[10] J. H. Stathis, D. J. DiMaria, *IEDM*, 1998, p. 167

[11] M. T. Takagi, Y. Toyoshima, *IEDM*, 1998, p. 575

[12] B. Maiti, P. Tobin, V. Misra, *IEDM*, 1997, p. 651

[13] T. Hori, *Trans. Electron Devices*, 1990, v. 37, p. 2058

[14] R. B. Fair, *J. Electrochem. Soc*, 1997, vol.144, p.709

ADVANCED GATE DIELECTRICS SYNTHESIZED BY JVD

T.P. Ma
Yale University
Center Microelectronic Structures and Materials, and
Department of Electrical Engineering
New Haven, Connecticut, 06520, USA

ABSTRACT

The principle of the Jet-Vapor Deposition (JVD) technique for thin dielectric deposition will be introduced, the properties of ultra-thin JVD silicon nitride (designated SiN in this paper) as an advanced MOS gate dielectric will be presented, and recent results on the JVD TiO_2/SiN gate stack will also be reported.

INTRODUCTION

The JVD process relies on supersonic jets of a light carrier gas such as helium to transport depositing vapor from the source to the substrate, where films are formed.

Figure 1 shows the top view of a JVD system used to deposit the silicon nitride films reported here. One can see that the wafer is scanned with respect to the vapor jet to achieve film uniformity. A microwave cavity surrounds a portion of the jet nozzle and generates a gaseous discharge in the nozzle.

Figure 2 shows a close-up view of the coaxial dual-nozzle jet-vapor source used for silicon nitride deposition, in which He is used as the carrier gas to rapidly transport the depositing species (silane in the inner nozzle and reactive nitrogen in the outer nozzle) to the substrate where they undergo reactions to form the silicon nitride film. The pressures in both nozzles are significantly higher than that in the chamber so that a supersonic flow occurs at the exit of both nozzles. It is believed that the high speed of the jet leads to two important consequences which contribute to the high quality of the deposited films: (i) the high impinging energy of the depositing species allowed the use of room-temperature substrates, and (ii) the short transit time of the depositing species minimizes the possibility of gas-phase nucleation.

The deposition rate is controlled by adjusting the flow rates of silane, nitrogen, and helium, the microwave power, and the wafer scan speed. For thin gate dielectric applications, a deposition rate lower than 0.5 nm/min is typically used, although it is possible to increase the rate significantly.

A post-deposition annealing (PDA) step normally follows the silicon nitride deposition to densify the film, and the PDA temperature ranges from 600 to 800 °C for up to 30 minutes in nitrogen. A post-metal anneal (PMA) step is also normally part of the processing sequence, and the PMA is typically performed in a forming gas ambient (5% H_2 + 95% N_2) at 400 °C for 5 to 30 min.

A more detailed description of the JVD process can be found in ref [1].

PROPERTIES OF JVD SILICON NITRIDE FILMS

Figure 3 shows the FTIR spectrum of a JVD silicon nitride film deposited on Si, which exhibits the expected SiN bonds. The hydrogen concentration, as manifested by the NH and SiH signals, is relatively low as compared to the plasma-enhanced CVD silicon nitride.

Figure 4 shows the Auger depth profile of a JVD silicon nitride film deposited n Si. One can see the expected silicon and nitrogen signals. There is also an oxygen signal throughout the thickness of the film, although no oxygen gas was intentionally introduced into the deposition chamber. This unintentional oxygen is believed to come from the residual gas in the chamber, which is evacuated by mechanical pumps to a base pressure of ~ 1 mTorr. The effects of the oxygen on the film properties are not entirely clear, although it is known to lower the dielectric constant (it changes from 7.8 for pure nitride to 6.5 with 20% of oxygen) and thus increases the gate tunneling current for a given EOT [2].

Figure 5 shows a histogram of the breakdown fields for a set of MNS capacitors with EOT of 4.2 nm. These data are comparable to high-quality thermal SiO_2.

Figure 6 shows a pair of high-frequency and quasi-static C-V curves for an MNS capacitor with a JVD nitride film of 4.2 nm of equivalent oxide thickness (EOT). There is no discernible hysteresis in the high-frequency C-V curve when swept back and forth along the voltage axis. These results are comparable to quality MOS capacitors, and certainly much better than any other MNS capacitors made of conventional CVD silicon nitride reported in the literature.

Figure 7 shows the result from a constant-current stress experiment. Based on the saturation voltage shift (of order of 30 mV), one may estimate a trap density in the $10^{11}/cm^2$ range, which is comparable to the values for quality thermal SiO_2, but much lower than conventional CVD silicon nitrides, which typically contain at least one order of magnitude higher densities of traps [3-9].

The I-V characteristics of the MNS capacitors fit Fowler-Nordheim(F-N) plot very well, as exemplified by the data in Fig.8, where J is the current density and E is the equivalent oxide field. Using an effective mass of 0.5 times the free electron mass, a barrier height of 2.1 eV is calculated from the slope of the straight line. Other MNS capacitors made with the JVD silicon nitride show similar barrier height values from their I-V characteristics.

Since in the conventional CVD silicon nitride the dominant current component is due to the Frenkel-Poole (F-P) conduction mechanism, it is very interesting that the JVD nitride shows primarily F-N tunneling mechanism. It is believed that the lack of traps in the JVD nitride is probably responsible for this, because the F-P conduction requires a high density of traps.

The fact that the dominant conduction mechanism in the JVD nitride is tunneling is also confirmed by its relatively weak temperature dependence, as shown in Fig.9. As a comparison, the CVD nitride shows a much stronger temperature dependence.

Theoretical calculations indicate that, if tunneling is the dominant current transport mechanism rather than Frenkel-Poole process, then the gate leakage current should be substantially lower for the nitride when compared with thermal SiO_2 of the same equivalent thickness [1, 10]. Figure 10 shows an example where the dielectric constant of the nitride is assumed to be 6.5, which is a reasonable assumption based on the experimental results. Note that the kink in each curve corresponds to the transition from direct tunneling to Fowler-Nordheim tunneling as the dominant transport mechanism.

One can see that, for a given equivalent oxide thickness, the leakage current is substantially lower in the nitride, especially in the direct tunneling regime, where the difference is several orders of magnitude. By assuming a higher dielectric constant for the nitride, the calculation showed even lower nitride current [10, 11].

Figure 11 shows the leakage currents measured in three JVD nitride films of 2.1, 2.9, and 3.9 nm of EOT, respectively. For comparison, the data for thermal SiO_2 of comparable thicknesses are also included. The data clearly indicate a lower leakage current through the nitride for a given oxide equivalent field, in qualitative agreement with the theoretical prediction. In the low-field regime, the leakage currents in the nitride films are not as low as what the tunneling calculations have predicted. It is suspected that the Frenkel-Poole conduction mechanism has not been entirely eliminated in these nitride films.

The reason that the tunneling current is much lower in silicon nitride for a given EOT, despite its lower barrier height than SiO_2, is because of its larger physical thickness (nearly twice as thick as SiO_2) which more than offsets the effect of its lower barrier height.

Figures 12(a) and (b) show I-V characteristics for a set of ultra-thin JVD silicon nitride films measured before and after high-field stress in both polarities, and the SILC values are very small.

Both n- and p-channel field-effect transistors (FET's) with JVD silicon nitride as the gate dielectric exhibit good quality. Figure 13(a) shows a family of drain current curves of a N-channel MNS transistor with EOT of 3.8 nm, while Fig.13(b) shows the corresponding transconductance characteristics. The data for a control MOSFET are also included for comparison. Note that both sets of data have been normalized with respect to the gate EOT in order to have a fair comparison. One can see that the low-field transconductance is lower for the nitride sample, but its high-field transconductance is higher than that of the oxide sample. This behavior may be attributed to carrier trapping by the slow interface traps (or border traps [12]) near the nitride/Si interface, as described elsewhere [13].

The data for the P-channel devices are shown in Figs.14(a) and (b),where the transconductance for the MNS device is also lower than that for the MOS device at low fields, but higher at high fields. The current driveability is comparable between the two dielectrics.

The JVD silicon nitride film is highly resistant to boron penetration and oxidation [1], making it ideally suited as an interfacial buffer for high-k gate dielectrics which are possible successors to ultra-thin silicon nitride as future-generation gate dielectrics. Our preliminary results on the TiO_2/SiN dielectric stack have indeed lived up to our expectations [14].

JVD TiO_2/SiN GATE STACK

Figure 15 shows the high-low frequency C-V curves of a MIS capacitor with a TiO_2/SiN stack gate dielectric, indicating an excellent interface. Figure 16 shows that the gate leakage current through a TiO_2/SiN stack is several orders of magnitude lower than through its thermal oxide counterpart of similar EOT. The transistor characteristics of MISFETs made with TiO_2/SiN gate dielectric are also excellent, as shown in Fig.17. The corresponding transconductance curves show smaller low-field mobility but higher high-field mobility, when compared with the thermal oxide sample, as shown in Fig. 18. The robustness of the TiO_2/SiN stack may be illustrated by the data shown in Fig.19, where one observes practically no SILC after prolonged high-field stress. The breakdown distribution for either voltage polarity is rather tight, as shown in Fig.20

CONCLUSION

The JVD silicon nitride film and the JVD TiO_2/SiN stack exhibit properties that make them promising candidates to succeed thermal SiO_2 as future generations of advanced gate dielectrics.

ACKNOWLEDGEMENT

The author wishes to thank the JVD team at Yale University, Xiewen Wang, Ying Shi, Mukesh Khare, and Xin Guo, for all the results reported here, Takashi Tamagawa and Guang-Ji Cui at JPC for their technical support, and SRC, NSF, ONR, and DARPA for supporting this work.

REFERENCES

1. T.P. Ma, IEEE Trans. Electron Dev., vol.45, no.3, pp.680-690, (1998).
2. Xin Guo and T.P. Ma, IEEE Electron Dev. Letts. Vol.19, No.6, pp.207-209, (1998).
3. B. Deal, E. MacKenna, and P. Castro, J. Electrochem. Soc., vol.116, pp.997-1005 (1969).
4. A. Ginovker, V. Gristsenko, and S. Sinitsa, Phys. Stat. Sol.(a), vol. 26, pp.489-495, (1974).
5. V.J. Kapoor, R. S. Bailey, and H.J. Stein, J. Vac. Sci. Technol., vol. A1, no., pp.600-603, (1983).
6. S. Fujita, T. Ohishi, T. Toyoshima, and A. Sasaki, J. Appl. Phys., vol.57, no.2, pp.426-431, (1985).
7. M. Maeda and H. Nakamura, J. Appl. Phys., vol.58, no.1, pp.484-489, (1985).
8. K. Alloert, A. Van Calster, H. Loos, and A. Lequesne, J. Electrochem. Soc., vol.132, pp.1763-1766, (1985).

9. W.R Knolle and J.W. Osenbach, J. Appl. Phys., vol. 58, no.3, pp.1248-1254, (1985).
10. X.W. Wang, Y. Shi, T.P. Ma, G.J. Cui, T. Tamagawa, J. Golz, B. Halpern, and J. Schmitt, in 1995 Symp. VLSI Technol. Dig. Tech. Papers, pp.109-110.
11. Mukesh Khare, X.W. Wang, T.P. Ma, G.J. Cui, T. Tamagawa, B. Halpern, and J. Schmitt, in 1998 Symp. VLSI Technol. Dig. Tech. Papers, pp.218-219.
12. D.M. Fleetwood, IEEE Trans. Nucl. Sci., Vol.39, p.269 (1992).
13. M. Khare and T.P. Ma, IEEE Electron Dev. Letts. Vol. 20, No.1, pp.57-59, (1999).
14. Xin Guo, T.P. Ma, T. Tamagawa, and B. Halpern, IEDM Tech. Digest, pp.377-380 (1998).

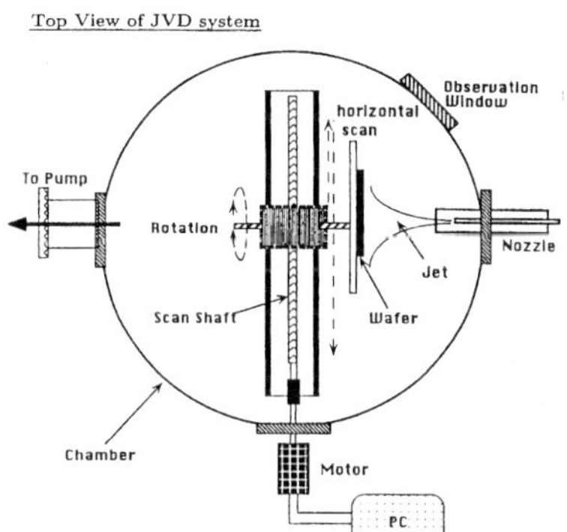

Fig.1 Schematic top view of a JVD system.

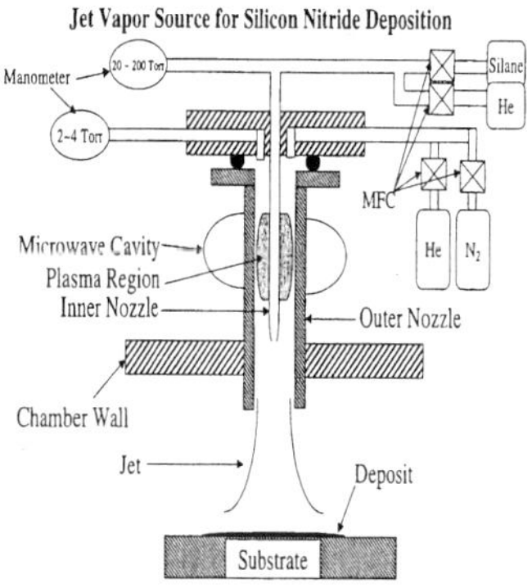

Fig.2 Dual-nozzle jet-vapor source for silicon nitride deposition.

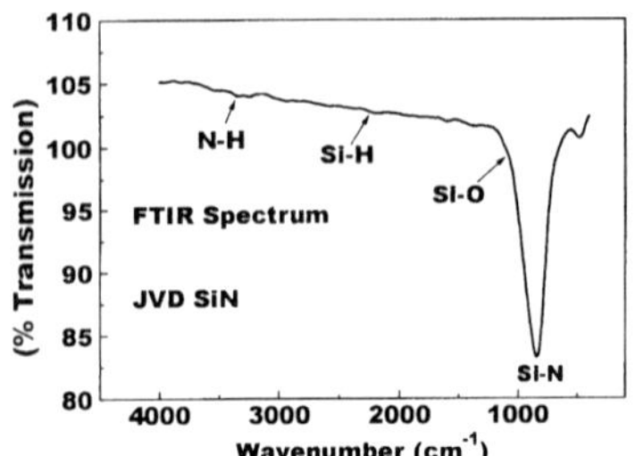

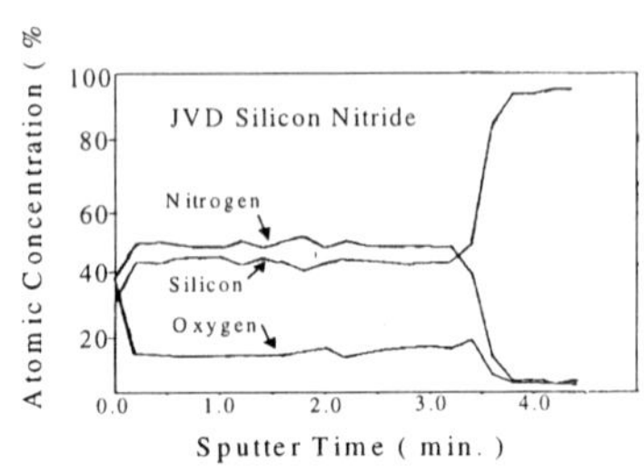

Fig.3 FTIR spectrum of a JVD silicon nitride film.

Fig.4 Auger depth profile of a JVD silicon nitride film.

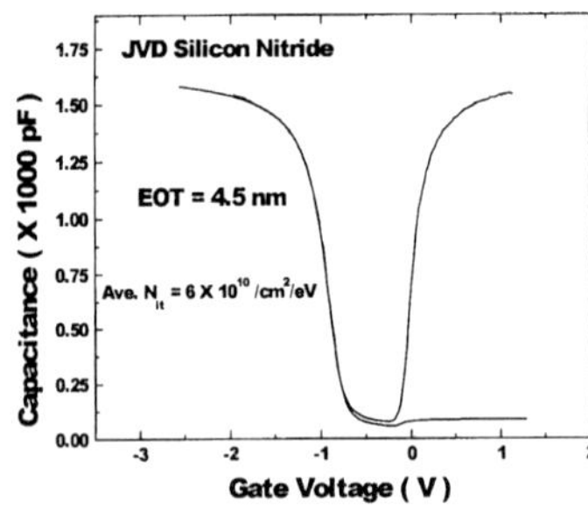

Fig.5 High-frequency and low-frequency C-Vcurves of a MNS capacitor with JVD silicon nitride as the gate dielectric.

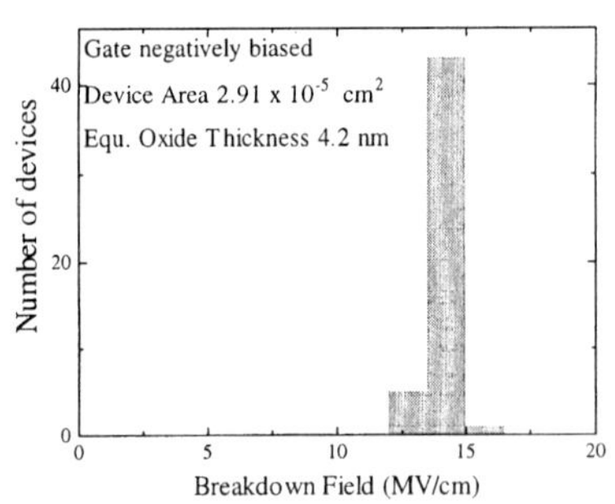

Fig.6 Breakdown field histogram of a set of MNS capacitors made of JVD silicon nitride as the gate dielectric

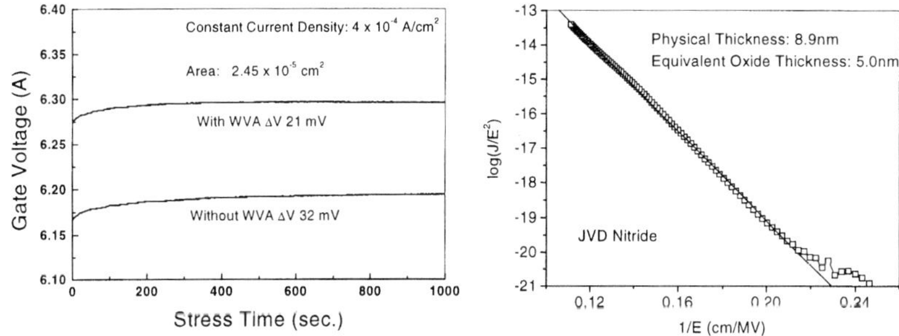

Fig.7 Gate voltage required to maintain a constant current. The upper sample had received water-vapor anneal (WVA)

Fig.8 Fowler-Nordheim plot from the I-V characteristic of a JVD MNS capacitor.

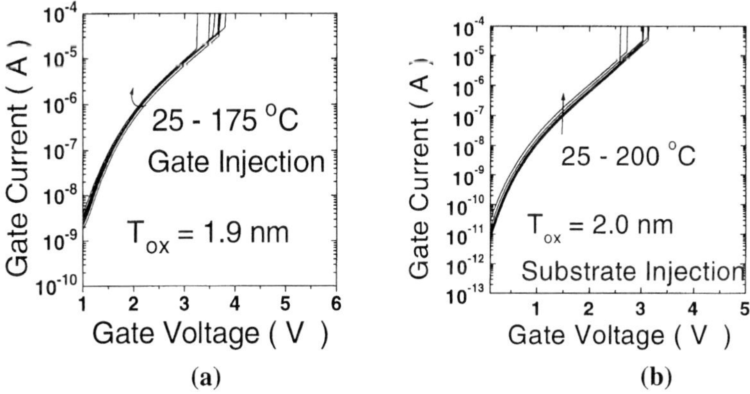

Fig.9 Relative weak temperature dependence of the I-V characteristics for a set of MNS capacitors made of JVD silicon nitride as the gate dielectric: (a) electron injection from the gate electrode, (b) electron injection from the silicon substrate.

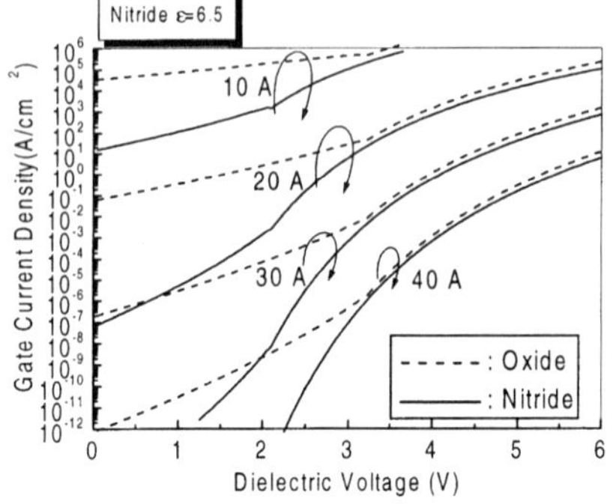

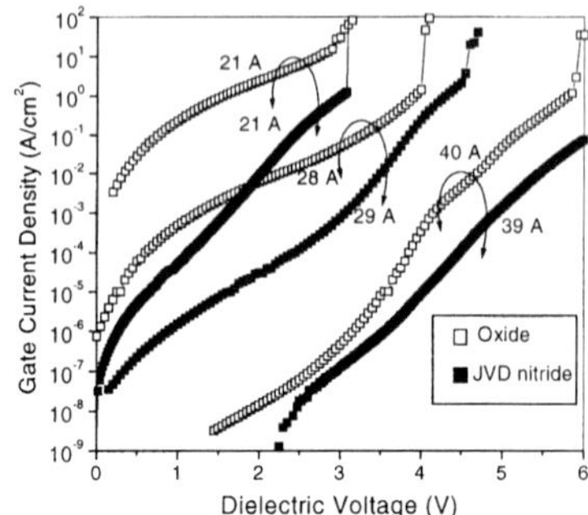

Fig.10 Theoretical gate current as a function of dielectric voltage for silicon oxides and nitrides of various thicknesses.

Fig.11 Comparison of measured gate currents between oxide and and nitride samples.

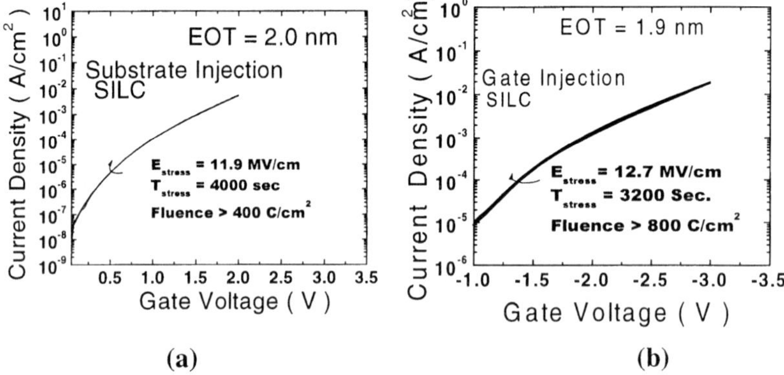

Fig.12 Stress-induced leakage current (SILC) is relatively small for MNS capacitors made with JVD silicon nitride as the gate dielectric: (a) electron injection from the silicon substrate, and (b) electron injection from the gate electrode.

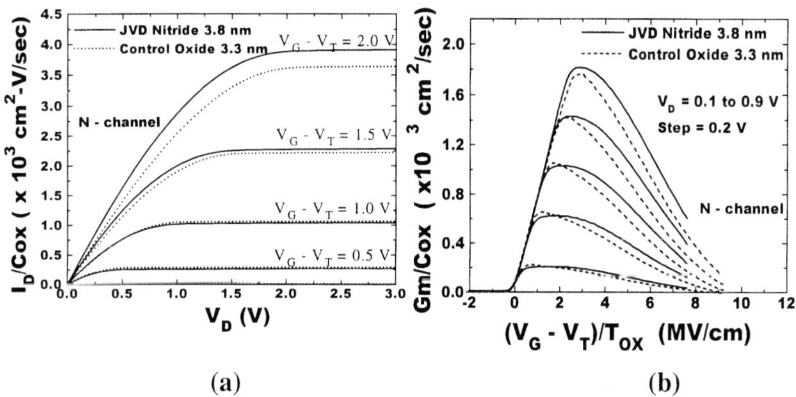

(a) (b)

Fig. 13 Comparison of N-channel transistors made of oxide gates with those made of JVD nitride gates: (a) family of I-V curves; (b) corresponding transconductance curves.

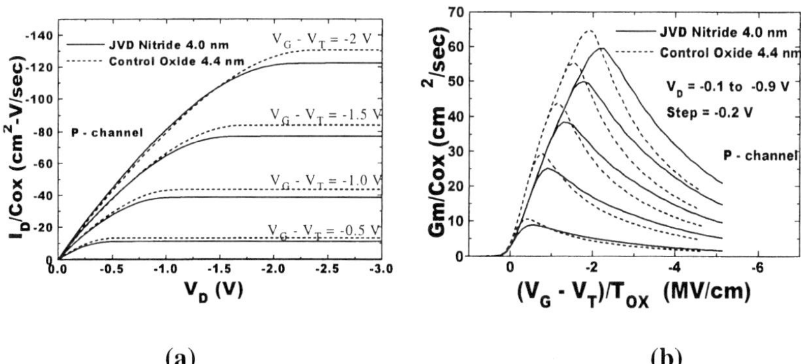

(a) (b)

Fig.14 Comparison of P-channel transistors made of oxide gates with those made of JVD nitride gates: (a) family of I-V curves; (b) corresponding transconductance curves.

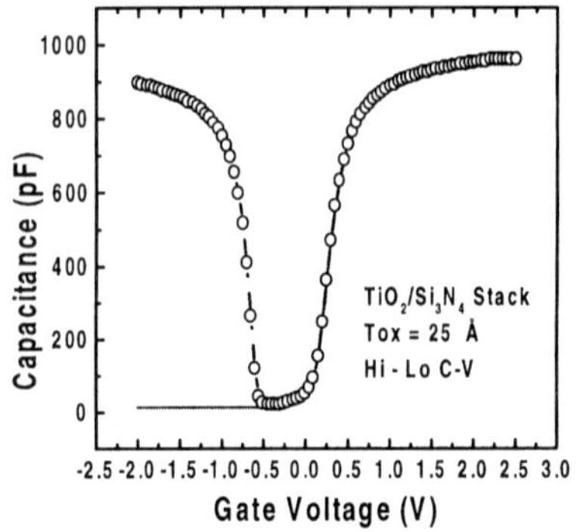

Fig.15 High-Low frequency C-V curves of a MIS capacitor with TiO2/SiN stack dielectric.

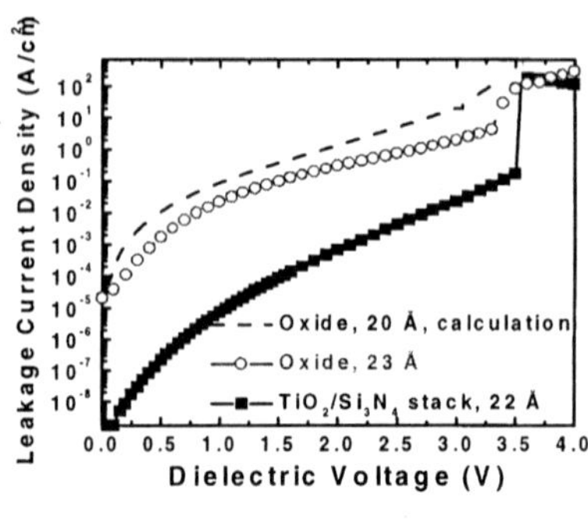

Fig.16 I-V characteristics of a TiO2/SiN stack and its thermal oxide counterpart of a similar EOT.

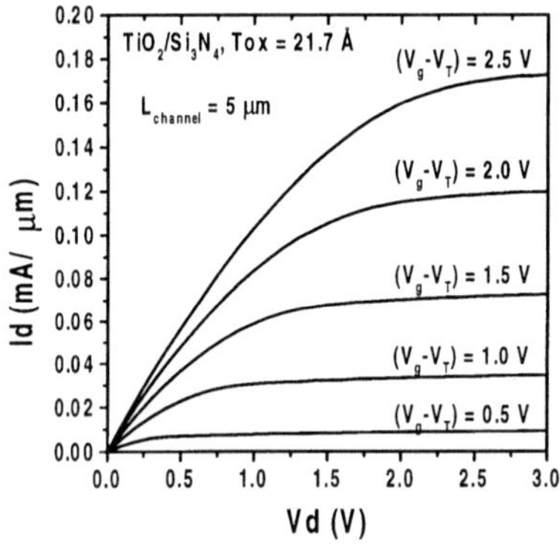

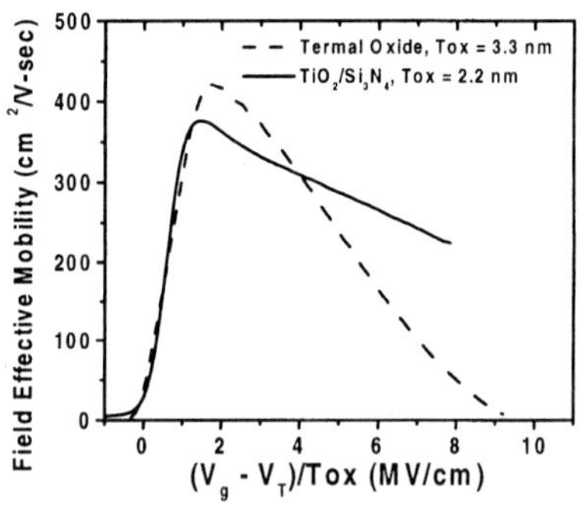

Fig.17 N-channel characteristics of a MOSFET with TiO_2/SiN stack-gate dielectric.

Fig.18 The TiO_2/SiN stack-gate transistor shows lower low-field mobility but higher high-field mobility when compared to its thermal SiO_2 counterpart.

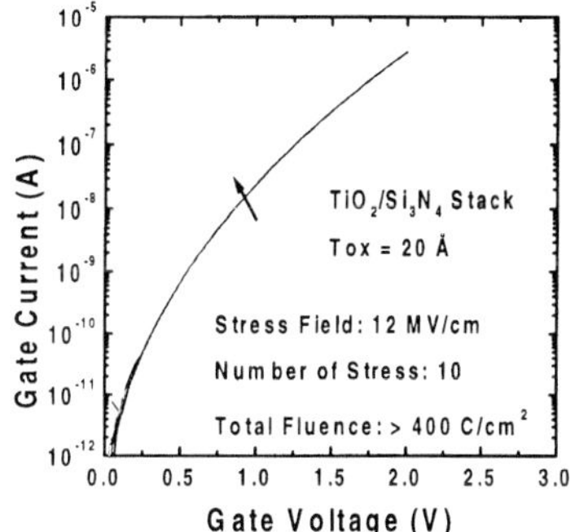

Fig.19 I-V characteristic of a TiO_2/SiN stack before and after prolonged high-field stress.

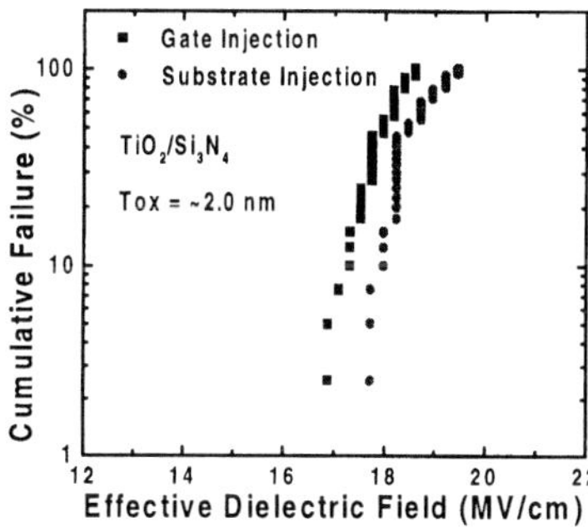

Fig.20 Breakdown voltage distribution for a set of TiO_2/SiN samples.

INTEGRATION OF ALTERNATIVE HIGH-K GATE DIELECTRICS INTO AGGRESSIVELY SCALED CMOS Si DEVICES: CHEMICAL BONDING CONSTRAINTS AT Si-DIELECTRIC INTERFACES

Gerald Lucovsky

Departments of Physics, Electrical and Computer Engineering
and Materials Science and Engineering,
NC State University Raleigh, NC 27695-8202

Optimization of performance and reliability for aggressively-scaled CMOS FETs with high capacitance dielectrics ($t_{ox-eq} < 2$ nm) can proceed in two steps i) first scaling to t_{ox-eq} ~1.2-1.3 nm using stacked oxide/nitride gate dielectrics with nitrogen profiles controlled at the atomic layer level, and ii) then to $t_{ox-eq} < 1$ by using other alternative high-K gate dielectrics such as Ta_2O_5, $Zr(Hf)O_2$-SiO_2 alloys and Al_2O_3 in stacked structures with *hyper-thin (< 0.5 nm)* nitrided oxide interfaces. This paper addresses three aspects of chemical bonding that are important for the implementation of alternative gate dielectrics: i) the balance between interfacial electronic and nuclear charge, ii) mechanical constraints imposed by average interfacial bonding coordination, and iii) interfacial band offset energies, primarily between the conduction bands of Si and the gate dielectric. These bonding *constraints* are applied to alternative gate dielectrics including i) oxide/nitride gate stacks, ii) Ta_2O_5, iii) $Zr(Hf)O_2$-SiO_2 alloys and iv) Al_2O_3, each requiring hyper-thin nitrided SiO_2 interfaces.

INTRODUCTION

Aggressive scaling of CMOS devices will eventually require the oxide-equivalent thickness of FET devices, t_{ox-eq}, to be reduced to < 1 nm to match decreased lateral dimensions required by increased levels of speed and integration, and decreased power consumption. Direct tunneling limits use of thermally-grown or deposited oxides to t_{ox-eq} ~ 1.5 to 1.7 nm. At these ~t_{ox-eq}'s direct tunneling at biases of about 1 V can exceed 1 A-cm^{-2} [1], and issues have been raised regarding the reliability of devices with tunneling leakage currents in this regime [2]. Technology pathways have been proposed to extend CMOS technology beyond the limits imposed by these ultrathin oxide gate dielectrics. The SIA roadmap [1] calls for thermally nitrided oxides to be replaced by physically-thicker stacked dielectrics based on alternative insulators with dielectric constants greater than SiO_2. These include i) silicon nitride and/or oxynitride alloys, and ii) other alternative high-K gate dielectrics such as Al_2O_3 [3], Ta_2O_5 [4], transition metal binary oxides such as $ZrSiO_4$ [5], as well as iii) transition metal oxide/SiO_2 alloys such as TiO_2/SiO_2, each used in conjunction with hyper-thin (< 0.5 nm) nitrided Si-SiO_2 interfaces.

This paper first discusses chemical bonding interface constraints that require introduction of hyper-thin nitrided Si-SiO_2 interfaces, and then focuses on different representative alternative gate dielectrics. The factors that limit t_{ox-eq} in silicon nitride and/or silicon oxynitride gate stacks to values of about 1.2 nm are first addressed. The paper then identifies the aspects of bonding that will limit t_{ox-eq} in the implementation of other high-K alternative dielectrics in which the bulk film is an amorphous wide band-gap oxide. Finally, the paper briefly addresses the factors that are important in implementing epitaxial oxides such as $SrTiO_3$ [6].

CHEMICAL BONDING INTERFACE CONSTRAINTS

Three chemical interface bonding issues that must be taken in account for the implementation of alternative gate dielectrics into aggressively-scaled CMOS devices are i) electron/nuclear charge balance in interface bonds [7], ii) mechanical bonding constraints related to the average number of interfacial bonds/atom, N_{av} [8]. and iii) asymmetries and reductions in band offset energies that apply primarily in transition metal (TM) oxides, and derive from a combination of a) the d-state origins of conduction band states, coupled with b) increases in TM-oxygen bond ionicity [9].

Charge Balance in Interface Bonding

The bonding between Si *surface atoms* and O-atoms of an SiO_2 dielectric at Si-SiO_2 interfaces is isovalent in the context of the definitions presented in Ref. 7. Interface bonds are partially-ionic (~50%) two electron pair bonds in which one of the electrons comes from the interfacial Si-atom, and the other from an O-atom of the SiO_2 network. The electron pair in these bonds is *exactly balanced* by positive charge on the respective Si- and O-atom nuclei. Similar considerations also apply to the bonding of N-atoms i) at Si-dielectric interfaces, ii) in homogeneous oxynitride alloys, and/or iii) at internal interfaces between oxide and nitride layers in stacked ON gate dielectrics.

However, the bonding between crystalline Si and the majority of the proposed alternative high-K transition metal oxides will not be ideal covalent bonding involving interfacial Si-O two electron pair bonds, primarily because of the increased bonding coordination of O-atoms in the transition metal oxides that results from increased bond ionicity. There are three different classes of high-K materials that will be addressed where interfacial bonding will differ in detail: i) transition metal oxides such as Ta_2O_5, ii) transition metal oxide-SiO_2 alloys or compounds such as TiO_2-SiO_2 and $ZrSiO_4$, respectively, as well as iii) group III, non-transition metal oxides such as Al_2O_3.

The local bonding in non-crystalline transition metal oxides such as Ta_2O_5 has not been determined to same degree as the bonding in Si-oxide, Si-nitride and Si-oxynitride alloys. However, Raman scattering [10] and extended X-ray absorption fine structure (EXAFS) [11] measurements indicate that Ta atoms are predominantly six-fold coordinated consistent with the Ta-O bonds having considerably more ionic character than Si-O bonds. For example, based on the Pauling electronegativity (X) scale ΔX(Si-O) = 1.54 , whereas ΔX(Ta-O) is increased to 1.94 [12]. Interface bonding of Ta_2O_5 to Si will not be by partially-ionic two-electron isovalent pair bonds as at Si-SiO_2 interfaces, but may also involve formation of Si-Ta bonds to provide the balance between nuclear and bond charge. The interface bonding is not easily resolved due to interfacial Si-SiO_2 and Ta_2O_5-SiO_2 bonding, both of which can occur during thermal CVD deposition of Ta_2O_5, or during post deposition anneals in oxidizing ambients [13,14]. These anneals are need to fully oxidize transition metal oxides deposited by thermal CVD, or to oxidize metal films prepared by PVD. Attempts to deposit Ta_2O_5 directly on the clean (e.g., H-atom terminated Si surfaces) generally result in Ta-Si bond formation as identified by X-ray photoelectron spectroscopy, and interfacial layers of silicon nitride have been deposited prior to the Ta_2O_5 depositions to passivate the Si substrate [13].

In contrast, interfacial bonding at interfaces between Si and group IV(Ti, Zr, and Hf) transition metal oxide-SiO_2 pseudo-binary alloys, or silicate compounds can occur through isovalent two-electron pair bonds for low alloy atom concentrations of Ti, and for $Zr(Hf)O_2$-SiO_2 alloys with $Zr(Hf))O_2$ concentrations up to the respective compound silicate compositions [5]. These represent compositions in which the group IV transition metal atoms retain a four-fold coordination, so that local bonding at the Si-atoms is by sp^3 hybrids, and at the Ti-, Hf- or Zr-atoms by sd^3 hybrids. Note that similar tetrahedral sd^3 bonding occurs in metal-organic molecules such as Zr(IV) t-butoxide. However, kinetic effects can promote formation of silicide bonds during the initial stages of film deposition. This would also require passivating interfacial nitrided SiO_2 layers prior to film deposition. Finally, the limitation on the TiO_2-content in the TiO_2-SiO_2 alloy system is associated with concentration-dependent bonding interactions between nearest-neighbor Ti-O bonds. At sufficiently high TiO_2 content, these will nucleate formation of a second phase, i.e., c-TiO_2, in which the bonding of the Ti atoms becomes more ionic, and the bonding coordination is increased from four to six.

The interfacial bonding between Si and Al_2O_3 will be primarily two electron pair bonds, but the interface may be required to have additional charged bonding arrangements to balance other aspects of bonding in the more ionic Al_2O_3 [16] The atomic structure of non-crystalline Al_2O_3 has two charged bonding components in a 3:1 ratio: *three* tetrahedral $[AlO_{4/2}]^-$ groups that form a covalently-bonded network for *each* Al^{3+} ion that occupies an *approximately* octahedral interstitial site within that network structure. Like group IV transition metal oxides deposited by thermal CVD, interfacial Si-SiO_2 bonding, and/or interfacial mixed oxide, Al_2O_3-SiO_2, bonds are likely to occur in thermal CVD deposition, or post deposition anneals. Si-Al bond formation is less likely. Bonding at internal Al_2O_3-SiO_2 interfaces or within Al_2O_3-SiO_2 alloys will be isovalent, with neutral $[SiO_{4/2}]^o$ and charged $[AlO_{4/2}]^-$ groups forming interface bonds analogous to the $[SiO_{4/2}]^o$–$[AlO_{4/2}]^-$ bonds in zeolites, with charge compensation by Al^{3+} ions [17].

From this discussion of charge balance, we can conclude that interfacial Si-SiO_2 or nitrided Si-SiO_2 will be required for i) group IV transition metal oxides such as Ta_2O_5, as well as other ii) group III and group IV transition metal oxides. Interfacial Si-SiO_2 or nitrided Si-SiO_2 may also be required at the $Zr(Hf)O_2$- and TiO_2-SiO_2 alloy interfaces, primarily to prevent TM-silicide formation during film deposition. Finally, the partially ionic character of Al_2O_3 may also require Si-SiO_2 or nitrided Si-SiO_2 to reduce effects of the ionic bonding in the Al_2O_3 layers on transport within the channel regions of FETs.

Mechanical Bonding Constraints

Constraint theory is based on the idea that all of the bonding forces (stretching, bending, etc.) in a covalently bonded network can be arranged in a hierarchy from strong to weak [17-21]. The constraining effects of these forces are a linear function of the average coordination number, N_{av}. If both bond-bending and stretching forces are present, the optimal average coordination, N_{av}*, that matches constraints to degrees of freedom is 2.4, as in $As_2S(Se)_3$ bulk glasses and thin films. However, for SiO_2, $N_{av}^* = 2.67$ is optimal because bond-bending forces at O-atoms are too weak to function as significant constraints at typical growth or annealing temperatures [17-20]. For over-constrained networks such as Si_3N_4 ($N_{av} = 3.43$), Si-atom stretching constraints are stronger than bending constraints, so that strain energy accumulates along the bending constraints. The average Si-N-Si bond angle θ_{ij} is distorted from the local value θ_{ij}^* by an amount $\delta\theta$ proportional to $\delta N_{av}^* = N_{av} - N_{av}^*$. Since total strain energy scales as $(\delta\theta)^2$ [21], the bonding defect concentration, D, e.g., for Si- and/or N-dangling bonds in Si_3N_4, will be proportional to $\{N_{av} - N_{av}^*\}^2$.

A structural model for the extension of constraint theory to crystalline-Si-dielectric-interfaces is illustrated in Fig. 1 and identifies three interfacial contributions to N_{av}: i) the

Si substrate represented by one-half a Si atom, ii) 1.5-2.0 molecular layers of a hyper-thin oxide or nitride interfacial layer (0.5 – 0.6 nm), and iii) the bulk dielectric represented by one-half a molecular layer [8]. Using the structural model of Fig. 1, a value of N_{av} ~3, is in excellent agreement with experimental data for separating device-quality intefaces from defective interfaces [2,11]. Experiments performed on i) Si-Si_3N_4, and ii) Si-Si_3N_4-SiO_2 and Si-SiO_2-Si_3N_4 interfacial stacks have demonstrated that D ~ $\{N_{av} - N_{av}^*\}^2$, where N_{av}^* is the average coordination for an ideal Si-SiO_2 interface (see Fig. 2) [8]. The model demonstrates that Si-SiO_2 interfaces with N_{av}~2.8 display excellent interface properties, whereas Si-Si_3N_4 interfaces with N_{av}~3.5 do not. Equally significant, the calculations demonstrate that interposition of ultrathin SiO_2 layers between Si and Si_3N_4 results in device-quality interfaces with N_{av} □ 3, whereas interposition of ultrathin Si_3N_4 layers between Si and SiO_2 results in defective interfaces with $N_{av} > 3$ (see Table I).

Fig. 2 demonstrates that defect scaling as noted above and previously established for bulk films, also holds at interfaces [8]. However, the situation is somewhat complicated by the nature of the interfacial defects. The electron occupancy and charge state of defects in n-channel and p-channel field effect transistors, FETs, determines their effect on device performance. Experiments performed on n-channel and p-channel FETs with Si-Si_3N_4 interfaces indicate that defects at these interfaces are predominantly donor-like and located in the lower half of the Si band-gap [22]. They are neutral when occupied and do not adversely effect channel transport in NMOS FETS, but are charged positively when empty and adversely effect channel formation and channel transport in PMOS FETS.

Table I includes calculated values of N_{av} for the interfaces discussed above, as well as interfaces with *emerging* high-K candidate materials such as TiO_2, Ta_2O_5 and Al_2O_3. Since N_{av} values for these oxides are > 3, interfacial SiO_2 or nitrided SiO_2 alloy layers are required to eliminate defects that can arise from over-constrained interface bonding. The interfaces between Si and binary SiO_2-transition metal oxides and compounds will not be over-constrained, however, Si-SiO_2 interfacial regions may be a natural consequence of the deposition technologies used to form these dielectrics.

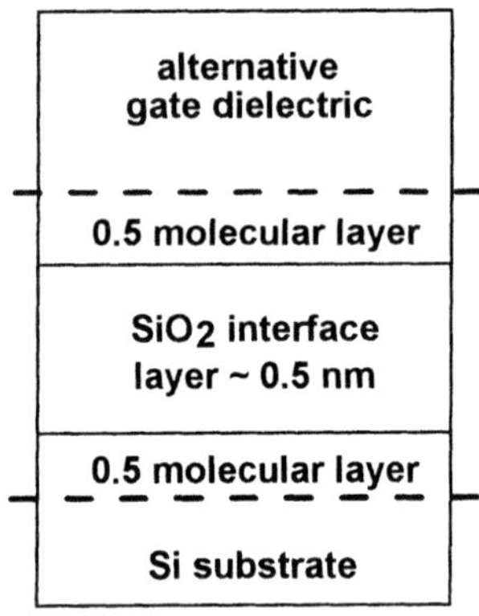

Figure 1. Schematic representation of interface model for application of constraint theory to semiconductor dielectric interfaces.

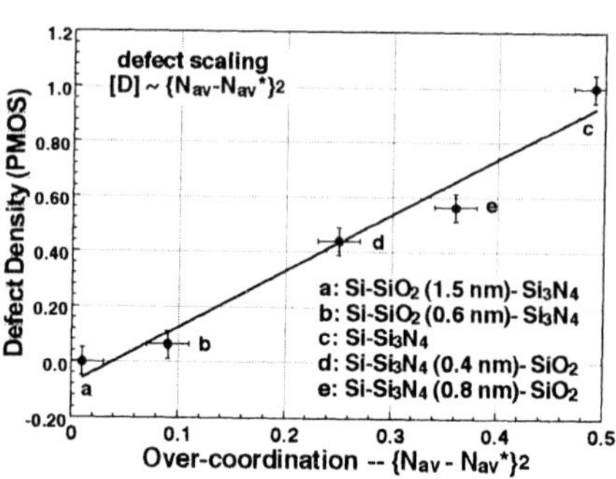

Fig. 2. Application of constraint theory to c-Si-dielectric interfaces. Verification of the defect scaling relationship [8].

Table I Average bonding coordination at Si-dielectric interfaces [8]

Interface System	Average Coordination (N_{av})	Electrical Quality
$Si-SiO_2$ (1.5 molecular layers)	2.8	excellent
$Si-Si_3N_4$ (1.5 molecular layers)	3.5	very poor
$Si-\{SiO_2\}(t)-Si_3N_4$	t = 0.6 nm: 3.0	very good
t = oxide layer thickness	t = 1.5 nm: 2.9	excellent
$Si-\{Si_3N_4\}(t)-SiO_2$	t = 0.4 nm: 3.3	poor
t = oxide layer thickness	t = 0.8 nm: 3.4	poor
$Si-N-SiO_2$ {1 monolayer (ML)}	2.8	excellent [Ref. 23]
$Si-TiO_2\}$[a] (1.5 molecular layers)	4.0	unreported
$Si-Ta_2O_5\}$[b] (1.5 molecular layers)	3.5	unreported
$Si-Al_2O_3\}$[c] (1.5 molecular layers)	3.6	unreported

[a] average coordination [Ti] = 6, [O] = 3.0: N_{av} = 4 [as in rutile/anatase crystrals]
[b] average coordination: [Ta] = 6, [O] = 2.4: N_{av} = 3.4 [Refs. 10,11]
[c] average coordination: Al = [4.5], [O] = 3.0: N_{av} = 3.6 [Ref. 15]

Band Alignment Energies

The valence and conduction band offset energies for Si with respect to both SiO_2 and Si_3N_4 are large and positive; e.g., the conduction band offset energy is ~ 3.1-3.2 eV for $Si-SiO_2$, and ~ 2.1-2.2 eV for $Si-Si_3N_4$ [24]. The conduction band offset energies between Si and elemental and binary transition metal oxides are generally less because of two contributing factors: i) increased ionic bonding which moves the valence band to lower energies than in SiO_2, and ii) reduced band-gaps due to conduction band states being derived from unoccupied transition metal d-states, rather than s-states as in non-transition metal oxides [9]. This is shown schematically in Fig. 3. Based on the calculations of Ref. 8, as well as the arguments presented above, it will be necessary to bridge crystalline Si substrates and transition metal oxide dielectrics with hyper-thin nitrided SiO_2 interface layers (see Fig. 3). Based on comparisons between Si-Al and $Si-SiO_2-Al$ interfacial electronic structures, the insertion of SiO_2 interface layers should promote an increase in conduction band offset energies from those calculated in Ref. 9. Taking this into account the band offset energy between Si and Ta_2O_5, with an SiO_2 interface layer can be as high as ~1eV, but it is still in a range where leakage current may involve hopping conduction through the Ta_2O_5 films by a Poole-Frenkel mechanism [4].

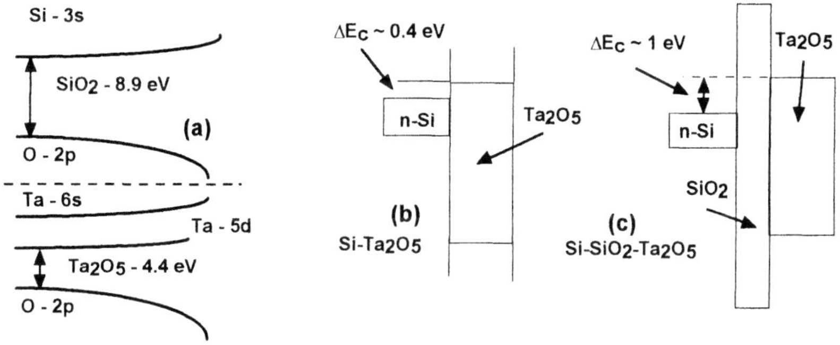

Figure 3. (a) Schematic representation of band gap electronic states in SiO_2 and Ta_2O_5. (b) Band offset energies at $Si-Ta_2O_5$ interfaces. (c) Band offset energies at $Si-SiO_2-Ta_2O_5$ heterointerface structures.

In addition to the transition metals oxides, other studies have addressed non-transition metal oxides. In contrast to interfaces involving the group IV and group V transition metal oxides, the band offset energies at Si-Al_2O_3 interfaces are expected to be comparable to those at Si-SiO_2 interfaces because the valence and conduction band states of Al_2O_3 are also derived respectively from O-atom 2p states and Al-atom 3s states.

Finally, the band offset energies for Si-SiO_2-(transition metal oxide-SiO_2 alloy) heterostructures such as TiO_2-SiO_2 and Zr(Hf)SiO_4, will depend on alloy composition and the bonding coordination of the transition metal. To a first order approximation, the band gaps of these alloys are expected to scale linearly between the end member oxide values due to the hybridized nature of the conduction band states as discussed above. This would predict band-gaps of about 6-7 eV for the compound silicates, $ZrSiO_4$ and $HfSiO_4$, with conduction and valence band offset energies of approximately 2 eV.

Incorporation of Hyper-thin Nitrided Oxide Interfacial Layers

Based on consideration of i) electron/nuclear charge balance in interface bonds, ii) mechanical bonding constraints related to the average number of interfacial bonds/atom, N_{av}. and iii) estimates of band offset energies, as presented above, there are at least four different classes of alternative dielectric structures with qualitatively different interface properties. However, each of these will require the incorporation of hyper-thin interfacial SiO_2 or nitrided SiO_2 layers between the Si substrate and the high-K oxide. These four classes are i) oxide/nitride stacks, ii) Si-SiO_2-transition metal oxide stacks (e.g., Ta_2O_5), iii) Si-SiO_2-(transition metal oxide-SiO_2) alloy stacks, and iv) Si-SiO_2-Al_2O_3 stacks. Each of these is discussed in the next section with emphasis on just which aspects of the interface constraints apply and force the incorporation of the interface buffer layers.

The inclusion of the hyper-thin interface layers will limit the aggressive scaling of stacked gate dielectrics. Table II includes a calculation of the limiting values of t_{ox-eq} for two conditions: i) for a physical thickness, t_{phys}, of 2.5 nm, and ii) a physical thickness of 2.0 nm. The limiting values of t_{ox-eq} assume 0.5 nm of nitrided SiO_2 with a dielectric constant of 3.8, so that $t_{ox-eq} = 0.5\ nm + (t_{phys} - 0.5\ nm)\bullet(3.8/k)$, where k is the dielectric constant of the alternative gate oxide material.

Table II Limiting values of t_{ox-eq} for stacked dielectrics

a) physical thickness = 2.5 nm
0.5 nm SiO_2* and 2.0 nm high-k*

k – dielectric constant	8	10	15	20	25
t_{ox-eq} (nm)	1.45	1.26	1.02	0.88	0.80

b) physical thickness = 2.0 nm
0.5 nm SiO_2* and 1.5 nm high-k

k – dielectric constant	8	10	15	20	25
t_{ox-eq} (nm)	1.21	1.07	0.88	0.79	0.73

* Si-SiO2 interfaces should have one monolayer of nitrogen to reduce tunneling and improve reliability

APPLICATION TO ALTERNATIVE GATE DIELECTRIC STACKS

There are four significant criteria that alternative gate dielectrics must have to meet the demands of aggressive device scaling: i) the gate dielectric capacitance must correspond to $t_{ox\text{-}eq} < 1$ nm, ii) the direct tunneling leakage current at the operating voltage, ~0.6 to 1.0 V must be significantly less than 1 A/cm^2, iii) the extrapolated time to failure must exceed 10 years, and preferably 20 years, and iv) the drive current/micron of channel length must be maintained at oxide device levels for both NMOS and PMOS devices (this translates to channel mobilities not being degraded by more than 10% in advanced gate stack structures). This paper focuses on the first two criteria; other studies have addressed the remaining two criteria, and where relevant to the arguments developed in this paper they will be introduced and discussed as well.

Si-SiO_2/Si-Nitride and Oxynitride Stacks

Only one of the three bonding constraints presents a substantive issue for these stacked dielectrics. This is the mechanical interface bonding constraint which requires hyper-thin SiO_2 or nitrided SiO_2 interfacial layers between the nitride or oxynitride alloy and the Si substrate. Since the dielectric constant of silicon nitride is at most 7.6, this places a severe restriction on the limiting values of $t_{ox\text{-}eq}$ that can be obtained in this way. Based on Table II, this type of gate stack can be expected to be aggressively scaled to about 1.2 to 1.3 nm, where the limitation will be based primarily on the tunneling current at operating bias levels. Experiments discussed below have yielded tunneling leakage currents between 10^{-3} and 10^{-2} A/cm^2 for PMOSFETs with $t_{ox\text{-}eq} \sim 1.6$ nm.

There are two factors that are related to the way nitrogen atoms are incorporated into the stacked gate structure that are important for meeting the targeted goals for tunneling current reduction for these gate stacks. There are two independent mechanisms for reduction of direct tunneling current associated with i) interface nitridation at the mono-

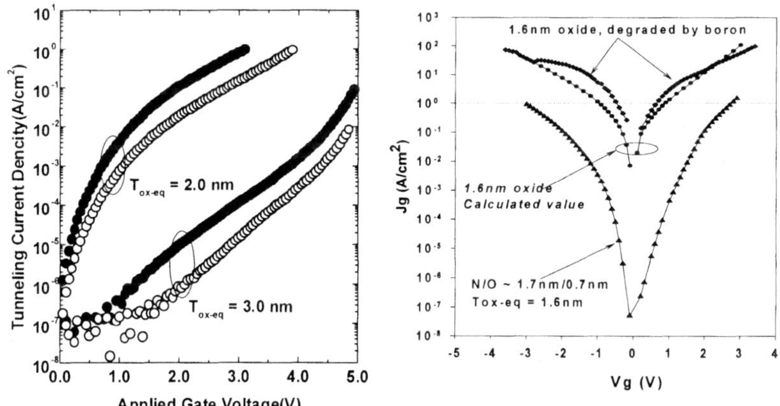

Figure 4. (a) Direct tunneling in RPECVD PMOS devices ($t_{ox\text{-}eq} \sim 2$ nm and 3 nm) with and without interface nitridation. (b) Direct tunneling in an aggressively-scaled oxide-nitride stack ($t_{ox\text{-}eq} \sim 1.6$ nm) with monolayer interface nitridation.

layer level, and ii) physically thicker oxide/nitride (or oxynitride) composite layers. Figure 4 demonstrates direct tunneling current reductions due to (a) interface nitridation (a factor of ~ 10 at ~ 1V), and (b) the combined effect of interface nitridation and physically thicker stacks (a factor > 100). Tunneling in nitride stacks without monolayer interface nitridation yields a factor of ~ 10 in tunneling current reduction for ~ 1.9 nm, independent of the nitride/oxide thickness ratio. Since these two reductions in tunneling are independent, they can be *combined* in stacked dielectrics with nitrided Si-SiO_2 interfaces. It is further assumed that the nitrided interfaces can be combined with other alternative gate dielectrics to achieve similar reductions in tunneling.

There is an additional aspect of bonding to consider in oxide/nitride stacks: the bonding constraints at the internal oxide/nitride interface. The average coordination at this interface is 3.05, which is at the *boundary* between device-quality and defective interfaces. Experiments performed on O/N/O [25] and O/N [8] stacks indicate fixed positive charge levels in the mid 10^{11} cm^{-2} range at these interfaces. Analysis of drive currents in NMOS and PMOS FETs with ~ 0.6 nm oxide layers and thicker nitride layers, ~1.5 to 2.0 nm, indicate channel mobility reductions of the order of 5 to 10% with respect to control devices with oxides that may be due, in part, to the fixed interfacial charge [8].

Si-SiO_2/Transition-Metal-Oxide Stacks

All of the interface bonding issues discussed above generally apply at Si-transition-metal-oxide interfaces, so that hyper-thin SiO_2 or hyper-thin nitrided SiO_2 buffer layers must be used. Figure 5 includes two different types of heterostructures with (a) an abrupt interface between the SiO_2 and Ta_2O_5 layers, and (b) a graded interface between the SiO_2 and Ta_2O_5 layers. The structure in (a) is more effective in reducing t_{ox-eq} and increasing capacitance, whereas the structure in (b) is more effective in i) reducing direct tunneling and ii) eliminating any temperature dependent contributions to the leakage current that involve hopping transport (e.g. Poole-Frenkel mechanism) through the transition metal oxide film [4]. Comparing experimental results for devices with thermal CVD Ta_2O_5 suggests that devices of Ref. 13, which show a temperature

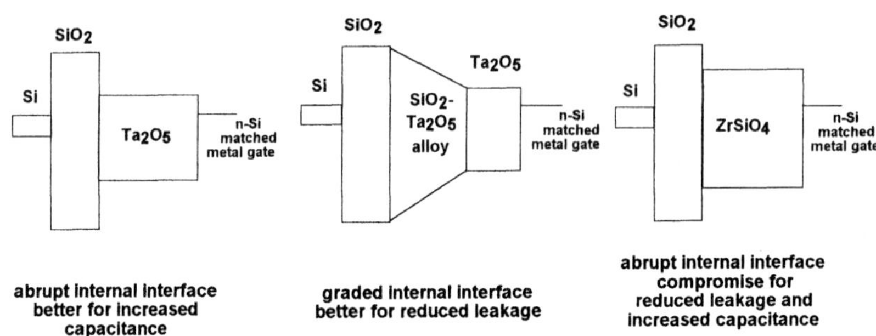

Figure 5. (a) Schematic representation of Si-SiO_2-Ta_2O_5 heterostructure. (b) Schematic representation of Si-SiO_2-graded SiO_2-Ta_2O_5 heterostructure. (c) Schematic representation of a Si-SiO_2-(Zr(Ti)O_2-SiO_2 alloy) heterostructure.

independent leakage current, have a band diagram corresponding to (b), and those of Ref. 4 with temperature dependent leakage current have a band diagram corresponding to (a). Clearly additional characterizations of these devices are needed to confirm this proposed explanation for the qualitatively different leakage current behaviors.

Similar to the oxide/nitride stacks, the *internal dielectric interfaces* in these stacks will also be influenced by chemical bonding *mismatches* with respect to both charge balance and bond coordination induced defects. Increased levels of fixed charge are expected with respect to the internal oxide/nitride interfaces, and additionally silicide bond formation may take place to balance nuclear and electronic charge. Experimental results to date for gate stacks that include Ta_2O_5 have indicated channel mobility degradations in NMOSFETs of the order of a factor of two, whereas channel mobility in PMOSFETs is degraded by less than 5 % [4]. This suggests acceptor like defects in the upper half of the Ta_2O_5 bandgap at the internal dielectric interface. This explanation is consistent with a crossover point in Ta_2O_5 between acceptor-like and donor-like defects being closer to the conduction band than the valence band due to the difference in conduction and valence band density of state functions [9]. Stated differently, the charge neutrality level, CNL, is in the upper half of the Ta_2O_5 band gap, and this is where a cross-over in interface induced defect states is expected occur. Similar considerations will also be expected to apply to other transition metal oxides, since their conduction bands are also derived from transition metal d-states.

Si-SiO_2/Transition-Metal-Oxide/SiO_2 Alloy Stacks

There are two types of alloys to consider i) those with a chemically ordered compound in the phase diagram as for example Zr(Hf)O_2-SiO_2, where the compounds are silicates with a one to one ratio of each oxide - Zr(Hf)SiO_4, and ii) those where there is no compound in the pseudo-binary phase diagram, such as TiO_2-SiO_2. The chemically-ordered alloys should retain tetrahedral bonding at each of the constituent atomic sites up to the silicate compound composition. In contrast. tetrahedral coordination in the TiO_2-SiO_2 system will be a function of the TiO_2 fraction. As long as the concentration of TiO_2 is significantly less than 50%, the Ti atoms will remain in a tetrahedral coordination. However, there will be a critical concentration of TiO_2 at which six-fold coordination of Ti becomes favored over tetrahedral coordination and this will lead to chemical phase separation into a diphasic material comprised of TiO_2 crystallites in an SiO_2 matrix. This concentration must be determined from alloy experiments which are currently underway in our research group.

For concentrations at which teterahedral coordination is maintained, these transition-metal-oxide alloys are expected to have band structures intermediate between SiO_2 and the transition metal oxides of Fig. 3(a). For example at the compound composition in the Zr(Ti)O_2-SiO_2 alloy system, $ZrSiO_4$, the conduction band states will be hybridized Zr(sd^3) and Si(sp^3) states, corresponding the local tetrahedral symmetry, so that band structure will be markedly different from the Ta_2O_5 structure shown in Fig. 3(a). As such these materials represent yet another pathway for reducing tunneling and increasing capacitance. Based on the assumptions of a linear variation of band energies and an electron tunneling mass essentially equal to that of SiO_2, these materials are expected to have dielectrics constants of the order of 10-15, and therefore function as CMOS gate dielectrics down to $t_{ox\text{-}eq} \sim 1\text{–}1.2$ nm with tunneling leakage currents below 10^{-2} A/cm^2. It is difficult to predict the limiting concentrations of TiO_2 and/or Ta_2O_5 in non-chemically ordered alloys with SiO_2, and therefore impossible to make reliable judgements relative to ultimate performance limitations. Finally, there may be advantages in exploring more complex ternary oxide gate dielectrics; e.g., studies based on sol-gel preparation have indicated amorphous films can be formed between ZrO_2 and Y_2O_3 [28]. There are numerous issues regarding leakage current, dielectric constants and band alignments in these binary alloys. There are equally interesting questions regarding what benefits could be derived from consideration of ternary systems involving SiO_2, ZrO_2 and Y_2O_3. In this regard both ZrO_2 and Y_2O_3 form compound silicates, so that improvements in properties in the ternary alloys could be driven by increased chemical order.

Si-SiO_2/Al_2O_3 Alloy Stacks

The local atomic structure in Al_2O_3 is qualitatively different from that in the group IV transition metal/SiO_2 alloys. As discussed above, Al_2O_3 is effectively amphoteric with local bonding arrangements that are both acidic, $[AlO_{4/2}]^-$, and basic, Al^{+3} [15] Bonding issues associated with the average coordination of Al_2O_3 indicate the necessity for interfacial mechanical buffer layers. However, as noted above the charged nature of the network constituents introduces additional bonding considerations. For example, do charged bonding sites play a role in defining the bonding at the Si-dielectric interface, and/or at the internal interface between Al_2O_3 and SiO_2 in a stacked dielectric?

Since band offset energies in Al_2O_3 are expected to be comparable to those in SiO_2, the band offset energy profiles are expected to be similar to those in Si-SiO_2 structures. Neglecting the more ionic bonding, tunneling should be the dominant leakage mechanism; however, transport may be modified by the presence of the charged species, e.g., the Al^{+3} interstitials which could act as intermediate states in a two step tunneling process, or transport states in a hopping conduction, depending on their electronic energies relative to Si. Experiments in which Zr and Si were introduced in the Al_2O_3 network indicated that these species reduced the leakage current [3]. If these dopants are incorporated in a +4 valence state, then this would reduce the concentration of Al^{+3} sites by a 3:4 ratio, paralleling charge compensation in zeolites [16]. This suggests that transport through the Al_2O_3 films may involve localized electronic states associated with the positively charged ionic species. Additional experiments are needed to clarify this issue.

Si-epitaxial Dielectric Interfaces

The group at Oak Ridge National Laboratory, ORNL, has recently demonstrated heteroepitaxial growth of $SrTiO_3$ onto Si(100) substrates [6]. Their first publication included a C-V trace for a structure in which the t_{ox-eq} was estimated to be less than 1 nm. More recently, they have reported successful fabrication and operation of NMOSFETs, and have estimated electron mobilities of about 200 $cm^{-2}/V\text{-}s$ [27]. The epitaxial growth requires a complex interfacial buffer layer comprised of SrSi and Sr,BaO, and the details of the bonding within this layer may be responsible for the large positive shift (~ 1 V) of the flatband voltage in the C-V trace of Ref. 6, and the approximately two-fold reduction of the electron mobility in the NMOSFETs. The initial results are clearly encouraging and additional studies are necessary, before any definitive judgements can be made regarding heteroepitaxially grown dielectrics.

Finally, the necessity to introduce a SrSi layer between the crystalline Si substrate and the $SrTiO_3$ dielectric is consist with other attempts to grow dielectrics on Si, as for example CaF_2, wherein the interface bonding is by Si-Ca, rather than Si-F bonds [28]. The effects on monolayer interfacial silicide bonds in Si-gate stacks need additional study to fully evaluate their potential effects on electrical performance and/or reliability.

SUMMARY

A consideration of chemical bonding constraints involving the substitution of alternative high-K gate dielectrics for SiO_2 in aggressively scaled CMOS devices identifies the need for interfacial SiO_2 or nitrided SiO_2 buffer layers. These clearly impact of the anticipated reductions in $t_{ox,eq}$ limiting reductions as indicated in Table II. In addition, inherent differences between the nature of the conduction band states in transition metal oxides and SiO_2 (and Al_2O_3) place additional restrictions on composition profiles that must simultaneously address capacitance increases and reductions in leakage current. This identifies major characterization challenges of obtaining i) compositional and ii) energy band gap profiling. These will be required to develop device simulators, and also to provide a quantitative relationship between device fabrication metrics and on-line/in-situ characterizations, and targeted device performance.

At the present point in time, there is no obvious high-k ($k > 10$) winner in the race to find a replacement for SiO_2 in gate stacks. It is however almost certain that the transition from SiO_2 to an alternative high-k dielectric will proceed in two steps, i) first to $t_{ox\text{-}eq} \sim$ 1.2 to 1.3 nm, by introduction of a stacked structure that includes either a nitride or heavily nitrided oxynitride alloy component, and ii) then to $t_{ox\text{-}eq} < 1$ nm through the introduction of another alternative oxide material. The two classes of alternative dielectrics that are least constrained by interfacial chemical, physical and electronic structure are the chemically-order transition metal alloys such as $Zr(Hf)O_2$-SiO_2 [5], and doped Al_2O_3 [3]. However, the dielectric constants of these materials, and the necessity for hyper-thin interfacial SiO_2 or nitrided SiO_2 are likely to limit $t_{ox\text{-}eq} \sim 0.9$ nm. As noted above, the field of dielectrics can be increased substantially by considering ternary oxides use as the SiO_2:ZrO_2:Y_2O_3 system identified above. These ternaries could provide higher dielectric constants and thereby extend the limiting value of $t_{ox\text{-}eq}$ to even less than 0.9 nm.

When one, or more of these alternatives dielectrics are eventually integrated into CMOS devices, another, and equally challenging set of issues will arise in the choice of gate electrodes. If depletion regions in doped poly-Si or poly-Si,Ge prove to be too severe a constraint on the available scaled V_{dd}, then dual-gate metals will be required. This in turn raises another set of interface bonding chemistry issues that in many ways are the same as those we have already addressed at the Si-dielectric interface [29].

The replacement of the Si-SiO_2-poly-Si gate stack materials to achieve higher levels of capacitance and therefore enable aggressively scaling of device dimensions is the greatest challenge to date in the materials package for CMOS devices. Alternative materials will be identified, but implementing them into manufacturing will provide yet another set of challenges that embrace many aspects of materials engineering, and in-situ characterization at the monolayer scale. As mentioned above, the chemical profile of the entire stack will be critical in achieving the four targeted goals for gate stack performance: i) increased capacitance with $t_{ox-eq} < 1$ nm, ii) low leakage current, iv) high reliability, and v) maintenance of drive currents in both NMOS and PMOS devices.

ACKNOWLEDGEMENTS

This research has been supported in part by the National Science Foundation, the Office of Naval Research and The Air Force Office of Scientific Research. The author acknowledges the contributions of his graduate students, Hiroake Niimi, Hanyang Yang, David Wolfe and Yi-Der Wu, and his post doctoral fellows, Bruce Claflin and Robert Therrien. The author further acknowledges a productive collaboration with Jim Phillips at Lucent Bell Laboratories which led to the application of constraint theory to Si-dielectric interfaces. Finally, the author wishes to thank Professors John Hauser, Greg Parsons and Veena Misra, all of NC State University for their helpful discussions and constructive criticisms.

REFERENCES

[1] National Roadmap for Semiconductor Technology, (SIA, Santa Clara, CA, 1997).
[2] J.H. Stathis and D.J. DiMaria, IEDM Tech. Dig. 167 (1998).
[3] L. Manchanada et al. IEDM Tech. Dig. 605 (1998).
[4] A. Chatterjee et al, IEDM Tech. Dig. 777 (1998).
[5] G. Wilk and W.A. Wallace Appl. Phys. Lett. 74 (1999); D.M. Wolfe and G. Lucovsky, MRS Symp. Proc. (1999) in press.
[6] R.A. McKee, F.J. Walker and M.F. Chisholm, Phys. Rev. Lett, **81**, 3014 (1998).
[7] W.A. Harrison, E.A. Kraut, J.R. Waldrop and R.W. Grant, Phys. Rev. **B 18**, 1402 (1978).
[8] G. Lucovsky et al., Appl. Phys. Lett. **74**, 2005 (1999).
[9] J. Robertson and C.W. Chen, Appl. Phys. Lett. **74**, 1168 (1999).
[10] F.L. Galeener, W. Stutius and G.T. McKinley, in *The Physics of MOS Insulators*, ed by G. Lucovsky, S.T. Pantelides and F.L. Galeener (Pergamon, NY, 1980), p. 77.
[11] H. Kimura, J. Mizuki, S. Kamiyama and H. Sizuki, Appl. Phys. Lett. 66, 2209 (1995).
[12] J.E. Huheey, *Inorganic Chemistry*, 2nd Ed., (Harper & Row, New York, 1978), Chap. 4.
[13] H.F. Luan et al., IEDM Tech. Dig. 609 (1998).
[14] G.B. Alers, D.J. Werder, Y. Chabal, H.C. Lu, E.P. Gusev, E. Garfunkel, T. Gustafson and R.S. Urdahl, Appl. Phys. Lett. **73**, 1517 (1998).
[15] G. Lucovsky, A. Rozaj-Brvar and R.F. Davis, in *The Structure of Non-Crystalline Materials 1982*, edited by P.H. Gaskell, J.M. Parker and E.A. Davis (Taylor and Francis, London, 1983), p. 193.
[16] F.A. Cotton and G. Wilkinson, Advanced Inorganic Chemistry, 3rd Ed. (Interscience, New York, 1972), Chap. 11.
[17] J.C. Phillips, J. Non-Cryst. Solids **34**, 153 (1979); J. Non-Cryst. Solids **47**, 203 (1983).
[18] H. He and M. F. Thorpe, Phys. Rev. Lett. **54**, 2107 (1985).
[19] J.C. Phillips, in *Rigidity Theory and Applications*, ed. by M.F. Thorpe and P. Duxbury, (Michigan State University Press, East Lansing, 1999) to be published.
[20] G. Lucovsky and J.C. Phillips, J. Non-Cryst. Solids **227**, 1221 (1998).
[21] J. H. Van der Merwe, J. Appl. Phys. **34**, 123 (1963).
[22] V. Misra et al., submitted to IEEE Electron Device Trans. (1999).
[23] G. Lucovsky, J. Vac. Sci. Technol. **A 16**, 356 (1998).
[24] H.Y. Yang, H. Niimi and G. Lucovsky, J. Appl. Phys. **83**, 2327 (1998).
[25] Y. Ma et al., J. Vac. Sci. Technol. **B 11**, 1533 (1993).
[26] G. Antionioli et al., J. Non-Cryst. Solids **177**, 179 (1994).
[27] R.A. McKee, in Symposium R at MRS Spring Meeting 1999, San Francisco, CA.
[28] M. A. Olmstead et al., J. Vac. Sci. Technol. **B 4**, 1123 (1986).
[29] B. Claflin and G. Lucovsky, J. Vac. Sci. Technol. **A 16,** 1757 (1998).

EVALUATION OF ULTRA-THIN GATE STACK DIELECTRICS FOR 0.1 μm PMOSFETs

A. Srivastava*, C.M. Osburn, K.F. Yee, H.H. Heinisch, E.M. Vogel, K.Z. Ahmed, Z. Wang, K. Min, B. Timberlake, C. Parker, J.J. Wortman and J.R. Hauser
Dept. of Electrical and Computer Engineering, North Carolina State University, Raleigh, NC 27695.
* Motorola, 3501 Ed Bluestein Blvd., MD: K31, Austin, TX 78721.

Ultra-thin 2.0 - 3.0 nm dielectrics were examined in this study as candidates for a 0.1 μm technology. Dielectric evaluation was conducted with respect to boron penetration, interface state densities and gate tunneling currents for furnace oxides, RTO oxides, RTCVD oxides, RTO oxide /RTCVD oxynitride stacks and RPE oxide /RPE nitride stacks. Based on the evaluation, RPE oxide /RPE nitride stacks were determined to best meet CMOS dielectric needs for 2 nm thick gate dielectrics by providing excellent resistance to boron penetration, while exhibiting interface densities and gate leakages comparable to mainstream furnace oxides.

INTRODUCTION

The scaling of conventional furnace gate oxides in silicon CMOS devices is limited by the exponential increase in gate leakage beyond 3.0 nm due to direct tunneling, and by the increase in boron penetration from P^+ gates into P-MOSFET channels at low oxide thicknesses, resulting in device instabilities (1,2). Several alternatives to thermal gate oxides are currently under scrutiny; in particular, oxynitride gate dielectrics and gate stacks incorporating nitride or oxynitride layers have shown considerable promise in restricting boron penetration (3,4). In this paper, we examine several dielectric formation techniques as potential candidates for the 0.1 μm technology regime. 2.0 nm - 3.0 nm thick furnace oxides, Rapid Thermal Oxides (RTO) (5), Rapid Thermal Chemical Vapor Deposition (RTCVD) oxides (6), RTO oxide/RTCVD oxynitride stacks (7) and Remote Plasma Enhanced (RPE) oxide/nitride stacks (8,9) are investigated with either an in-situ or an implant doping scheme for the gate. Dielectric performance is evaluated with respect to boron penetration, gate tunneling leakage and interface state densities.

EXPERIMENTAL

Device fabrication was carried out using a 0.1 μm PMOS technology, featuring LOCOS isolation, i-line lithography and heavily doped extensions at the MOSFET source/drain junctions. The process flow followed the sequence described in (10). Two different gate formation and doping approaches were examined: in-situ doping of an RTCVD deposited gate (Split A), and implant doping of an LPCVD deposited gate (Split B). Several different dielectrics were investigated for each gate doping strategy. Table I lists the dielectrics considered for each case, together with the range of thicknesses examined for each dielectric.

Furnace oxidation was performed at 700°C in dry O_2 and 4.5 % HCl (for dielectric conditions A1,B1,B2 in Table I); rapid thermal oxidation was carried out in an O_2 ambient at 1000°C and 735 Torr (dielectrics A2,A4,B3); and the RTCVD oxides were deposited using a Si_2H_6 /N_2O chemistry at 800°C and 3 Torr (dielectric A3). An additional 900 °C, 15 sec post-deposition anneal in N_2O was applied to the RTCVD oxide samples. Nitrogen incorporation into bulk oxide has been reported for the RTCVD process due to the presence of N_2O (7). A 3 % oxynitride layer was deposited over varying RTO oxide thicknesses at 800°C and 3 Torr using a SiH_4/Ar/N_2O/NH_3 reaction chemistry (dielectrics A4,B3). The

RPE oxides were grown thermally at 300°C and 300 mTorr in an O_2 ambient (dielectrics A5,B4), while the RPECVD nitride depositions were carried out in-situ immediately afterward, using SiH_4 and N_2 (dielectrics A5,B2,B4). In-situ boron doped gates (120 nm) were deposited by RTCVD using $B_2H_6/Si_2H_6/Ar$ at 550 °C and 4 Torr. A 950°C, 25 sec Rapid Thermal Anneal (RTA) was applied after gate deposition to achieve dopant activation (ρ_{sh} = 230 Ω/sq after activation). Implant doped gates were obtained using a 5 keV, 5e15 cm^{-2}, B implant into LPCVD deposited polysilicon (110 nm); a 950°C, 45 sec RTA was used following the implant for gate dopant diffusion (ρ_{sh} = 205 Ω/sq after activation).

Table I. Gate stack dielectrics and dielectric thicknesses considered for in-situ and implant doped polysilicon gates.

GATE	GATE DIELECTRIC	EQUIVALENT OXIDE DIELECTRIC THICKNESS
1. In-Situ doped RTCVD polysilicon gate (Split A)	1. Furnace Oxide (A1)	~ 2.0 nm
	2. RTO oxide (A2)	2.4 nm - 2.9 nm
	3. RTCVD (N_2O) oxide (A3)	1.9 nm - 2.7 nm
	4. RTO oxide + RTCVD oxynitride {A4)	2.0 nm - 2.7 nm
	5. RPE oxide + RPECVD nitride (A5)	2.0 nm
2. Implant doped LPCVD polysilicon gate (Split B)	1. Furnace Oxide (B1)	2.4 nm
	2. Furnace Oxide + RPECVD nitride (B2)	2.3 nm
	3. RTO oxide + RTCVD oxynitride (B3)	2.0 nm
	4. RPE oxide + RPECVD nitride (B4)	1.5 nm - 2.5 nm

Well and channel implants were employed prior to gate oxidation to produce a dopant concentration of approximately 2.5 x 10^{18} cm^{-3} at the Si-SiO_2 interface. Enhanced gate linewidth resolution down to 0.1 μm was obtained using i-line lithography together with an organic Anti-Reflective Coating (ARC), followed by photo-resist trimming. The etched gates were thermally re-oxidized to grow a 4.0 nm oxide along the exposed polysilicon edges and on the substrate. 40 nm sidewall spacers (oxide spacers-split A, nitride spacers-split B) were then formed along the gate edges. A drain extension structure was used for the junctions, with low energy BF_2 implants employed to produce shallow extension ($X_{j\ ext}$ = 30 nm) and contacting ($X_{j\ cont}$ = 60 nm) junctions. A common RTA cycle of 950°C, 10 sec was used to activate both implants. After passivation and contact hole etching, a bi-layer of Ti /Al was evaporated onto the wafer substrates and patterned using lift-off. A final 400°C, 20 min anneal in forming gas was then carried out to complete the processing.

ANALYSIS AND DISCUSSIONS

A. Boron Penetration (for in-situ and implant doped gates)

Extensive CV measurements were performed to characterize the magnitude of boron penetration in the different dielectrics. The measured CV profiles were post-processed using an extraction program (11) to extract the electrical equivalent thickness of the gate-dielectric ($t_{ox,el}$), the dopant concentration in the channel (N_{ch}), the MOS flatband voltage (V_{FB}) and the MOSFET threshold voltage (V_T). Estimates of the penetrated boron were then determined from the extracted flatband voltage shift, using a model described in (32,33), which considers the penetrated boron in the channel to form a shallow, fully-depleted, p-type layer close to the Si-SiO_2 surface in the substrate. The CV extraction program used during analysis included models to correct for quantum mechanical effects associated with channel quantization. The extracted values of the dielectric thickness and the channel concentration showed good agreement with ellipsometric measurements and SIMS-determined concentrations respectively.

For devices processed with in-situ doped gates (except dielectric A5, but including B1 with an implant doped gate), the total thermal budget seen after gate deposition was 950 °C, 35 sec (950 °C, 25 sec - gate anneal RTA + 950 °C, 10 sec - S/D anneal RTA), while for devices with implant doped gates (except B1, but including A5), the combined thermal budget was 950 °C, 55 sec (950 °C, 45 sec gate anneal RTA + 950 °C, 10 sec S/D anneal RTA). Normalized high frequency CV characteristics for the different dielectrics with in-situ and implant doped gates are shown in Figs. 1 and 2 respectively. Close agreement is found in the normalized accumulation capacitances of the different dielectrics, as expected. The extracted parameters are listed in Table II. No extraction could be performed for dielectric B3 (from Table I), due to distortions in the measured CV characteristics caused by high gate leakage. An independent extraction of V_T from IV measurements in this case revealed extremely high levels of boron penetration, a likely cause of the high leakage.

Table II. Flatband voltages for the different gate dielectrics with in-situ and implant doped polysilicon gates

DIELECTRIC	GATE THERMAL CYCLE	THICKNESS (nm) (Extracted)	SUBSTRATE CONC. (cm-3) (Extracted)	FLATBAND VOLTAGE (V) (Extracted)	$\Delta V_{FB} = V_{FB} - Ø_{ms}$ (V)
(A1) FURNACE Oxide	950°C, 35 sec	2.1	2.3×10^{18}	1.24	0.20
(A2) RTO Oxide	950°C, 35 sec	2.4	2.3×10^{18}	1.23	0.19
(A3) RTCVD Oxide	950°C, 35 sec	2.6	2.2×10^{18}	1.05	0.01
(A4) RTO Oxide + RTCVD Oxynitride	950°C, 35 sec	2.0	2.1×10^{18}	1.26	0.21
(A5) RPE Oxide + RPECVD Nitride	950°C, 55 sec	2.0	2.3×10^{18}	0.73	-0.31
(B1) FURNACE Oxide	950°C, 35 sec	2.3	1.9×10^{18}	1.42	0.38
(B2) FURNACE Oxide + RPECVD Nitride	950°C, 55 sec	2.3	2.4×10^{18}	0.61	-0.43
(B4) RPE Oxide + RPECVD Nitride	950°C, 55 sec	2.1	2.7×10^{18}	0.72	-0.32

Boron penetration is observed for all dielectrics in Fig. 1, except for the oxide/nitride stack, where the flatband voltage shift relative to $Ø_{ms}$ is negative (i.e. the measured value of V_{FB} is less than the value of $Ø_{ms}$), indicating a positive fixed charge in the dielectric. Here, $Ø_{ms}$ was calculated to be 1.04 V. Of the dielectrics processed with in-situ doped gates, similar levels of penetration, as estimated from ΔV_{FB}, were found for the furnace and RTO oxides and the RTO oxide/RTCVD oxynitride stack, although lower penetration was seen for the RTCVD oxide. The relative similarity in penetration resistance between the RTO oxide and RTO oxide /RTCVD oxynitride stack samples is a little surprising, since the presence of the oxynitride is expected to reduce penetration. It is possible that the thickness of the oxynitride film used in this experiment (0.5 nm) was too small to be effective. By comparison, a smaller flatband voltage shift is seen for the RTCVD oxide. The reduced penetration observed in this case is attributed to the nitrogen incorporated into the film by the Si_2H_6/N_2O deposition process (12). Although the Si_2H_6/N_2O process incorporates a lower level of nitrogen than the NH_3 process (7) in sample A3, it is seen to be more effective in controlling boron penetration. A possible reason is the greater thickness of the RTCVD oxide (2.7 nm), relative to the oxynitride film (0.5nm), leading to a larger cumulative N content.

The higher resistance of the RTCVD oxide to boron penetration is also evidenced by the observed variation in flatband shift with dielectric thickness. Fig. 3 shows the normalized high frequency CV plots for three thicknesses of RTCVD oxide. A progressive reduction in the flatband voltage is seen as the oxide thickness increases, suggesting a decrease in boron penetration with increasing oxide thickness. The negligible flatband voltage shift for $t_{ox(RTCVD)}$ = 2.5 nm suggests that RTCVD oxide thicknesses larger than 2.5 nm will be effective in controlling boron penetration.

The highest resistance to boron penetration was obtained with the oxide/nitride stacks (dielectrics B2 and B4). No change in flatband voltage was observed in RPE oxide/nitride stacks, even after these stacks were subjected to progressively larger thermal cycles. The absence of penetration was confirmed in this study for a thermal budget of 950°C, 130 sec, (no shift in V_{FB} seen at the higher thermal budget in Fig. 4) but has been reported in other literature for thermal budgets as high as 1000°C, 120 sec (4).

The negative V_{FB} shift encountered in oxide/nitride stacks with both in-situ and implant doped gates is attributed to sheet fixed charge at the oxide/nitride interface (13). From the data available in this experiment, the value of this sheet charge density was determined to be ~ 5 x 10^{12} cm^{-2}. It is suspected, however, that the high fixed charge may have been a consequence of non-optimal process conditions during the RPE nitride deposition. Fixed charge densities lower by more than an order of magnitude than those seen in this study were measured for oxide/nitride stack devices when the nitride was deposited using a *reduced plasma power* RPE process (9). Consequently, it is believed that enhanced nitridation occurred at the oxide/nitride interface due to the excessively high plasma power setting used in our depositions, resulting in a higher fixed charge.

B. Density of Interface States

The density of interface states in fabricated devices was measured using a charge-pumping scheme (14). Fig. 5 presents D_{it} values for different gate dielectrics. Each value represents an average over 5 measurements made at devices with the same gate area. Data from splits A (in-situ doped gates) and B (implanted gates) are presented separately. Excellent, low interface state densities were measured for the furnace oxide/RPE nitride and RPE oxide/nitride (conditions B2 and B4) dielectric stacks with implanted polysilicon gates. In both cases, the low D_{it} values are partially attributed to a pile-up of nitrogen at the $Si\text{-}SiO_2$ interface during the 900°C, 15 sec anneal following nitride deposition. A small nitrogen peak at the interface following the anneal was confirmed from SIMS analysis in (9) for a furnace oxide/ RPE nitride stack. Nitrogen atoms are believed to replace weak Si-O bonds at the interface, relieving strain in the process due to the smaller effective size of the nitrogen atoms (15). Consequently, lower interface state densities may be expected.

High values of D_{it} were found for all the oxides and oxynitrides processed with in-situ doped gates (except A5, for which D_{it} data were not available) relative to those with implanted gates. It is interesting to note that for dielectrics with in-situ doped gates, the aggravated interface state density coincides with a higher measured boron penetration, as seen in section (A). While no direct correlation between boron penetration and interface state densities is known, it is conceivable that the penetrated charge present in the thin oxides may either alter the D_{it} levels or otherwise influence their measurement. With reference to the control furnace oxides (in Split A), the D_{it} was ~35 % lower for RTO oxides, ~80 % higher for RTCVD oxides and 15 % higher for the RTO oxide /RTCVD oxynitride stacks. The reduced D_{it} in the RTO may be a consequence of the higher temperature of the RT oxidation (1000°C) relative to the furnace oxides (700°C). The higher D_{it} observed for the RTCVD oxides and the oxide-oxynitride is consistent with previous literature on D_{it} characterization in deposited oxides (16), and in oxynitrides deposited using NH_3 (17).

C. Gate leakage

Dielectric evaluation was also conducted by comparing tunneling leakage magnitudes for the different dielectrics at a fixed accumulation gate voltage. Measurements were made for V_G = 1.2 V on both capacitor and MOSFET (S/D shorted for higher carrier injection) sites. The measured gate leakage is plotted as a function of dielectric thickness in Fig. 6 for the different gate stack dielectrics. In each case, the equivalent oxide thickness was extracted from a low frequency CV characteristic measured at the same site, and analyzed using the CV extraction program (11) described in Section (A).

Tunneling data were not available for all of the dielectrics in overlapping thickness regimes. While the leakage variation with thickness does suggest a trend that may be used in evaluating the different dielectrics, extrapolation of the trend to lower thicknesses must be undertaken with caution, since the oxide barrier height has been reported to decrease at low oxide thicknesses (7) due to image forces and the conservation of crystal momentum (18). An examination of the leakage variation with thickness for the different dielectrics confirms the expected inverse exponential dependence of the leakage on the insulator thickness. When plotted on a logarithmic scale, the slope of the leakage variation with thickness is inversely dependent on the barrier height (19); consequently, the increase in the leakage slopes observed at lower thicknesses for some of the dielectrics may be interpreted to result from a lowering in barrier height at these thicknesses.

Of the different dielectrics examined, RTO oxides showed the lowest leakages down to the smallest available RTO thickness of 2.4 nm. Moderately higher leakage was observed for the RTO oxide /RTCVD oxynitride stack and the RPE oxide/nitride stack; in both cases, the leakage was comparable to that found for furnace grown thermal oxides. It is interesting to note that relative to the leakage trend for the RTO oxide /RTCVD oxynitride stacks, the furnace oxides showed a slightly higher leakage, while the furnace oxide /RPE nitride stacks showed a moderately smaller leakage. The improvement in leakage obtained with the furnace oxide /RPE nitride stack in this case is due to the increase in total dielectric thickness when an oxide/nitride stack is used instead of a pure oxide. The nitride layer, with a relative dielectric constant of approximately 7.0, contributes only half its thickness to the value of $t_{ox,eq}$. Of all the dielectrics considered, the leakage was found to be highest for the RTCVD oxides, between a factor of 3 to 6 times larger than that seen for the RTO oxide/ RTCVD oxynitride samples over the same thickness range.

CONCLUSIONS

Ultra-thin 2.0 - 3.0 nm gate dielectrics formed using several different growth and deposition techniques were examined in this study as candidates for a 0.1 μm technology. Dielectric evaluation was conducted with respect to interface state densities, gate tunneling currents and dielectric resistance to boron penetration. Furnace oxides, RTO oxides, RTCVD oxides, RTO oxide /RTCVD oxynitride stacks and RPE oxide/nitride stacks were examined with either an in-situ or an implant doping scheme for the gate.

RTO oxides exhibited the lowest gate leakages, moderately improved D_{it} and comparable resistance to boron penetration relative to the furnace grown oxides. The addition of a thin RTCVD deposited oxynitride to the RTO oxide increased the dielectric resistance to boron penetration a little, although some degradation in interface state density was also observed. RTCVD oxides, deposited with Si_2H_6/N_2O, showed enhanced protection against boron penetration, which improved as the dielectric thickness was increased. However, the RTCVD oxides also showed higher gate leakages and larger interface state densities, relative to the other dielectrics. Of all the dielectrics considered, only the RTCVD oxides and the oxide-nitride stacks demonstrated acceptable resistance to boron penetration, and, of these two dielectrics, only the oxide/nitride stacks were able to contain boron penetration for $t_{ox,eq} < 2.5$ nm. The level of penetration in the oxide/nitride stacks was found to be negligible for $t_{ox,eq} = 2.0$ nm even at the largest thermal budgets examined in this study (950°C, 120 sec). The oxide/nitride stacks also had very low interface state densities, and comparable gate leakage to furnace oxides. A high fixed charge density was encountered at the oxide-nitride interface in this study; however, it appears that optimization of the nitride deposition process, as demonstrated by other researchers, may considerably reduce the magnitude of the fixed charge. With such an optimized nitride deposition process, oxide/nitride stacks appear uniquely equipped to address CMOS dielectric needs for thicknesses in the 2.0 nm regime, by their capability to contain boron penetration while exhibiting comparable dielectric response and device performance relative to conventional furnace oxides.

ACKNOWLEDGMENTS

The authors are grateful to the staff of the Microelectronics Laboratory at the North Carolina State University for their assistance in wafer preparation and to Sung Han for his help in measurements. This work was partially supported by the Semiconductor Research Corporation under grant SRC 94-MC-509 and by the NSF Engineering Research Centers program through the Center for Advanced Electronic Material Processing, under grant CDR-8721545. This work was also performed in part at the Cornell Nanofabrication Facility (a member of the National Nanofabrication Users Network) which is supported by the National Science Foundation under grant ECS-9319005, Cornell University and industrial affiliates.

REFERENCES

1. T. Kuroi *et al, 1996 VLSI Tech. Symp. Dig.*, 210 (1996)
2. Y. Taur *et al, IEEE Proceedings*, **85**, 484 (1997)
3. R.B. Fair, *Proceedings of the 6th International Symposium on ULSI Science and Technology*, Electrochemical Society, PV **97-3**, 247 (1997)
4. Y. Wu and G. Lucovsky, *1998 Proc. IRPS,* 70 (1998)
5. J. Nulman *et al, IEEE Electron Device Lett.* , **6**, 205 (1985)
6. X. Xu *et al, Applied Phys Lett.,* **60,** 3063 (1992)
7. E.M. Vogel, *PhD Thesis*, North Carolina State University (1998)
8. C.R. Parker *et al*, **19**, 106 (1998)
9. Y. Wu and G. Lucovsky, *IEEE Electron Device Lett.*, **19**, 367 (1998)
10. A. Srivastava *et al, Proceedings of the Rapid Thermal and Integrated Processing Conference,* Materials Research Society, San Francisco, April (1998)
11. J.R. Hauser and K. Ahmed, *1998 International Conference on Characterization and Metrology for ULSI Technology,* NIST, Gaithersburg, March (1998)
12. H. Fukada *et al, Japan J. of Appl. Phys.,* **29**, 2333 (1990)
13. J.R. Hauser, *Private communication.*
14. Khaled Ahmed, *PhD Thesis*, North Carolina State University (1998)
15. M.L. Green *et al, Appl. Phys. Lett.*, **65**, 848 (1994)
16. T. Hori *et al, IEEE Trans. Electron. Devices* , **36**, 340 (1989)
17. Y. Okada *et al, J. Electrochem. Soc.*, **140**, 487 (1993)
18. S-H Lo *et al, IEEE Electron Device Lett.*, **18**, 209 (1997)
19. K.F. Schuegraf *et al, 1992 Symposium on VLSI Tech. Dig.*, 18 (1992)

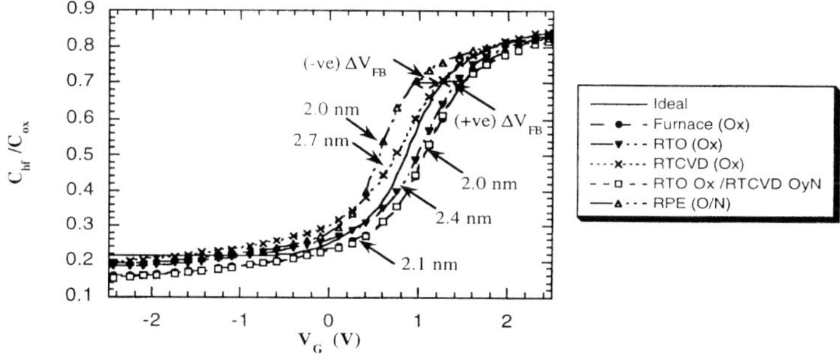

Figure 1. *High frequency CV characteristics for the different gate stack dielectrics fabricated with in-situ boron doped gates*

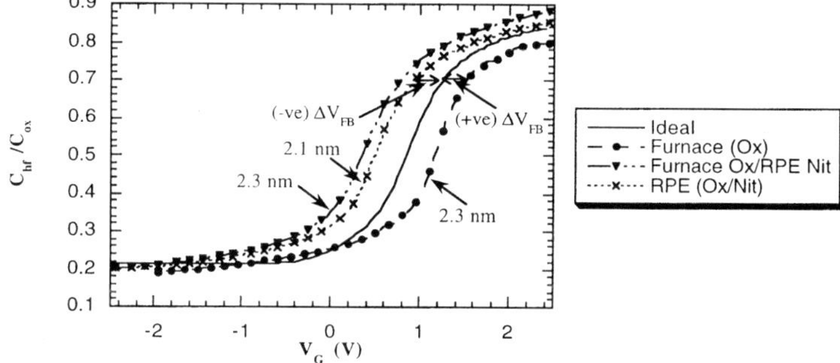

Figure 2. *High frequency CV characteristics for the different gate stack dielectrics fabricated with implant doped gates*

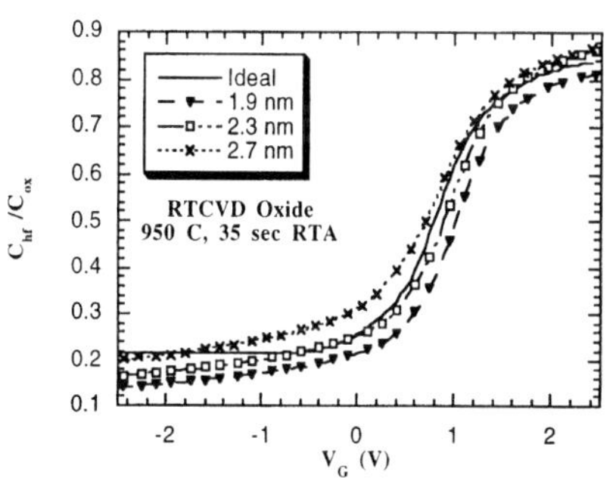

Figure 3. *High frequency CV characteristics for three different RTCVD oxide thicknesses. The gates were fabricated with the in-situ process*

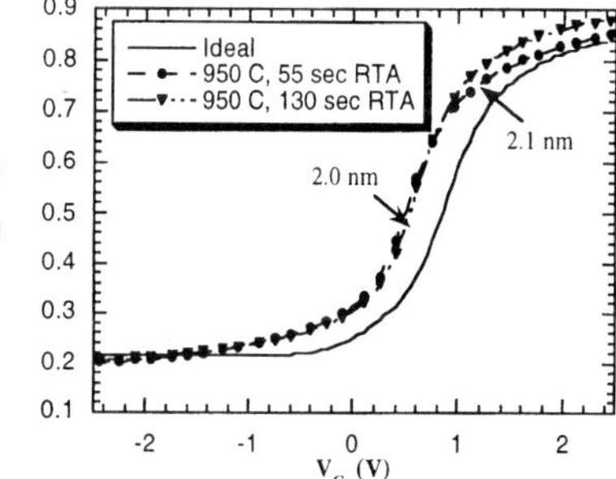

Figure 4. *High frequency CV characteristics for RPE oxide/nitride stacks annealed using different thermal cycles. The gates were doped by implantation*

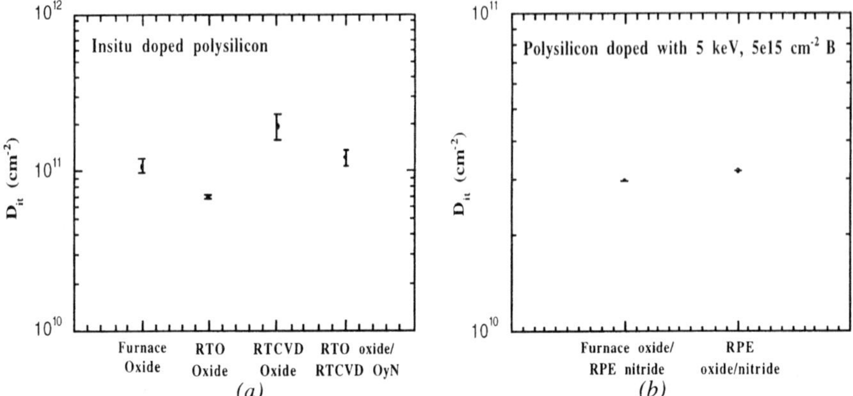

Figure 5. *Measured D_{it} distributions for the different dielectrics conditions fabricated with (a) in-situ doped gates on oxide or oxynitride dielectrics, and (b) implant doped gates on oxide/nitride dielectrics*

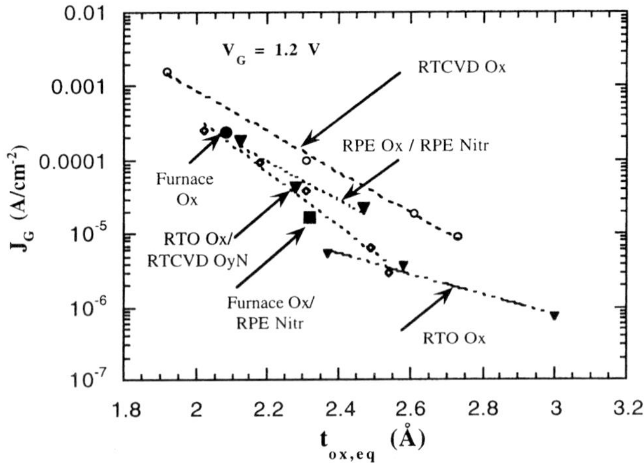

Figure 6. *Variation of gate tunneling current vs dielectric thickness for the different dielectrics.*

GROWTH KINETICS AND MODELING OF DIRECT OXYNITRIDE GROWTH WITH NO-O_2 GAS MIXTURES

Rahul Sharangpani[1], Sing-Pin Tay[1], Randhir Thakur[2],
Sarah Everist[3] Jerry Nelson[3] Paul Martin Smith[3]
[1]Steag RTP Systems, San Jose, CA 95134
[2]Steag Electronic Systems, San Jose, CA 95134
[3]Sandia National Laboratories, Albuquerque, NM 87185

We have modeled growth kinetics of oxynitrides grown in NO-O_2 gas mixtures from first principles using modified Deal-Grove equations. Retardation of oxygen diffusion through the nitrided dielectric was assumed to be the dominant growth-limiting step. The model was validated against experimentally obtained curves with good agreement. Excellent uniformity, which exceeded expected values, was observed.

INTRODUCTION

The reliability advantages of oxynitrides over conventional SiO_2 for gate and tunnel dielectric applications have been well established. Various nitrogen sources have been investigated for this purpose, such as NH_3, N_2O, and NO. Nitrous oxide (N_2O) provides high growth rates but has inadequate nitrogen incorporation and the nitrogen distribution is severely non-uniform (1). Ammonia (NH_3) is a promising alternative (2) but requires a post-nitridation reoxidation step to eliminate hot carrier degradation caused by hydrogen incorporation (3). Nitric oxide (NO) is being studied because of its ability to provide good uniformity, high nitrogen concentrations, a robust interface, and excellent reliability characteristics (4). Rapid thermal processing (RTP) for oxynitride growth can provide advantages over furnace processing, including better uniformity, improved dielectric quality, and reduced thermal budget (5,6). Further advantage is gained by direct oxynitridation in a 1-step process using NO-O_2 mixtures at temperatures over 950°C because of the increased throughput compared to multi-step processing, although the effect on dielectric quality is unknown. Since incorporated nitrogen inhibits oxygen diffusion through the dielectric, reduced growth rates are observed with gas mixtures that produce oxynitride films as compared with rapid thermal oxidation (RTO) (7). In this paper, we have modeled the growth kinetics by modifying the diffusion dependent parameters of the modified Deal-Grove equations. Excellent thickness uniformity was obtained for typical growth conditions.

EXPERIMENTAL

All experiments were performed in a Heatpulse 8800 rapid thermal processor equipped with a ceramic shield and non-contact, closed loop temperature control. The system was also equipped with gas sensors, alarms, and other safety features required for toxic gas handling. The substrates were 200 mm, p-type, Si <100> wafers. All wafers were ramped at a constant rate of 60°C/s to the processing temperatures of 950°C or

1050°C. Cooling occurred radiatively. Mixtures of 25% or 50% NO in O_2 were flowed over the wafers at atmospheric pressure from the start of the ramp step until the process cycle was complete. After processing, the wafer cooled to 350°C in a constant flow of 20 slm of nitrogen before it was removed from the oven. The oxynitride film thicknesses were measured using a single wavelength, Gaertner Scientific Corp. ellipsometer (λ = 632.8 nm), assuming a constant refractive index of 1.4. Uniformity data was obtained by 49 point mapping with a UV Tencor 150 SE spectroscopic ellipsometer (3 mm edge exclusion).

RESULTS AND DISCUSSION

The classical Deal-Grove linear parabolic equations define thermal oxide growth kinetics using a set of diffusion and temperature dependent parameters. The validity of these equations has been well established for thicker (>500 A) oxides (8). However, it has been observed that growth rates of SiO_2 in the thin regime are higher than those predicted by the Deal-Grove model (9). Massoud, *et al.*, have formulated a new equation, which is presented below for the case of lightly doped substrates (9).

$$\partial x / \partial t = B/(2x + A) + C\exp(-x/L) \qquad [1]$$

Where x is the oxide thickness after time, t. The term C represents a departure from the classical Deal-Grove linear parabolic model and is included for thin oxides to account for increased oxidation sites below the surface (9). L is constant for a given wafer orientation. Typical values of L for RTO are 10-15 A (10). It is seen that for thicker oxides, the last term on the right hand side of equation (1) approaches zero. When the resulting differential equation is solved it yields the standard Deal-Grove linear-parabolic model. Therefore, Deal-Grove kinetics can be viewed as a special case of equation 1 for thick oxides.

The excess growth rate (C) depends on the incorporated nitrogen (N), temperature (T), and oxide thickness (x). Term A is a function of the reaction rate constant (k) and the diffusion coefficient (D). Term B is a function of the diffusion coefficient, which in turn depends on the temperature, thickness, and incorporated nitrogen. Finally, the incorporated nitrogen is a function of partial pressure of NO (P), temperature, and time. These relationships are expressed quantitatively in equations 2-6 below.

$$B=2DC_0/C_i \qquad [2]$$
$$A=2D/k \qquad [3]$$
$$D=f(N,T,x) \qquad [4]$$
$$C=g(N,T,x) \qquad [5]$$
$$N=h(T,t,P) \qquad [6]$$

C_o and C_i are constants that are independent of wafer orientation and are defined in reference (8). Equations 1-6 can now be combined into a single unified differential equation expressed as:

$$\partial x / \partial t = (2f(h(T,t,P),T,x)(Co/Ci)) / (2x + 2f(h(T,t,P),T,x)/k) + g(h(T,t,P),T,x)\exp(-x/L) \quad [7]$$

A Fourth order Runge-Kutta method is used to solve this equation and plot the theoretical thickness/time curves. In order to avoid complications due to different ramp rates and bring in times, t = 0 is defined as the end of the ramp step. Initially, L (Equation 1) and C were obtained from curve fitting with unnitrided (P = 0) growth curves as described in our earlier work (10). The nature of the dependence of the diffusion coefficient (D) on the incorporated nitrogen (N), and N on temperature, NO partial pressure, and time was determined from published data (11-13) and is not reproduced here. Experimental and theoretical curves are compared in Figures 1 and 2.

This model assumes that the dominant growth limiting step is the diffusion of oxygen across the dielectric layer, which challenges work that attributed the reduction in growth rate to nitrogen occupation of oxidation sites at the Si surface (14). We believe this assumption is valid because NO processing results in films with relatively high nitrogen concentrations (4) and Okada, et al., have shown growth retardation due to limited diffusion at nitrogen concentrations above 0.9% (7). With our model, nitrogen incorporation (atoms/cm^2) can also be calculated, increasing the usefulness of the model since resistance to boron penetration and hot carrier effects depend on nitrogen incorporation. Further, using a similar set of equations it is possible to predict growth rates and nitrogen incorporation for 2- and 3-step dielectric film growth.

Discrepancies between the model and the experiment arise because there is no unambiguous way to quantify nitrogen incorporation in the dielectric as a function of growth parameters. In addition, there is no straightforward way to relate the oxygen diffusion coefficient to incorporated nitrogen. Equations of the form published for boron diffusion through oxynitrides (13) were used with suitable modifications. This extrapolation may lead to errors since the growth kinetics depends on the oxygen diffusion coefficient. Nonetheless, there is good agreement in the regime of practical interest for the formation of ultrathin gate and tunnel dielectrics. Therefore, this model provides a good basis for the development of thin oxynitride dielectric films for advanced applications.

Figure 3 shows a thickness contour map for a wafer processed using 50% NO in O_2 at 1050°C for 30 s. It should be noted that the wafer uniformity (1 sigma = 0.8%) is better than what is expected from an analysis of the Heatpulse 8800 specifications and the temperature sensitivity of the observed oxynitride process.

CONCLUSIONS

A model is proposed that shows good agreement between experimental and theoretical growth curves. This model provides insights into the oxynitride growth process and can be used to theoretically determine growth kinetics for dielectric thicknesses of practical importance to advanced gate and tunnel dielectric applications. A user-friendly version of this model is being implemented for more convenient computation of film thicknesses.

ACKNOWLEDGEMENTS

The authors would like to thank the personnel at Sandia's Microelectronics Development Laboratory (MDL) for wafer processing. Sandia is a multiprogram laboratory operated by Sandia Corporation, a Lockheed Martin Company, for the United States Department of Energy under Contract DE-AC04-94AL85000.

REFERENCES

1. T.Y. Chu, W.T. Ting, J.Ann and D.L. Kwong, *J. Electrochem. Soc.*, **138** L13 (1991).
2. W. Cho, K. S. Balakrishnan, T. F. Edgar, I. Trachtenberg, 5th Int. Symp. on Process Systems Engineering, Seoul, Korea, p. 1235-1239, 1994.
3. I.J.R. Baumvol, F.C. Stedile, J.J. Ganem, I. Trimaille, S. Rigo, *J. Electrochem. Soc.*, **143** 1426 (1996)
4. H.B. Harrison, Z.Q. Yao, S. Dimitrijev, D. Sweatman and Y.T. Yeow, Mat. Res. Soc.
   Proc. **387**, 233, 1995.
5. J. Nulman, J.P. Krusius and A. Gat, *IEEE Electron Device Letters* EDL-6, 205 (1986).
6. A.B. Joshi and D. Kwong, *IEEE Transactions on Electron Devices*, 39(9), 2099, 1992.
7. Y. Okada, P.J. Tobin, V. Lakhotia, S.A. Ajuria, R. Hegde, J.C. Liao, P.R. Rushbrook and L.J. Arias, *J. Electrochem. Soc.*, **140**, L87, 1993
8. S.P. Murarka and M.C. Peckerar *Electronic Materials, Science and Technology*, p. 102-106. Academic Press, Inc., San Diego (1989).
9. H.Z. Massoud, J.D. Plummer and E.A. Irene, *J. Electrochem. Soc.*, **132**, 2693, 1985.
10. R. Sharangpani, J. Das, S. Tay, R.P.S. Thakur, T.C. Yang and K.C. Saraswat, Proc. Mat. Res. Soc. Meeting, vol. 525, 1998, p. 143.
11. I. Sagnes, D. Laviale, M. Regache, F. Glowacki, L. Deutschmann, B. Blanchard and F. Martin, Mat. Res. Soc. Symp. Proc. Vol. 429, 1996, p. 251.
12. G. Weidner, D. Kruger, M. Weidner, K. Tittelbach-Helmrich, Mat. Res. Soc. Symp. Proc. Vol. 387, 1995, p. 265.
13. R.B. Fair, *J. Electrochem. Soc.* **144**, 708 (1997).
14. S. Dimitrijev, D. Sweatman and H.B. Harrison, *Appl. Phys. Lett.*, **62**, 1539, 1993.

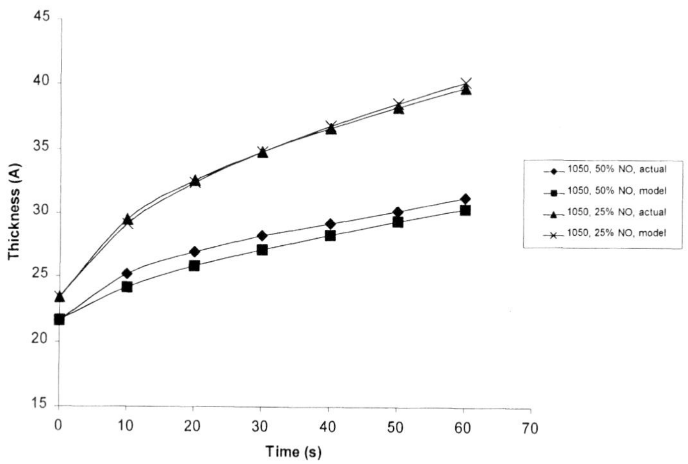

Figure 1: Experimental vs. predicted growth curves at 1050°C.

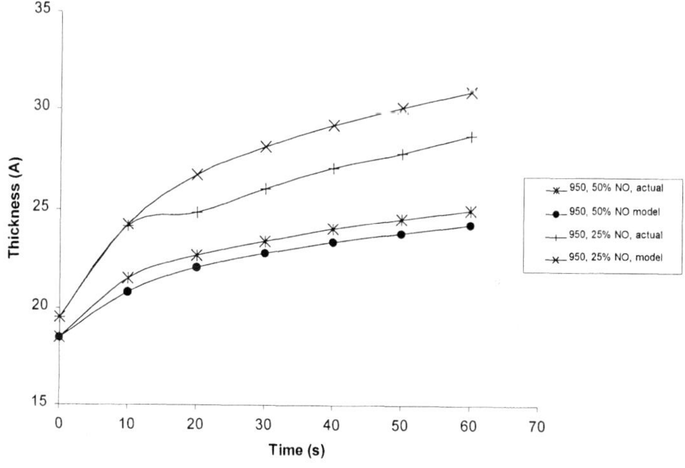

Figure 2: Experimental vs. predicted growth curves at 950°C.

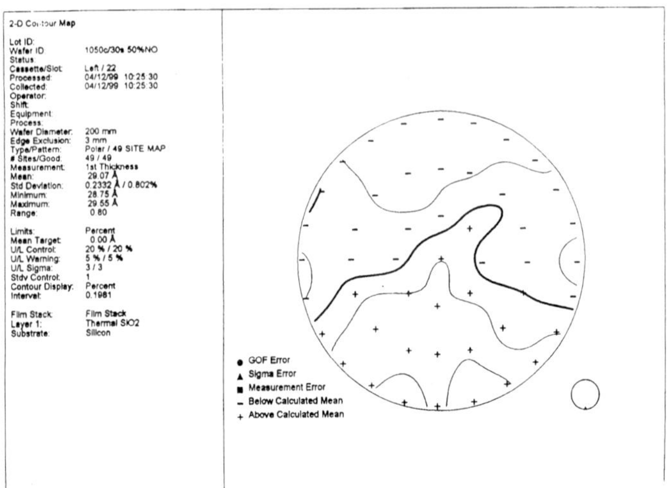

Figure 3: Thickness contour map for 1050°C / 30s process carried out using 50% NO showing uniformity $1\sigma = 0.8\%$.

INTERFACIAL PROPERTIES of $Si\text{-}Si_3N_4$ FORMED BY REMOTE PLASMA AND RAPID THERMAL PROCESSING

H. Lazar, V. Misra, Z. Wang, M. Kulkarni, W. Li, M. Mahler and J.R. Hauser
Department of Electrical and Computer Engineering,
North Carolina State University, Raleigh, NC 27695, USA

This paper discusses the interfacial properties of $Si\text{-}Si_3N_4$ in NMOSFETs and PMOSFETs. Silicon nitride, formed by remote plasma/rapid thermal processing, was found to display severely degraded interfacial properties, in which the PMOS interfaces were significantly more degraded than NMOS interfaces. This is believed to be indicative of a relatively high density of interface traps located below the Si mid-gap that inhibit hole channel formation. Bonding constraint theory was applied to conclude that the $Si\text{-}Si_3N_4$ interface is over-constrained compared to the $Si\text{-}SiO_2$ interface and consequently results in a higher intrinsic defectivity. A systematic study of the oxygen and hydrogen content in the silicon nitride film and its effect on electrical properties is also presented. Based on the electrical results it is concluded that the presence of oxygen either as a monolayer at the interface or within the silicon nitride film can produce high quality interfaces suitable for aggressively scaled CMOS devices.

INTRODUCTION

The national technology road map for semiconductors projects that gate dielectrics with oxide equivalent thicknesses to ~1.0 nm will be required as channel lengths decrease towards 50 nm. This will require new gate dielectric materials with dielectric constants (K) higher than that of SiO_2 to provide the increased capacitance without compromising gate leakage currents [1]. However, many of the high-K dielectric candidates are thermodynamically unstable on Si and result in undesired growth of SiO_2. Therefore, ultra thin advanced interface layers with good diffusion barrier properties are needed to facilitate the formation of high capacitance gate stacks. Silicon nitride is an excellent candidate for an interface layer since it has approximately twice the dielectric constant of SiO_2, reasonable band gap and additionally is effective in blocking oxygen diffusion from subsequent high-K dielectrics processing.

Silicon nitride films have been studied extensively in the past for their potential use in non-volatile memory devices and as gate insulators for thin-film transistors. Typically, low pressure chemical vapor deposition techniques were used to generate these films, however, the characteristics of this deposition process can leave the film hydrogen rich, which in the form of Si-H and N-H bonds, can act as precursors for defect generation [2]. A complete optimization of the film chemistry, deposition process and post treatment is needed to obtain good quality silicon nitride films. Recently, a few encouraging processes such as Jet Vapor Deposition and rapid thermal processing have emerged that demonstrate good gate quality silicon nitride behavior [3,4]. These films have shown lower leakage conduction than SiO_2 at the same oxide equivalent thickness and have also displayed excellent interface characteristics. However, the step sequences required to create these high quality films typically results in a oxygen rich layer near the Si interface or oxygen in the nitride film.

In order to fully obtain the benefits of a higher dielectric constant interface layer, the investigation of pure stoichiometric Si_3N_4 on Si is required. Therefore, the goal of this work is to evaluate the interface of Si and pure Si_3N_4 formed by remote plasma enhanced chemical vapor deposition (RPECVD) and to obtain a fundamental understanding of this interface at the atomic level. The electrical behavior on both NMOSFETs and PMOSFETs was investigated. The asymmetric electrical characteristics of NMOS and PMOS observed in this work will be linked to an atomic level model that links the defect characteristics to the intrinsic nature of the Si-silicon nitride interface.

EXPERIMENTAL

Silicon nitride films were formed in a cluster tool in which the entire gate stack was completed before exposure to atmosphere. Prior to gate stack formation, a 200 nm field oxide was grown on 100 mm substrates followed by patterning of the active areas. A 10nm sacrificial oxide was then grown and removed with a 2 % HF solution immediately before insertion into the cluster tool. Nitride layers were formed by RPECVD at 400 W, 300 mTorr and 300 °C using SiH_4+NH_3(N_2) [5]. Single layer nitrides varying in thickness from 1.5 to 3.0 nm were formed on both n- and p-type substrates. In addition to forming single layer nitride films, nitride/oxide stacks were also formed to study the properties of ultra thin nitrides (0.5-2 nm) at the interface without encountering characterization problems typically associated with ultra thin layers. After deposition, films were rapid thermal annealed at 900 °C in vacuum to reduce the hydrogen content and promote formation of additional Si-N bonds [8-10]. Gate dielectrics were then capped by rapid thermally-deposited polysilicon in the cluster tool system. Control oxides of varying thickness were also prepared by thermal oxidation. Surface channel NMOS and PMOS devices were formed by conventional techniques, and a forming gas (10 % H_2 in N_2) anneal at 400 °C was performed after metallization.

RESULTS AND DISCUSSION

Secondary Ion Mass Spectroscopy analysis results of RPECVD nitride films are shown in Fig.1. The hydrogen content is less than 15 % throughout the film. The oxygen concentration is also very low (< 3 %) with a slight increase at the interface. This slight increase may be related to surface contamination from being exposed to the clean room before being loaded in the RPECVD chamber. Transmission electron microscopy indicated that the dielectric constant was ~7.4. The Si-2p XPS spectra (not shown) indicated the presence of Si-N bonds and the absence of Si-O bonds indicating that the Si-interface represents a Si-pure Si_3N_4.

Current voltage measurements were made on NMOSFETs and PMOSFETs with control oxide and silicon nitride. The thicknesses of the nitride and control oxide,as extracted from the CV data using a least square fit method [6], were 2.1 and 2.3 nm, respectively. The Gm*Tox vs Electric Field is shown is Fig. 2. The nitride devices displayed a ~50 % degraded peak mobility, but an enhanced high-field mobility. The NMOSFETs with nitride dielectrics also had a lower V_t compared to the control devices. The behavior of reduced peak degradation, enhanced high field mobility and lower V_t is consistent with what has been observed in NMOSFETs with heavily nitrided oxides [7,8] and is attributed to fixed positive charge caused by the nitrogen incorporation. The PMOS data with nitride dielectric, shown in Figure 2b, displays markedly different characteristics compared to NMOS devices. Firstly, the channel conduction is significantly reduced, and secondly, there was a very large V_t-shift of the order of ~1.5 V. The degraded I_{ds} and large V_t-shift make the PMOS devices effectively non-functional.

In an effort to better understand the cause of the degraded PMOS behavior, capacitance voltage (CV) measurements were performed on transistors to obtain both accumulation and inversion characteristics. As shown in Fig. 3a, the NMOS devices showed degraded characteristics near the accumulation region, whereas the inversion CV curves were normal. The PMOSFET CV data for silicon nitride, shown in Fig. 3b, indicated normal characteristics in the accumulation region. However, no inversion layer formation was observed which is consistent with the severely degraded drive currents obtained from the I_d-V_g measurements. This also suggests that only holes in the channel were being affected. Although the degradation in the NMOS devices was lower than in the PMOS, these results demonstrated that the nitride dielectric was affecting channel holes significantly more than channel electrons.

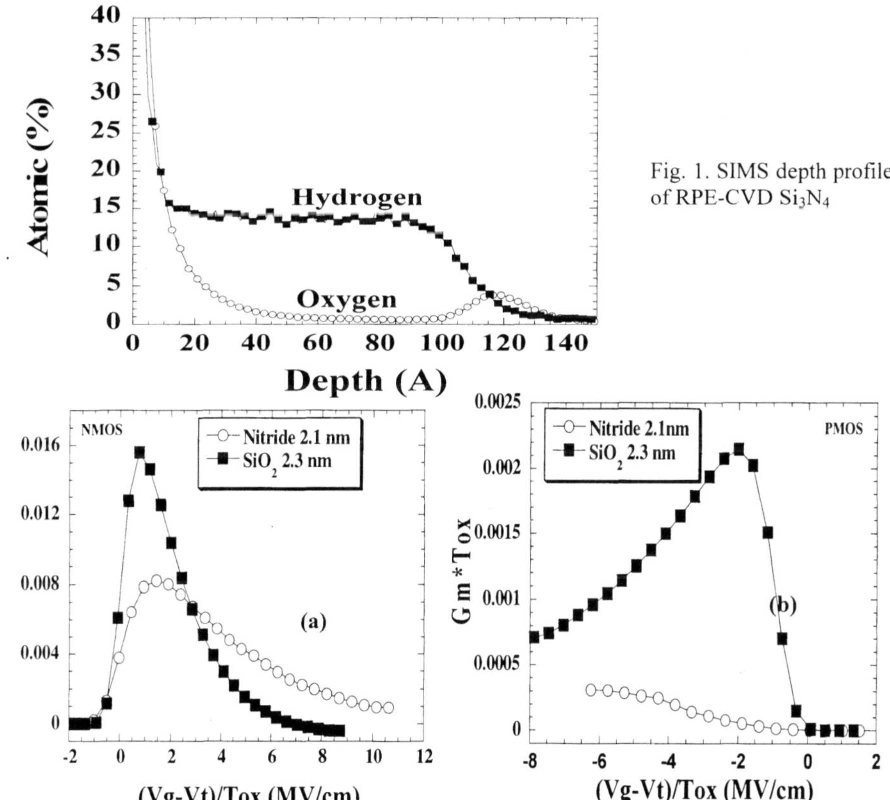

Fig. 1. SIMS depth profile of RPE-CVD Si_3N_4

Fig. 2. G_m* T_{ox} *vs.* oxide field for silicon nitride and SiO_2 for 0.6 μm channel length (a) NMOSFETs and (b) PMOSFETs. The oxide field was calculated as (Vg-Vt)/Tox.

Gate tunneling current was measured both in accumulation and inversion for both NMOS and PMOS devices (area = 30 μm^2) and is shown in Fig. 4. The voltage drop across the oxide was obtained using V_g-V_t in inversion and V_g-V_{fb} in accumulation. The silicon nitride shows slightly lower tunneling current than SiO_2 in the positive region for both NMOS (Fig. 4a) and PMOS (Fig. 4b) devices. However, at negative bias the gate current through the nitride layers is excessive compared to SiO_2 even though the nitride layers are physically thicker.

In an effort to understand the mechanisms responsible for the degraded hole behavior, a carrier separation technique was used to separately measure the hole and electron currents [9]. Measurement of gate, channel and substrate current provides a measure of the electron and hole current components. The results for SiO_2 in both NMOS and PMOS are shown in Fig. 5a. For both transistors, the gate current is nearly equal to the channel current for all voltages implying that the majority of the carriers that tunnel to the gate come from the channel, i.e. electron current dominates in the NMOS and hole current dominates in the PMOS. The substrate current is a

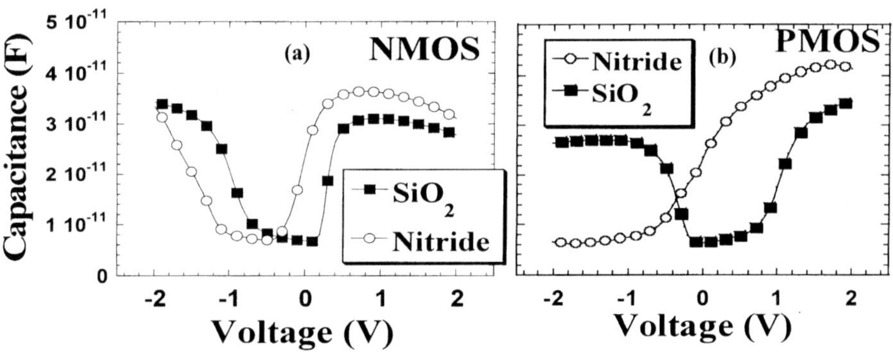

Fig. 3. CV on 50 x 50 μm^2 (a) NMOSFETs (2.1 nm) and (b) PMOSFETs (2.3 nm) at 1 MHz.

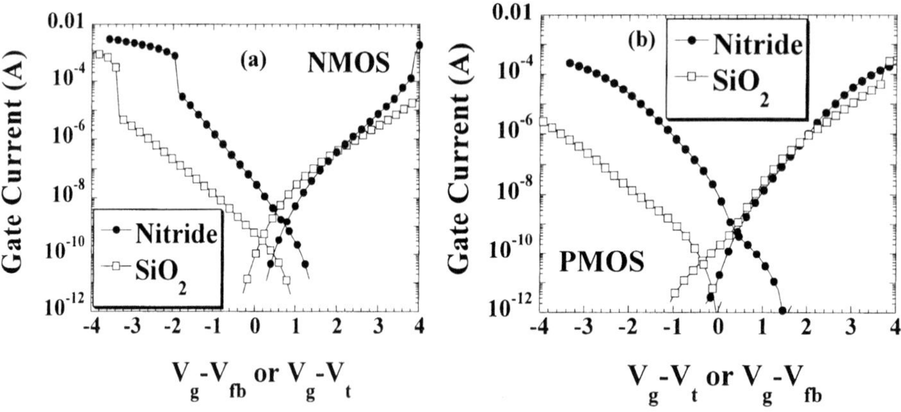

Fig. 4. Gate current for (a) NMOSFET (2.1 nm) and (b) PMOSFET (2.3 nm).

very small percentage of the gate current. The measurements for Si_3N_4 for NMOS and PMOS are shown in Fig. 5b. The NMOS behavior with nitride dielectrics is similar to the control oxide, i.e., the gate current is dominated by source/drain channel current suggesting that electron current is the dominating process in this regime. However, the PMOS with silicon nitride displays markedly different behavior than its SiO_2 counterpart. Firstly, at smaller gate biases the drain current is very low and is not equal to the gate current suggesting that the channel holes are not participating in the carrier conduction in this regime. Secondly, the substrate current is very high in this regime and equals the gate current. This implies that the electron conduction from the gate is the dominating process. It should also be noted that after V_g exceeds ~ −1.5 V, the channel current starts increasing and becoming equal to the gate current.

There are several possible reasons that can be responsible for the poor Si_3N_4-Si interfacial behavior. To understand the responsible mechanism the following effects need to be carefully examined a) role of hydrogen in Si_3N_4, b) role of oxygen in Si_3N_4 and c) intrinsic stress in the Si_3N_4 films due to the higher bond constraints in Si_3N_4 compared to SiO_2. These mechanisms will be discussed next.

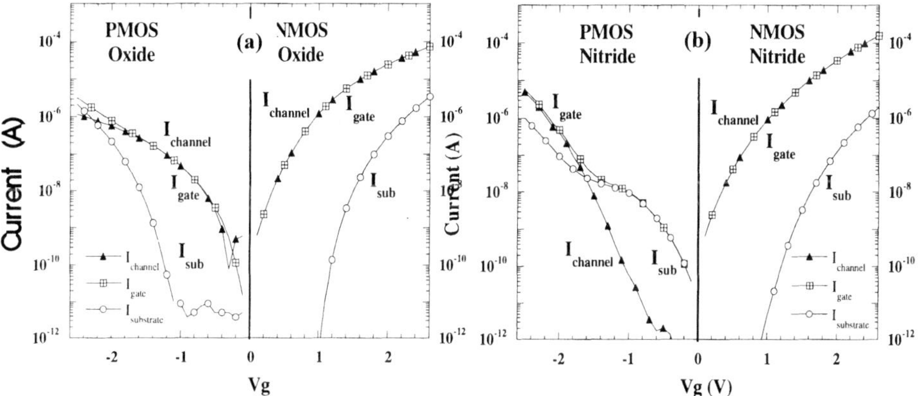

Fig. 5. Carrier separation data on N- and P-MOSFETs for (a) SiO_2 and (b) silicon nitride.

The presence of hydrogen in the Si_3N_4 films is unavoidable if the films are formed thermally since most gases used to form Si_3N_4 contain hydrogen. However, remote plasma processing can be used to form hydrogen free Si_3N_4 films by using direct nitridation of the Si surface in a N_2 plasma. Ultra-thin interfaces, (5 Å, 10 Å and 15Å) were grown by direct nitridation of the surface using N_2. The films were covered by a 30 Å low temperature remote plasma SiO_2 and Si_3N_4 films in order to obtain a thick enough gate stack to allow for conventional electrical measurements. Control SiO_2 layers and deposited Si_3N_4 layers were also included. The results are shown in Fig. 6. The N_2-grown samples indicate a large flatband shift compared to the deposited nitride. The same result was observed in n-type substrates. These results suggests that hydrogen may act as a defect passivator in Si_3N_4 films and may bond with dangling N and Si bonds to create N-H, and Si-H bonds [10]. However, even though the presence of hydrogen can act to reduce the initial defect density, it is detrimental to the device reliability since both the N-H and the Si-H bonds are strong defect precursors.

In an effort to determine the role of oxygen at the interface, oxygen was introduced at the interface using three techniques: a) growing an Si_3N_4 interface in N_2 followed by O_2-plasma

Fig 6. CV characteristics of grown Si_3N_4 in N_2, control oxide, and deposited Si_3N_4.

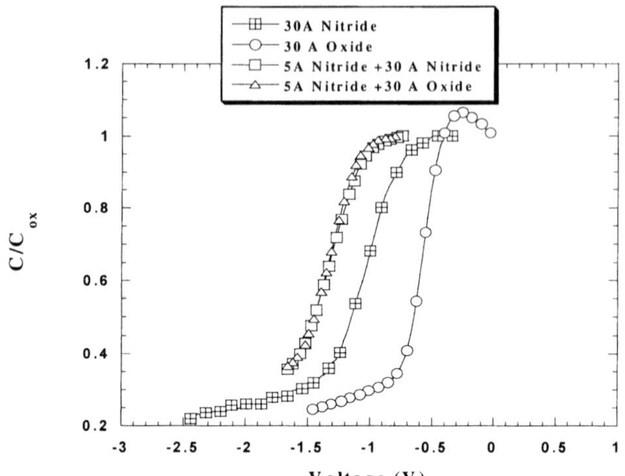

treatment, b) growing a ultra-thin SiO_2 layer (< 5 Å) followed by N_2 plasma treatment, and c) growing interfaces in O_2/N_2 mixtures, where O_2 varied from 15 to 85 %. All films were covered by 30 Å of RPECVD Si_3N_4 to enable conventional CV measurements. The results for processes (a) and (b) are shown in Fig 7. If the Si_3N_4 interface is grown prior to any oxidation process, the

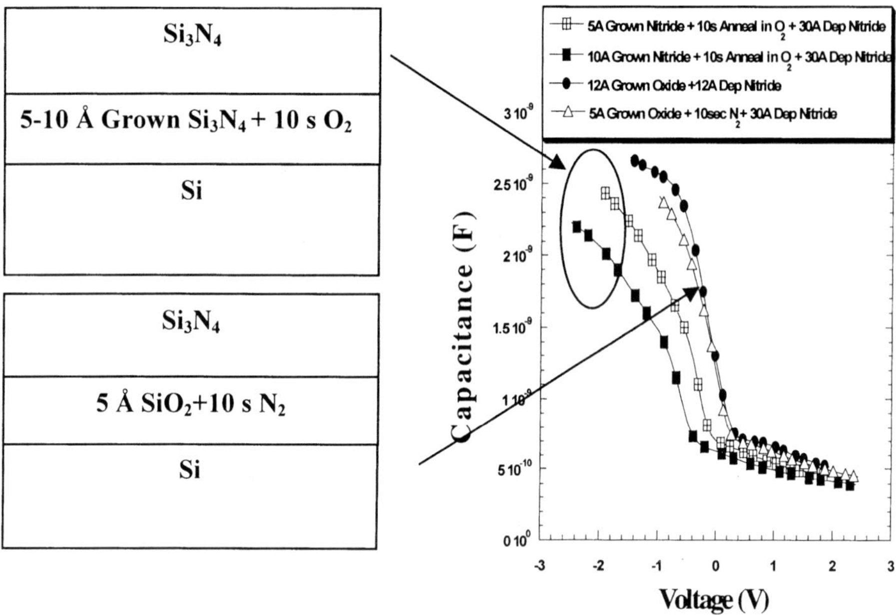

Fig 7. Effect of the sequence of interface formation on CV characteristics.

interfacial properties remain degraded, i.e. the oxygen is unable to repair the interface properties. However, if the oxygen is grown prior to nitridation, the interfacial properties are drastically improved and resemble the control samples. This indicates that once the Si-N bonds are created at the interface they are unable to be modified under a remote plasma environment.

The role of oxygen during nitridation of the Si substrate was also evaluated. The results indicated a systematic relationship between the oxygen content and the V_{fb}-shifts. Moreover, good quality films can be obtained using O_2 during the interface formation even in the presence of N_2. This indicates that the presence of O_2 species can produce a high quality interface during the growth an oxynitride alloy.

A possible physical-chemical explanation for the Si- Si_3N_4 interface defectivity may lie in the bonding constraints at the Si-Si_3N_4 interface. It has been reported that bonding constraint theory calculations can relate defect formation to bonding coordination at the interface of two materials [11]. Bonding constraints arise due to differences between the average number of bonds per atom on each side of the interface. The calculations for various Si based dielectrics has been performed and reported [12,13]. It was found that when the average coordination was < 3.0, good quality interfaces were obtained as indicated by the Si-SiO_2 interface, which has an average coordination of 2.8. However, when the average coordination was > 3, defective interfaces are obtained as indicated by the Si-Si_3N_4 interface, which has an average coordination of 3.5. It is also interesting to note that if a monolayer of SiO_2 is interposed between the Si and the Si_3N_4, then the average coordination is lowered to 3. To verify this experimentally, a thin layer of oxide (< 0.6 nm as measured by in-situ Auger) was interposed between the nitride and the Si substrate. Transconductance values, shown in Fig. 8, now indicate normal characteristics for both P and N devices suggesting that ~ 0.6 nm of interfacial oxide is sufficient to reduce the interface state density both near the conduction and valence band edges. In this work it is believed that this over-coordination of the Si_3N_4-Si bonds compared to the Si-SiO_2 interface can propagate various types of interface traps that have been reported in literature as the cause of interface defectivity.

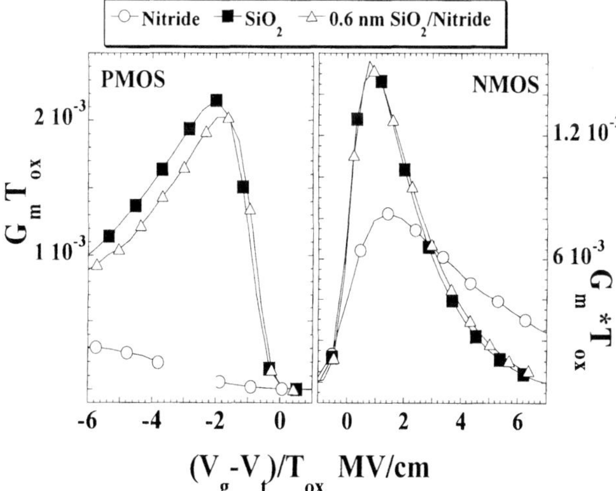

Fig. 8. G_m* T_{ox} vs. oxide field for silicon nitride, SiO_2 and stacked O/N dielectrics for 0.6 μm channel length PMOSFETs and NMOSFETs. The oxide field was calculated as $(V_g-V_t)/T_{ox}$.

CONCLUSIONS

The data presented in this paper demonstrate an asymmetric degradation of PMOS and NMOS devices with pure Si_3N_4 as gate dielectrics. This degradation originates from interfacial defects where a markedly different distribution of interfacial traps near the valence band vs. the conduction band edges results in the inability to form a PMOS channel. The observation that these traps are not present in PMOS devices in which an ultra thin oxide layer < 0.6 nm is interposed between the Si and the nitride film demonstrates that these defects are a property of the Si-Si_3N_4 interface. Experiments performed to study the role of hydrogen and oxygen suggested that both are beneficial to the properties of the Si-Si_3N_4 interface. However, the hydrogen acts as a defect precursor. The role of oxygen was found to be very beneficial and the results obtained here indicate that oxynitride alloys can provide reasonable interface performance. Bonding constraint theory calculations were used to show that the Si-Si_3N_4 interface is heavily over-constrained. This over-coordination can be the basis for propagation of various types of interface traps that have been reported in literature as being the cause of interface defectivity. The addition of oxygen as an ultra thin interfacial oxide layer and/or in the nitride film to form an oxynitride alloy, may be necessary if nitride interface layers are to be integrated into aggressively-scaled CMOS devices.

ACKNOWLEDGEMENT

This research is supported by the NSF, SRC, SEMATCH and ONR.

REFERENCES

1. The SIA National Technology Roadmap for Semiconductors, 1997 edition.
2. W.S. Lau, *Jpn. J. Appl. Phys.*, **29** (1990) 690.
3. T.P. Ma, *IEEE Trans. Elec. Dev.*, **45** (1998) 680; T.P. Ma, *These proceedings*, p. 57.
4. B.Y. Kim, H.F. Luan, D.L. Kwong, *IEDM Tech. Dig.*, (1997) 463.
5. Y. Ma, T. Yasuda, G. Lucovsky, *Appl. Phys. Lett.*, **64** (1994) 2226.
6. J.R. Hauser, private communication.
7. H. Momose, S. Kitagawa, K. Yamabe, H. Iwai, *IEDM Tech. Dig.*, (1989) 267.
8. T. Hori, Y. Naito, H. Iwasaki, H. Esaki, *IEEE Elect. Dev. Lett.*, **7** (1986) 669.
9. A.S. Ginovker, V.A. Gristsenko, S.P. Sinitsa, *Phys. Status Solidi (a)*, **26** (1974) 489.
10. Y. Saito, K. Sekine, M. Hirayama, T. Ohmi, *Jpn. J. Appl. Phys. 1*, **38** (1999) 2329.
11. J. Philips, *J. Non-Cryst. Solids*, **34** (1979) 153.
12. G. Lucovsky, Y. Wu , H. Niimi, V. Misra J.C. Phillips, *App. Phys. Lett.* **74** (1999) 2005.
13. G. Lucovsky, *These proceedings*, p. 69.

Section II

MOSFET Source/Drain Engineering: Ultrashallow Junctions

SHALLOW JUNCTION FORMATION BY LOW ENERGY IMPLANT AND HIGH RAMP-UP RATE RTA PROCESS

Seiichi Shishiguchi, Akira Mineji, Tomoko Yasunaga Matsuda and Hiroshi Kitajima
ULSI Device Development Laboratory, NEC Corporation
1120 Shimokuzawa, Sagamihara, Zip 229-1198, Japan, shishi@lsi.nec.co.jp

Implant and RTA conditions are investigated for source/drain (S/D) formation process. In deep sub-quarter micron CMOS-FETs, ultra shallow junction for S/D-extensions as well as TED reduction process for deep-S/Ds are required. Boron or phosphorus δ-doped super-lattices were used for implant-damage evaluation. Both a single boron implant and a germanium pre-amorphized boron implant process were studied for PMOS-FET fabrication. The TED is minimized by boron implant without pre-amorphization if the boron energy is below 1keV, and by germanium pre-amorphized process if the boron energy is above 1keV. This shows that ultra low energy boron implantation without pre-amorphization should be used for S/D-extension formation and the germanium pre-amorphized process should be used for deep-S/D formation in PMOS-FETs. Arsenic or phosphorus implant process is studied for NMOS-FETs. Low energy arsenic implantation can reduce both the TED and the junction-depth; therefore, this process should be used for shallow S/D-extension formation. Deeper junctions are required for deep-S/Ds. A low energy arsenic and phosphorus mixed implantation can increase the junction depth with slightly increase in TED. This process should be used for deep-S/D formation in NMOS-FETs. Optimization of RTA conditions is key for ultra shallow junction formation. Effects of ramp-up rate or soak time were investigated. Higher ramp-up rates for short soak time conditions are effective to obtain shallow and low sheet resistance junctions. The effects of ambient or cap-film were studied. It is shown that boron diffusivity becomes sensitive to the oxygen content or the existence of cap-films, when the boron implant energy is reduced down to the sub-keV region. A pure nitrogen ambient without cap-film is recommended for shallow junction formation. The effects of LPCVD grown nitride cap films were also investigated. Large TED and boron out-diffusion are observed for nitride cap conditions.

INTRODUCTION

For realizing deep sub-quarter micron CMOS-FETs, ultra shallow source/drain (S/D) with low sheet-resistance is required to suppress short channel effects and to obtain high current drivability. To meet these requirements, drain-extension structures and processes using pre-amorphized boron implant or sub-keV ultra-low energy implant technology have been developed. In these technologies, activation annealing as well as implant conditions should be optimized.

As for implant conditions, key technology is to suppress dopant enhanced diffusion, such as TED or BED caused by point defects during the following activation annealings. Implant energy reduction has been reported to decrease the point-defect concentration, resulting in suppressing the transient enhanced diffusion (TED) of boron [1-3]. In addition, recent studies have demonstrated another boron enhanced diffusion mechanism, called BED, caused by the point defect injection due to highly-doped boron [4,5]. This implies that the use of an appropriate implant dose (not too high) is important for ultra-shallow junction formation. As for activation annealing conditions, the rapid thermal annealing (RTA) process has been widely studied. High ramp-up rate and short soak-time conditions have been developed for reducing the thermal budget [6-8]. Effects of RTA ambient or cap-films on dopant diffusivity or out-diffusion have also been reported [9-12]. In addition, effects of film formation process should be studied, since oxide- or nitride-films are used as gate side-wall spacers or etch-stop layers in device process.

In CMOS fabrication, TED affects not only the dopant diffusion itself but also the diffusion of dopant located in other regions, as shown in Fig.1. The flow of point-defects from S/D-extensions or deep-S/Ds enhances the re-distribution of channel dopant during S/D activation RTA. The point-defects from deep-S/Ds enhance the dopant diffusion in S/D-extension regions. The re-distribution of channel dopant can cause fluctuating threshold voltage (Vt), resulting in reverse short channel effects (RSCE) [13]. The diffusion enhancement in S/D-extensions can cause short channel effects (SCE). The effects of TED are listed in Table 1.

In this paper, TED reduction processes are presented for NMOS- or PMOS-FETs by decreasing point-defects due to implant damage. A pre-amorphization process by germanium implantation is proposed for PMOS deep-S/D implantation. Arsenic and phosphorus mixed implant technology is applied to NMOS deep-S/D formation. Next, the RTA conditions for ultra-shallow junction formation are optimized. The effects of ramp-up rate, oxygen content and cap-films are demonstrated.

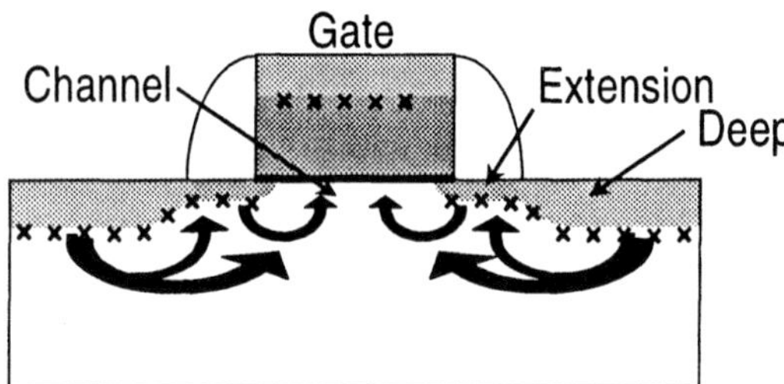

Figure 1. Point defect diffusion during S/D annealing.

Table 1. Device effects of TED caused by point defects during annealing.

Point defect		TED region	TED effects
Extension I/I	⇒	Channel region	Vt, RSCE
Deep-S/D I/I	⇗⇒	Extension region	Xj, Leff

RESULTS AND DISCUSSION

TED reduction process for S/D formation

Boron or phosphorus δ-doped super-lattices were formed as marker layers for implant-damage evaluation, by ultra high vacuum chemical vapor deposition (UHV-CVD) on (100) Si substrates. These δ-doped layers were 5nm-thick and spaced by 100nm. Dopant ions were implanted into the substrates, and the implant damage was evaluated by the diffusion length (FWHM) of the δ-doped marker layers as shown in Fig.2.

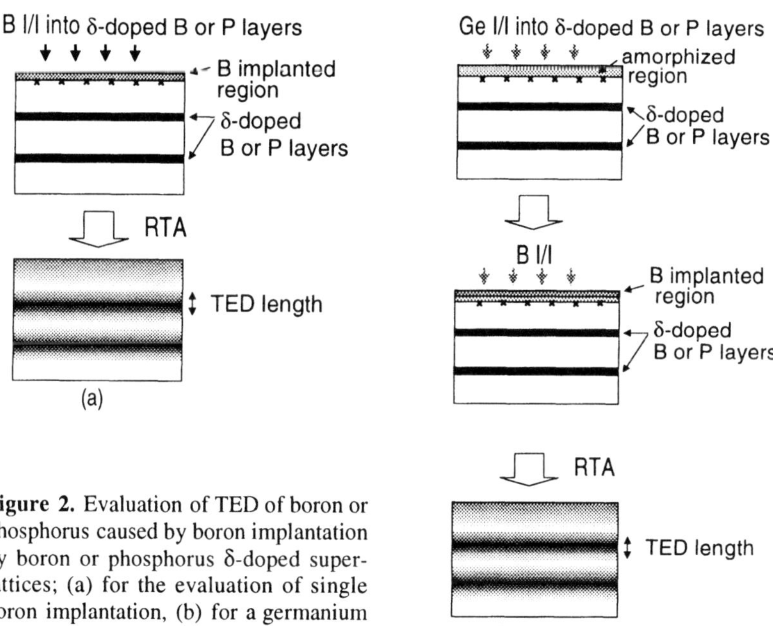

Figure 2. Evaluation of TED of boron or phosphorus caused by boron implantation by boron or phosphorus δ-doped super-lattices; (a) for the evaluation of single boron implantation, (b) for a germanium pre-amorphized boron implant process.

As for evaluating S/D implant damage in PMOS-FETs, a boron or a germanium pre-amorphized boron implantation process was investigated. Boron ions of $1\times10^{15}/cm^2$-dose were implanted at energies from 0.2keV to 2keV, and annealed at 950°C/10 sec for the single boron process. For the germanium pre-amorphization process, germanium ions were implanted with an energy of 5keV and a dose of $1\times10^{15}/cm^2$ before boron implantation. Figure 3 shows the boron diffusion depth-profiles for the single boron and the germanium pre-amorphized boron process.

The boron TED length dependence on boron implant energy is plotted in Fig. 4 (a). The TED was increased with the implant energy for single boron process. On the contrary, the TED for the pre-amorphized process was constant for the boron energy range below 2keV. These results show that the point-defect density due to the implant damage is proportional to the implant energy for the boron process and the density is

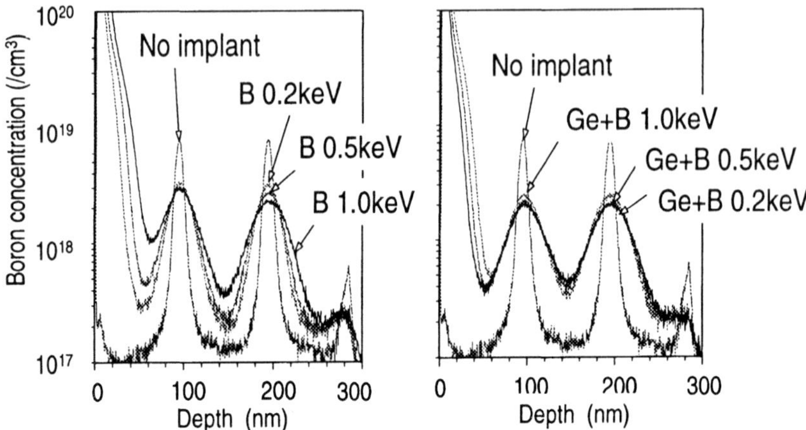

Figure 3. B diffusion profiles for the boron and the germanium pre-amorphized boron implant process. The diffusion width of δ-doped boron layers was increased with the boron implant energy for the single boron process. The width was constant for the germanium pre-amorphized process.

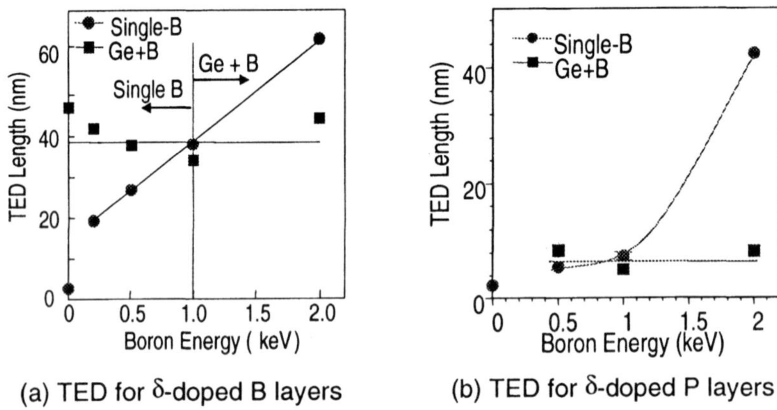

Figure 4. TED length for δ-doped (a) boron or (b) phosphorus layers after RTA. TED length was determined as FWHM of δ-doped layers. TED was minimized by boron implantation if the implant energy was below 1keV, and by germanium pre-amorphized boron implantation process if the boron implant energy was above 1keV.

constant for the germanium pre-amorphized process. It is explained that the point-defect density for pre-amorphized process is determined by the germanium implant damage for pre-amorphization, because most of the boron ions are implanted in the amorphous region. By the same way, the TED of phosphorus by boron implant damage was evaluated. The trend of TED for phosphorus was the same as that for boron, as shown in Fig. 4 (b). A single boron process below 1keV and a germanium pre-amorphized process above 1keV minimized the TED of boron and phosphorus. Therefore, it is recommended to use sub-keV single boron for shallow S/D-extension implantation and to use boron implantation above 1keV with germanium pre-amorphization for deep-S/D formation.

Arsenic or phosphorus ions were implanted into boron δ-doped super-lattices, in order to evaluate the S/D implant damage in NMOS-FETs. The implant dose for arsenic or phosphorus was $5\times10^{15}/cm^2$ and the RTA conditions were 1000°C for 10 sec.

Figure 5 shows the TED length dependence on the implant energy. The TED was increased proportionally to the energy of arsenic or phosphorus implantation. In addition, the TED by phosphorus was longer than that by arsenic implantation. This demonstrates that arsenic should be used for S/D implantation in NMOS-FETs from the TED point of view. The junction-depth (Xj) was estimated as the depth for the ion concentration of $1\times10^{18}/cm^3$ of SIMS profiles. Figure 6 shows the TED vs. Xj map for the arsenic or the phosphorus implantation. This map indicates the junction depth by arsenic process was too shallow for the application to deep-S/Ds for 0.18μm to 0.13μm technology node. This is due to low arsenic diffusivity. The low energy phosphorus condition (2keV) could decrease the TED length; however, the junction depth was too deep. Sub-keV ultra-low energy phosphorus conditions may decrease both the TED and the junction depth; however, these conditions have a problem of too low beam current; resulting in low throughput.

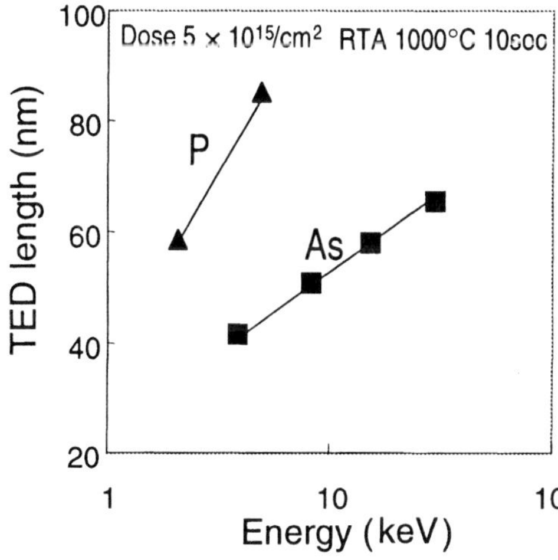

Figure 5. TED length for δ-doped boron layers after RTA. TED was decreased with decrease in implant energy. TED by phosphorus implantation was longer than that by arsenic implantation for the same energy.

An arsenic and phosphorus mixed implant process was evaluated. First, arsenic ions were implanted to amorphize silicon surfaces. Then phosphorus ions were implanted. The mixing rates of arsenic and phosphorus were varied as the total dose was kept constant as $5\times10^{15}/cm^2$, as listed in Table 2. Conditions of A to D were mixing of arsenic/4keV with phosphorus/5keV, and condition E was mixing of arsenic/4keV with phosphorus/2keV. The mixed implant process gave TED and junction depth values in between those of a single arsenic and a single phosphorus implantation with dose $5\times10^{15}/cm^2$. Only the phosphorus ions, implanted out of the amorphized region, could increase the TED. The deeper junctions were realized by the high diffusivity of phosphorus. It is demonstrated that TED and junction depth can be controlled by the arsenic and phosphorus mixed implant process. Consequently, a low energy arsenic implantation should be used for shallow S/D-extension formation and an arsenic plus phosphorus mixed implantation should be used for deep-S/D formation.

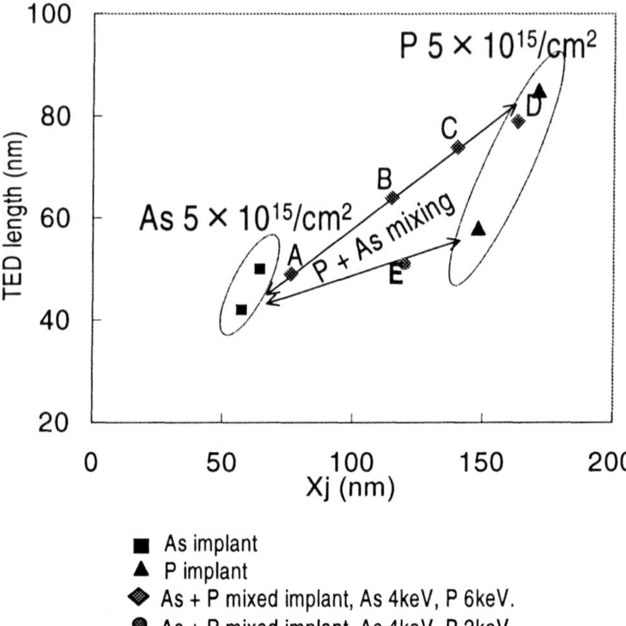

Figure 6. TED vs. Xj map for the arsenic plus phosphorus mixed implant process. Low energy arsenic implantation reduced both TED and junction depth. Low energy arsenic implantation should be used for shallow S/D-extensions. The arsenic and phosphorus mixed implant process could increase the junction depth with a slight increase in TED. The mixed implant process should be used for deep-S/Ds.

Table 2. Arsenic and phosphorus mixed implant conditions.

Condition	Arsenic implant	Phosphorus implant
A	As 4keV, $4.5\times10^{15}/cm^2$	P 5keV, $5\times10^{14}/cm^2$
B	As 4keV, $3.5\times10^{15}/cm^2$	P 5keV, $1.5\times10^{15}/cm^2$
C	As 4keV, $1.5\times10^{15}/cm^2$	P 5keV, $3.5\times10^{15}/cm^2$
D	As 4keV, $5\times10^{14}/cm^2$	P 5keV, $4.5\times10^{15}/cm^2$
E	As 4keV, $3.5\times10^{15}/cm^2$	P 2keV, $1.5\times10^{15}/cm^2$

Reduction of boron implant energy was effective in decreasing P+/N junction depth. Figure 7 shows the effect of boron energy reduction on the junction depth for the process with or without germanium pre-amorphization. The junction depth was decreased with decreasing boron energy. The junction depth was shallower with the pre-amorphization process included. However, the difference between the process with or without pre-amorphization became small, if the boron implant energy was reduced down to 1keV. Pre-amorphization suppresses boron channeling, but boron diffusion is enhanced by TED due to the germanium implantation for the pre-amorphization. These results demonstrate that sub-keV boron conditions are effective for ultrashallow junction formation.

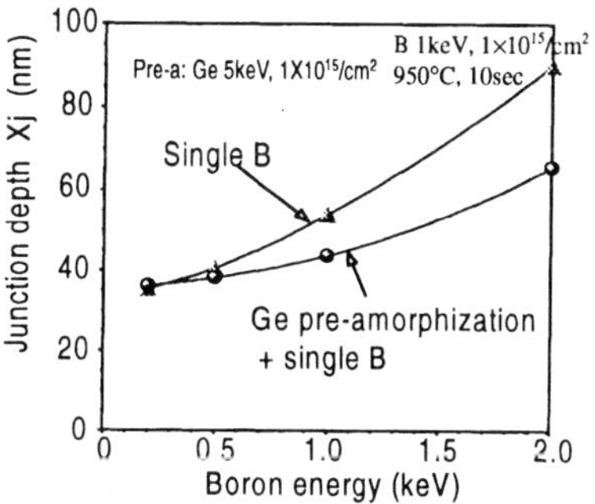

Figure 7. Junction depth dependence on boron implant energy for the boron implant process with or without pre-amorphization. Shallow junctions could be obtained without pre-amorphization process for sub-keV energy region.

RTA conditions were optimized to realize ultrashallow junctions by ultralow energy implantation. Effects of ramp-up rate, ambient or cap-films were investigated.

Point defects are generated by transient enhance diffusion (TED) and oxidation enhanced diffusion (OED), if oxygen is introduced during RTA. Figure 8 shows the point defect density during a thermal sequence of a RTA process. Point defect generation of TED is completed during the initial stage of the thermal sequence. On the contrary, OED is continued during the entire thermal sequence. The point defect density generated by TED or OED was evaluated by boron δ-doped super-lattices, in the same way as described in the previous section. For the TED evaluation, boron ions were implanted into boron δ-doped super-lattices, followed by RTA at 1000°C/10 sec in pure nitrogen ambient. For the OED evaluation, boron δ-doped super-lattices were annealed at 1000°C/10 sec in 10% oxygen content ambient without boron implantation. Figure 9 shows the enhanced diffusion length of the boron δ-doped layers by TED or OED. The TED length was decreased with decreasing boron implant energy. The OED length was compared to the TED length in Fig. 9. The OED length by 10% oxygen ambient was comparable to the TED length by 0.5keV boron implantation. This shows that OED is not negligible for ultra shallow junction formation by ultra low energy sub-keV boron implantation. In order to reduce the OED effects, a decrease of oxygen content is effective. It has been reported that the precise control of the oxygen concentration down to several tens of ppm is necessary to obtain shallow and low sheet-resistance junctions [9]. OED effects can also be reduced by short time RTA conditions (spike anneal conditions) as shown in Fig. 8, because point defect generation is continued during the soak time. A soak anneal at 1000°C/10 sec and higher temperature spike anneal at 1050°C were compared. It was found that the OED length evaluated by boron δ-doped super-lattices was reduced by the spike condition, as shown in Fig. 9. Figure 10 shows the boron diffusion profiles for the soak anneal and the spike anneal conditions for 0.5keV boron implantation. The profiles were measured by SIMS after the removal of surface oxide. The diffusion profiles in pure nitrogen ambient were almost the same for both the soak and the spike conditions. On the other hand, in oxygen ambient, the diffusion profile for the spike condition was shallower than that for the soak condition. These results indicate that higher temperature spike anneal condition is effective for reducing OED.

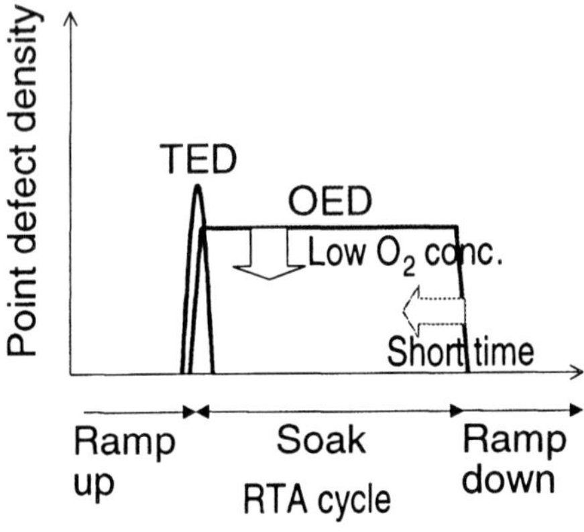

Figure 8. Point defect density during RTA. TED is completed in the initial stage of the RTA cycle. OED is continued during soak time. Reduction of oxygen content or short soak time condition was effective to reduce OED.

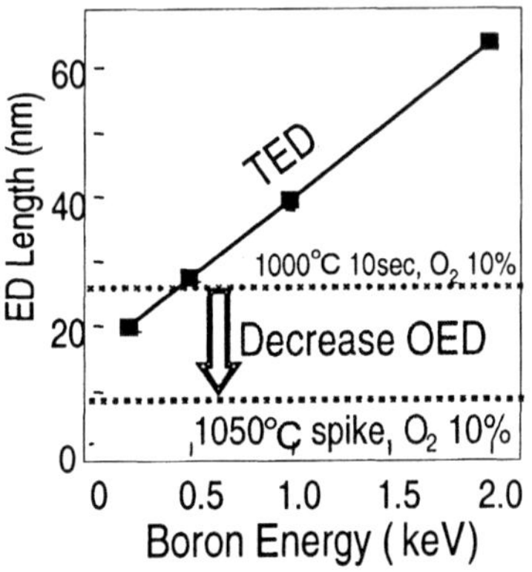

Figure 9. Enhance diffusion (ED) length due to TED or OED. OED for soak anneal was comparable to TED for sub-keV boron implantation. Spike conditions could reduce OED.

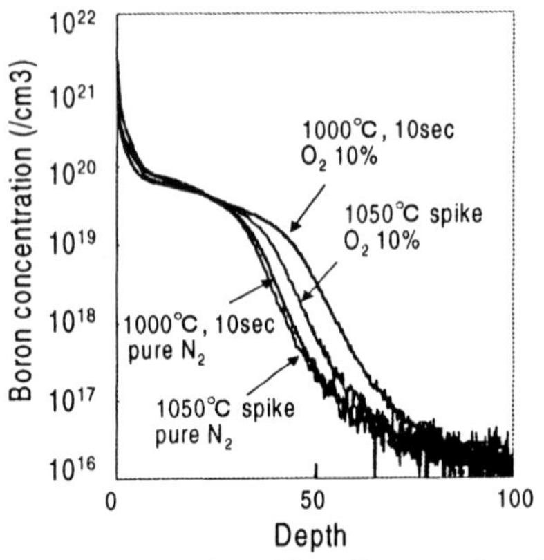

Figure 10. Boron diffusion profiles for soak or spike RTA. The profiles for RTA in nitrogen ambient were almost the same. The profile for spike annealing was shallower for RTA in oxygen.

Next, surface film effects on junction formation were investigated. Rp of boron implantation becomes so shallow for ultra low energy implantation that the surface effects on boron diffusion become strong. Figures 11 and 12 show the effects of surface films. The boron depth profiles were measured by SIMS after the removal of the surface films. Figure 11 shows the effects of pre-treatment conditions before the 0.5keV boron implantation. The implant profile for HF-dip before the implantation was the deepest, and the profile for the implantation through 2.5nm-thick screen-oxide was the shallowest. However, the diffusion profile for the HF-dip pre-treatment condition was the shallowest, and the profile for 2.6nm-thick through implant was the deepest. These results show that the screen-oxide or even the native-oxide enhances the boron diffusivity. Figure 12 shows the effects of cap-films on boron diffusion during RTA. The boron was implanted through native oxide. The oxide- or nitride-cap was grown by LPCVD at 700°C or 630°C after a 0.5keV boron implantation, respectively. The boron profile for the LPCVD nitride or the without cap-film condition was shallower than that for the LPCVD oxide-cap condition. TED for these RTA conditions was evaluated by boron δ-doped super-lattices. TED was increased if the surface was covered by the cap-oxide, as shown in Fig.13. The deeper profile caused by the oxide-cap is explained by the TED increase. In contrast, the effect of nitride-cap was not so simple. The effect of the nitride cap was stronger than that of the oxide-cap from a TED point of view, as shown in Fig. 13. However, the boron diffusion depth for the nitride-cap condition was almost the same for the native-oxide

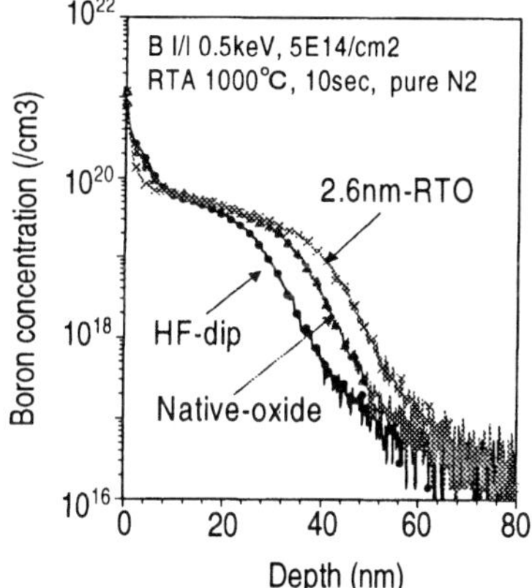

Figure 11. Effects of pre-treatment before boron implantation. The profile for screen-oxide through implant was deeper.

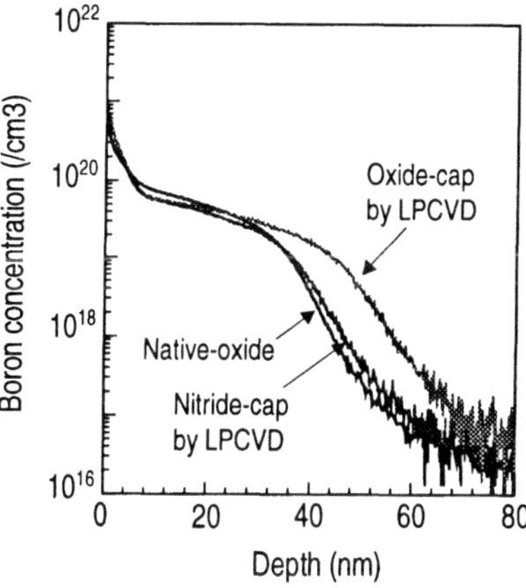

Figure 12. Effects of cap-films during RTA. The profile for the cap-oxide was deeper.

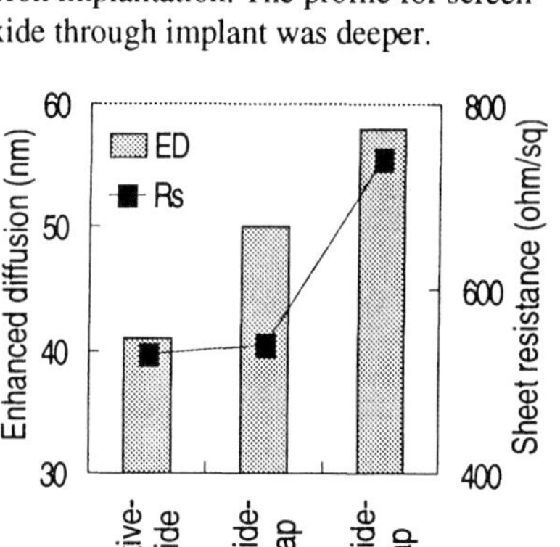

Figure 13. Effects of cap-films for diffusion enhancement and sheet-resistance. Nitride-cap enhanced boron diffusivity and increased sheet-resistance.

condition. Figure 12 shows that the near surface boron concentration for the nitride-cap condition is lower than that for the native oxide condition. The retained dose calculated by the SIMS profile for the nitride-cap condition was decreased. Therefore, the shallow profile for the nitride-cap condition was not due to the TED suppression, but due to the large out-diffusion of boron during RTA. This resulted in the large increase in sheet-resistance for the nitride-cap condition, as shown in Fig. 13.

The effects of the thermal budget on shallow junction formation were investigated. Figure 14 illustrates a boron depth profile after RTA. When boron profile becomes shallower, the peak depth is expected to be closer to the surface; therefore, the ions in inactive regions (over the solid solubility at the RTA temperature) increase. This is why the sheet-resistance increases when a junction becomes shallower. Annealing at higher temperature is effective to expand the active ion region and to obtain lower resistance. Shorter soak-time and higher ramp-up rate are required for suppressing boron diffusion and realizing shallower junctions in higher temperature conditions. Figure 15 shows the effects of soak-time and ramp-up rate on 1keV implanted boron diffusivity with germanium pre-amorphized process. The investigated conditions were soak time from 50msec to 10sec and ramp-up rate from 25°C/sec to 400°C/sec. Shorter soak-time was effective to reduce the boron diffusivity, as shown in Fig. 15(a). The soak time setting for the spike anneal was 50msec. The effects of the soak time reduction from 5sec to 1sec was large. On the contrary, the effects from 1sec to 50msec was limited. Therefore, the effective soak time for the spike condition may be around several hundred msec. This is due to the lower ramp-up rate at the temperature close to the soak temperature, or the low ramp-down rate. Higher ramp-up rate for the spike anneal conditions were also effective to reduce the boron diffusivity, as shown in Fig. 15 (b). These results show that the reduction of thermal budget can decrease the junction depth. Figure 16 shows junction depth (Xj) for various RTA conditions (soak temperature of 950°C~1150°C, soak time of 50msec~10sec, and ramp-up rate of 25°C/sec ~400°C/sec). The junction depth was reduced along each equi-temperature line when samples were annealed in shorter time and/or higher ramp-up rate. Minimum junction-depth for each RTA temperature was obtained by 400°C/sec ramp-up rate with spike condition. However, trade off exists between junction-depth and sheet-resistance, because the boron activation is limited by the solid solubility at the RTA temperature. By RTA of higher-temperature conditions, the equi-temperature line proceeded closer to the ideal line, indicated in broken line in Fig. 16. The ideal line corresponds to the junction-depth and the sheet-resistance for BOX carrier profile. The temperature over 1100°C was not effective for the boron implant condition in this work. The sheet-resistance for the 50nm-depth junction was reduced by using higher temperatures; however, the sheet-resistance was saturated to the lowest value when the temperature was higher than 1100°C. This demonstrates that there

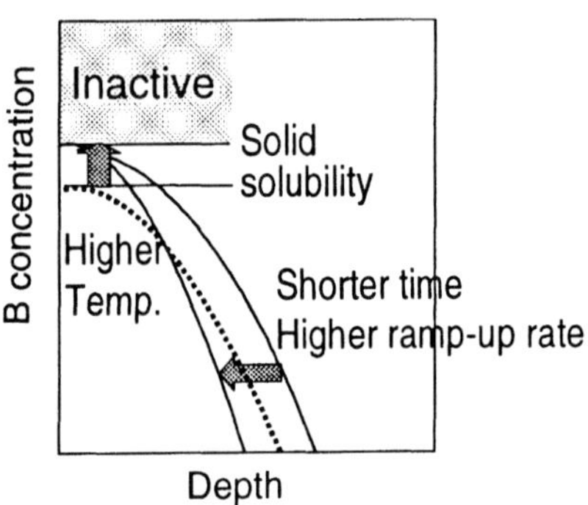

Figure 14. Effects of short time/high ramp-up rate process for boron diffusion layer formation.

exists a lower limit for the sheet-resistance and the junction-depth by implantation and RTA, as shown in Fig. 16. Around 30nm a junction-depth with 1kΩ/sq sheet resistance can be achieved by this process. For realizing shallower junctions with lower sheet-resistance, a new activation technology should be developed.

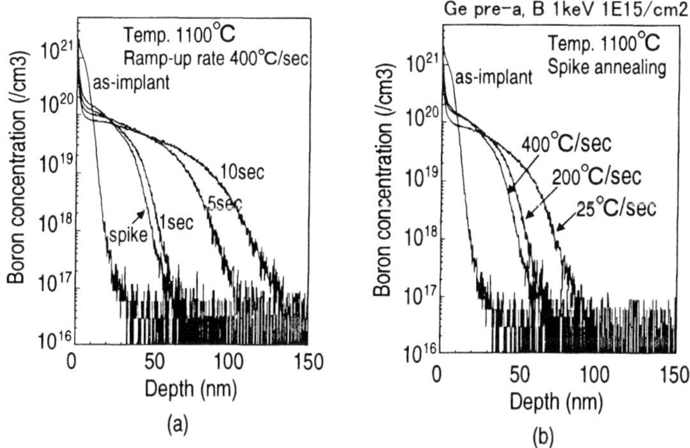

Figure 15. Effects of short time/high ramp-up rate (spike) process on boron diffusivity. (a) Soak time dependence. (b) Ramp-up rate dependence.

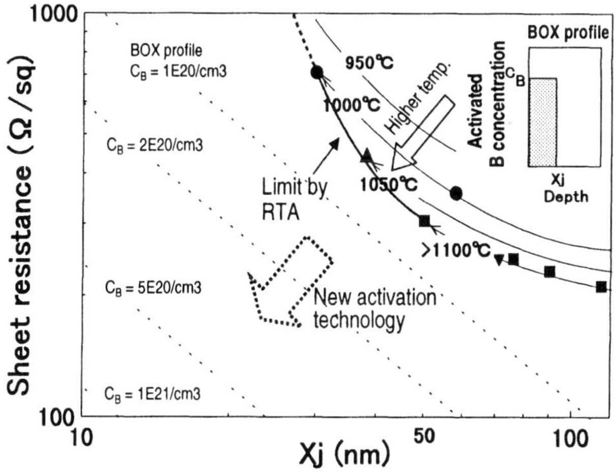

Figure 16. Limit of ultra-shallow junctions by I/I and RTA. A trade-off exists between Xj and Rs, because the boron activation is limited by the solid solubility.

CONCLUSION

Implant and RTA conditions were studied for the S/D formation process in CMOS-FETs. TED due to S/D implantation was evaluated by δ-doped super-lattice structures formed by UHV-CVD. It was found that TED was minimized by boron implantation without pre-amorphized process if the boron energy was below 1keV, and by a germanium pre-amorphized process if the boron energy was above 1keV. It was recommended to use sub-keV boron implant for S/D-extensions and a germanium pre-amorphized process for deep-S/Ds for PMOS-FET fabrication. Low energy arsenic implantation could reduce both junction depth and TED. Arsenic and phosphorus mixed implant process was effective to increase the junction depth with slightly increase in TED. For NMOS-FET fabrication, a low energy arsenic implant process for S/D-extensions and a mixed implant process for deep-S/Ds were recommended. For ultra shallow S/D extension formation, a low energy boron process was investigated. An ultra-low energy (sub-keV) single-boron process could reduce the junction depth without pre-amorphization. It was shown that cap-films enhanced the boron diffusivity. Nitride cap-film by LPCVD decreased the retained dose after RTA, resulting in high sheet-resistance. RTA conditions of annealing without cap- or screen-oxide in nitrogen ambient were recommended for ultra shallow junction formation. The reduction of thermal budget was effective to reduce junction depth. High temperature/high ramp-up rate/short time conditions were promising process parameters. The existence of trade-off between junction-depth and sheet-resistance was demonstrated. Junction-depth and sheet-resistance were limited by the solid solubility. 30nm-depth with 1kΩ/sq junctions can be obtained by implant and RTA process. New activation technologies should be developed for shallower or lower sheet-resistance junctions.

ACKNOWLEDGEMENTS

The authors would like to thank S. Saito for fruitful discussions, O. Kudo, N. Endo and K. Ikeda for their encouragement during this work.

REFERENCES

[1] H.-J. Gossmann et al., Appl. Phys. Lett. **71** (1997) 3862.
[2] P.A. Stolk et al., J. Appl. Phys. **81** (1997) 6031.
[3] A. Agarwal et al, Appl. Phys. Lett. **71** (1997) 3141.
[4] A. Agarwal et al., IEDM 1997 Tech. Digest, p.467.
[5] A. Agarwal et al., Appl. Phys. Lett. **74** (1999) 2435.
[6] S. Shishiguchi et al., VLSI 1997 Tech. Symp., p.89.
[7] A. Agarwal et al., "*Ultra-shallow junction formation by rapid thermal annealing in a furnace-based RTP system*", Ion Implantation Technology-**98**, to be published by IEEE in 1999.
[8] D.F. Downey, S.D. Marcus, and J.W. Chow, J. Electron. Mat., **27** (1998) 1296.
[9] D.F. Downey, S.L. Daryanani, M. Meloni, and K.M. Brown, Mat. Res. Soc. Symp. Proc. **470** (1997) 299.
[10] S. Marcus, W. Lerch, D.F. Downey, S.Todorov , J. Electron. Mat., **27** (1998) 1291.
[11] T. K. Mogi et al., Proc. 5th Int. Symp. ULSI Sci. Technol., p. 145, 1995
[12] S. Shishiguchi et al., VLSI 1998 Tech. Symp., p.134.
[13] A. Kamgar et al., IEDM 1997 Tech. Digest, p. 695.

SPIKE ANNEALS IN RTP: KINETIC ANALYSIS

E. G. Seebauer
Department of Chemical Engineering, University of Illinois, Urbana, IL 61801

The relative importance of kinetic factors in process and equipment design should continue to increase as device dimensions shrink, thereby offering less "wiggle room" for undesired effects like interface degradation. Through considerations of rate selectivity, this paper shows that spike annealing techniques in RTP make kinetic sense only when the activation energy for the desired transformation exceeds that for the undesired. Increasing the heating rate improves performance, but only up to a certain point. Beyond that point one can expect only rapidly diminishing returns. Spike anneals can have process windows defined in the same way as for conventional RTP with a soak period.

INTRODUCTION

The use of rapid thermal processing (RTP) in microelectronic fabrication has expanded over the past decade to include steps such as defect annealing, chemical vapor deposition (CVD), oxidation, nitridation, and silicidation. From the processing perspective, considerable attention has been paid to issues of metrology, temperature uniformity and control, cost of ownership, and the like. Considerably less attention has been paid to putting the kinetic analysis of RTP on a sound theoretical footing, even though in some applications (like defect annealing), the kinetic advantages afforded by high ramp rates and short processing times are key drivers for using RTP. The relative importance of kinetic factors in process and equipment design should continue to increase as device dimensions shrink, leaving less "wiggle room" for undesired effects like interface degradation, bridging, unwanted dopant diffusion, and silicide agglomeration.

This research group has undertaken to provide such a theoretical footing, most notably in pointing out deficiencies in the concept of thermal budget when designing heating programs [1-3]. However, these criticisms and the replacement framework we have proposed [4,5] concerns RTP in its standard form, in which the heating sequence proceeds through three phases: a ramp-up, a constant-temperature "soak", and a cool-down. Recently, interest has grown in a two-phase temperature program: the so-called "spike anneal." In this approach, the soak phase disappears. The entire heating sequence becomes transient, with the temperature T either rising or falling the entire time. Sometimes this strategy is employed concurrently with attempts to push the ramp rate as high as possible. For example, Shishiguchi *et al.* [6] have reported ultrashallow junction formation using spike anneals with ramp rates of 400°C/s.

Analysis of the complicated phenomenon of transient-enhanced diffusion in a case like this presents a formidable problem for detailed kinetic analysis, which undoubtedly will require direct numerical simulation as an important tool. Nevertheless, it appears that a coherent kinetic analysis of spike anneals remains lacking even for processes with relatively simple kinetics. Ideally, such a framework would be analytical rather than numerical, and would answer the following questions with relatively little effort:

(1) Under what conditions should a spike anneal be preferred to conventional RTP with a soak period?

(2) For a given process, what process window can be established for the ramp rate β and maximum temperature T_M?

(3) What values of β and T_M constitute the optimum?

This paper provides answers to these questions from a purely kinetic perspective. Such a perspective of course suffers from important limitations, as other process considerations like temperature measurement and control can exert large effects on the design of equipment and heating programs. Nevertheless, developing a precise representation of kinetic limitations and ideals represents an important step forward in process optimization.

PROBLEM DEFINITION

Quantitative kinetic analysis requires precise definition of what is being examined and the assumptions employed. Before we proceed further, it makes sense to lay out these things explicitly. We idealize a spike anneal in terms of a ramp-up phase with a constant heating rate β from an initial temperature T_0 to a maximum temperature T_M, followed by a cool-down phase back to T_0 in which virtually all heat loss occurs by radiation. A schematic diagram of this heating program appears in Fig. 1. Of course, in real systems a constant heating rate can be difficult to maintain all the way to T_M. However, for variable heating rates the conclusions of the paper continue to hold if one employs the heating rate near T_M because that is where the physical and chemical transformations that concern us occur most rapidly. For cooling, there exist three mechanisms of heat transfer: radiation, conduction, and convection. The rate of radiative cooling increases as T^4, while the rates of conduction and convection increase only linearly in T [7]. Thus, our idealization of pure radiative cooling works best near the peak of the spike, and improves as T_M increases.

For simplicity, we will restrict our attention to just two rates proceeding on the wafer. One we will designate as "desired," with examples like CVD, oxidation, silicide phase change, and defect annealing. The other we will designate as "undesired," with examples like dopant diffusion and interface degradation. We will assume that we have suitable data for these rates, especially for their temperature dependencies. Analytical expressions are preferred, but scattered data points in graphical form may suffice in some cases.

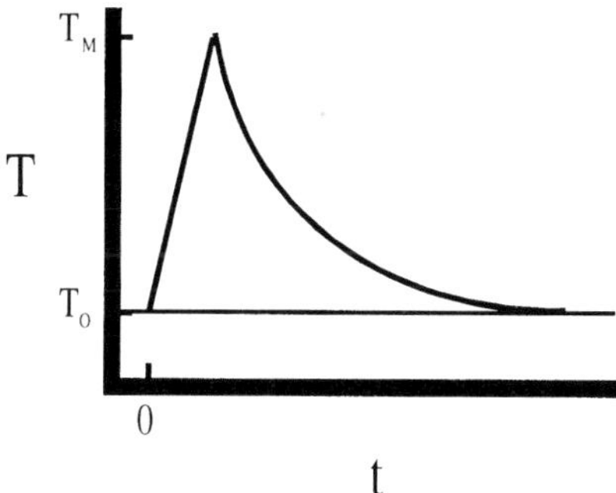

Fig. 1: Schematic temperature-time diagram of an idealized spike anneal with linear heating and pure radiative cooling.

For concreteness, we will assume our desired rate r_d has the following form:

$$R_d = A \exp(-E_d/kT) \quad (1)$$

where k denotes the Boltzmann's constant; E_d denotes the activation energy; and A incorporates a pre-exponential factor together with any other dependence of system variables other than T (like pressure in CVD). We will also assume that our undesired rate describes a diffusional process, and obeys a random-walk form for the mean square displacement x^2:

$$x^2 = 6Dt \quad (2)$$

where the diffusivity D obeys:

$$D = D_o \exp(-E_u/kT) \quad (3)$$

The subscript "u" in the activation energy E of Eq. (3) implies "undesired," while D_o denotes a pre-exponential factor.

DEFINING A PRELIMINARY KINETIC WINDOW

Analysis of Ramp-up

With the assumption of constant heating rate β discussed earlier, the temperature varies with time t during ramp-up according to:

$$T = T_o + \beta t \quad (4)$$

We seek the integrated amount of transformation for both the desired and undesired phenomena. For the desired phenomenon, this integrated rate R_{up} during the ramp phase becomes:

$$R_{up} = \int_{T_O}^{T_M} A\exp(-E_d/k(T_o + \beta t))dt \quad (5)$$

For unwanted diffusion, x_{up}^2 itself represents the integrated unwanted transformation. However, to calculate x_{up}^2 we must insert a temperature-dependent diffusivity into Eq. (2):

$$x_{up}^2 = \int_{T_O}^{T_M} 6D_o \exp(-E_u/k(T_o + \beta t))dt \quad (6)$$

After a change of variable under the integral signs in Eqs. (5) and (6) from t to T (*i.e.*, t = T/β), both integrals can be manipulated to assume the form:

$$I(y) = \int_a^b F(T)\exp(y\phi(T))dT \quad (7)$$

where $b = T_M$ and $a = T_o$. Evaluation of these integrals using the Laplace asymptotic approximation [8] with b > a in the limit $y \to \infty$ yields:

$$I(y) \approx F(a)\frac{\exp(y\phi(a))}{y\phi'(a)} \quad (8)$$

This limit is almost always satisfied in practical cases, where $y = E/kT_o$ typically lies near 30 or more. Application of the limit represented by Eq. (8) to Eqs. (5) and (6) yields:

$$R_{up} = \frac{AT_M}{\beta}\left(\frac{kT_M}{E_d}\right)\exp(-E_d/kT_M) \quad (9)$$

and

$$x_{up}^2 = \frac{6D_o T_M}{\beta}\left(\frac{kT_M}{E_u}\right)\exp(-E_u/kT_M) \quad (10)$$

Analysis of Cool-down

During cool-down by radiation, the wafer temperature obeys the following differential equation:

$$dT/dt = -C(T^4 - T_o^4) \quad (11)$$

where C is a constant that incorporates the emissivity (assumed constant [9]), view factors, and the like. In Eq. (11) we have set the temperature of the surroundings at T_o, which usually falls far below the temperature where significant transformations take place. Thus, to a good approximation we can neglect T_o^4 in Eq. (11), thereby yielding a simple relation between T and t:

$$dT/T^4 = -Cdt \tag{12}$$

The integrated rates describing cool-down look just like those of Eqs. (5) and (6) with the limits reversed. Changing variables from t to T using Eq. (12) and applying the Laplace approximation yields:

$$R_{down} = \frac{A}{CT_M^3}\left(\frac{kT_M}{E_d}\right)\exp(-E_d/kT_M) \tag{13}$$

and

$$x_{down}{}^2 = \frac{6D_o}{CT_M^3}\left(\frac{kT_M}{E_u}\right)\exp(-E_u/kT_M) \tag{14}$$

Creating the Preliminary Window

Putting together the integrated expressions for the ramp up and cool-down phases, we obtain the following equations for the total desired and undesired transformations:

$$R_{total} = \left[\frac{1}{\beta} + \frac{1}{CT_M^4}\right] AT_M\left(\frac{kT_M}{E_d}\right)\exp(-E_d/kT_M) \tag{15}$$

and

$$x_{total}{}^2 = \left[\frac{1}{\beta} + \frac{1}{CT_M^4}\right] 6D_oT_M\left(\frac{kT_M}{E_u}\right)\exp(-E_u/kT_M) \tag{16}$$

The only variables over which a designer has much control in these equations are β and T_M. In fact, setting one or the other of these quantities fixes the value of the other through Eq. (15). Put another way, for a fixed amount of desired transformation R_{total}, Eq. (15) describes a curve on a plot of $1/\beta$ vs. T_M whose average slope increases with activation energy E_d. An example of such a plot appears in Fig. 2a. (A more detailed discussion of drawing plots of this form appears in Ref. [5].) The curve slopes down and to the right because the needed maximum temperature increases as the heating rate goes up (so that there is less time for transformation). Fig. 2a represents a process step like silicidation in which we worry less about the precise extent of the transformation than about whether it exceed some minimum level. Examples include defect annealing or conversion of the C49 phase of titanium silicide to the C54. In cases like these, the desired transformation could take place anywhere within the shaded region. An analogous diagram for the undesired transformation appears in Fig. 2b. Again, the key concern is whether unwanted diffusion or interface degradation remains below some maximum level, so that operation anywhere within the shaded region is permissible.

Superimposing Figs. 2a and 2b leads to Fig. 2c, which shows the process window that satisfies both requirements. High values for both β and T_M are recommended. Constraints on these quantities imposed by equipment limitations or other factors can be incorporated to modify the preliminary process window in Fig. 2 in exactly the same way as we have described in previous publications [4,5].

Interestingly, Eqs. (15) and (16) differ only modestly in form from the equations that describe R_{total} and x_{total}^2 during a constant-temperature soak [5]. By comparing the forms of the equations, we can identify $1/\beta$ with the soak time t, and T_M with the soak temperature T. The similarity in form becomes particularly pronounced when the heating rate is chosen to be low enough so that $\beta << CT_M^4$, and $1/\beta$ dominates the sum within brackets.

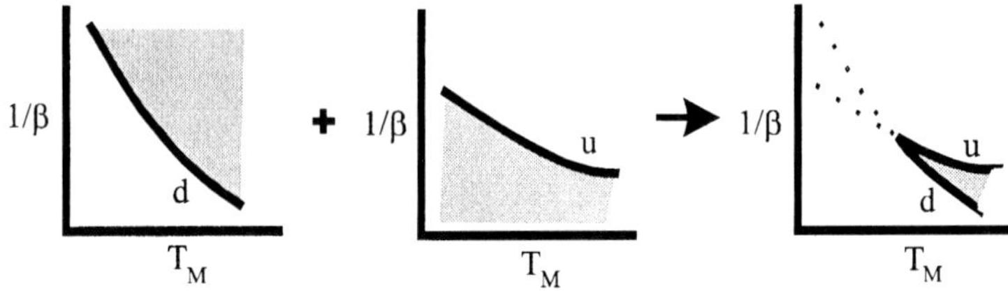

Fig. 2: Steps for defining a kinetic window for the case $E_d > E_u$. Overlaying (a) the kinetics of the desired transformation with (b) the kinetics of the undesired transformation and finding the intersection (c) yields the preliminary operating window (that ignores other process constraints).

However, something very different happens when the heating rate is high enough so that $\beta >> CT_M^4$, and $1/CT_M^4$ dominates the sum within brackets. In this case the integrated rates of both the desired and undesired phenomena become largely independent of β. This behavior reflects the fact that most transformation now takes place during cooling. From a practical point of view, we conclude that for a given value of T_M, continuing to increase the heating rate (using more lamps and bigger power supplies) brings diminishing returns. Of course, in real equipment operated near the design limit, β and T_M tend to increase together. The relative behavior of R_{total} and x_{total}^2 (that is, the selectivity) tends to depend on the exponential terms in Eqs. (15) and (16), with higher T_M being beneficial when $E_d > E_u$. However, other constraints on T_M from wafer damage or the like then bring the diminishing returns of increased β into play.

SPIKE ANNEAL vs. CONVENTIONAL RTP WITH SOAK

Fig. 2 shows that for $E_d > E_u$, both the heating rate and maximum temperature should be as high as possible. Now consider the reverse case: $E_d < E_u$. Here, kinetic analysis akin to that of Fig. 2 suggests that β and T_M should remain as low as possible in a spike anneal. As a practical matter, such a strategy would draw out the process time so much that it would be better to operate with conventional RTP including a soak period. In fact, it's not difficult to show that with $E_d < E_u$, the heating program should have a fast ramp to the lowest possible soak temperature that is consistent with limitations on the process time (from throughput, for example) [3-5], followed by the fastest possible cool-down.

INCORPORATING PROCESS CONSTRAINTS

So far we have considered the process window from the perspective of kinetics only; we have ignored process constraints. These constraints arise from factors like limitations imposed by equipment (lamp power, robustness of temperature control, etc.) or by physical behavior of the wafer (strain, thermodynamics of dopant activation, etc.) In general we can construct a window of such constraints on a plot of $1/\beta$ vs. T_M just as we did for the kinetic window. Fig. 3 shows an example [9].

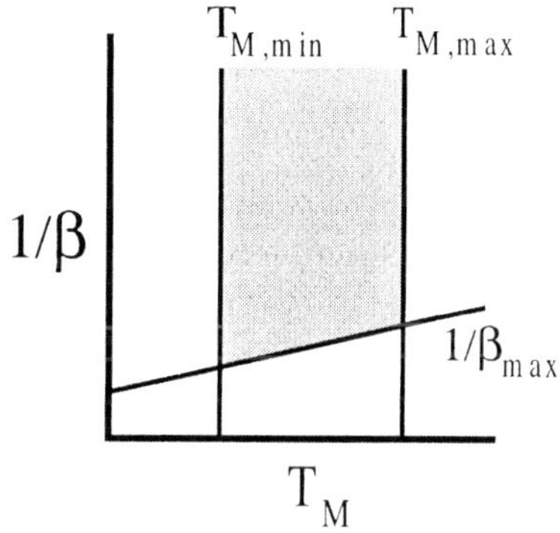

Fig. 3: An example of a window of process constraints. Here we have assumed for concreteness that β_{max} decreases with T_M (due to decreased robustness of temperature control at high T, for example).

Fig. 4: The complete process window developed from the intersection of the windows for kinetics and constraints taken from Figs. 2 and 3. The heavy dot represents the kinetic optimum because it lies furthest from the undesired curve.

The figure shows limits on the maximum and minimum peak temperature $T_{M,max}$ and $T_{M,min}$ that might arise from the physical behavior of the wafer, and implicitly assumes for concreteness that these limits are independent of β. This assumption may not always hold, especially for $T_{M,max}$, where constraints imposed by wafer damage may depend on time spent near $T_{M,max}$ and therefore on β.

The figure also shows a limit for β_{max} that does depend on T_M. The mechanism assumed is that adequate temperature control is harder to accomplish at higher peak temperatures – a problem that is sometimes mitigated by using lower heating rates. Thus, $1/\beta_{max}$ increases with increasing T_M. Of course, other couplings of β_{max} and T_M can be imagined. An example comes from consideration of Eq. (15) for the desired process: when $\beta >> CT_M^4$, little additional benefit from increasing β is gained. This condition creates a practical upper limit for β. Here, as T_M increases, the value of $1/\beta_{max}$ where this condition is fulfilled decreases.

Fig. 3 shows no limitation on β_{min}. As we discussed above, small heating rates in a spike anneal are desirable only when $E_d < E_u$. When this condition holds, standard RTP with a soak should be chosen over a spike anneal.

The complete window for the process can be obtained by superimposing the preliminary kinetic window with the window of constraints. Fig. 4 shows a complete window developed from the windows of Figs. 2 and 3. Operation can take place anywhere within the darkly shaded region representing the intersection of the two subsidiary windows.

OPTIMIZATION WITHIN THE "COMPLETE" WINDOW

The complete window represented by Fig. 4 shows the maximum area in (T_M, β) parameter space where operation can take place consistent with both kinetics and other known limitations on T_M and β. From a kinetic standpoint, however, some regions in this window offer better process performance than others. For example, it makes sense to operate as far away from the curve for the undesired phenomenon as possible (in other words, to minimize diffusion). This point appears as the heavy dot in Fig. 4.

ACKNOWLEDGMENTS

This work was supported in part by NSF (CTS 98-06329) and the DOE through the Materials Research Laboratory at the University of Illinois (DEFG02-91ER45439).

REFERENCES

[1] R. Ditchfield and E. G. Seebauer, *Rapid Thermal and Integrated Processing V*, p. 138 (MRS Vol. 429, 1996).

[2] R. Ditchfield and E. G. Seebauer, *Rapid Thermal and Integrated Processing VI*, p. 313 (MRS Vol. 470, 1997).

[3] R. Ditchfield and E. G. Seebauer, *J. Electrochem. Soc.*, **144**, 1842 (1997).

[4] R. Ditchfield and E. G. Seebauer, *Sol. St. Technol.*, **40**, 111 (1997).

[5] R. Ditchfield and E. G. Seebauer, *Rapid Thermal and Integrated Processing VII*, p. 57 (MRS Vol. 525, 1998).

[6] S. Shishiguchi, A. Mineji, T. Hayashi and S. Saito, *VLSI Tech. Symp.* 1059 (1997).

[7] F. Kreith and W. Z. Black, *Basic Heat Transfer*, Harper & Row, New York (1980).

[8] C. M. Bender and S. A. Orszag, *Advanced Mathematical Methods for Scientists and Engineers*, p. 261, McGraw-Hill, New York (1978).

[9] This approximation is quite poor in some cases, but does not affect the chief conclusions of the paper.

[10] The method shown here follows closely a similar method for conventional RTP this laboratory has published in Ref. [5].

TRANSIENT ENHANCED DIFFUSION AND OSTWALD RIPENING OF ION-IMPLANTATION GENERATED DEFECTS IN SILICON

N.E.B. Cowern[a], G. Mannino[a,b], F. Roozeboom[a], P.A. Stolk[a], H.G.A. Huizing[a], J.G.M. van Berkum[a], N.N. Toan[c], P.H. Woerlee[a,c], F. Cristiano[d] and A. Claverie[d]
[a] Philips Research Laboratories, Prof. Holstlaan 4, 5656 AA Eindhoven, The Netherlands
[b] INFM and Dept. of Physics, University of Catania, Corso Italia 57, 95129 Catania, Italy
[c] Dept. of Electrical Engineering, Twente University, Enschede, The Netherlands
[d] CEMES/CNRS, BP 4347, 31055 Toulouse, France

We report a study of transient enhanced diffusion arising from Ostwald ripening of Si implant-generated defects. Early during annealing, small interstitial clusters with low binding energy give rise to a large interstitial supersaturation, $S(t) \sim 10^7$, which drops to a nearly constant level $\sim 10^4$ as the clusters ripen into $\{113\}$ defects. Inverse modelling of Ostwald ripening yields the dissociation energy, E_{diss}, as a function of size. Based on this model we predict trends in TED as a function of implant dose and RTA ramp rate.

INTRODUCTION

Transient enhanced diffusion (TED) is a significant limitation for the fabrication of ultra-large scale integrated circuits. It broadens out dopant profiles and drives the deactivation of electrically active dopants in critical device regions. In recent years the role played by extended defects ($\{113\}$ defects [1] and dislocation loops [2]) in TED has become reasonably well understood. Under most conditions, $\{113\}$ defects are formed in the early stages of annealing, and thereafter act as the main source of silicon self interstitials driving TED. The structure and annealing kinetics of these defects have been well characterized [3, 4], and the associated interstitial supersaturation, $S(t)$, has been studied using marker-layer diffusion experiments [4, 5, 6, 7]. At high implant doses, dislocation loops form in competition with $\{113\}$ defects and can lower the supersaturation of interstitials by growing at the expense of the less stable $\{113\}$ defects. At longer annealing times, when all the $\{113\}$ defects have gone, the remaining dislocation loops are the only net source of interstitials, and they drive a long-timescale, low level of TED.

An analogous behavior has been seen in the case of low-dose Si implantations, where extremely rapid TED has been observed in the absence of $\{113\}$ defects. As the Si dose is increased, $\{113\}$ defects are able to nucleate and the amount of ultrafast TED is reduced [5]. At longer times, the $\{113\}$ defects no longer act as a sink of interstitials, but as a net source [5, 4]. In general, therefore, we might best try to understand TED in terms of a hierarchy of interstitial defects, ranging from small precursor clusters with low dissociation energies, via $\{113\}$ defects with intermediate dissociation energies, through (at high implant doses) to large dislocation loops with high dissociation energies.

Unfortunately, very little is known about the precursor clusters involved in the nucleation and ripening of $\{113\}$ defects. Theoretical study of these clusters is a major challenge, and only the very smallest, the di-interstitial [8] and the 4-cluster [9], have been investigated in detail. Experimentally even less is known, although Benton *et al.* have reported the growth and decay of cluster-related deep levels, consistent with an Ostwald ripening process [10]. The absence of data on these precursor defects is a significant problem when modelling TED at low temperatures and short annealing times, when {113} defects have not yet been formed. This is a particularly important issue for the understanding of RTA ramp-rate effects on TED.

The present paper reports results from a novel experimental approach to determine the dissociation energies E_{diss} as a function of cluster size, n. The approach relies on diffusion measurements to determine the transient interstitial supersaturation, $S(t)$, followed by inverse modeling to deduce E_{diss} as a function of n.

EXPERIMENT

A 2.5 μm intrinsic silicon layer containing thin, lightly boron-doped marker layers at depths of 0.9 and 1.3 μm, was epitaxially grown on a lightly n-type doped substrate using an ASM Epsilon One CVD reactor. This reactor produces highly pure material with an extremely low density of interstitial traps [11]. Boron markers were used since boron diffuses by an interstitial-assisted mechanism, driven by kick-out reactions with self interstitial atoms, and its diffusivity enhancement remains proportional to S up to very high values [12]. After growth, the wafer was implanted with $2 \times 10^{13}/cm^2$ 40 keV ^{28}Si ions, cut into samples and cleaned. Samples were annealed in dry N_2 at 600, 700 and 800°C for times in the range 1 s to 20 hours. Rapid thermal annealing (RTA) was used for anneals up to 5 hours, and furnace annealing was used for the time range 10-20 hours. RTA time-temperature cycles were measured by optical pyrometry, using a primary thermocouple calibration, and, at 600 and 700°C, confirmed to better than 4°C accuracy by measuring the regrowth rate of Ge-amorphized Si. Overshoot at the start of each RTA anneal was limited to less than 1.5°C by using a two-step ramp-up, with a set heating rate of 100°C/s up to 50°C below the set temperature, and 50°C/s for the remaining 50°C increment. A graphite susceptor was used to achieve satisfactory radiation coupling in the low-temperature range of the RTA experiments. Cooling at the end of the RTA anneals was facilitated by an increased N_2 gas flow rate.

In order to take proper account of diffusion effects occurring during thermal heating and cooling, which may be important for short-time RTA anneals, the actual temperature-time cycle for each RTA anneal measured by the pyrometer was recorded digitally. This turned out to be a key point in later analysis of the diffusion results, where it was possible to include the complete temperature-time cycle in an Ostwald ripening analysis of TED. Examples of the time-temperature profiles used in this experiment are shown in Fig. 1. The Ge-regrowth data used to cross-calibrate the RTA temperature at 600°C are shown in Fig. 2.

After annealing, boron concentration profiles were measured by secondary-ion mass spectrometry (SIMS). Typical results are shown in Fig. 3. Significant broadening appears within the first seconds of annealing at 600°C, corresponding to a diffusion enhancement of order 10^7. The enhancement decreases as a function of time, but is still $\sim 10^4$ after 20h at 600°C.

EVOLUTION OF THE INTERSTITIAL SUPERSATURATION

The time dependence of S is extracted from the SIMS data as follows. The as-grown profile is convolved with the B intermittent diffusion function $f(x,t)$ described in Ref. [13], and fitted to the measured diffused profile to extract the diffusive broadening $w =$

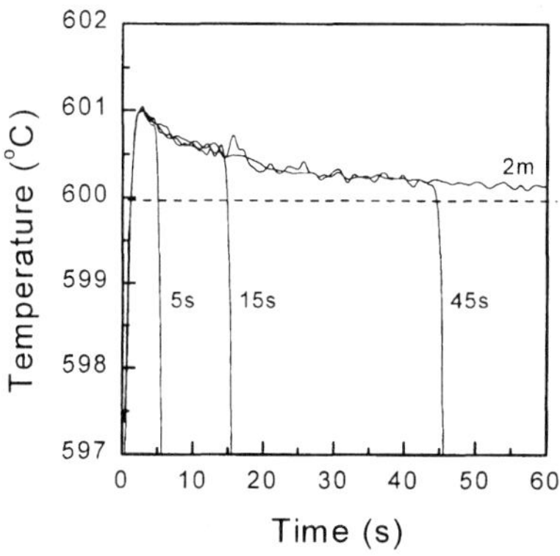

Figure 1: *Pyrometer data showing the close agreement between the RTA set temperature and the actual temperature-time cycle.*

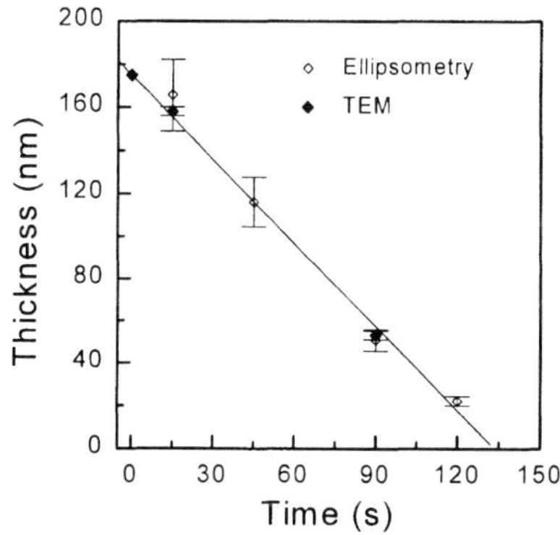

Figure 2: *Thickness of a Ge-preamorphized layer as a function of annealing time, used to determine the Si sample temperature at 600°C from published data on amorphous-layer regrowth rates.*

$\sqrt{(2\int D.dt)}$. The average diffusivity D_B over the period t_{i-1} to t_i is obtained from the relation $2D_B(t_i - t_{i-1}) = w_i^2 - w_{i-1}^2$, and finally $S(t)$ is estimated from $S = D_B/D_B^{eq}$ using Fair's value for the equilibrium diffusivity of B, $D_B^{eq} = 0.76\exp(-3.46eV/kT)$ [14].

The extracted values of S are shown in Fig. 4. The data show two phases of enhanced diffusion. An initial phase of ultrafast TED is followed by a sharp drop in S and a lower 'plateau' with near-constant S up to time τ, where the transient diffusion rather abruptly ends. The ultrafast phase persists for a much longer time than it takes for free interstitials, I, to diffuse through the measurement structure ($\ll$ 1s at 600°C). As we shall see, the ultrafast phase reflects ripening of very small interstitial clusters – precursors in the nucleation of $\{113\}$ defects. The plateau region closely matches previous observations of TED driven by $\{113\}$ defects [4, 7, 1]. The timescale τ, which represents the survival time of the $\{113\}$ defect band against loss of I to the silicon surface, is determined by the areal density of I present in the $\{113\}$ defects, and by the capacity of the silicon crystal to transport them to the surface. Assuming the surface is a perfect sink for I, as evidenced by previous studies [15], the flux of I to the surface is given by $D_I C_I^* S/L$ where D_I and C_I^* are the diffusivity and equilibrium concentration of I in silicon and L ($\approx r_p$, the ion projected range) is the depth of the defect band.

OSTWALD RIPENING ANALYSIS

To explain and model the supersaturation data in Fig. 4 we have to consider the Ostwald ripening of the interstitial clusters during the implant damage annealing process. This has to be done in a compact way, since we want to use the model to extract physical parameters by inverse modelling (fitting). We therefore use a system of ordinary differential equations that describe the Ostwald ripening of the defects and the loss of interstitials to the surface, without explicitly modelling the spatial and time-dependent diffusion of interstitials. The model accounts for capture and emission of I to and from all significantly populated cluster sizes ($n < 250$ in this study). It is assumed that, once the clusters have nucleated, the number of free I in the wafer is always much smaller than the number of clustered I. In

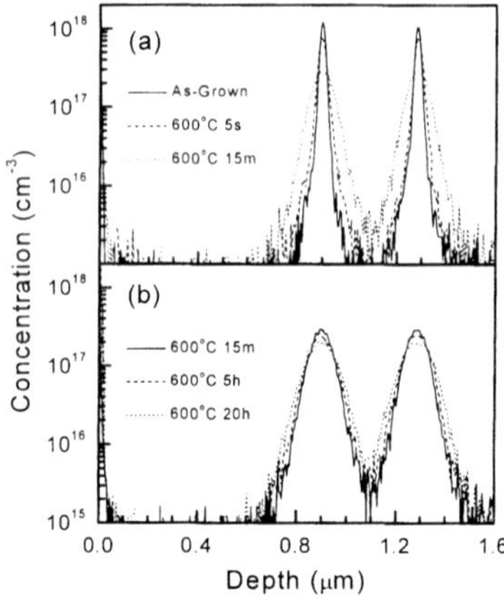

Figure 3: *SIMS profiles of boron-doped marker layers before and after rapid thermal annealing for a range of times at 600°C.*

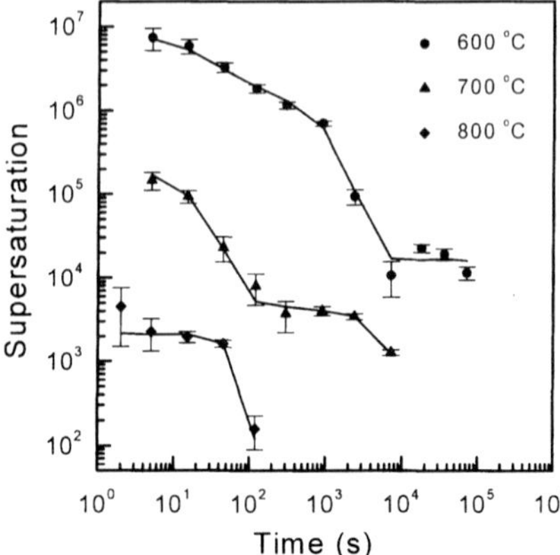

Figure 4: *Interstitial supersaturation versus time. Symbols represent data averaged over successive time periods. Curves are model fits as described in the text.*

other words, after an interstitial is emitted from a cluster, it is very soon captured again, either by another cluster or by the silicon surface. In this case, $S(t)$ quickly reaches a quasi steady state with respect to the cluster size distribution, the actual number of free I can be eliminated from the analysis, and S can be calculated directly. Atomistic Monte Carlo simulations [16, 17] confirm the validity of this picture. The evolution of the cluster size distribution N_n and the supersaturation S are then given by [18]:

$$\frac{dN_n}{dt} = F_{n-1}N_{n-1} - F_nN_n - R_nN_n + R_{n+1}N_{n+1} \quad (1)$$

$$S = \frac{1 + (r_p/D_IC_I^*)\sum_{n=2}^{\infty}\beta_nR_nN_n}{1 + 4\pi r_p\sum_{n=2}^{\infty}a_nN_n} \quad (2)$$

where

$$F_n = 4\pi D_IC_I^*a_nS \quad (3)$$

$$R_n = (6D_{0n}a_n/\lambda^3)\exp-(E_{diss}(n))/kT \quad (4)$$

are forward and reverse reaction rates describing the capture and emission of I from clusters of size n. The quantity $E_{diss}(n)$ is the cluster dissociation energy, D_{0n} is the emission prefactor, $a = a_n$ is the capture radius for I at clusters of size n, and β is the number of I generated by the emission reaction ($\beta = 2$ when $n = 2$, otherwise $\beta = 1$). For current purposes, the capture radius is assumed proportional to cluster size ($a_n = n\lambda/2$), as discussed in Ref. [16]. The problem can now be solved using equations (1-4).

In order to fit the experimental data in Fig. 4 we embed the Ostwald ripening solver in a least-squares optimization routine and extract fitted values for the product $D_IC_I^*$ and the variation of E_{diss} as a function of n. The initial conditions for the simulation are $n = 2$ and $N_2 = \Phi/2$ where Φ is the Si implant dose [16]. To test the consistency of the extracted parameters we fit the data for each temperature separately. The resulting fits are shown in Fig. 4. It proves possible to obtain excellent agreement ($\chi^2 \sim 1$) for all three temperatures, including the very steep drop in S in the 600°C curve at $t \sim 10^3$s and the less steep decrease

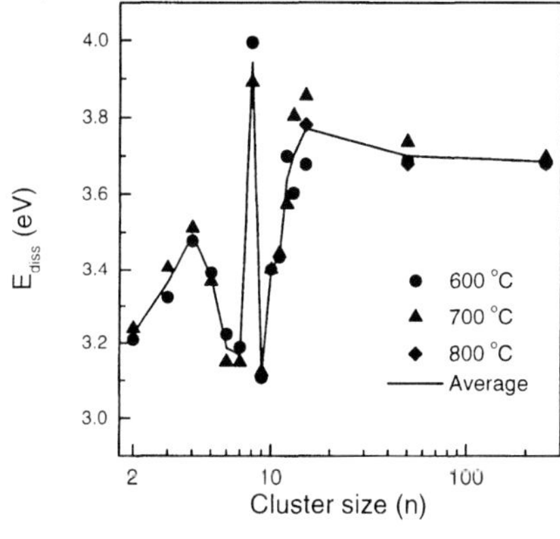

Figure 5: *Cluster dissociation energies extracted from the Ostwald-ripening analysis. Symbols are values fitted at 600, 700 and 800°C, and the curve is the averaged result.*

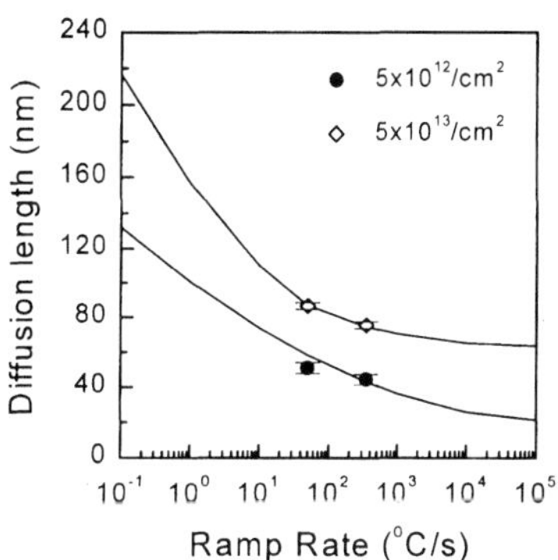

Figure 6: *TED diffusion length as a function of heating rate, up to a set temperature/time of 950°C 10s. Symbols represent experimental data and curves are simulation results.*

in the 700°C curve.

Our extracted values of E_{diss} are plotted in Fig. 5. The values for small clusters oscillate dramatically as a function of n, before settling down to higher, near-constant, values for $n > 10$ which are consistent with previous estimates for the stability of {113} defects [1]. The oscillatory structure at small n is essential for describing the shape of $S(t)$ at 600 and 700°C, shown in Fig. 4. In particular, the sharp drop in $S(t)$ in the range 15m - 2h at 600°C is related to the strong oscillation in $E_{diss}(n)$ at around $n = 8$. The annealing kinetics associated with this feature are discussed in detail in Ref. [16]. The dissociation energies of the smaller clusters in Fig. 5 are much higher than previous theoretical estimates based on the energetics of {113} defect chains [19]. We speculate that the oscillatory energies of these defects reflect local bonding 'opportunities' that depend on the number of atoms in the cluster. Recent calculations by La Magna *et al.* [20] are consistent with this view, although their predicted energies do not agree in detail with our results. It is also interesting to note the recent proposal by Arai *et al.* of a stable size-4 interstitial cluster which might form the basis of larger clusters [9]. This prediction seems consistent with our designation of sizes ~ 4 and ~ 8 as most-stable clusters in the size range $n < 10$.

RAMP-RATE EFFECTS ON TED

Having determined a curve of dissociation energy versus size for interstitial clusters, it should be possible to predict trends in TED as a function of temperature and time. This provides us with an opportunity to estimate the behavior of TED in future RTA systems, where it is expected that higher ramp rates will deliver lower amounts of TED and thus shallower, more highly-activated dopant profiles. To investigate this issue we have carried out preliminary experiments using RTA ramp rates of 50 and 350°C/s up to a peak temperature of 950°C, and compared the results with simulations over a much wider range of ramp-rates, based on the above-derived model. The simulated diffusion length is scaled by a factor 1.4 to obtain optimal agreement with the TED data from the ramp-rate experiment.

Fig. 6 shows the amount of diffusion plotted as a function of the RTA ramp-up rate, for doses of 5×10^{12} and 5×10^{13} Si/cm^2. For low ramp rates, all the TED occurs during the ramp-up part of the temperature cycle – the lower the rate, the lower the temperature at which the majority of TED occurs, and the larger the absolute amount of diffusion. The simulation demonstrates the major benefits of RTA in reducing TED when compared with furnace annealing. However, because of the extremely high supersaturation at very short anneal times, significant diffusion still occurs even during typical RTA ramp-up cycles. This problem persists even up to much higher ramp rates than are used in current RTA systems. Thus, although the potential for reducing TED by fast ramp-rate RTA is important, very large increases in ramp-rate are needed to actually suppress Si implant/TED. To simulate this effect for the case of other implant species such as B and P, the equivalent information on cluster dissociation energies will be also be needed for impurity-interstitial clusters. For example, for B implantation one can expect that the curve in Fig. 6 will shift to the left, owing to the relatively high dissociation energies of B-I clusters.

The simulations shown in Fig. 6 also show why the amount of TED is not simply proportional to the $\sqrt{(\text{dose})}$, as the "+1" model would predict in the case of diffusion at a constant temperature. The reason is that at low doses, a larger proportion of TED occurs in the low-temperature part of the ramp-up, where the coupling between interstitials and boron dopant diffusion is strongest. This effect is clearly reproduced in the TED data in Fig. 6. Similar ramp-rate effects also apply to the energy dependence of TED [18].

CONCLUSIONS

An Ostwald ripening model has been used to extract basic data on interstitial cluster energetics from Si implant/TED experiments. The resulting model has then been used to simulate the dose and ramp-rate dependence of TED. The trend of the simulation results is in good agreement with ramp-rate experiments. Finally, we have shown that further improvements in TED suppression are possible using even higher ramp rates than are achievable in current RTA systems. Similar trends, though with different binding energies and time scales, also apply to the case of dopant implantation and annealing.

ACKNOWLEDGEMENTS

We thank P.H.L. Bancken for carrying out the Si implantations and J. Jans for spectral ellipsometry measurements. S. Coffa, A. La Magna and M. Jaraíz are acknowledged for valuable discussions. We are also grateful to Steag-AST Elektronik GmbH for assistance with absolute calibration of the RTA system. This work has been partially supported by the ESPRIT Long Term Research project RAPID.

References

[1] P.A. Stolk, H.-J. Gossmann, D.J. Eaglesham, D.C. Jacobson, C.S. Rafferty, G.H. Gilmer, M. Jaraiz, J.M. Poate, H.S. Luftman, and T.E. Haynes, *J. Appl. Phys.* **81**, 6031 (1997).

[2] A. Claverie, in '*Silicon Front-End Technology: Materials Processing and Modeling*, Mat. Res. Soc. Spring Meeting, San Francisco, 1999.

[3] M. Kohyama and S. Takeda, *Phys. Rev. B* 46, 12305 (1992); *ibid.*, *Phys. Rev. B* **51**, 13111 (1995).

[4] D.J. Eaglesham, P.A. Stolk, H.-J. Gossmann, and J.M. Poate, *Appl. Phys. Lett.* **65**, 2305 (1994).

[5] N.E.B. Cowern, G.F.A. van de Walle, P.C. Zalm, and D.W.E. Vandenhoudt, *Appl. Phys. Lett.* **65**, 2981 (1994).

[6] H.G.A. Huizing, C.C.G. Visser, N.E.B. Cowern, P.A. Stolk, and R.C.M. de Kruif, *Appl. Phys. Lett.* **69**, 1211 (1996).

[7] H.S. Chao, P.B. Griffin, J.D. Plummer and C.S. Rafferty, *Appl. Phys. Lett.* **69**, 2113 (1996).

[8] Y.H. Lee, *Appl. Phys. Lett.* **73**, 1119 (1998).

[9] N. Arai, S. Takeda, and M. Kohyama, *Phys. Rev. Lett.* **78**, 4265 (1997).

[10] J.L. Benton, K. Halliburton, S. Libertino, D.J. Eaglesham, and S. Coffa, *J. Appl. Phys.* **84**, 4749 (1998).

[11] K.J. van Oostrum, P.C. Zalm, W.B. de Boer, D.J. Gravesteijn, and J.W.F. Maes, Appl. Phys. Lett. **61**, 1513 (1992).

[12] N.E.B. Cowern, G.F.A. van de Walle, P.C. Zalm, and D.J. Oostra, *Phys. Rev. Lett.* **69**, 116 (1992).

[13] N.E.B. Cowern, K.T.F. Janssen, G.F.A. van de Walle, and D.J. Gravesteijn, *Phys. Rev. Lett.* **65**, 2434 (1991).

[14] R.B. Fair, in *Impurity Doping Processes in Silicon*, edited by F.F.Y. Wang, (North-Holland, 1981), p. 315.

[15] D.R. Lim, C.S. Rafferty and F.P. Clemens, *Appl. Phys. Lett.* **67**, 2303 (1995).

[16] N.E.B. Cowern, G. Mannino, P.A. Stolk, F. Roozeboom, H.G.A. Huizing, J.G.M. van Berkum, F. Cristiano, A. Claverie, and M. Jaraíz, *Phys. Rev. Lett.* **82**, 4460 (1999)

[17] M. Jaraíz, L. Pelaz, E. Rubio, J. Barbolla, G.H. Gilmer, D.J. Eaglesham, H.J. Gossmann, and J.M. Poate, *Mat. Res. Soc. Symp. Proc.* **532**, 43 (1998).

[18] N.E.B. Cowern, G. Mannino, P.A. Stolk, F. Roozeboom, and H.G.A. Huizing, submitted to *Mat. Sci. in Semiconductor Processing*.

[19] N. Cuendet, T. Halicioglu, and W.A. Tiller, *Appl. Phys. Lett.* **68**, 19 (1996).

[20] A. La Magna, S. Coffa, and S. Libertino, in '*Silicon Front-End Technology: Materials Processing and Modeling*, Mat. Res. Soc. Spring Meeting, San Francisco, 1999, paper S5.4.

ELECTRICAL MEASUREMENTS OF ANNEALED BORON IMPLANTS FOR SHALLOW JUNCTIONS

A. T. Fiory,[1] K. K. Bourdelle,[2] M. E. Lefrancois,[3] D. M. Camm,[3] and A. Agarwal[4]

[1]Bell Laboratories, Lucent Technologies, Inc., Murray Hill, NJ
[2] Bell Laboratories, Lucent Technologies, Inc., Orlando, FL
[3]Vortek Industries Ltd., Vancouver, BC, Canada
[4]Eaton Semiconductor Equipment Operations, Beverly, MA

Ultra-low energy, 0.5-keV, boron implants were annealed at ramp-up rates of 150 °C/s in an incandescent-lamp system, or up to ≈1500 °C/s in an arc-lamp system, to promote activation and minimize diffusion. Annealing treatments covered a range of temperatures and included "spike anneal" methods, which result in effective annealing times of ≈1 s for incandescent-lamps and ≈0.3 s for an arc-lamp. Results for the electrical activation and junction depths were derived from measurements of sheet Hall van der Pauw electrical transport and from secondary ion mass spectroscopy. Isothermal annealing shows the dopant diffusion to have a square-root dependence on the effective annealing time for short annealing cycles. Accordingly, arc-lamp annealing leads to a shallower junction than incandescent-lamp annealing. Similar relationships between sheet resistance and junction depth are observed for the two annealing methods.

INTRODUCTION

Formation of shallow p+/n junctions in silicon with ultra-low energy boron implants and anneals with the shortest possible cycle times had been investigated several years ago by Shishiguchi and co-workers at NEC (1). Here, "ultra-low" energy (ULE) pertains to implant energies in the range of 0.5 keV. Annealing to electrically activate ULE boron implants at doses of 1×10^{15} cm^{-2} or greater creates junctions that are significantly deeper than the as-implanted range. The junction depth is affected by mechanisms of boron enhanced diffusion (BED), in which Si interstitials are created in the high boron concentration layer produced by the implant (2). A "spike anneal" method, which uses the fastest available rates of heating and cooling, and nominally "0 s" dwell at maximum temperature, has advantages in promoting electrical activation with less diffusion, owing to the differing activation energies for the equilibrium diffusivities of boron and Si interstitials, 3.5 and 4.9 eV, respectively (3,4). Previous studies employing rapid thermal processing (RTP) equipment with incandescent lamp heating have shown advantages from reduced thermal anneal cycles (1,4,5). The present work investigates processing with incandescent and arc lamp RTP systems. Substantially shorter cycle times are achievable with an arc lamp system, which emits a stronger flux of radiation below 1 μm and responds more rapidly to changes in excitation power (6).

EXPERIMENT

Samples were prepared from 150-mm n-type wafers implanted with ^{11}B at 0.5 keV and doses of $1x10^{15}$ and $2x10^{15} cm^{-2}$ in an Eaton ULE2 implanter. Junctions were formed directly in lightly doped n-type bulk silicon ($n=2x10^{14}$ cm^{-3}) and in wafers with an $n = 5x10^{17}$ cm^{-3} tub prepared by phosphorus implants that were diffused and activated prior to the B implant. The annealing studies used several whole wafers and a variety of 9.5-, 10-, and 15-mm square samples cut from unpatterned wafers. An AG Associates Heatpulse 8108 (HP) tool with a 2.5 μm ripple pyrometer calibrated against thermocouples (SensArray type 1530) was used for wafer and sample annealing. The maximum heating rate in the HP system is about 150 °C/s and a typical cooling rate is 80 °C/s. Other samples were heated at rates up to 1500 °C/s and maximum cooling at 125 °C/s in a developmental arc-lamp system at Vortek Industries, similar to an earlier design (6). A 0.9-μm or 1.5-μm pyrometer calibrated against the melting point of silicon was used for control at 1-ms resolution. Measurements on the samples included Hall van der Pauw sheet electrical transport (HvdP), junction I-V and C-V curves, and B depth profiling by secondary ion emission spectroscopy (SIMS). Junction depths were estimated from fits to HvdP data that assumed an erfc model function for the carrier profile and from ^{11}B SIMS profiles (7). An ellipsometer was used to measure oxide film thickness.

SPIKE THERMAL CYCLES

RTP systems with incandescent lamps as the heating source typically take advantage of chamber reflections to boost efficiency and uniformity in the rapid heating of silicon wafers. The thermal cycle time generally has three contributions, a heating rate dictated by maximum lamp power and reflector efficiency, the relaxation time for the tungsten filament temperature, and cooling by radiative transfer from the wafer to the chamber. The analogous contributions may be reduced in an arc-lamp RTP system, owing to greater available lamp power, faster response of an arc relative to a tungsten filament, and reduced back reflection of wafer emission radiation. Figure 1 shows pyrometer recordings for spike thermal cycles at 1050 °C peak temperature in the HP and Vortek RTP systems.

A cumulative effective time is defined by integrating time weighted by an Arrhenius factor, $\exp [E_A (T_P^{-1} - T^{-1}) / k_B]$, where T is the instantaneous temperature and T_P is the maximum temperature. The effective cycle times are 1.0 s for curve H and 0.33 s for curve V, taking an activation energy of 5 eV for the electrical activation of shallow boron implants (4). The temperature and time region shown in Fig. 1 encompasses over 95% of the effective cycle times. Figure 2 shows the temperature vs cumulative effective time computed for the spike anneal with the Vortek system. The arc lamp produces a 0.1-s transition from heating at 365°C/s to cooling at 115°C/s. This rapid transition accounts for a major part of the factor of 3 reduction in cycle time for the arc lamp method.

DIFFUSED BORON PROFILES

Boron concentration profiles obtained by SIMS are shown in Fig. 3 for several spike anneals in N_2 ambients of 0.5 keV $2\times10^{15}cm^{-2}$ implants. Two samples annealed in a Vortek system with a 0.9-µm pyrometer are shown by curves A (1326 °C/s ramp, 1058 °C peak, R_S = 468 Ω/Sq) and B (543 °C/s ramp, 1063 °C peak, R_S = 402 Ω/Sq). Junction depths X_J = 35 nm are obtained at 10^{18} cm^{-3} concentration. An anneal in the HP system is shown by curve H (158 °C/s ramp, 1050 °C peak, R_S = 332 Ω/Sq, X_J = 44 nm). Curves Af and Bf show the model erfc functions that were used to fit the HvdP measurements on the samples corresponding to Vortek anneals of A and B, respectively. The electrical measurements are consistent with a shorter effective thermal cycle for A produced with the faster ramp (although SIMS shows nearly equivalent diffusion depths). The faster ramp is achieved with approximately 2.4 times more heating power. The trends in these results are opposite to possible photo-assisted activation or diffusion of the boron, since the larger radiation flux produces higher R_S and diffusion that is either unchanged (according to SIMS) or reduced (according to HvdP). Shallower diffused profiles are obtained with the Vortek system, compared to the HP system, largely because of the reduced cycle time. Additional SIMS profiles for Vortek anneals of 1×10^{15} cm^{-2} dose give X_J = 33 nm for 358 °C/s ramp, 1049 °C peak, and R_S = 536 Ω/Sq; and X_J = 37 nm for 459 °C/s ramp, 1070 °C peak, and R_S = 459 Ω/Sq.

TIME DEPENDENCE

Dependence of sheet resistance on effective annealing time, defined for a 5 eV activation energy, is shown in Fig. 4 for a 0.5 keV 2×10^{15} cm^{-2} boron implant annealed in the HP and Vortek (with 1.5-µm pyrometer) at 1000 °C and 1100 °C in N_2 ambients. The shortest time in HP (1 s) is obtained for spike anneals with 150 °C/s heating and up to 80 °C/s cooling; longer times used the same transition rates with plateau temperatures of 1 s to 800 s. The two Vortek anneals used 330 °C/s heating to 1000 °C and 351 °C/s to 1100 °C. Time dependence for the sheet carrier density is shown in Fig. 5. The electrical activation has an initial square-root time dependence characteristic of a diffusion mechanism. Junction depths, computed from the HvdP data with the erfc model function are shown for isothermal annealing at 1000 °C in Fig. 6 as a function of the square-root of the effective time. Electrical activation at this temperature is sufficiently slow to reveal the initial time dependence associated with the diffusion of the boron profile. A comparison between Figs. 4 and 6 shows more pronounced time dependence for R_S than for X_J at the shorter annealing times.

SPIKE ANNEALING

Samples of a 0.5 keV 2×10^{15} cm^{-2} implant were cut from the same wafer and spike annealed in the HP and Vortek systems at various peak temperatures in N_2 ambients. Ellipsometry measurements, assuming an SiO_2 surface film, show 2 nm of oxide after HP anneals and 3 nm of oxide after Vortek anneals. Figure 7 shows the temperature dependence of R_S and Fig. 8 shows the temperature dependence of X_J from HvdP analysis. Shorter annealing cycles (Vortek relative to HP) produce higher R_S and smaller X_J.

However, different temperature measurement and calibration methods were used in the two RTP systems. Figure 9 shows a plot of R_S as a function of X_J, which removes the uncertainty in temperature scale (apart from differences in temperature uniformity). The HvdP measurements for anneals in the Vortek system are represented by crosses and triangles, corresponding to 0.9-μm and 1.5-μm pyrometry methods, respectively. The plot includes selected samples measured by SIMS (from Fig. 1 and additional data from Ref. 5). The dashed curve shows the relationship computed for a carrier profile with an assumed erfc functional form and fixed surface concentration.

JUNCTION LEAKAGE

Current-voltage curves were recorded at 1 to 100 Hz using contacts applied to the activated boron layer and the n-type substrate of annealed cut samples. Capacitance was determined from the displacement current. The surface oxide grown during the activation anneal serves to passivate the exposed cleaved edges. Figure 10 shows current densities at a reverse bias of –1V for anneals in the HP and Vortek systems.

Samples for anneals in the HP system used B implanted into a 150-nm deep tub doped to $n = 5\text{x}10^{17}\ cm^{-3}$ by diffusing previous P implants of 30 keV, $2\text{x}10^{12}\ cm^{-2}$ and 90 keV, $6\text{x}10^{12}\ cm^{-2}$. A 0.1% O_2 ambient was used for annealing to suppress boron out diffusion and which produced surface oxides of 1.7 to 1.9 nm. Results were obtained for spike anneals at "0 s" and anneals with the same ramp rates and a 10-s plateau at the maximum temperature. Reverse bias leakage shows a maximum in the mid $10^{-4}Acm^{-2}$ range near 1000 °C for 10-s anneals and 1100 °C for 0-s anneals. Junction depths are estimated to be about 65 nm for these annealing conditions. Carrier concentration gradients at the junctions were computed from the zero-bias capacitance C(0), as $6\text{x}10^{25}\ cm^{-4}\ (\mu Fcm^{-2})^{-3}\ C^3(0)$, and found to vary from $7\text{x}10^{22}$ to $5\text{x}10^{23}\ cm^{-4}$, with the higher concentration gradient and C(0) correlating with lower leakage current.

Samples for annealing in the Vortek system used B implanted into uniformly and lightly-doped $n = 2\text{x}10^{14}\ cm^{-3}$ silicon. These junctions have much lower built-in electric field and show reverse-bias leakage under 30 μAcm^{-2}.

CONCLUSIONS

Results from sheet electrical transport, SIMS, and junction I-V curves are presented for annealing shallow boron implants by RTP systems using incandescent and arc lamps for radiant heating. Isothermal annealing shows that the dopant diffusion has a square-root time dependence at short annealing cycles. Accordingly, arc-lamp annealing, with a 0.3 s thermal cycle time, leads to a shallower junction than incandescent-lamp annealing with a 1.0 s cycle time. Inverse relationships between sheet resistance and diffusion depth have similar behavior for the two annealing methods. Reverse bias leakage of p+/n junctions was measured on annealed samples with several n doping levels.

ACKNOWLEDGEMENTS

The authors acknowledge the technical contributions of M. Buonanno, S. McCoy, and M. Walsh; and stimulating discussion with C. Rafferty and H.-J. Gossmann. SIMS analyses were performed at Evans East.

REFERENCES

1. S. Shishiguchi, A. Mineji, T. Hayashi, and S. Saito, VLSI Symp. Tech. Digest, 89 (1997); S. Saito, S. Shishiguchi, A. Mineji, and T. Matsuda, Mat. Res. Soc. Symp. Proc. **532**, 3 (1998); S. Shishiguchi, A. Mineji, and T. Matsuda (*these proceedings*).
2. A. Agarwal, D. J. Eaglesham, H.-J. Gossmann, L. Pelaz, S. B. Herner, D. C. Jacobson, T. E. Haynes, Y. E. Erokhin, and R. Simonton, IEDM Tech. Digest, 467 (1997); A. Agarwal, H.-J. Gossmann, D. J. Eaglesham, S. B. Herner, and A. T. Fiory, and T. E. Haynes, Appl. Phys. Lett. **74**, 2435 (1999); A. Agarwal, H.-J. Gossmann, D. J. Eaglesham, Appl. Phys. Lett. **74**, 2331 (1999).
3. R. B. Fair and J. C. C. Tsai, J. Electrochem. Soc. **124**, 1107 (1997).
4. A. T. Fiory and K. K. Bourdelle, Appl. Phys. Lett. **74** (3 May 1999 issue, *in press*).
5. A. Agarwal, A. T. Fiory, H.-J. L. Gossmann, C. S. Rafferty, P. Frisella, Mat. Sci. in Semicond. Proc. **1**, 237 (1998).
6. M. E. Lefrancois, D. M. Camm, and B. J. Hickson, Mat. Res. Soc. Symp. Proc. **429**, 321 (1996).
7. A. T. Fiory and K. K. Bourdelle, J. Electron. Mater. (*to be published*).

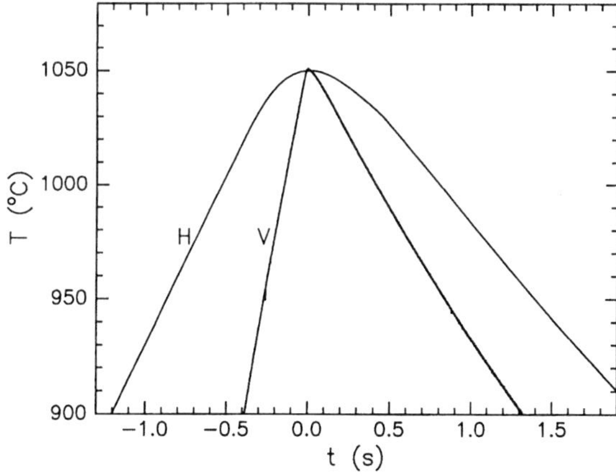

FIG. 1. Temperature vs. time (relative to peak temperature) from pyrometer recordings for spike annealing. Curve H: Heatpulse 8108 with incandescent lamps; curve V: Vortek system with arc lamp and close-coupled reflector.

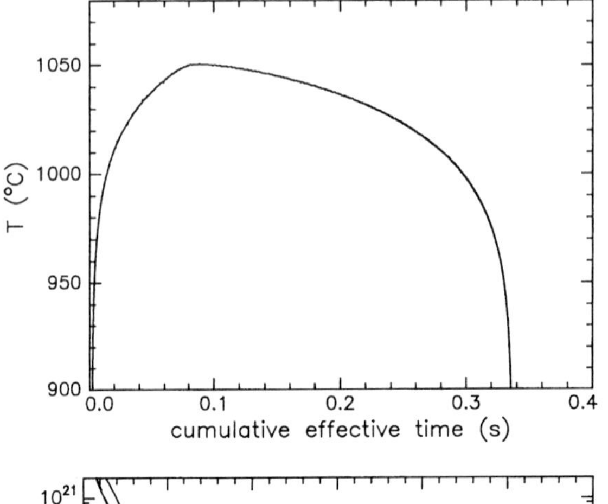

FIG. 2. Temperature vs. cumulative effective time, computed with activation energy of 5 eV, for spike annealing in Vortek system with arc lamp and close-coupled reflector.

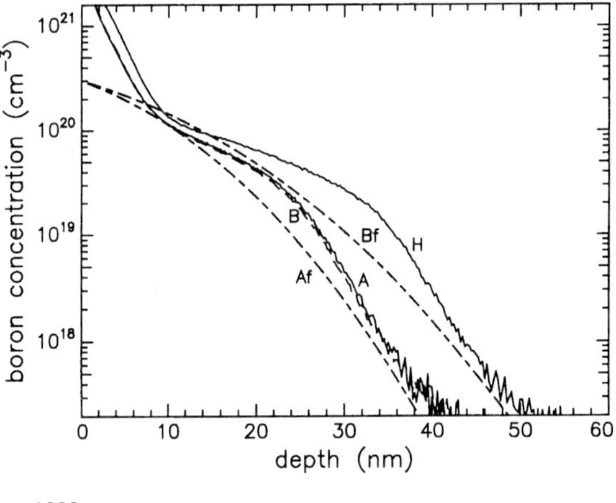

FIG. 3. SIMS ^{11}B profiles for annealed 0.5 keV, $2x10^{15}$ cm^{-2}: sample **A** (solid curve), Vortek, 1326 °C/s, 1058 °C; sample **B** (dashed curve), Vortek, 543 °C/s, 1063 °C; sample **H**, Heatpulse 150 °C/s, 1050 °C. Fits to Hall van der Pauw data with erfc model function: dashed curve **Af** for sample A, dashed curve **Bf** for sample B.

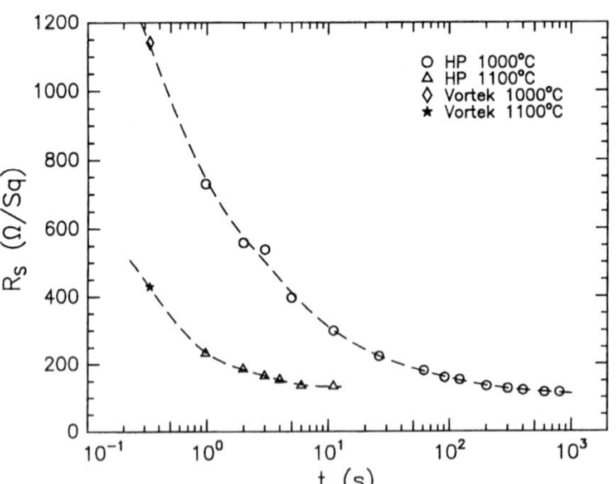

FIG. 4. Sheet resistances of boron implant 0.5 keV, $2x10^{15}$ cm^{-2} as function of effective annealing time at 1000 °C and 1100 °C for samples annealed in Heatpulse and Vortek RTP systems. (Curves are spline guides to the eye.)

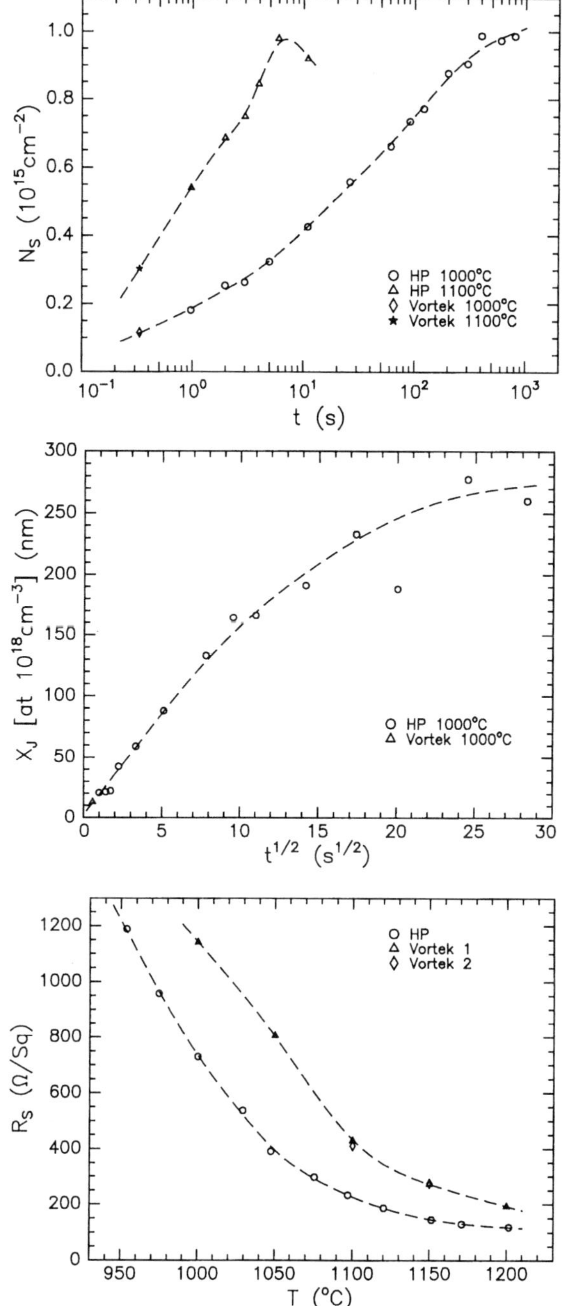

FIG. 5. Sheet carrier densities of boron implant 0.5 keV, 2E15 cm^{-2} as function of effective annealing time at 1000 °C and 1100 °C for samples annealed in Heatpulse and Vortek RTP systems.

FIG. 6. Junction depths of boron implant at 0.5 keV and 2x10^{15} cm^{-2}, from analysis of Hall van der Pauw data with erfc model function, as function of the square-root of effective annealing time at 1000 °C, for samples annealed in Heatpulse and Vortek RTP systems.

FIG. 7. Sheet resistances for boron implant at 0.5 keV and 2x10^{15} cm^{-2} as functions of peak temperature of spike thermal annealing in Heatpulse (heating at 150 °C/s) and Vortek systems (Vortek 1, heating at 306 to 351 °C/s; Vortek 2, heating at 126 and 140 °C/s).

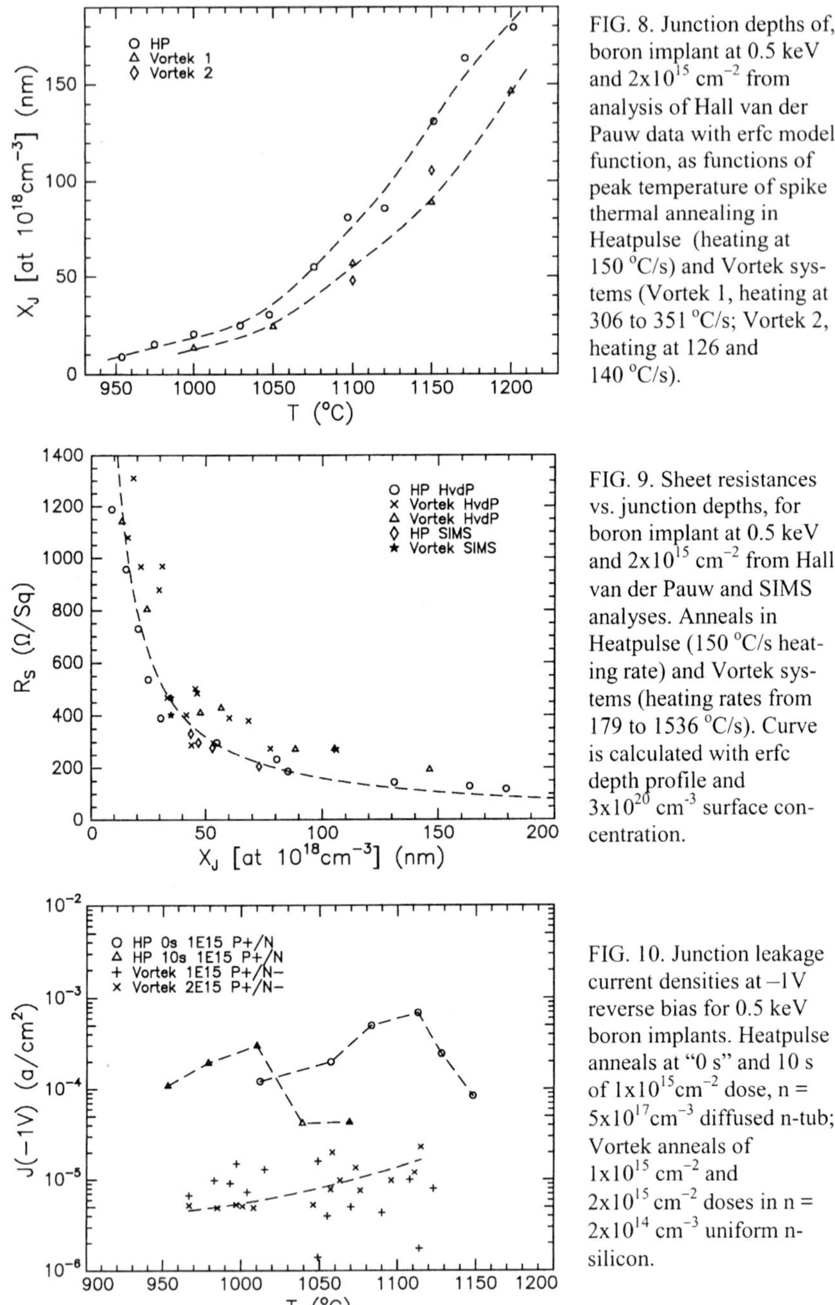

FIG. 8. Junction depths of, boron implant at 0.5 keV and $2x10^{15}$ cm^{-2} from analysis of Hall van der Pauw data with erfc model function, as functions of peak temperature of spike thermal annealing in Heatpulse (heating at 150 °C/s) and Vortek systems (Vortek 1, heating at 306 to 351 °C/s; Vortek 2, heating at 126 and 140 °C/s).

FIG. 9. Sheet resistances vs. junction depths, for boron implant at 0.5 keV and $2x10^{15}$ cm^{-2} from Hall van der Pauw and SIMS analyses. Anneals in Heatpulse (150 °C/s heating rate) and Vortek systems (heating rates from 179 to 1536 °C/s). Curve is calculated with erfc depth profile and $3x10^{20}$ cm^{-3} surface concentration.

FIG. 10. Junction leakage current densities at −1V reverse bias for 0.5 keV boron implants. Heatpulse anneals at "0 s" and 10 s of $1x10^{15}cm^{-2}$ dose, n = $5x10^{17}cm^{-3}$ diffused n-tub; Vortek anneals of $1x10^{15}$ cm^{-2} and $2x10^{15}$ cm^{-2} doses in n = $2x10^{14}$ cm^{-3} uniform n-silicon.

INFLUENCE OF THERMAL NITRIDATION ON THE DIFFUSION OF ARSENIC DURING RAPID THERMAL ANNNEALING

W. Lerch[1], N. A. Stolwijk[2], S. D. Marcus[3], D. F. Downey[4], M. Schäfer[5]
[1] STEAG AST Elektronik GmbH, Dornstadt, Germany
[2] Institut für Metallforschung, Westfälische Wilhelms-Universität Münster, Germany
[3] STEAG AST Elektronik, Tempe (AZ), USA
[4] Varian Ion Implant Systems, Gloucester (MA), USA,
[5]CADwalk, Almendingen, Germany

Nitridation of bare silicon surfaces in an ammonia-containing ambient is known to introduce vacancies in silicon at much lower temperatures than in a pure nitrogen ambient. The assumed vacancy excess during direct nitridation may enhance the diffusion of arsenic in silicon. After ultra-shallow implantation of As^+ with an energy of 1 keV to a dose of $\Phi \approx 1 \cdot 10^{15}$ cm^{-2} the change of the junction depth due to rapid thermal annealing (RTA) in an argon ambient with a varying but controlled concentration of ammonia (1000 ppm up to 100 %) is studied. Isochronal 10 s anneals are performed at 1000°C, 1050°C and 1100°C. Additionally, isothermal anneals at 1000°C, 1050°C and 1100°C for various times are performed as well. Concentration-depth profiles, measured by secondary ion mass spectroscopy were simulated within the SSUPREM IV computer code allowing for arsenic diffusion via neutral and negatively charged vacancies and silicon self-interstitials. From these experiments the RTA process for the formation of ultra-shallow junctions with arsenic can be optimized and the diffusion broadening of these near-surface implantations under non-equilibrium conditions better understood.

INTRODUCTION

The continued downscaling of CMOS devices to the 0.1 μm gate length feature and the increasing circuit complexity involves several process improvements and additions. In the critical section of the device fabrication process long-time furnace annealing steps have gradually been replaced by rapid thermal annealing (RTA) processes (1).

One of the main difficulties is the formation of low-resistance silicide contacts for the source and drain junctions of these small geometry devices while simultaneously maintaining low leakage currents and good transistor turn-off performance. This challenges junction annealing for controlling diffusion while fully activating the dopant. The formation of shallow n-junction using As implantation involves less problems than that of shallow p-junctions formed by boron or BF_2, both concerning ion-implant channeling (in particular boron) and diffusion broadening (2).

Previous publications (3), (4) of the present authors address the effect of RTA ambient oxygen on the diffusion of ultra-shallow $^{75}As^+$, $^{11}B^+$ and $^{49}BF_2^+$ implants. These works have shown a direct correlation between oxygen concentration and enhanced diffusion for B and

BF_2. We have also proposed a model based on the supersaturation of silicon self-interstitials caused by surface oxidation and on the kick-out diffusion model (3).

In a similar way the present work attempts to explain and model the diffusion characteristics of ultra-shallow 1 keV $^{75}As^+$ implants at a dose of $1 \cdot 10^{15}$ cm^{-2} as a function of ambient conditions. Since As diffusion in contrast to B diffusion is dominated by the vacancy mechanism, oxidizing ambients have little effect on the broadening of implanted profiles (4). Nitridizing ambients, however, may introduce vacancies in the bulk. Therefore we have annealed our implants for 10 s at 1000°C, 1050°C and 1100°C in argon ambients containing ammonia concentrations varying from 0~1 ppm to 100 %. Simulations of the resulting secondary ion mass spectroscopy (SIMS) profiles using the SSUPREM IV computer code, implemented in the desktop framework Athena (5), (6), reveal not only a direct nitridation-induced enhancement of As diffusion but also more indirect effects which appear to be due to capping of the silicon surface.

EXPERIMENTAL

For the experiments p-type, 200 mm Wacker prime Si wafers, (100)-orientation, 10-20 Ωcm, were used either for ion implantation or to monitor the silicon-oxynitride thickness growth during annealing independent of damage enhanced growth effects (7).

The arsenic implants (1 keV, $1.0 \cdot 10^{15}$ cm^{-2}) were performed on a Varian VIISion-80 PLUS low-energy high-current implanter at a tilt and twist angle of 0°. The beam current densities have been measured using an in-situ 2D beam profiler. A key issue for these ultra-shallow implantations of 210 Å depth (@ $1.0 \cdot 10^{18}$ cm^{-3}) is the variability of the native oxide. To avoid any influence prior to the implantation process a 30 s wet-chemical etch was performed in a HF (49 %):H_2O (1:40) solution. After implantation for each ammonia concentration a set of two wafers (a bare and an implanted wafer) were annealed in a STEAG AST Elektronik AST2800ε rapid thermal processing system (RTP). The recipes consist of a prolonged purge step for stabilization of the gas ambient, a prestabilization at 650°C, 10 s, followed by a final 10 s isochronal anneal at 1000°C, 1050°C and 1100°C, respectively. The ramp up rate to the soak time as well as the ramp down rate are 50 K/s. An average post-anneal uniformity (1σ) of less than 2 % on a 200 mm wafer was found for these ultra-shallow junctions.

The ammonia concentrations during the annealing step in Ar have been varied in a range of 0~1 ppm to 100 % within this work. Argon was used as inert carrier gas for avoiding unintentional and uncontrolled nitridation of silicon by nitrogen at higher temperatures (8). For the purpose of these experiments a low-flow ammonia mass flow controller (20 sccm/200 sccm) in conjunction with a standard (30 slm) argon mass flow controller were used to ensure the correct ammonia concentration.

The silicon-oxynitride thickness growth within the annealing time of 10 s was measured using a Plasmos SD3200 single wavelength ($\lambda = 633$ nm) ellipsometer (refractive index $n = 2.01$) with a relative measurement accuracy of 1 Å and a repeatability of 0.1 Å. Subsequent to ellipsometric characterization the sheet resistance was probed through the silicon-oxynitride on a KLA-Tencor RS100.

The implanted profiles have been analyzed using SIMS. The depth profiling was performed at Evans East Inc. using a Physical Electronics Φ 6600 SIMS spectrometer. The near-surface concentration accuracy has intentionally been improved by the use of an O-leak technique. The penetration depth for the SIMS measurements throughout this publication is defined as the depth where the total concentration, sum of electrically and non-electrically active As atoms falls to a level of $3.0 \cdot 10^{18}$ cm^{-3}.

RESULTS AND DISCUSSION

Characterization of nitridation

The gaseous ambient during RTA plays a crucial role for the formation of ultra-shallow junctions because it may change the Si point defect equilibrium. For boron the injection of Si self-interstitials due to surface oxidation leads to an oxidation enhanced diffusion (OED) (3).

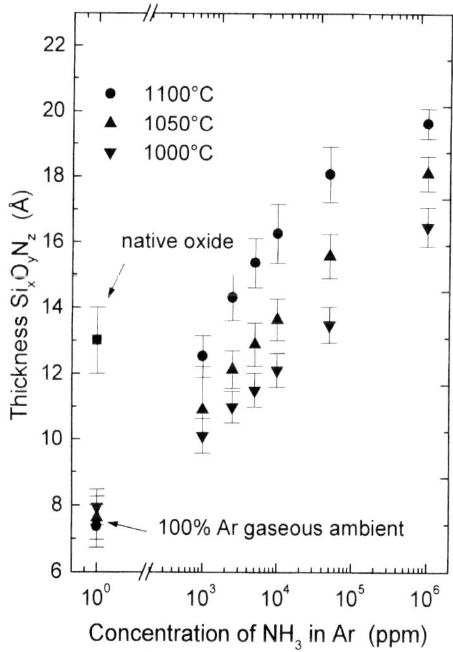

Figure 1: Silicon-oxynitride ($Si_xO_yN_z$) thickness versus relative ammonia concentration for various temperatures after isochronal annealing (10 s) of bare Si wafers. The refractive index used for the native oxide was $n = 1.465$, for all other measurements $n = 2.01$ of Si_3N_4 was used

Similar experiments were performed with arsenic ($^{75}As^+$, 2 keV, $1 \cdot 10^{15}$ cm^{-2}) (4). At 1050°C the junction depth was not much affected by the variation of the O_2 concentrations between 0~1 ppm (505 Å) and 50000 ppm (540 Å). The sheet resistance was almost constant. So, there was no pronounced OED effect; rather an oxidation retarded diffusion was observed. From the literature is known that the contribution of self-interstitials to arsenic diffusion is equal or less than 30 % (9). As a consequence, for the tailoring of as-implanted arsenic profiles during the RTA we have to use, apart from the thermal budget (10), thermal nitridation since it has been established that this process injects vacancies into the bulk (11).

Jacob et al. (8) report that RTA annealing in N_2 at 1200°C to 1250°C resulted in an introduction of vacancies. However, annealings at and below 1100°C in pure nitrogen even for prolonged times did not result in measurable vacancy concentrations. Therefore the presented experiments were conducted in an ambient of NH_3 (reactive gas) and Ar (inert gas).

In Fig. 1 the dependence of the silicon-oxynitride thickness on the relative NH_3 concentration is displayed for bare silicon wafers which were annealed for 10 s at 1000°C, 1050°C and 1100°C. A monotonic increase of the silicon-oxynitride thickness with the NH_3 concentration is visible. However, the small thickness observed for 100 % Ar ambient (0~1 ppm NH_3) compared to the width of the native oxide layer reveals the effect of thermal etching.

Profile analysis

It is important to note that the prestabilization anneal at 650°C for 10 s did not change the as-implanted profile. In Fig. 2 an as-implanted profile together with an annealed one is displayed. The gaseous ambient was 100 % NH_3.

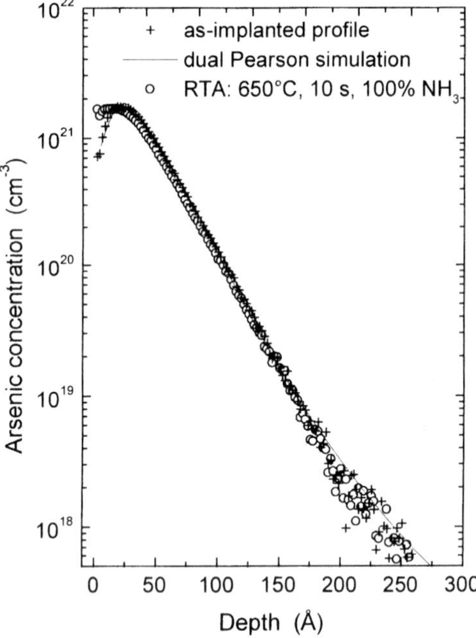

Figure 2: SIMS profile comparison of the As distribution after implantation to that after additional annealing (1 keV, $^{75}As^+$, $1 \cdot 10^{15}$ cm^{-2})

Even for this most reactive ambient there is no significant broadening of the profile after this low temperature process. Both implanted and annealed profiles are well described by a dual Pearson distribution.

In Fig. 3 As concentration depth profiles measured by SIMS after 1100°C diffusion / activation annealing are displayed together with the numerically simulated profiles. It is obvious, that the penetration depth increases monotonically i. e. (ignoring the nominally inert case) from 734 Å to 884 Å with NH_3 concentration running from 1000 ppm to 50000 ppm. Similar results were obtained for 1000°C and 1050°C. The shallowest penetration under nitridizing ambient at 1000°C is observed for 1000 ppm NH_3 and characterized by a depth of 354 Å. The corresponding depth at 1050°C is 492 Å.

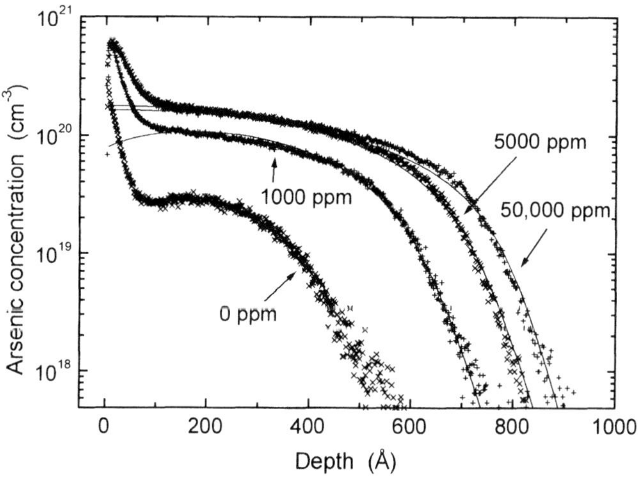

Figure 3: Arsenic SIMS profiles (x, +) after annealing for 10s at 1100°C for various concentrations of ammonia. The computer-simulated solid lines for nitridizing ambients show excellent agreement with the experimental data

Modeling of the As profiles was performed with the SSUPREM IV software package taking the as-implanted distribution (Fig. 2) as starting profile. For arsenic the local actual diffusivity D is given by

$$D = h\left(D_i^0 + D_i^-\left(\frac{n}{n_i}\right)\right) \qquad [1]$$

where D_i^0 denotes the As diffusivity via neutral point defects, i.e. Si vacancies (V^0) and Si self-interstitials (I^0), D_i^- the As diffusivity via singly negatively charged point defects (V^-, I^-), n the electron concentration which may exceed the intrinsic carrier density at the diffusion temperature $n_i(T)$, and h the electric-field enhancement factor which may approach the value 2 for high As concentrations. In the computer simulations all surface reactions with ambient gases were switched off so that vacancies and self-interstitials were maintained at their thermal equilibrium values. Only a possible outflux of arsenic atoms through the Si surface was taken into account via the surface transport coefficient v_{surf} (TRANS parameter in SSUPREM IV). The pile-up of As observed in the region up to 100 Å beneath the Si surface is most probably due to segregation to the Si/Si-oxynitride interface. For arsenic as well as phosphorous and antimony this is described in the literature. Segregation coefficients of approximately 10 are usually quoted (12). This pile-up is not taken into consideration in the simulations.

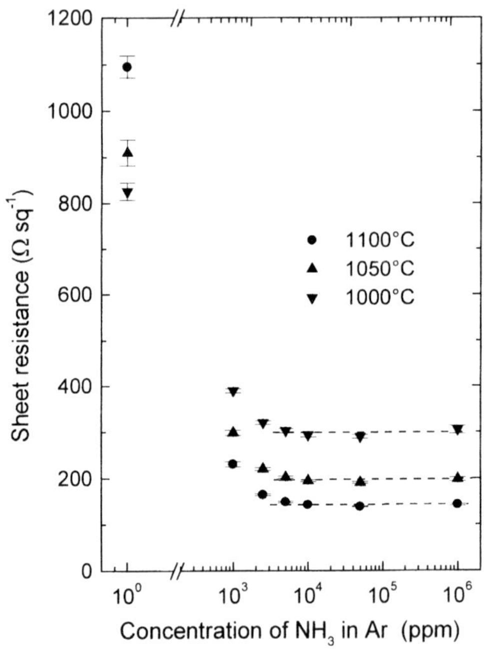

Figure 4: Sheet resistance versus ambient ammonia concentration as measured for the present As implants (1 keV, $1 \cdot 10^{15}$ cm^{-2}) annealed for 10 s at the indicated temperatures. The dashed lines represent saturation values

As illustrated in Fig. 3 for the 1100°C profiles, good or even excellent fits were obtained by optimizing D_i^0, D_i^- and v_{surf} being the only free parameters in the simulations. Moreover, the adjusted values of D_i^0 and D_i^- turned out to be constant within narrow error limits in the concentration range of 5000 ppm to 100 % of ammonia in Ar. In contrast, v_{surf} is orders of magnitude larger for 0~1 ppm NH_3 (100 % Ar) than for high NH_3 ambient concentrations. This reflects the substantial loss of As from the substrate for the lower NH_3 ambient concentrations as seen in Fig. 3, and hardly any loss for 100 % NH_3.

The same phenomenon is also apparent in Fig. 4, i. e. from the monotonic decrease of the sheet resistance with increasing NH_3 concentration which saturates at about 5000 ppm. The high sheet resistance for 0~1 ppm NH_3 is not only caused by extreme As loss but also by surface erosion due to thermal etching. Obviously, these unwanted effects make it difficult to obtain a satisfactory fit to the 0~1 ppm SIMS profile in Fig. 3.

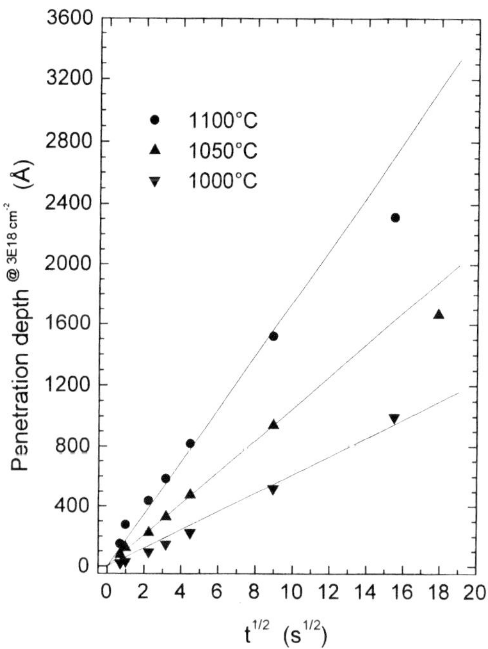

Figure 5: Penetration depth at $3 \cdot 10^{18}$ cm^{-3}, x_p, versus $\sqrt{t}$ after annealing for various times at three temperatures in 100 % NH_3 gaseous ambient. The solid lines represent theoretical predictions for a timely constant effective diffusivity

These findings imply that the increase of the As penetration depth with increasing NH_3 concentration shown in Fig. 3 is *not* caused by a *direct* nitridation-enhanced diffusion effect as resulting from vacancy injection. Rather the present results can be rationalized by a more *indirect* surface capping effect which correlates with the growing thickness of the $Si_xO_yN_z$ layer exhibited by Fig. 1. Concomitantly, the ever better prevention of As loss from the substrate leads to an increase of the As concentration in the plateau region beneath the surface. In turn, these increased concentrations give rise to higher electron densities and therefore to a stronger self-doping enhancement of the diffusivity (see Eq. [1]). In Fig. 5 the netto penetration depth, x_p, at $3 \cdot 10^{18}$ cm^{-3} versus the annealing time is displayed for the anneals with 100 % ammonia. In x_p a correction is made for the penetration depth of the as-implanted profile. There is a fair linear dependence up to a time of 100 s. For longer times x_p tends to be shallower than the extrapolation of the linear fit.

Diffusion coefficients

Fig. 6 reveals the intrinsic As diffusion coefficients that were obtained from computer modeling of the measured profiles, together with literature data (13). The present data are averages over NH_3 ambient concentrations from 5000 ppm to 100 % NH_3 for which no significant differences were observed. Not only the individual diffusivities D_i^0 and D_i^- are displayed in Fig. 6 but also their sum $D_i = D_i^0 + D_i^-$ which is the total As diffusivity under intrinsic conditions. Fig. 6 also includes D_i^- estimated from the time dependence of the penetration depth x_p shown in Fig. 5. This estimation is based on the relationship (see Eq. [1] with $h = 2$)

$$x_p \approx 2 \cdot \sqrt{D \cdot t} \approx 2 \cdot \sqrt{2 \cdot D_i^- \cdot \left(\frac{n}{n_i}\right) \cdot t} \qquad [2]$$

which approximately holds in high concentration regions $(D_i^-(n/n_i) >> D_i^0)$. It is seen that the two ways of calculating D_i^- lead to a reasonably good agreement for all temperatures.

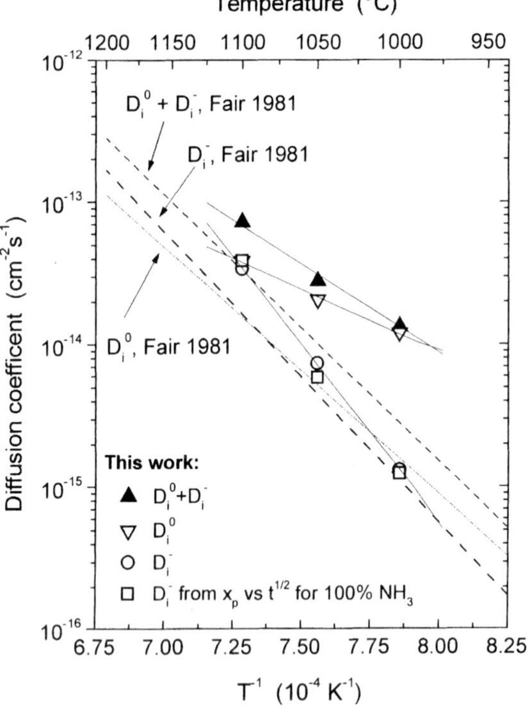

Figure 6: Intrinsic diffusion coefficients of As under NH_3 ambients, present work (symbols) compared to literature data for inert ambient diffusion (13) (dashed lines)

Comparing the total intrinsic diffusivity $D_i^0 + D_i^-$ under NH_3 ambient (this work) with that for inert conditions (13) we see a diffusion enhancement of a factor of 3.5 at 1000°C and of nearly 2 at 1100°C. This picture does not change much if we take other established data for intrinsic As diffusion from the literature (14), (15). Enhancements of similar magnitude have been found in earlier As diffusion experiments carried out in NH_3 ambient at 1100°C (9). This lends credit to the view that we are dealing in the present study with nitridation-enhanced diffusion.

However, some caution seems appropriate since residual implantation damage usually gives rise to strong transient enhanced diffusion (TED) effects in the case of B and P in Si. For As, several studies report pertinent TED effects (e.g.(16), (17)), whereas others explicitly state the absence of TED (e.g. (18), (19)). Downey and Jones (20) state that during the implantation of 1 keV $^{75}As^+$ at a dose of $1 \cdot 10^{15}$ cm^{-2} no extended defects or trapped interstitials are formed and that the amorphous layer is about 120 Å thick. Therefore, for the present implantation followed by recovery annealing at 650°C the TED effect is likely to be of minor importance. This conclusion is supported by the observation that x_p versus $\sqrt{t}$ holds for annealing times in the range 0-100 s.

CONCLUSIONS

Annealing of implanted $^{75}As^+$ (1 keV, $1 \cdot 10^{15}$ cm^{-2}) was performed in an RTP system for various times and temperatures under different concentrations of ammonia in argon. The nitridation of the surface leads to capping of the surface which influences the retained dose of the implanted ion and therefore indirectly enhances As penetration via an increased self-doping effect. Nevertheless from the simulation of the SIMS profiles under various ammonia concentrations also a direct nitridation-induced enhancement, probably due to vacancy injection, was found from the comparison with inert ambient diffusivities (13), (14), (15). Both influences on As diffusion are almost constant for ammonia concentrations in argon varying from 5000 ppm to 100 %. From the processing efficiency point of view 5000 ppm of ammonia seems to be an optimal choice.

ACKNOWLEDGMENTS

The authors would like to thank Dr. G. Roters for helpful discussions and critical reading of the manuscript. We are indebted to Dr. Z. Nenyei, the Applications Lab staff of STEAG AST Elektronik GmbH, the VARIAN IIS team, and co-authors of our previous publications.

REFERENCES

(1) F. Roozeboom in Advances in Rapid Thermal and Integrated Processing, edited by F. Roozeboom, NATO ASI Series E, Vol. 318, (Kluwer 1996)

(2) D. F. Downey, S. L. Darayanani, M. Meloni, K. Brown, S. B. Felch, B. S. Lee, S. D. Marcus, J. C. Gelpey, *Mat. Res. Soc. Symp. Proc.* **470**, 299 (1997)

(3) W. Lerch, M. Glück, N. A. Stolwijk, H. Walk, M. Schäfer, S. D. Marcus, D. F. Downey, J. W. Chow, H. Marquardt, *Mat. Res. Soc. Symp. Proc.* **525** 237 (1998)

(4) D. F. Downey, J. W. Chow, W. Lerch, J. Nieß, S. D. Marcus, *Mat. Res. Soc. Symp. Proc.* **525** 263(1998)

(5) Computer code SSUPREM IV developed at the Integrated Circuits Laboratory Stanford University by R. W. Dutton and J. D. Plummer

(6) ATHENA trademarked and commercially available process simulation framework developed by SILVACO International comprising SSUPREM IV

(7) J. F. Götzlich, K. Haberger, H. Ryssel, *Radiation Effects* **47** 203 (1980)

(8) M. Jacob, P. Pichler, H. Ryssel, R. Falster, M. Cornara, D.Gambaro, M. Olmo, M. Pagani, GADEST'97, Diffusion and Defect Data, Part B (*Solid State Phenomena*) Vols. 57-59 349 (1997)

(9) P. Fahey, G. Barbuscia, M. Moslehi, R. W. Dutton, *Appl. Phys. Lett.* **46**(8) 784 (1985)

(10) Z. Nenyei, C. Grunwald, W. Lerch, J. Nieß, D. F. Downey, R. Ostermeir, 6th International Conference on Advanced Thermal Processing of Semiconductors, RTP'98, Sep. 9-11, 1998; Kyoto, Jp

(11) P. M. Fahey, P. B. Griffin, J. D. Plummer, *Rev. Mod. Phys.* **61**(2) 289 (1989)

(12) S. M. Sze, VLSI Technology, 2nd Edition, (McGraw-Hill 1988), pp. 332-340

(13) R. B. Fair, in Impurity Doping, edited by F. F. Y. Wang (North Holland, Amsterdam) p. 315 (1981)

(14) D. A. Antoniadis, A. M. Lin, R. W. Dutton, *Appl. Phys. Lett.* **33** 1030 (1978)

(15) C. Hill, *Proc. Electrochem. Soc.* **81-5** 988 (1981)

(16) Y. Kim, H. Z. Massoud, R. B. Fair, *J. Electron. Mater* **18** 143 (1989)

(17) R. B. Fair, *J. Electrochem. Soc.* **137** 667 (1990)

(18) T. O. Sedgwick, T. O., A. E. Michel, S. A. Cohen, V. R. Deline, G. S. Oehrlein, *Appl. Phys. Lett.* **47** 949 (1985)

(19) H. Park and M. E. Law, *Appl. Phys. Lett.* **58** 733 (1991)

(20) D. F. Downey and K. S. Jones, presented at the 12th International Conference on Ion Implantation Technology, June 22-26, 1998, Kyoto Jp

DOPING AND ANNEALING REQUIREMENTS TO SATISFY THE 100 nm TECHNOLOGY NODE

Daniel F. Downey, Susan B. Felch, and Scott W. Falk
Varian Semiconductor Equipment Associates
35 Dory Road
Gloucester, MA 01930

The achievement of ultra-shallow source/drain extensions that satisfy the requirements for 100 nm technology requires not only reproducible, low energy, low contamination doping, but also well controlled, 'fast' rapid thermal anneals. This paper reviews the doping requirements for these junctions, providing specific examples for doping via ion implantation and also by Plasma Doping (PLAD). The effects of ramp-rates, time and temperature of anneal, and anneal ambient are discussed in detail to clarify the important contributions that these parameters play in the overall production of ultra-shallow junctions. It is shown that ramp-rates alone do not matter at higher energies where transient enhanced diffusion (TED) is known to be present. However, below an energy threshold, where no extended damage ever forms and when anneals are performed in a low ppm O_2 in N_2 ambient, the observed diffusion is totally dominated by thermal diffusion. It is under these conditions that ramp-rates do have an effect. Examples of anneals with ramp-rates up to 425°C/s are provided.

INTRODUCTION

This paper reviews combined doping (ion implant and Plasma Doping) and anneal processes required to achieve the 1997 National Technology Roadmap requirements at the 100 nm node for the source/drain extensions employing single implants (or doping steps) of either $^{11}B^+$, $^{49}BF_2$ (BF_3 for PLAD), or $^{75}As^+$ (for the n and p-extensions). The National Technology Roadmap predicts that at the 100 nm node, source/drain extension junction depths (x_j) will be between 20 nm to 40 nm (for a uniform channel concentration of 2-3e18/cm^3) with a minimum activated surface dopant concentration level of 1e20/cm^3. Many device manufacturers, however, are even more aggressive with respect to their desired dopant concentration levels and sheet resistance (R_s) values, demanding the lowest achievable R_s.

Several publications from our laboratory (1-10) demonstrate that low energy implants (down to 0.25 keV) in conjunction with annealing in low ppm concentrations of O_2 in N_2 (3-10) minimize oxidation enhanced diffusion (OED), and in the case of Boron eliminate/minimize boridation enhanced diffusion (BED) (11). BED has previously been considered to be a limiting factor in achieving shallow junctions with boron. Phosphorus implants have also been shown to behave similarly to B producing a phosphoridation enhanced diffusion effect (PED) (8). Under the appropriate anneal environment, employing low ppm levels of O_2 in N_2 during RTA, the PED effect can be similarly well controlled. In another communication from our laboratory (3), it has been shown that fast ramp-up and cool-down rates (240°C/s, and 87°C/s) for spike anneals at 1050°C, in a low ppm O_2 in N_2 ambient, provide highly activated ultra-shallow junctions at energies of 1.0 keV and above. Another publication (7) reported that, for implants below an energy dependent threshold where loops and extended-defects never formed and hence were not available to produce TED, the effective diffusion is dominated by thermal and concentration enhanced diffusion. In these cases slow, controlled, 1050°C spike anneals (50°C/s ramp-up and 87°C/s cool-down rates with a 750°C, 10s thermal stabilization step to optimize uniformity) in an ambient of 33 ppm O_2 in N_2 were performed. An additional paper (9) addresses fast ramp-rates and spike anneals for implants below the energy threshold where defects never occur. In this range all enhanced diffusion was eliminated (including TED), and the faster ramp-rates did yield shallower junctions because of the reduction in thermal diffusion (which is now not being dominated by the initial TED).

The challenge in producing ultra-shallow (20 nm to 40 nm) source/drain extensions is to do so simultaneously with a low sheet resistance. These (shallow junctions with low sheet resistance) requires balancing competing phenomena, since as the layer becomes shallower, the retained dose after anneal tends to decrease, the amount of dopant below the solid solubility limit decreases, and the electrical mobility increases from surface scattering as well as an increase in impurity scattering (from the increased dopant concentration). In addition to the above competing phenomena, anneal temperatures of 1000°C to 1050°C are required to achieve sufficient electrical activation to yield low sheet resistance. It is at these temperatures where the control of dopant diffusion is the most difficult. An economical solution to this problem is reviewed in this paper, where single implants (dopant step) of a well-controlled and shallow doping profile are obtained using high beam current, low contamination implants or PLAD at energies below the threshold where extended defects occur (7). After all enhanced diffusion effects are eliminated/minimized (by the sub-threshold energy implant and an RTA in low ppm O_2 in N_2), the effects of thermal diffusion can be minimized by faster ramp-rates, cool-down rates and "spike" anneals. Dopant loss during the anneal is minimized by the correct selection and control of ppm levels of O_2 in N_2.

EXPERIMENTAL

The implants reported in this review were performed on either a Varian VIISta-80 (12), or the VIISion-80LE (13). In the case of PLAD a research test stand was employed (14). 200 mm, <100>, n-type or p-type wafers with resistivities of 10-20 Ω-cm were used as the substrate material. Implants were performed at doses ranging from 2.5e14/cm^2 to 5e15/cm^2 with energies down to 0.25 keV and tilt angles of 0°. To eliminate the variability of native oxide thickness, all bare wafers were pre-stripped in a 40:1 H_2O:HF (49%) acid for 30s within 1hr of implant. After implant, the wafers were annealed using a STEAG AST SHS-2800∊ or SHS-3000 RTA system employing a controlled O_2 in N_2 ambient (4,5). After anneal, all wafers were probed on a Prometrix OmniMap RS35c four-point probe system following a BOE strip (1). Secondary ion mass spectrometry (SIMS) depth profiling was performed at Evans East Inc. using a Physical Electronics Φ 6600 quadrupole SIMS instrument (using the O_2 leak technique) (2). Select wafers were also analyzed by spreading resistance profilometry (SRP) (15), tapered groove profilometry (TGP) (16), transmission electron microscopy (TEM) (7) and/or the Hall Effect (1).

Ion implantation is the dominant technique for doping of silicon wafers. The dopant ions are created in a plasma in the ion source, then mass-analyzed through a magnet, accelerated to the desired energy in a beamline, and finally implanted in a scanned fashion into the wafer in a process chamber. The VIISion-80LE implanter is a batch, high current ion implanter that uses a dual Einzel lens and modified extraction optics for increased low energy beam currents. The VIISta-80 is a serial process, high current implanter which produces an ion beam that is a ribbon which is 50 mm wider than the implanted wafer. In this implanter, the wafer is scanned vertically through the beam. As is true for all semiconductor equipment used to manufacture devices, these implanters have been optimized to produce high wafer throughput, energy purity, high uniformity and excellent repeatablility, minimal contamination, and minimal charging damage.

Plasma doping (PLAD) is regarded as an attractive, lower cost, alternative to conventional beamline implantation techniques. In the PLAD process, the silicon wafer is placed directly in a plasma containing the desired dopant ions (17). The wafer is then pulse-biased to a negative potential to accelerate positive ions toward and into the silicon surface. This technique differs from conventional ion implantation in that the wafer is placed directly in the plasma source and the accelerating bias is applied to the wafer. The other parts of a conventional implanter (acceleration columns, mass-analyzing magnets, ion optics, and scanning systems) are missing, resulting in a much simpler, smaller, and less costly tool. For the data reported in this paper, a BF_3 source gas and wafer biases of –0.14 kV to -5.0 kV were used to implant boron into 200 mm wafers to form ultra-shallow p^+-n junctions (20). Typical bias pulses were 20 μs long with a pulse repetition rate of 500 Hz. For a description of the PLAD research test stand, see Ref. (14).

The as-implanted junction depth for plasma-doped wafers is determined by the wafer bias. As Fig. 1 shows, the junction depth decreases as the wafer bias decreases. For this data, a BF_3 plasma was used to implant n-type, (100) Si wafers with 10-20 Ω-cm resistivity (20). The approximate doses for all implants were 5×10^{14} B cm^{-2}. The junction depth (x_j) was determined from SIMS profiles as the depth where the boron concentration is 1×10^{18} B cm^{-3}. This value was used as a measure of x_j because it approximates the well doping level under the source/drain region of CMOS devices. The data is well represented by a linear function for biases ranging from - 0.14 kV to - 5 kV, with a least-squares fit producing good correlation. This fit suggests that the as-implanted junction depth decreases 12.5 nm for every 1 kV reduction in bias. The as-implanted junction depth for – 0.14 kV is only 7.3 nm, which should result in annealed junctions that are shallow enough to meet the requirements of the 100 nm technology node.

RESULTS AND DISCUSSION

Effects of Low ppm O_2 in N_2 RTA Ambient

The effects of anneals in ambients with low ppm levels of O_2 in N_2 have been extensively reported by us for ^{11}B, $^{49}BF_2$, ^{75}As (4,5) and ^{31}P (8). In all of these cases an oxidation enhanced diffusion and other oxygen related diffusion effects were observed and eliminated by a controlled reduction in the unintentional background of O_2 in the anneal. Small controlled amounts of O_2, however, were required to prevent surface etching and severe out-diffusion of dopant (5). For 1e15/cm^2, 1.0 keV ^{11}B and 5.0 keV $^{49}BF_2$, a concentration of about 33 ppm of O_2 in N_2 optimized the anneal. For the case of ^{11}B annealed at 1050°C 10s, junctions 32 nm shallower than previously reported were obtained at this O_2 concentration level with improved R_s and across-the-wafer uniformity. At 1000 ppm the junction depth increased 32 nm to those levels previously reported elsewhere (11). Figures 2 and 3 illustrate the effects of oxygen on lower energy implants. In Figure 2 the effect of varying the O_2 ppm levels in N_2 for an RTA anneal at 1000°C, 10s for a 0.25 keV $^{11}B^+$ at 2.5e14/cm^2 dose is shown. In this figure it is observed that the O_2 content has a significant effect on the annealed junction depth (x_j). The shallowest junction occurs for a 33 ppm O_2 in N_2 ambient. The 0-1 ppm case, as mentioned previously, results in an increased R_s from surface etching by SiO formation and dopant evaporation [5]. The result of this etching process also yields deeper junctions than observed at 33 ppm, perhaps due to interstitial injection. Similar effects of the O_2 content during anneal are seen for a 1.1 keV, 5e14/cm^2 BF_2 implant (Figure 3). In Figure 3 it can be seen that 33 ppm O_2 in N_2 still yields the shallowest junction. At 0-1 ppm of O_2 again surface etching and out-diffusion is observed, and at 1000 ppm, dopant loss into the RTA grown oxide becomes significant (6). For the case of the BF_2, it is believed that this reduction in x_j is a result of decreasing oxidation enhanced diffusion (OED), while in the case of ^{11}B the lowering of the O_2 level not only decreases OED but now also prohibits the formation of an SiB_x layer which causes BED. This SiB_x layer does not appear in BF_2 ion implanted or PLAD doped (21) wafers (which contain F), most likely because of the F chemistry with B and Si.

Defect Free Energy Threshold at 1e15/cm^2 doses

A comprehensive investigation was conducted (7) to determine for the four species of primary doping interest (^{11}B, $^{49}BF_2$, ^{31}P and ^{75}As) the energy threshold (at 1e15/cm^2) below which extended defects are never observed to form. To determine this threshold, 750°C 15 min furnace anneals (and in some cases lower T anneals were also studied) were performed for a variety of implant energies. Plan-view TEM was employed to quantify the extended defects. For the case of $^{49}BF_2$, this threshold was determined to be between 3.0 and 3.5 keV. Above this threshold, type II loops formed, whose size and density increased with implant energy. For the ^{11}B implants the threshold was between 0.25 keV and 0.50 keV. Above this threshold, type I loops formed, whose size and density were about the same at 0.5 and 1.0 keV. At 2.0 keV, however, wide 30 nm long {311}s were also observed. For the case of As^+ this threshold was between 3.75 and 4.0 keV. Above this threshold, type II loops formed whose density increased with increasing energy, at least over this energy range. These loops were extremely small, being only 3-4 nm in diameter. For the P^+ implants (which were amorphizing), the threshold was between 1.0 and 2.0 keV. Above this threshold, type II loops formed, whose density increased with implant energy. It was also demonstrated in that work (7), that below the energy threshold, TED was not observed, and that when RTA was performed in a low O_2 ppm in N_2 ambient, all enhanced diffusion effects were eliminated and that the diffusivity factor was smaller than or equal to 1.0 (the case of pure thermal diffusion).

Effects of Ramp-up Rates, 'Spike' Anneals and Cool-down Rates

The effects of ramp-up rates, 'spike' anneals and cool-down rates have been extensively researched for the four primary dopant species at energies above (3, 8) and below the energy threshold for extended defect formation (8, 9). Above the energy threshold, cool-down rates and 'spike' anneals did have an effect which could be explained by the reduction in thermal diffusion (3). In this regime, however, ramp-rates did not matter since the initial diffusion was dominated by TED. Below the energy threshold, TED was not observed and increased ramp-rates did indeed have an effect of yielding shallower junctions as a result of the reduction in thermal diffusion. This is clearly illustrated (for the case of ^{11}B and $^{49}BF_2$) in Figure 4, which is a SIMS overlay of 1e15/cm^2 ^{11}B at 1.0, 0.5 and 0.25 keV with a 1.1 keV BF_2 annealed with two different ramp-rates, 50°C/s and 425°C/s. For the 1.0 keV ^{11}B implant the SIMS profiles for the two different ramp-rates overlay each other, while for the other three cases a reduction in junction depth with increased ramp-rate is observed. Figure 5 clearly illustrates that even with fast 'spike' anneals that the O_2 ambient concentration is still critical. In this case of a 1e15/cm^2, 0.25keV ^{11}B implant, a 2.7 nm reduction in x_j (at 3e18/cm^3) with a more abrupt junction was observed at 33 ppm vs. 1000 ppm. Similar reductions in x_j were also observed for $^{49}BF_2$, ^{31}P, and ^{75}As as a function of the O_2 content during fast 'spike' anneals (9, 18).

Diffusion in Equivalent Energy ^{11}B and $^{49}BF_2$ Above and Below Defect Energy Theshold

In Table I, the effects of extended defects and the presence of fluorine (for the case of BF_2) on the final annealed junction depths, are compared for equivalent effective energies of 1e15/cm^2 ^{11}B (2.0, 1.0 and 0.25 keV) and $^{49}BF_2$ (8.9, 5.0 and 1.1 keV) implants. The anneals were performed in a 33 ppm O_2 in N_2 ambient to eliminate any enhanced oxygen related diffusion effects. In this Table the junction depth differences calculated by thermal diffusion theory between the "as-implanted" profiles and the fast or slow 1050°C spike anneal is compared to the actual SIMS measured differences. The calculated Boron diffusivity factor for the measured diffusion is also listed in Table I. In the thermal diffusion calculation no concentration enhanced diffusion, TED, OED or any other oxygen related diffusion effects were assumed. These differences are then compared to the residual fluorine level, the initial defect densities after a 750°C furnace baseline anneal, and the calculated trapped interstitial concentrations (after the furnace baseline). Referring to Table I it can be seen that the 8.9 keV $^{49}BF_2$ annealed wafer has a 6.9 nm enhancement over that predicted by thermal diffusion, while the 2.0 keV $^{11}B^+$ sample has a 28.4 nm difference above that predicted by thermal diffusion. Since both implants had similar amounts of extended damage and trapped interstitials, the reduction in the Boron diffusion in the $^{49}BF_2$ case is attributed to the chemical effect of the residual F in the $^{49}BF_2$ reducing the TED (7, 19). The diffusivity enhancement factor for the ^{11}B implant was 12 while for $^{49}BF_2$ it was 2.5. As the energies of the implants were lowered to 1.0 keV for ^{11}B and 5.0 keV for $^{49}BF_2$, less extended defects were formed and the overall observed TED was reduced. Again the residual F in the $^{49}BF_2$ impeded the TED yielding a diffusivity enhancement factor of 2.3 for the $^{49}BF_2$ case vs. 5.0 for the ^{11}B implant. At 2.2 keV $^{49}BF_2$, an energy below the defect threshold, the observed diffusivity enhancement factor was 1.0, that predicted by thermal diffusion alone. In this case there is no enhanced diffusion since all oxygen related enhancements are eliminated by the low O_2 ppm ambient and the absence of trapped interstitials generating TED. As the energies of the implants were lowered even further to 0.25 keV for ^{11}B and 1.1 keV for $^{49}BF_2$, no extended defects were formed and the overall observed TED was eliminated in both cases. The diffusivity enhancement factor was less than 1.0 in each case indicating that all diffusion was thermal. In summary it can be concluded that the amount of diffusion in ^{11}B implants is greatly reduced from 40.0 nm at 2.0 keV to 7.0 nm at 0.25 keV and follows the trends of the trapped interstitial population. Similarly $^{49}BF_2$ samples diffuse less with decreasing energy but with less TED (than ^{11}B) because of the diffusion retardation effect of the fluorine. The effects of fluorine play an increasingly larger role above the threshold energy of 3.0 keV.

1.0 kV PLAD Results

Figure 6 is a SIMS overlay of three 1.0 kV, 5e14/cm^2 PLAD samples rapid annealed for various times and temperatures in a 33 ppm O_2 in N_2 ambient. For these anneal conditions the range of values obtained from 950°C, 30s to 1050°C, 10s is 412.5 Ohms/sq. with an x_j of 43 nm (at 1e18/cm^3) to 191.3 Ohms/sq. at 76 nm. Similar to the above examples for ion implantation it is expected that use of lower PLAD biases and fast "spike" anneals will yield even shallower junctions.

Source/drain Extensions for 100nm Node

Combining the above doping and post-doping process solutions, the requirements for source/drain extensions for the 100 nm technology node are satisfied. Figure 7 is a SIMS overlay of 0.25 keV $^{11}B^+$ implants at doses of 2.5e14, 8e14 and 1.4e15/cm^2 annealed at 1050°C with the slow spike anneal. These profiles are plotted with respect to the 100 nm junction depth requirement at 2-3e18/cm^3 bulk concentrations. All of these implants have active dopant levels (determined by SRP) above 1e20/cm^3 as required by the NTRS (10). It is observed that the choice of dose provides various ranges of R_s and x_j values. At 3e18/cm^3 concentration levels, the range is from an R_s of 1141 Ohms/sq. and x_j of 24.6 nm to an R_s of 434 Ohms/sq. and an x_j of 41 nm. Figure 8 is a SIMS overlay of 1.1 keV $^{49}BF_2$ implants at doses of 5e14/cm^2, 8e14/cm^2 and a 2.2keV implant at 8e14/cm^2 annealed at 1050°C with the slow spike anneal. These profiles are again plotted with respect to the 100 nm junction depth requirement at 2-3e18/cm^3 bulk concentrations. All of these implants also have active dopant levels above 1e20/cm^3 (10). It is again observed that the choice of dose or energy yields various ranges of R_s and x_j values. At 3e18/cm^3 concentration levels, the range is from an R_s of 1319 Ohms/sq. and x_j of 20.4 nm to an R_s of 467 Ohms/sq. and an x_j of 34.7 nm. In the above examples it can be seen that the 100 nm technology requirements are achieved for both $^{11}B^+$ and $^{49}BF_2^+$. To achieve the optimum results for the 180 nm to the 130 nm nodes requires either higher doses and/or slightly higher implant energies. To achieve even shallower junctions than shown above, fast spike anneals (ramp-rates of 425°C/s) can be used, as discussed earlier in this section. Similar results are also obtained for 1.0 to 2.0 keV, 1e15/cm^2 $^{75}As^+$ implants (10). For the 2.0 keV implant, the R_s is 303 Ohms/sq. with an x_j of 38.4 nm; and for the 1.0 keV implant the R_s is 423 Ohms/sq. with an x_j of 28.5 nm. Both of these implants satisfy the 100 nm technology requirements and have active surface dopant concentrations greater than 1e20/cm^3 as determined by SRP (10).

SUMMARY

Employing low energy implants or PLAD (below the extended defect producing, energy threshold) and controlled low ppm oxygen ambient conditions, the effects of OED, TED, PED and BED can be eliminated/minimized. Once these enhancing diffusion phenomena are controlled, thermal diffusion conditions become dominant, and can be minimized by "fast" RTA spike anneals. Low sheet resistance is obtained by anneals at 1000°C to 1050°C, where the electrical activation is high, and by anneals in a low ppm O_2 in N_2 ambient which minimizes out-diffusion and dopant loss into an RTA oxide layer. For sub-keV $^{11}B^+$ and 1.1 to 2.2 keV $^{49}BF_2^+$ in controlled oxygen ambient conditions (33 ppm) with spike anneals, the requirements outlined by the NTRS roadmap for p-extension implants at the 100 nm technology node have been achieved. In order to satisfy the n-extension requirements described by the NTRS roadmap, lower energy (below 2 keV) $^{75}As^+$ can be employed.

ACKNOWLEDGEMENT

The authors acknowledge Carl Osburn of North Carolina State University for many helpful discussions; Si Prussin of UCLA for the TGP measurements; Dave Look of Wright State University for the Hall Measurements; Charles Magee and Bill Harrington of Evans East Inc. for the SIMS work; Kevin Jones for TEM analysis; Jinning Liu, Matthew Goeckner, and Adam Bertuch of Varian for their assistance; the Varian VIP and Application Lab staff for pre and post-processing help; and Steven Marcus, Wilfried Lerch and the STEAG-AST applications staff.

REFERENCES

1. D.F. Downey, C.M. Osburn, S. Marcus; Solid State Technology, p.71, Dec. 1997
2. C. Magee, J. Shallenberger, M. Denker, D.F. Downey, M. Meloni, S. Cloherty, S.B. Felch, B.S. Lee, Proc. Fourth Intl. Workshop on Ultrashallow Junctions, p.8.1 (April 1997).
3. D. F. Downey, S.D. Marcus and J.W Chow, "Optimization of RTP Parameters to Produce Ultra-shallow, Highly Activated B^+, BF_2^+ and As^+ Ion Implanted Junctions", presented at the TMS conference Feb. 1998, published in J. of El. Mat., vol. 27, No. 12, 1998, pp. 1296-1314.
4. D. F. Downey; US patent pending application, "Methods for forming shallow junctions in semiconductor wafers using controlled, low level oxygen ambients during annealing", 1998
5. D.F. Downey, J.W. Chow, W. Lerch, J. Niess, S.D. Marcus, "The Effects of Small Concentrations of Oxygen in RTP Annealing of Low Energy Boron, BF_2 and Arsenic

Ion Implants", presented at MRS conference April 1998, published in those Proceedings.
6. D.F. Downey, D. Brown, J.J. Cummings, A. Bertuch, E. Ishida, C.H. Ng; "Advanced Processing Technique to Minimize Enhanced Diffusion in Sub-keV Boron Implants"; presented at IIT '98, Kyoto, Japan; June 1998. To be published in the Proceedings for that conference.
7. D.F. Downey, K.S. Jones; "The Role of Extended Defects on the Formation of Ultra-Shallow Junctions in Ion Implanted $^{11}B^{+}$, $^{49}BF_2^{+}$, $^{75}As^{+}$ and $^{31}P^{+}$; presented at IIT '98, Kyoto, Japan; June 1998. To be published in the Proceedings for that conference.
8. J.W Chow, D.F. Downey and S. D. Marcus; Rapid Thermal Annealing of Low Energy Phosphorus Implants; Presented at RTP'98, Kyoto, Japan, Sept. 1998 and published in those Proceedings.
9. D. F. Downey, S.W. Falk, A.F. Bertuch and S.D. Marcus , "Effects of "Fast" Rapid Thermal Anneals on Sub-keV Boron and BF_2 ion implants" presented at the TMS conference March 1999, to be published in J. of El. Mat., Dec. 1999.
10. A.F. Bertuch, Z. Zhao, D.F. Downey and S.W. Falk, "Process Advantages to Achieve Ultra-shallow Junctions for 0.10 micron Technology Requirements", presented at The App. of Accelerators in Res. & Ind. Conference, Denton Texas, Nov. 1998 and published in those Proceedings.
11. A. Agarwal, D.J. Eaglesham, H.J. Gossman, L. Pelaz, S.B. Herner, D.C. Jacobson, T.E. Haynes, Y. Erokhin, R.Simonton; Proc.of IEDM97, p.467.
12. G. Angel, E. Bell, D. Brown, J. Buff, J. Cummings, W. Edwards, C. McKenna, S. Radovanov and N. White; "A Novel Beam Line for Sub-keV Implants with Reduced Energy Contamination"; presented at IIT '98, Kyoto, Japan; June 1998. To be published in the Proceedings for that conference.
13. G. Gammel, D. Brennan, E. Rushton, P. Sullivan and Z. Zhao; "VIISion PLUS Performance Improvements", presented at IIT '98, Kyoto, Japan; June 1998. To be published in the Proceedings for that conference.
14. T. Sheng, S. B. Felch, and C. B. Cooper III, "Characterization of a plasma doping system for semiconductor device fabrication", J. of Vac. Sci. & Tech., vol. B12, pp. 969-972, March/April 1994.
15. W. Harrington, C. Magee, M. Pawlik, D.F. Downey, C.M. Osburn, S.B. Felch, Proc. Fourth Intl. Workshop on Ultrashallow Junctions, p.7.1 (April 1997).
16. S. Prussin, C.A. Bil, D.F. Downey, M. Meloni, C.M. Osburn, "Characterization of Ultr-shallow Junctions with Tapered Groove Profilometry and Other Techniques", presented at NIST, 1998 Int. Conf on Char. & Metr. for ULSI Tech., March, 1998, to be published in those Proceedings.
17. S. B. Felch, T. Sheng and C.B. Cooper III, "Characterization of a plasma doping system", in Ion Implant Technology-92; Elsevier Science Pub, 1993, pp. 687-690.
18. D.F. Downey, S.W. Falk, A. F. Bertuch and J. Liu; "The effects of fast-ramp rates and spike anneals on ultra-low energy 11B, $^{49}BF_2$, ^{75}As and ^{31}P implants" to be presented at RTP'99 in Colorado; Sept 1999 and published in those proceedings.
19. D. F. Downey, J. W. Chow, E. Ishida and K.S. Jones, APL Vol.73, no.9 Aug. 31, 1998, pp. 1263-1265.
20. M.J. Goeckner, S.B. Felch, Z. Fang, D. Lenoble, J. Galvier, A. Grouillet, G.C-F. Yeap, D. Bang, and M-R. Lin, "Plasma Doping for Shallow Junctions" submitted to J Vac. Sci. Technol. A (1999).

21. D. Lenoble, M.J. Goeckner, S.B. Felch, Z. Fang, J. Galvier and A. Grouillet, "Evaluation of Plasma Doping for sub-0.18 μm Devices," presented at IIT '98, Kyoto, Japan; June 1998. To be published in the Proceedings for that conference.]

Table I.

Change in Junction Depth between "as-implanted" and 1050°C spike anneals in 33 ppm O_2 in N_2 ambients for ^{11}B and $^{49}BF_2$ above and below the energy threshold for extended defect formation

	2 keV, B	8.9 keV, BF_2	1 keV, B	5 keV, BF_2	2.2 keV, BF_2	*0.25 keV, B	*1.1 keV, BF_2
ΔX_j (Thermal Diffusion)	11.6nm	11.6nm	11.6nm	11.6nm	11.6nm	16.2nm	16.2nm
ΔX_j at 1e17/cm^2	40.0 nm	18.5nm	26.0nm	17.8nm	13.0nm	7.0nm	10.0nm
Boron Diffusivity Enhancement Factor	12	2.5	5	2.3	1.0	<1.0	<1.0
Residual Fluorine Levels	0	18.30%	0	7.90%	0.21%	0	<0.1% Surface Only
Defect type, size & Conc. 750°C, 15 Min.	Type I {311} 30nm 1.3e10 Loops 10nm 1.5e10	Type II 16nm 3.1e10	Type I 10nm 2.2e10	Type II 16nm 2e10	None	None	None
Trapped Interstitials from 750°C, 15 Min.	5e13	1e14	3e13	6.5e13	None	None	None

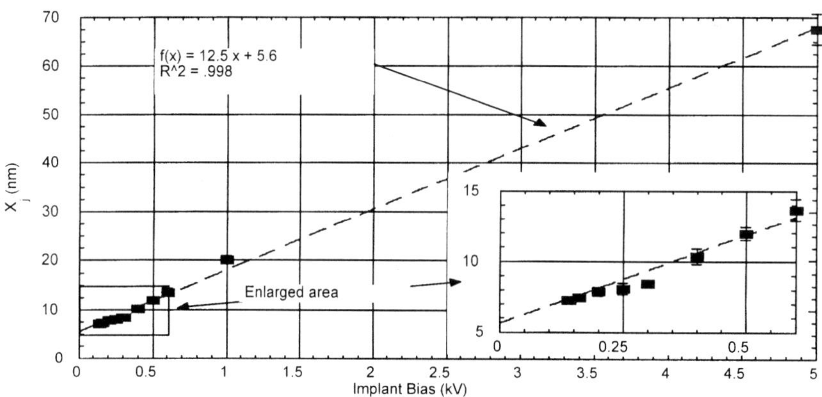

Figure 1. "as-implanted" junction depths for plasma-doped wafers from 0.15 to 5.0 kV.

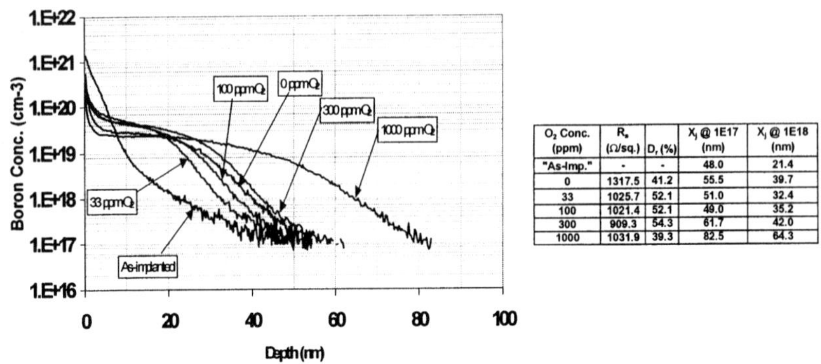

O_2 Conc. (ppm)	R_s (Ω/sq.)	D_r (%)	X_j @ 1E17 (nm)	X_j @ 1E18 (nm)
"As-Imp."	-	-	48.0	21.4
0	1317.5	41.2	55.5	39.7
33	1025.7	52.1	51.0	32.4
100	1021.4	52.1	49.0	35.2
300	909.3	54.3	61.7	42.0
1000	1031.9	39.3	82.5	64.3

Figure 2. The effects of varying the O_2 ppm level in N_2 anneal ambient for an ^{11}B, 0.25 keV, 2.5e14/cm^2 implant annealed at 1000°C for 10s.

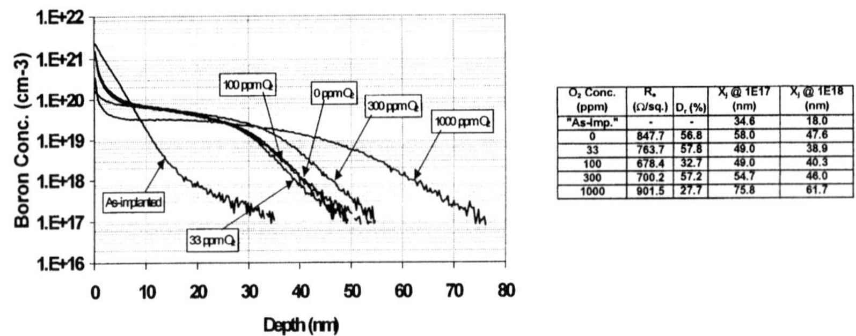

O_2 Conc. (ppm)	R_s (Ω/sq.)	D_r (%)	X_j @ 1E17 (nm)	X_j @ 1E18 (nm)
"As-Imp."	-	-	34.6	18.0
0	847.7	56.8	58.0	47.6
33	763.7	57.8	49.0	38.9
100	678.4	32.7	49.0	40.3
300	700.2	57.2	54.7	46.0
1000	901.5	27.7	75.8	61.7

Figure 3. The effects of varying the O_2 ppm level in N_2 anneal ambient for an $^{49}BF_2$, 1.1 keV, 5.0e14/cm^2 implant annealed at 1000°C for 1 0s.

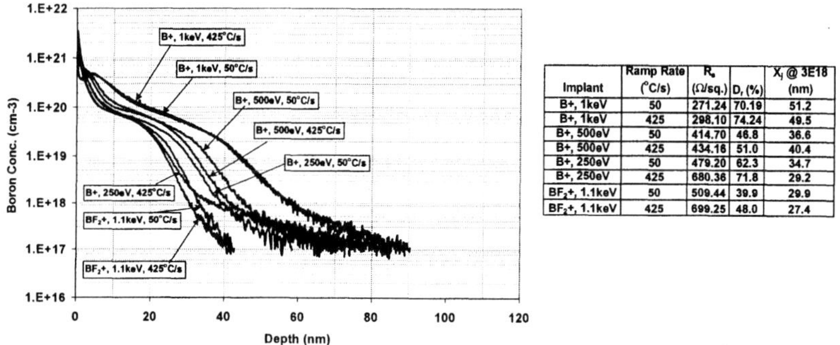

Implant	Ramp Rate (°C/s)	R_s (Ω/sq.)	D_r (%)	X_j @ 3E18 (nm)
B+, 1keV	50	271.24	70.19	51.2
B+, 1keV	425	298.10	74.24	49.5
B+, 500eV	50	414.70	46.8	36.6
B+, 500eV	425	434.16	51.0	40.4
B+, 250eV	50	479.20	62.3	34.7
B+, 250eV	425	680.36	71.8	29.2
BF_2+, 1.1keV	50	509.44	39.9	29.9
BF_2+, 1.1keV	425	699.25	48.0	27.4

Figure 4. SIMS overlay of 1e15/cm² ^{11}B implants at 1.0, 0.5 and 0.25 keV with a 1.1 keV BF_2 implant. All wafers annealed with two different ramp-rates, 50°C/s and 425°C/s for a 1050°C spike anneal in 33 ppm O_2 in N_2 ambient

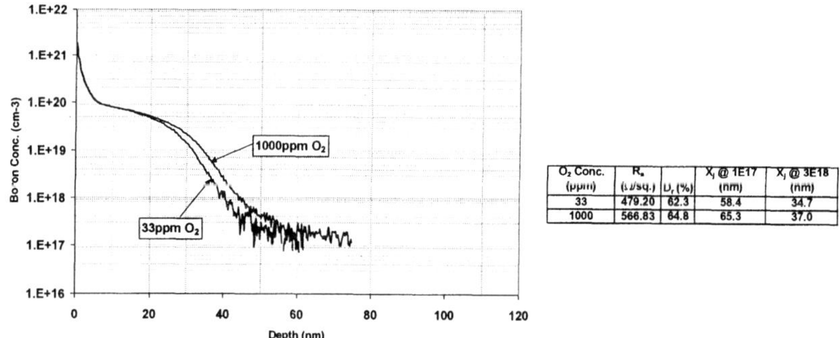

O_2 Conc. (ppm)	R_s (Ω/sq.)	D_r (%)	X_j @ 1E17 (nm)	X_j @ 3E18 (nm)
33	479.20	62.3	58.4	34.7
1000	566.83	64.8	65.3	37.0

Figure 5. The effects of varying the O_2 ppm level in N_2 anneal ambient from 1000 ppm to 33 ppm for an ^{11}B, 0.25 keV, 5.0e14/cm² implant annealed with a "fast" spike

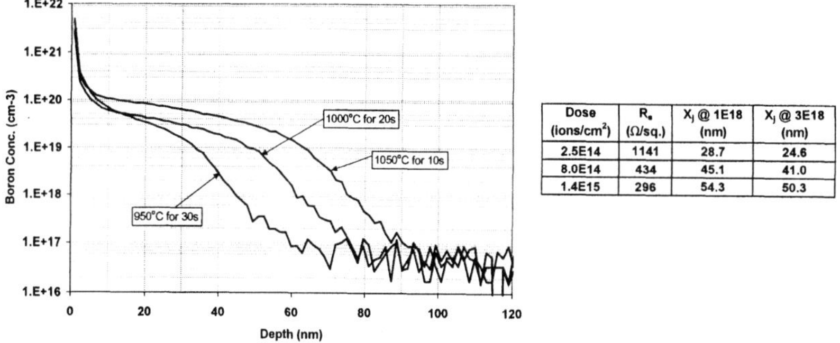

Dose (ions/cm²)	R_s (Ω/sq.)	X_j @ 1E18 (nm)	X_j @ 3E18 (nm)
2.5E14	1141	28.7	24.6
8.0E14	434	45.1	41.0
1.4E15	296	54.3	50.3

Figure 6. SIMS overlay of three 1.0 kV, 5e14/cm² PLAD samples rapid annealed for various times and temperatures in a 33 ppm O_2 in N_2 ambient.

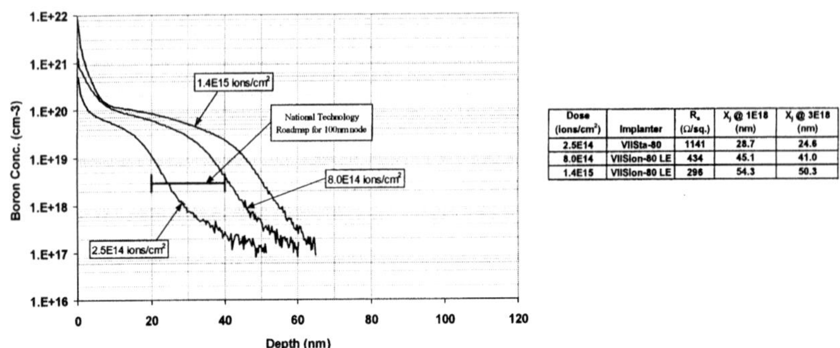

Dose (Ions/cm²)	Implanter	R_s (Ω/sq.)	X_j @ 1E18 (nm)	X_j @ 3E18 (nm)
2.5E14	VIISta-80	1141	28.7	24.6
8.0E14	VIISion-80 LE	434	45.1	41.0
1.4E15	VIISion-80 LE	296	54.3	50.3

Figure 7. 100 nm technology goals compared with various SIMS pro files for $^{11}B^{+}$ 0.25keV implants spike annealed at 1050C in 33 ppm O_2 + N_2 abient with a 50°C/s ramp rate, a 750°C-10s stabilization step and a 87°C/s cool down rate.

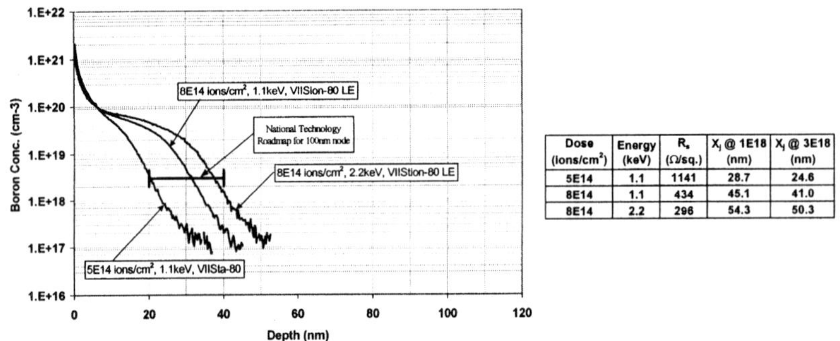

Dose (Ions/cm²)	Energy (keV)	R_s (Ω/sq.)	X_j @ 1E18 (nm)	X_j @ 3E18 (nm)
5E14	1.1	1141	28.7	24.6
8E14	1.1	434	45.1	41.0
8E14	2.2	296	54.3	50.3

Figure 8. 100 nm technology goals compared with various SIMS profiles for $^{49}BF_2^{+}$ implants spike annealed at 1050°C in 33 ppm O_2 + N_2 ambient with a 50°C/s ramp rate, a 750°C-10s stabilization step and a 87°C/s cool down rate.

DIRECT CORRELATION BETWEEN DEFECTS AND THERMAL STRESS IN RAPID THERMAL PROCESSING

V. Parihar, S. Venkataraman, R. Singh, K. F.Poole
Holcombe Department of Electrical and Computer Engineering
Center for Silicon Nanoelectronics
Clemson, SC 29634-0915, USA
and
R.P.S.Thakur
STEAG RTP Systems
San Jose, CA 95134-2000, USA

For the continued progress of silicon integrated circuit manufacturing, process simplification and defect reduction have become more important than ever. The use of thermal stress as an in-situ process control parameter can provide processed materials with built-in reliability. In this paper, we have studied diffusion as a front-end process and the deposition of low-dielectric constant material as a back-end process. Rapid Thermal Processing of both front -nd and back-end processes show a direct one-to-one correlation between defect density and the thermal stress.

INTRODUCTION

Process induced defects have a dramatic impact on the performance and yield as the wafer size increases and the device size decreases and this poses a critical challenge in sub-micron integrated circuit manufacturing. Defect reduction, isolation and control becomes more evident as the wafer size increases. As feature size is reduced, tolerances for the systematic defects will be smaller.. The defect size tolerance decreases as feature size decreases. As shown in the Fig. 1, extracting useful information out of huge amount of exponentially increasing data for integrated circuits with reducing dimensions will become very challenging. Therefore, we need to identify parameters that can be used for defect deduction, isolation and control. New advanced process controllers based on in-situ runtime approach are being developed to reduce process induced defects (1). This control technique involves collection, classification, processing, and decision making and feeding back of corrective measures. This complicated scenario will possibly solve the need of 0.07 μm device era. There is an immediate need for lower defects processing technique to obtain improved device performance and yield. Thermal processing is an integral part of a number of processing steps used in semiconductor manufacturing. We have studied the effect of thermal stress on the process-induced defects by employing both rapid thermal and rapid photothermal processing, RPP (2). Rapid photothermal processing is a novel processing technique that exploits the quantum effects associated with the ultraviolet (UV) and vacuum ultra violet (VUV) photons in the lower wavelength region of the spectrum.

The objective of this paper is to demonstrate that there is a direct correlation between defects and thermal stress both in front and back-end processing. In the front-end case we have employed the bulk minority carrier lifetime to establish a correlation between defects and thermal stress and in the back-end case we have correlated the leakage current of a low dielectric constant material in metal-insulator-metal (MIM) structure with thermal stress. In the case of front-end processing we have performed silicon diffusion. The effect of VUV photons has been analyzed by statistically designed experiments. In the case of back-end processing a direct liquid injection assisted CVD technique (3) has been used to deposit Teflon AF films. The effect of using VUV photons during the annealing of a fabricated Al/Teflon AF/Al -structure has been studied. Leakage current has been employed as the indicator of defects and the thermal stress obtained from the samples has been related to the leakage currents measured.

EXPERIMENTAL DETAILS

Czochralski (CZ) single-side polished p-type <100> silicon wafers with a resistivity of 1-10 Ωcm and a diameter of 2 inches were used in our diffusion study. The experiments have been performed with a home-built programmable rapid thermal processing system. We have used dedicated quartz reactors for different materials in order to avoid cross contamination. In lamp configuration used for these experiments, tungsten halogen lamps (THL) are placed on the polished (front) side of the wafer. The mercury lamps (source of UV photons) are placed directly on the unpolished (back) side of the wafer. Reflectors have not been used with the lamp banks in both the cases. For using vacuum ultra violet (VUV) photons, we designed an experiment where hot plate provided thermal energy and VUV lamp was placed very close to the hot plate facing the processing side of the wafer. During VUV experiments, system was kept in a non-enclosed condition and the experiments were carried out in a non-clean room environment.

The minority carrier lifetime of each sample was measured before processing. The samples are doped with a spin on source of phosphosilicite glass film (PSG) of density $10^{21}/cm^3$. Various lamp configurations discussed above were used to study the diffusion phenomena and minority carrier lifetime. After diffusion, the glass layer is removed and this process is repeated until the polished side of the sample becomes hygroscopic. This is followed by the sheet resistivity and minority carrier lifetime measurements. For blank wafers involving no diffusion phenomena are subjected to identical thermal cycles, as were used with the wafers used in the diffusion study. The initial lifetime is used as the reference for both samples.

The samples are subjected to diffusion with the following thermal cycle. The sample is heated to 400 °C in 30 s and is allowed to stay 400 °C for 30 s. Using a ramp rate of 10 °C/s., the wafer is heated to 870 ^{0}C and the hold time is 4 min The cooling cycle includes a cooling rate of 10 °C/s. up to 570 ^{o}C and below 570 °C natural cooling was used. In another thermal cycle for the VUV case, the temperature employed is 650 °C and the sample are exposed to the VUV photons for 5 min. The hotplate temperature of 650 °C reaches in 45 min and at this stage the VUV lamp is switched on. The sample and the hot-plate system are allowed to cool naturally.

In each case, the sheet resistivity of the samples is measured with a four-probe arrangement. The lifetime is measured by the photoconductive decay (PCD) technique The emitter layer is removed by wet etching in order to measure the bulk carrier lifetime. The method of sample preparation effective in passivating silicon surfaces, allowing a true measure of bulk lifetime. . The sample surface after the measurements are cleaned using a solution of H_2SO_4: H_2O_2: H_2O in the ratio1:1:1 followed by a dip in a diluted HF solution for a min. The wafer is then cleaned with HCl: H_2O_2: H_2O solution in the ratio 1:1:1 for 10 min. Each cleaning step is followed by a rinse in deionized water. Stress developed on the processed sample is measured using DEKTAK equipment.

An experimental setup similar to one described in our earlier work (3) was used to deposit the Teflon AF™ films on the test structure. The VUV source used in the back end experiments was 150-Watt Deuterium lamp model L 1835 manufactured by Hamamatsu. The system was calibrated using a SensArray's series 1501 wafer interfaced to a computer controlled temperature controller. The temperature variation across the wafer during the processing cycle was less than 1%. Teflon AF™ 1600 solution was dispensed in the system using direct liquid injection (DLI) technique (3).

For electrical characterization, metal-insulator-metal (MIM) structures were made. Ohmic contacts were formed by sputtering 500 nm of Al-Cu-Si on the back side of the silicon wafer followed by rapid isothermal annealing at 400 °C for 1 min in atmospheric pressure under nitrogen flow. Shadow masks were used for the deposition of the aluminum as the top contact. Finally, Al-Teflon AF™-Al plane structure capacitors with an area of 7.8×10^{-3} cm^2 were used for the measurements of electrical properties. Samples without electrodes were used for the structural characterization.

The leakage current density measurements are carried out by a Micro-manipulator probe station connected to a Hewlett-Packard 4140B PA meter that has the ability to resolving current to $+/-10^{-15}$A. The measurements were controlled by a personal computer working in Hewlett-Packard's visual engineering environment (HP VEE Version 4.0). Stress in the film was measured by FSM 128, using Non-Destructive Optilever™ Laser Scanning technique.

RESULTS AND DISCUSSION

In the diffusion part of the study, statistically designed experiments were carried out to investigate the relation between thermal stress and bulk minority carrier lifetime under different process conditions. The residual stress variation in case of processing samples in presence of VUV photons and in the absence of VUV photons is shown in Fig 2, whereas Fig. 3 shows the variation of the sheet resistance after the diffusion process with and without VUV photons. Minority carrier lifetime data for samples processed with and without VUV photons are presented in Fig. 4. From Figs. 2 and 3, we can observe that there is more residual stress in the samples with higher sheet resistance. In contrast, Figs. 2 and 4 show that the minority carrier lifetime is higher for the samples with lower residual stress. Residual stress and sheet resistance are reduced when the samples are processed in the presence of VUV photons, whereas the minority carrier lifetime is higher for the samples processed with VUV photons. The lower sheet resistance suggests that the dopants have diffused to a greater depth, indicating a higher diffusion coefficient for samples processed with VUV photons. This reiterates our earlier observation that the quantum photoeffects due to VUV photons lower the potential barrier (2) and lower the activation energy. A higher level of minority carrier lifetime suggests a lower number of defects generated during processing. The samples processed in the presence of VUV photons also show lower levels of residual stress, hinting towards a cause and effect relationship between defects, stress and minority carrier lifetime. This not only proves the fact that the VUV photons result in lower stress and enhanced minority carrier lifetime but also shows a trend indicating a direct correlation between thermal stress and minority carrier lifetime. Since the minority carrier lifetime reflects the defect levels present, we can say that we have observed a direct correlation between stress and process induced defects.

Statistically designed experiments were carried out to understand the relationship between residual stress of the Teflon AF film and leakage current through the film. The three thermal cycles used in this study are shown in Fig. 5. The stress variation for the samples processed with and without the presence of VUV photons is shown in Fig. 6. The leakage current density for the three different processing cycles is shown in Fig. 7. A bar chart representation of relevant stress and leakage current data at different voltage levels is presented in Fig. 8.

Fig. 6 shows that the average stress is lower in the samples processed in the presence of VUV photons. From Figs. 6, 7 and 8 we can observe that the leakage current values are low for the samples having lower stress levels. Lower levels of leakage current suggest a lower number of defects in the deposited films. It is also interesting to note that at low voltages, the leakage currents of the three samples are basically the same. However, at higher voltages the samples with more defects show a higher leakage current density. The measured stress varies in accordance with the number of defects present in the dielectric film. The presence of VUV photons during a process significantly reduces the stress in the deposited films and lowers the leakage current. Since leakage current reflects the defect levels present, we have observed a direct correlation between stress and process induced defects.

CONCLUSION

Based on the above results, we conclude that there is a trend showing a good correlation between the stress and performance of the materials. Since stress is observed to affect performance, reliability and yield of many materials used in front end and back end processing, we can use it as a numerical gauge to predict the robustness of a material/process. In-situ monitoring of stress can provide us an indirect indicator of defect generated during a process and alleviate the need to collect huge amount of data for defect reduction computations and lead to process simplification. We can also conclude that the presence of VUV photons during a process results in better device characteristics in both front end and back end processing and thereby RPP leads to improved performance.

REFERENCES

1. R. Singh, V. Parihar, K.F. Poole and K. Rajkanan, *Semicond. Fabtech* **9**, 223 (1999).
2. R. Singh, R. Sharangapani, K.C. Cherukuri, Y. Chen, D.M. Dawson, K.F. Poole, A. Rohatgi, S. Narayanan and R.P.S. Thakur, *Mat. Res. Soc. Symp. Proc.* **429** (1996) 81.
3. R. Sharangpani, and R. Singh, *Rev. Sci. Instrum.* **68**, 1564 (1997).

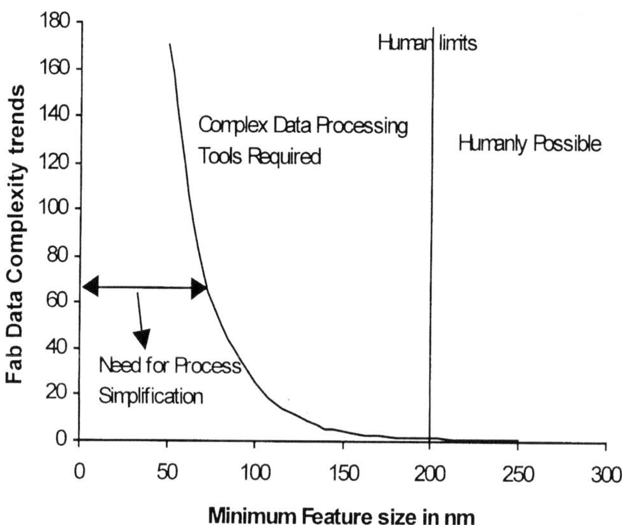

Fig. 1. Future fab data complexity and processing capability requirements as related to minimum feature size.

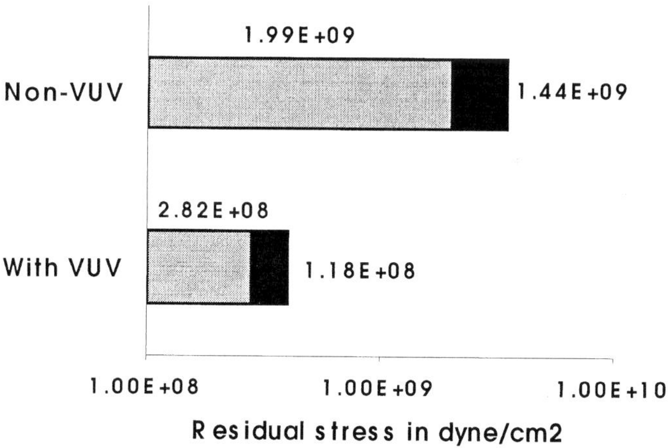

Fig.2. Bar chart showing the average residual stress and standard variation with and without VUV photons

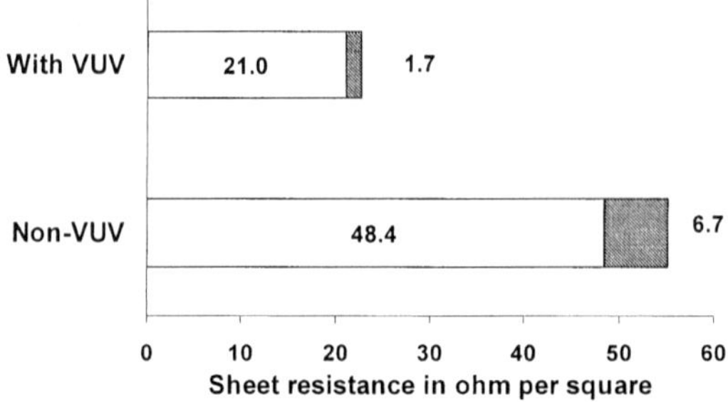

Fig.3. Bar chart showing the mean sheet resistance and standard variation with and without VUV photons

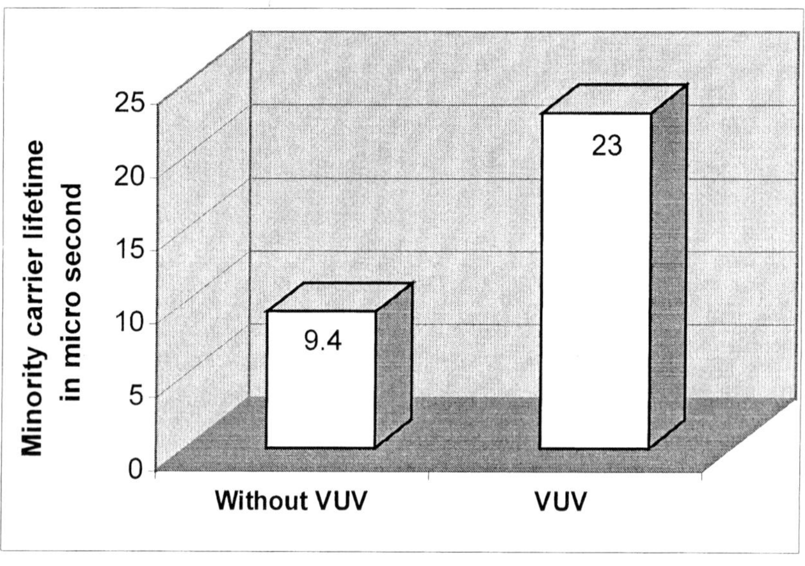

Fig.4. Minority carrier lifetime data for VUV and non-VUV based diffusion

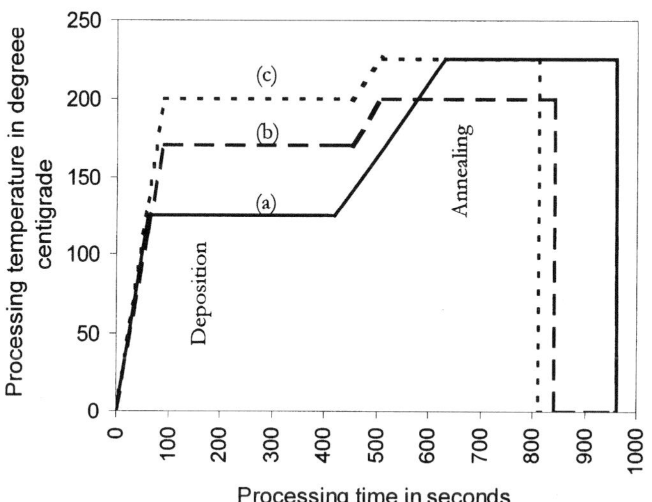

Fig. 5. The three temperature –time profiles used

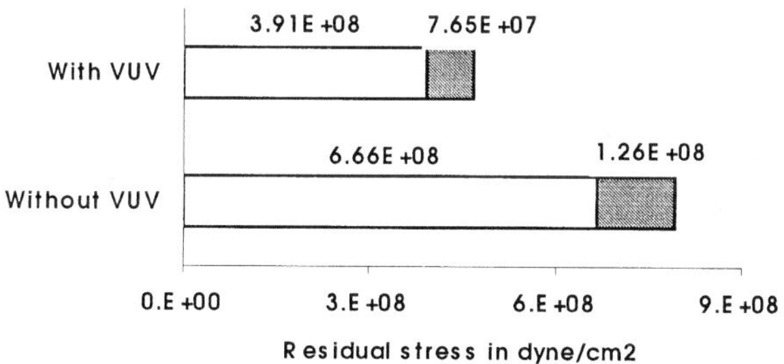

Fig.6. Bar chart showing mean stress and standard variation in VUV and non-VUV exposed Teflon AF samples

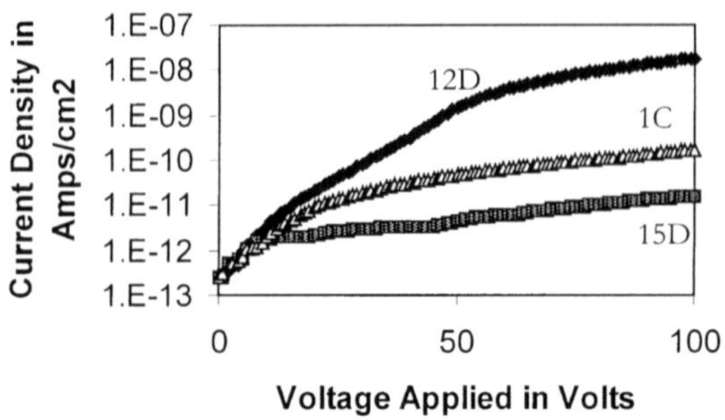

Fig.7. Leakage current density data

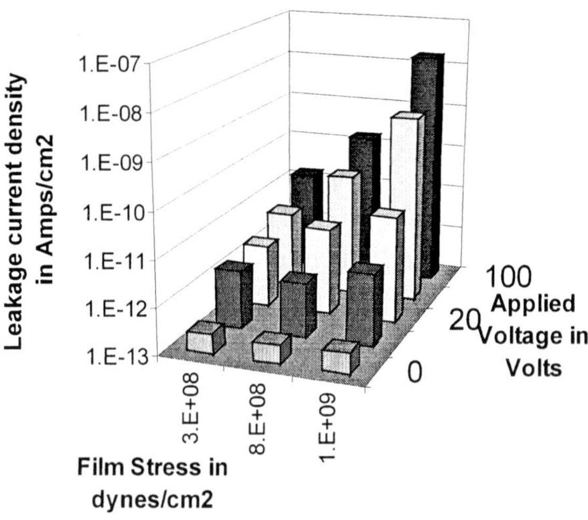

Fig.8. Stress and leakage current data for at different voltage levels

APPLICATION OF EXCIMER LASER ANNEALING IN THE FORMATION OF IMPLANTED SHALLOW JUNCTIONS

L.K. Nanver, E.J.G. Goudena, Q.W. Ren, M.v.d. Berg, R. Mallee, J. Slabbekoorn
ECTM, DIMES, TU Delft, NL-2628 CB Delft

Excimer laser annealing (ELA) is employed to activate implanted dopants in contacts where the p-n transition has already been formed by other means. In this manner, the thermal budget necessary to form a suitable junction with a low-ohmic contact resistance can be reduced. Elevated junctions, where the junction depth in the mono-silicon is reduced, are fabricated by depositing a thin polysilicon layer in the contact window before implantation and thermal annealing. The resulting high contact resistivity is reduced by ELA to about 10^{-7} - 10^{-6} Ωcm^2. The effect of the poly/mono interface on the junction quality and the implantation damage induced transient enhanced diffusion is also examined.

INTRODUCTION

The application of high-power excimer laser annealing (ELA) has in the past been proposed as a direct means of activating implanted dopants and forming shallow junctions without heating the bulk Si [1]. Previous work has also suggested that ELA can remove residual implantation damage and eliminate transient enhanced diffusion (TED) of dopants in the vicinity of the implant during subsequent thermal processing steps [2]. However, insufficient reproducibility and reliability have hampered the acceptance of ELA in IC-processing with respect to these applications. By ELA, a shallow surface region is melted and high electrical dopant activation is achieved, but the regrowth process will depend strongly on the overall situation giving variations in both the depth of the activated region and the crystallinity of the regrown layer. The regrowth velocity of the melted layer does not only depend on the amount of absorbed laser energy, but it is also very sensitive to the amount of heat transfer to the substrate. This is strongly influenced by the patterning of the wafer: the heat transfer from a small region with relatively large perimeter will be larger than that from a large region with relatively small perimeter. The proximity of any thermally isolating layers will also play a role. Due to these effects the processing window for junction formation by ELA is often quite small.

In the present work, it is shown that ELA can, however, be a very attractive means of making contact to diodes where the p-n transition has already been formed by other means [3]. The use of ELA to assure a high surface doping enhances the possibilities of forming shallow junctions at low processing temperatures. In many situations the ELA melts less than 20 nm of the silicon surface, so the contact resistance can be improved without influencing any of the other junction characteristics. This capability of ELA to reduce otherwise very high contact resistance has also been employed here to produce low-ohmic contact to elevated junctions. These are fabricated by depositing polysilicon over the contact windows before the diode implantation [4]. The junction depth in the mono-silicon is decreased, not only because the junction is elevated but also because the implantation damage induced transient enhanced diffusion (TED) is very significantly reduced by the presence of the poly/mono interface.

EXPERIMENTAL PROCEDURES

The laser system is a XeCl excimer laser ($\lambda = 308$ nm), the XMR5121, which is operated at an energy of around 500 mJ/pulse. The full width at half maximum is 60 ns and the repetition frequency is 5 Hz. The maximum spot size is 10×10 mm^2 and can be adjusted to obtain the desired energy density. The annealing process has been performed in a vacuum chamber at a pressure below 10^{-7} Torr and at room temperature. By using a beam homogenizer, the uniformity of the laser beam intensity is ± 10% within a 10×10 mm^2 beam area. The laser beam scans with 66% overlap in the lateral direction, so there are 3 shots at the same position. On patterned wafers, the laser annealing is performed in columns of dies in such a manner that the laser anneal energy density is adjusted per column so that comparisons can be made on the same wafer.

The fabrication of the implanted contacts is shown schematically in Fig. 1 for n-type contacts. A similar flow for p-type contacts is used with doping types interchanged. Basically a simple bipolar transistor structure is fabricated. The base and collector plug are implanted through a 30 nm thermal oxide and thermally annealed at 1000 °C. A stack of 100 nm oxide covered by 200 nm polysilicon is then deposited by LPCVD, at 700 °C and 675 °C, respectively. The latter is thick enough to absorb the range of laser beam energies used here without causing ablation. The contact windows for elevated contacts are plasma etched, a dip-etch in HF is performed to remove the native oxide and again a layer of

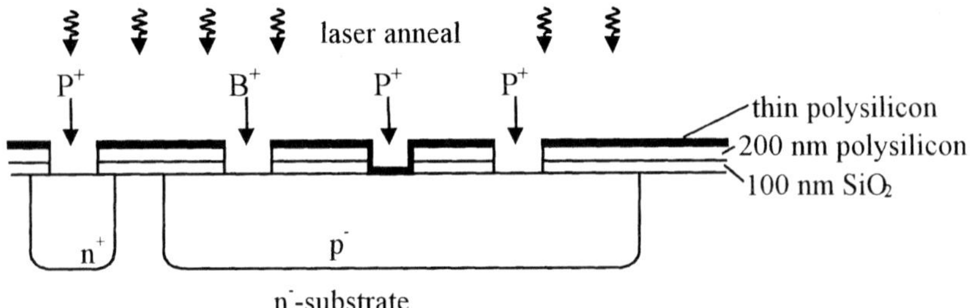

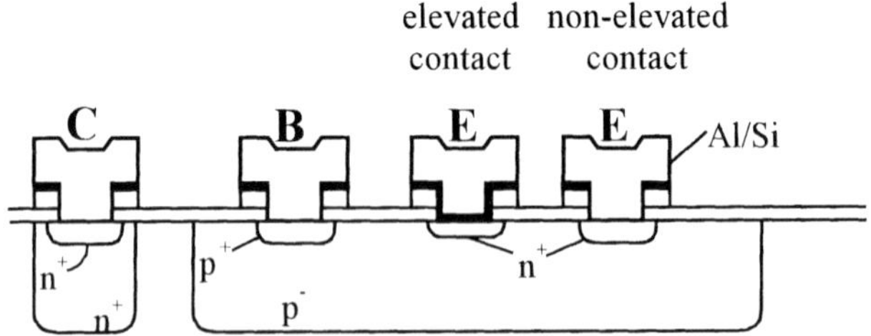

Fig. 1: Schematical process flow for fabricating contacts and electrical test structures.

polysilicon, this time 25 – 50 nm thick, is deposited. Now the contact windows for non-elevated contacts are plasma etched and all contact implantations can be masked with resist. Any desired sequence of thermal annealing, implantation and ELA can now be performed. A wet dip-etch step is used to remove the native oxide before the windows are contacted by sputtering Al/1%Si. A 400 °C 30 min alloying in forming gas completes the process. Contact dimensions down to 2 x 1 μm^2 have been studied.

In this manner, contacts in the mono-silicon and elevated contacts are fabricated on the same wafer. The p-type region, which is doped to $10^{17}/cm^3$, is used to evaluate the forward and reverse diode I-V characteristics. Further information on the residual damage at the junction is obtained when the diode is used as emitter with the p-region as base. The base current measured in a Gummel plot is orders lower than the forward diode current and gives a very sensitive means of evaluating the quality of the emitter-base junction.

The deep n^+ collector plug serves to form diffusion taps in the silicon for Kelvin contact resistance test structures. These have been specially developed for the direct and accurate measurement of the contact resistance to shallow junctions and/or doped regions that are self-aligned to the contact window [3]. The doping of the deep n^+ region is very light at the surface and does not contribute to the measured contact resistance, while the corresponding sheet resistance is so low (17 Ω/□) that parasitic resistances that otherwise may dominate the measured value are eliminated. In the elevated emitter junctions, the Kelvin measurement will not only give the poly-to-metal contact resistance, but will also include the resistance over the polysilicon and the poly/mono interface.

Sheet resistance measurements were performed by either four point probing on uniformly processed wafers or on wafers patterned with van der Pauw structures. Secondary ion mass spectrometry (SIMS) doping profiling was performed by Evans Europa and spreading resistance profiling by Solecon.

A number of the experimental results concern a highly doped phosphorus implantation (15 keV, $5\times10^{15}/cm^2$) preamphorized by a 40 keV, $2.5\times10^{14}/cm^2$ arsenic implantation. This is the implantation sequence that is used to form the implanted emitter in our high frequency SiGe HBT process [5]. In this process the stability of the SiGe base layer is safeguarded by keeping the thermal processing temperatures below 700 °C after the SiGe epitaxy.

RESULTS AND DISCUSSION

Contacts in mono-silicon

When an as-implanted region is laser annealed, high active doping levels can be obtained in a shallow region at the surface. An example is given by the spreading sheet resistance profiles in Fig. 2, where the 15 keV P^+ implantation is laser annealed at energies of 700, 1000 or 1200 mJ/cm^2. Both the dopant activation and the junction depth are increased considerably by increasing the laser energy density. The corresponding contact resistivity and sheet resistance is summarized Table I. SIMS profiles of the implantation before and after thermally annealing for 30 min at 700 °C are included in Fig. 3. Transient enhanced diffusion of the P increases the junction depth by 70 nm. The result of thermally

Fig. 2: Spreading resistance profiles of a 15 keV 5 x $10^{15}/cm^2$ P^+ implant preamorphized by 40 keV $5x10^{15}/cm^2$ As^+. Short dashes: after ELA; solid lines and long dashes: after subsequent thermal annealing at 700°C 30 min.

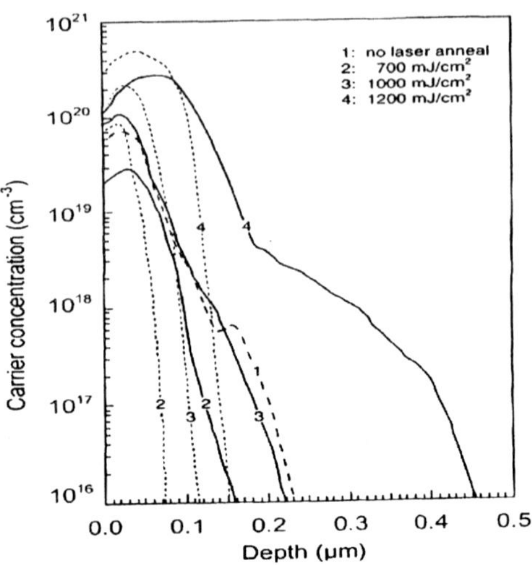

Table I: Contact resistivity and sheet resistance of implantations after thermal and/or laser annealing. The unit °C indicates a 30 min thermal anneal and J/cm^2 indicates ELA.

Implantation	Poly thickness (nm)	1st anneal	2nd anneal	Contact resistivity (10^{-7} Ωcm^2)	Sheet resistance ($\Omega/\Box$)
P^+ $5x10^{15}/cm^2$ 15 keV with pre-amorphization As^+ $2.5x10^{14}/cm^2$ 40 keV	0	0.7 J/cm^2		5.5 ± 1.5	45 ± 3
	0	1.0 J/cm^2		1.5 ± 0.3	28 ± 2
	0	1.2 J/cm^2		1.5 ± 0.3	24 ± 2
	0	700 °C		20 ± 10	100 ± 3
	0	700 °C	0.9-1.1 J/cm^2	2.0 ± 0.3	95 ± 5
	25	700 °C		100 ± 30	220 ± 5
	25	700 °C	0.9-1.0 J/cm^2	6.0 ± 1.0	160 ± 5
	0	650 °C		80 ± 20	
	0	650 °C	0.9-1.1 J/cm^2	2.0 ± 0.3	
As^+ $5x10^{15}/cm^2$ 40 keV	0	0.7–0.8 J/cm^2		5.0 ± 0.5	
	0	0.9–1.1 J/cm^2		1.5 ± 0.3	
	0	0.9–1.1 J/cm^2	700 °C	4.0 ± 0.6	
	0	700 °C		150 ± 50	
	0	700 °C	0.9–1.1 J/cm^2	0.7 ± 0.3	
	0	950 °C		3.0 ± 0.6	37.7 ± 0.5
	0	950 °C	0.9–1.1 J/cm^2	0.7 ± 0.3	30 – 38
BF_2^+ $5x10^{15}/cm^2$ 20 keV	0	0.7-1.0 J/cm^2		Contact failures	
	0	700 °C		25 ± 5	540 ± 10
	0	700 °C	1.0 J/cm^2	10 ± 1	75 ± 5
	25	700 °C			1650 ± 50
	25	700 °C	1.0 J/cm^2		150 ± 50

annealing after the laser anneal is also shown in Fig 2. The TED is strongly influenced by the laser anneal, giving a suppression of TED for low laser anneal energy and a strong enhancement for high energies.

The junctions formed by ELA of as-implanted regions are electrically less than perfect, and the uniformity and reproducibility is poor. On the same wafer the ideality factor can vary from n = 1.2 to 1.3 and the reverse breakdown voltage also shows a large spread. The contact resistance does, however, reach a reproducibly low value for laser energies above 900 mJ/cm^2. Equally low contact resistance is achieved for laser annealing of shallow as-implanted arsenic and boron regions, some results of which are included in Table I. On the other hand, ELA of as-implanted BF_2 contacts show a high frequency of failures. This is probably due to the presence of fluorine. However, thermally annealing these contacts before ELA, renders reliable results and for a 700 °C 30 min anneal the contact resistance is more than halved by the ELA.

In general, laser annealing of thermally annealed contacts improves the contact resistance. Some results are included in the Table I. Particularly for n-type contacts that have been thermally annealed at temperatures up to 700 °C, a very significant improvement is attained: contact resistivities well above the 10^{-6} Ωcm^2 are reduced to around 10^{-7} Ωcm^2. Low temperature thermal annealing of the contacts after ELA, will often give a significant increase in contact resistance due to dopant deactivation by clustering (e.g. the 700 °C anneal of the laser annealed As contact in Table I). Thus it is an advantage to perform ELA just before the metallization.

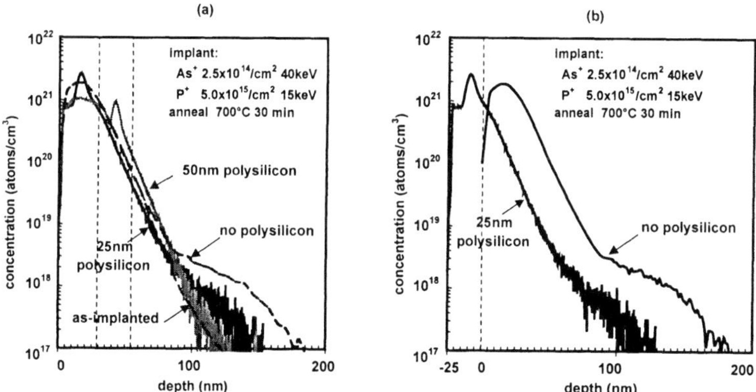

Fig. 3: SIMS profiles of phosphorus for implantation in 0, 25, or 50 nm polysilicon after 700 °C 30 min thermal annealing. In (b) the junction depth is compared for implantation through 0 and 25 nm polysilicon.

Elevated contacts

Elevated contacts have been fabricated by implanting 15 keV, 5 x $10^{15}/cm^2$ P^+ through either 25 or 50 nm polysilicon. The corresponding doping profiles after thermal annealing at 700 °C are compared in Fig. 3. The poly/mono interface modifies the dopant distribution in two ways: first there is a segregation of dopants in the poly near the interface, and second, the TED of the implantation tail is very significantly reduced. For the 50 nm poly case, the resulting junction characteristics are very poor, indicating that the dopants have not been diffused sufficiently far enough away from the implantation damage region and/or the interface. The measured contact resistance, which includes the resistance over the polysilicon layer and the poly/mono interface, is also very high, in the hundred ohms, and remains high even after laser annealing. This suggests that either the poly/mono interface oxide or a non-active poly layer near the interface is dominating the measured contact resistance. On the other hand, the diodes elevated by 25 nm poly are ideal with both diode and base current characteristics comparable to the corresponding non-elevated diodes with n = 1. The measured contact resistance, as a function of laser anneal energy, is plotted in Fig. 4, along with the contact resistance of non-elevated contacts. The laser anneal gives more than a factor 10 reduction in contact resistance, and the spread in the measured value is also very significantly decreased. A measured contact resistance of (6.0 ± 0.2) x 10^{-7} Ωcm^2 as compared to (2.0 ± 0.2) x 10^{-7} Ωcm^2 was achieved with and without poly, respectively. SIMS doping profiles of the laser annealed implantations do not show any dopant redistribution due to laser annealing. The dopant pile-up at the poly/mono interface is for example not modified. This suggests a melt depth of less than 20 nm, but enough heat must be transmitted to the rest of the poly to either activate dopants or to break up any native oxide at the interface.

The observed reduction in TED has been studied closer by placing epitaxially grown boron markers in the substrate. The thermal anneal behavior is shown in Fig. 5. The introduction of the poly/mono interface, which is know to act as a sink for interstitials, gives a small reduction of the TED of the boron markers. The effect is largest in the vicinity of the implantation and in the first stages of the thermal anneal. This is seen by extending the anneal time to 3 hr (Figs. 5c and 5e), after which time the whole TED process should be completed. Also the results of the 800 °C 30 min anneal (Figs. 5d and 5f) support these observations. In all cases the presence of the poly gives a slight reduction of both the clustering in the boron markers and the boron out-diffusion. The reduction in the junction depth of the P implant is more significant: in all 50 nm for the 700 °C 30 min anneal and 75 nm for the 800 °C anneal.

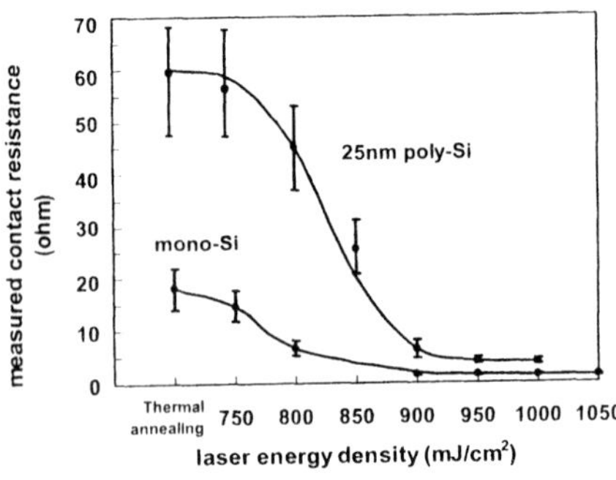

Fig. 4: The measured contact resistance of phosphorus implanted 4 x 4 μm^2 contacts versus laser energy density for mono-silicon and elevated contacts.

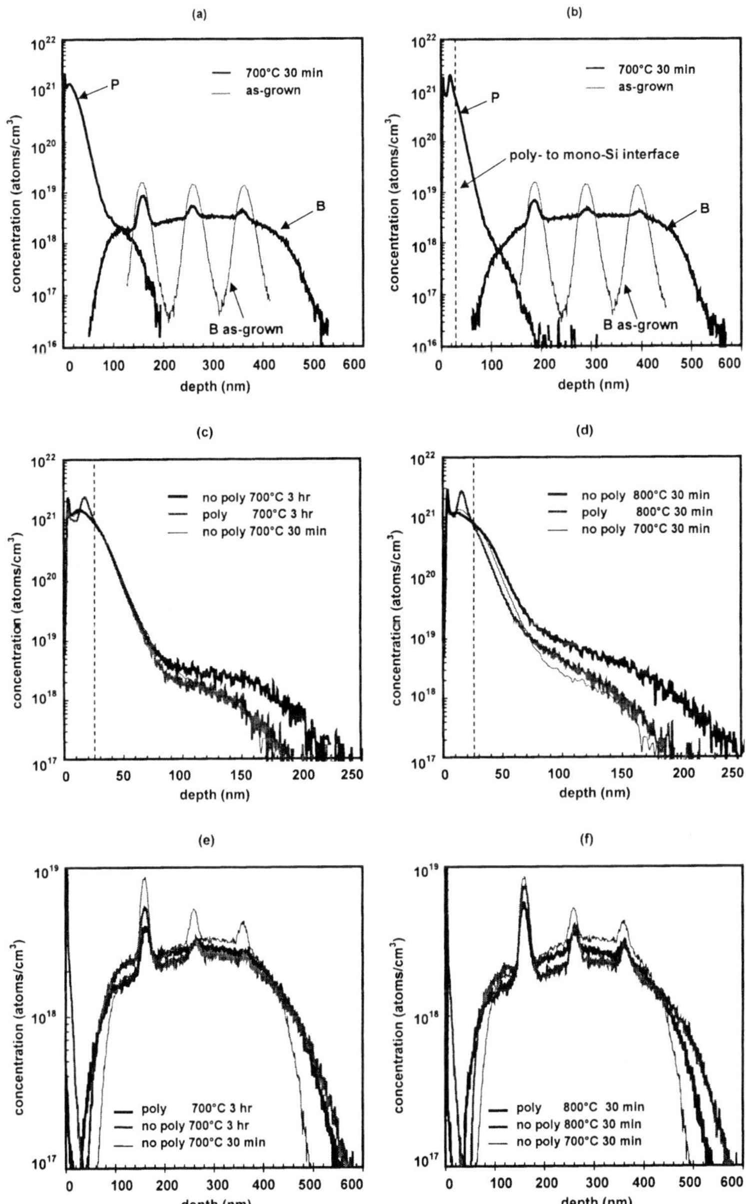

Fig. 5: SIMS profiles of a P+ implant (as Fig. 3) and three epitaxially grown boron markers for implantation in monosilicon (a) and 25nm polysilicon (b). Profiles for different anneal cycles are given for phosphorus in (c) and (d) and for boron in (e) and (f).

The reduction in sheet resistance of the thermally annealed P implant upon ELA is minimal (see Table I). This supports the above observation that the melt depth must be less than 20 nm. On the other hand, for BF_2 implants ELA results in a very large decrease in sheet resistance. In this case the SIMS profiles show that the pile up of dopants at the poly/mono interface due to thermal annealing is redistributed and flattened out by the laser anneal. This can be understood in terms of a large light absorption in the BF_2 contacts. The TED associated with the 20 keV BF_2 implantation is much lower than that of the P implantation, and the TED is only slightly influenced by the introduction of a poly/mono interface.

CONCLUSIONS

It has been demonstrated that excimer laser annealing is an efficient means of contacting shallow implantations in situations where the contact resistance otherwise would be very high, for example when the thermal budget is too low to provide high dopant activation. The laser anneal can electrically activates dopants in a thin surface layer of shallow junctions without altering the other junction characteristics. The processing window in which contact resistivities in the 10^{-7} Ωcm^2 range are achieved is large, usually from 800 - 1100 mJ/cm^2, making this application of ELA robust with respect to process variations.

The depth of shallow junctions can be reduced by using a thin 25 nm polysilicon layer to elevate the junction before implantation. The ELA then provides low-ohmic contact. For the high dose 15 keV P implant, the introduction of the poly/mono interface also decreases the TED and a total reduction of 50 nm in junction depth is realized. The TED of vicinal dopants is, however, only marginally reduced.

ACKNOWLEDGEMENTS

The authors would like to thank the whole staff of the DIMES IC processing group who have made this work possible, with special thanks to J. Bertens and T. Scholtes. Simulating discussions with R. Ishihara, N.E.B. Cowern, P.A. Stolk and H.G.A. Huizing are also gratefully acknowledged.

REFERENCES

1. H. Tsukamoto, H. Yamamoto, T. Noguchi and T. Suzuki, Jpn.J.Appl.Phys., **35**, pp.3810-3813: 1996.
2. T. Ghani, J.L. Hoyt, A.M. McCarthy, and J.F. Gibbons, J. El. Mat, **24**, pp. 999-1002: 1995.
3. L.K. Nanver, E.J.G. Goudena and J. Slabbekoorn, Proc. 1996 IEEE Int. Conf. Microelectronics Test Structures, **9**, pp. 241-245: March 1996.
4. Q.W. Ren, M.R.v.d. Berg, L.K. Nanver, and J. Slabbekoorn, Proc. ICSICT, pp. 102-105: Oct. 1998.
5. L.K. Nanver, C.C.G. Visser and A.v.d. Bogaard, J. Vac.Sc.Tech. B, **16**, p. 1533-1538: 1998.

ULTRA-SHALLOW P^+-N JUNCTIONS FOR 50-70 nm CMOS USING SELECTIVELY GROWN IN-SITU BORON-DOPED SILICON FILMS

*Ibrahim Ban and Mehmet C. Öztürk**

Department of Electrical &Computer Engineering, North Carolina State University, Centennial Campus, EGRC Building, Box 7920, Raleigh, NC 27695-7920

In-situ doped selective silicon films selectively grown in a Ultra-High Vacuum Rapid Thermal Chemical Vapor Deposition (UHV-RTCVD) reactor have been studied. Films with doping concentrations as high as $3x10^{21}$ cm^{-3} were grown using Si_2H_6, B_2H_6 and Cl_2 at 800 °C. Selectivity, surface morphology, and annealing behavior of these films have been studied for different boron concentrations. An anomalous diffusion enhancement was observed for very heavily doped films and have been attributed to self-interstitial injection during annealing. No Si-B phases were observed by x-ray and selected area diffraction measurements. Ultra-shallow, super abrupt p^+-n junctions with very high surface concentrations in the substrate (~$2x10^{20}$ cm^{-3}) were fabricated using in-situ doped Si as a solid dopant diffusion source. Reverse bias leakage measurements revealed excellent characteristics even for as-deposited junctions that did not receive out-diffusion anneals. Considering the fact that the deposition temperature was only 800°C resulting in practically no boron diffusion into the underlying substrate, these junctions can be referred to as *zero-depth* junctions.

INTRODUCTION

Source-drain junction depths less than 300 Å are projected for the 50 and 70 nm CMOS technology nodes (1) (see Table 1). Formation of such boron doped p^+-n junctions is extremely challenging due to high intrinsic and defect-enhanced diffusivity of boron. Recent advances in low-energy ion implantation along with low-thermal budget anneals such as spike annealing could enable formation of ultra-shallow junctions using ion implantation (2, 3). However, silicon consumption during silicidation is a problem, which requires an elevated buffer layer to realize junctions with low leakage.

Table 1: NTRS Roadmap requirements for junction depths for future generations

Technology	0.15 μm	0.13 μm	0.1 μm	70 nm	50 nm
Contact X_J (nm)	60-120	50-100	40-80	15-30	10-20
X_J @ channel	30-60	26-52	20-40	15-30	10-20

In this paper, we present a low-thermal budget in-situ doped selective silicon process which can be used to form elevated ultra-shallow or *zero-depth* junctions for the 70 nm and 50 nm CMOS technology nodes. The approach has three main advantages: a) it provides super abrupt, ultra-shallow/zero-depth junctions with surface concentrations

* Contact Author

as high as 10^{21} cm^{-3}, b) it provides a sacrificial layer which can be consumed during silicide formation, c) very high boron concentration in the film reduces the possibility of dopant depletion at the silicide/silicon interface during silicide formation which is an important advantage in forming junctions with low silicide/silicon contact resistance. Zero depth junctions with no out-diffusion anneals have been examined along with junctions formed by boron out-diffusion from in-situ doped layers. Boron diffusion during drive-in has been closely examined and linked to structural changes occurring during rapid thermal annealing (RTA) treatments (4, 5).

EXPERIMENTAL

In-situ boron-doped silicon films were selectively deposited in an Ultra-High Vacuum Rapid-Thermal Chemical Vapor Deposition Reactor (UHV-RTCVD) using disilane (Si_2H_6), diborane (B_2H_6) and chlorine (Cl_2) at 800 °C (6). Films with boron concentrations ranging from $\sim 10^{19}$ cm^{-3} to 3×10^{21} cm^{-3} have been investigated. Boron concentration and silicon growth rate were determined by Secondary Ion Mass Spectroscopy (SIMS). Atomic Force Microscopy (AFM) and Transmission Electron Microscopy (TEM) were used to study the microstructure of the deposited layers. Boron doped p^+-n junctions were fabricated on n-type Si (100) substrates. Active areas were defined by thermal oxide growth, photolithography and wet etching. In-situ doped silicon films were selectively deposited in the active areas. This was followed by low temperature SiO_2 deposition to avoid dopant out-diffusion during RTA used for junction drive-in. The samples were annealed at 1000 °C for 5, 10, and 15 s in an Argon ambient. The junction anneals were intentionally avoided for some of the samples to produce *zero-depth* junctions.

RESULTS AND DISCUSSION

Our previous reports on in-situ boron doping provide detailed information on boron incorporation mechanisms at low to medium levels ($\leq 10^{19}$ cm^{-3}) primarily intended for MOSFET channel engineering (6, 7). In this report, we focus on heavily doped films ($N \geq 10^{20}$ cm^{-3}) to form ultra-shallow source/drain junctions.

We have found that the film growth is highly selective with respect to the thermal oxide for all concentrations investigated. The process uses a relatively small amount of chlorine (Cl_2) (Cl_2 to Si_2H_6 ratio is 3/13) to achieve selectivity. AFM micrographs of selectively grown (with respect to the thermal oxide) films doped to 4×10^{19} cm^{-3} and $\sim 10^{20}$ cm^{-3} are shown in Fig. 1. Epitaxial growth was achieved for concentrations as high as 4×10^{20} cm^{-3}. These films were very smooth (RMS roughness=0.7 Å as measured by AFM) and free of visible structural defects up to a boron concentration of $\sim 10^{20}$ cm^{-3} (Fig. 1(a)). Above this level, pyramid-type defects were observed (Fig. 1(b)). Raising the boron concentration above this threshold resulted in 3-D islanding (Fig. 2(a) and 2(b)). The details of the surface processes that cause this 3-D island formation are not known at this moment. One possible mechanism is the lattice strain induced by the smaller boron atoms (8). In addition, it can be speculated that surface diffusion of species is impeded at the presence of such high concentrations of boron on the surface. The growth rate reduction observed at high B_2H_6 concentrations (not shown) suggests that some kind of surface poisoning takes place on the surface. It is possible that high density of these boron species hinder surface mobility of silicon ad-atoms resulting in more nucleation sites as opposed to step flow type growth.

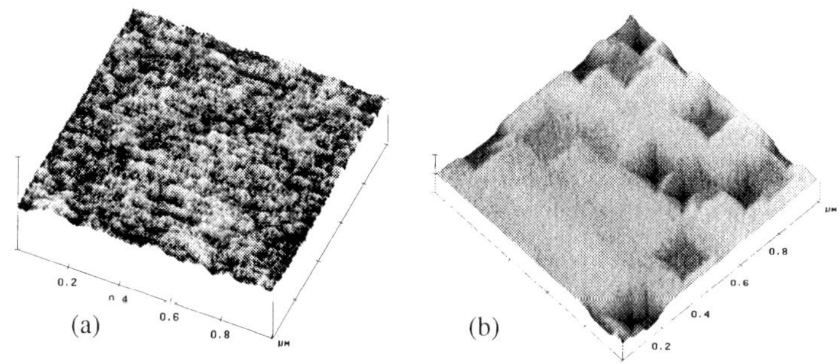

Figure 1: AFM micrographs of selectively grown in-situ doped epitaxial films: a) $N_B = 4x10^{19}$ cm^{-3}, b) $N_B = 1x10^{20}$ cm^{-3}.

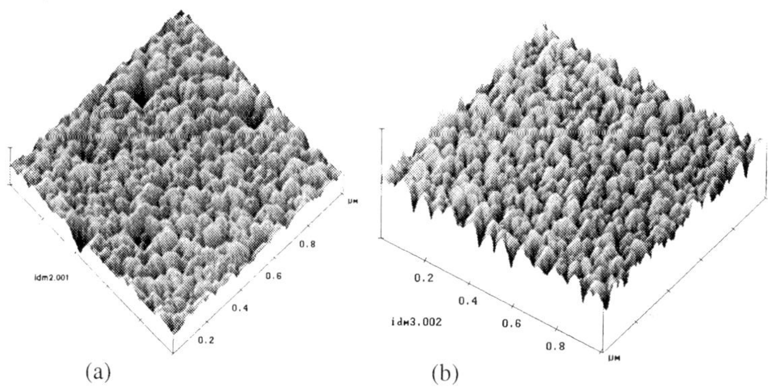

Figure 2: AFM micrographs showing 3-D film growth at high boron concentrations: a) $N_B = 4x10^{20}$ cm^{-3}, b) $N_B = 3x10^{21}$ cm^{-3}.

Boron diffusion from heavily doped silicon films was studied at concentrations ranging from 10^{19} cm^{-3} to $3x10^{21}$ cm^{-3}. Normal diffusion profiles were obtained with concentrations up to ~10^{20} cm^{-3} with no apparent diffusion enhancement (Fig. 3 (a)). Above this level, SIMS profiles indicate that enhanced-diffusion, which occurs within 5 sec of the RTA (1000 °C), limits the minimum junction depth achievable. As shown in Fig. 3(b), ($N_B = 1x10^{20}$ cm^{-3}) boron movement is fast and complete within 5 sec since the longer anneals do not result in appreciable diffusion. This diffusion enhancement is correlated with the increase in boron concentration between ~10^{20} cm^{-3} and ~10^{21} cm^{-3}; however, it becomes less severe in as-grown samples with boron concentrations higher than ~10^{21} cm^{-3} (Fig. 4 (a)). In this concentration regime, boron diffusion is more controllable and very abrupt, ultra-shallow junctions (X_j<300 Å for 1000 °C RTA, 10 sec) could be formed with high interface concentrations (~ $2x10^{20}$ cm^{-3}) at the silicon/substrate interface. It should be emphasized that the thermal budgets used here are much higher than what could typically be employed in future technologies. In addition, in forming these junctions we have used a conventional RTA tool without strict ambient or temperature control. Another phase of diffusion enhancement, more severe in magnitude, is observed when as-grown samples are annealed at a higher temperature (1050 °C) (Fig. 4(b)).

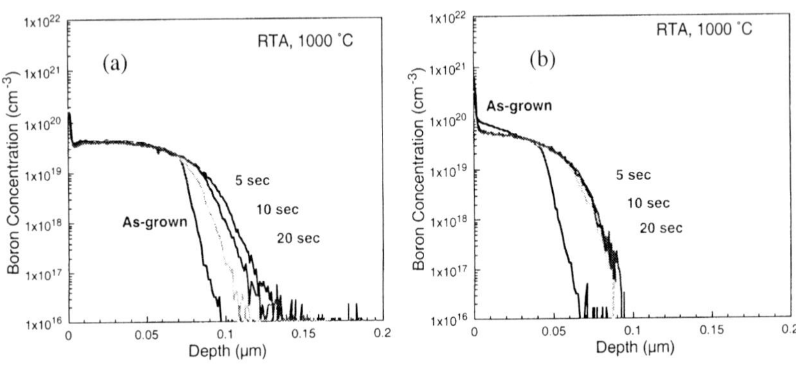

Figure 3: Boron profiles obtained by out-diffusion from selectively deposited silicon films. Boron concentrations in the as-grown films are : a) $4x10^{19}$ cm^{-3}, b) $1x10^{20}$ cm^{-3}. Note that abruptness of the as-grown profiles is limited by the SIMS cascading effect in (a) and surface roughness introduced by 3-D growth in (b).

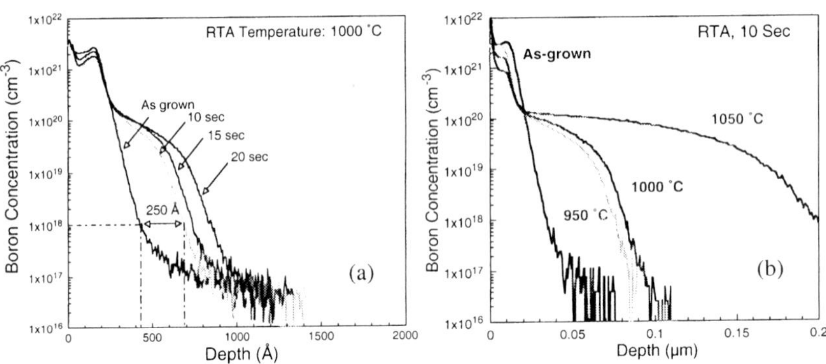

Figure 4: Boron profiles obtained by out-diffusion from selectively deposited silicon films doped to $3x10^{21}$ cm^{-3}. Annealing conditions: a) 1000 °C for 0, 10, 15, and 20 sec RTA, b) 10 sec RTA at 0, 950, 1000, and 1050 °C.

In addition to the diffusion enhancement, heavily-doped samples exhibited significant boron out-diffusion during RTA anneals which in some cases resulted in shallower junction depths for samples that underwent a higher thermal budget anneal (not shown). Figure 5 summarizes boron out-diffusion experiments for three different as-grown boron concentrations. As shown, the samples with the lowest boron concentration kept almost the same total dose during 20 s RTA anneals. However, boron out-diffusion enhances as the concentration is increased. It has been verified by samples with oxide caps that the dose loss is due to out-diffusion and that it can be avoided by a thin layer (experimentally only 400 Å was tried) of deposited oxide. It should be emphasized that the thermal budget considered in this particular experiment is well above the level practical for forming ultra-shallow junctions; hence, the out-diffusion observed is not a major concern for heavily-doped deposited silicon layers. However, our data demonstrates that significant dopant loss may occur for ion-implanted junctions since most of the dopant source is localized very close to the surface.

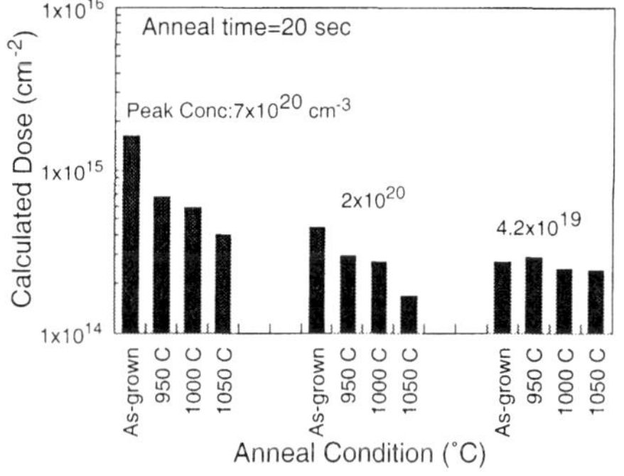

Figure 5: Calculated dose from the diffusion profiles for different anneal conditions. The differences in doses between as-grown and annealed samples are due to out-diffusion.

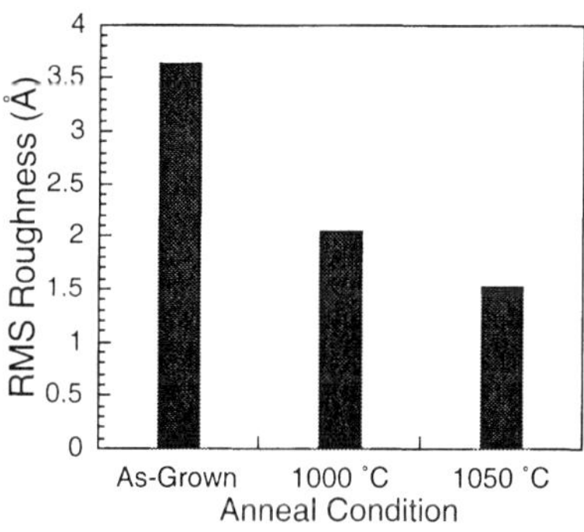

Figure 6: RMS roughness values obtain from AFM measurements before and after 10 sec anneals.

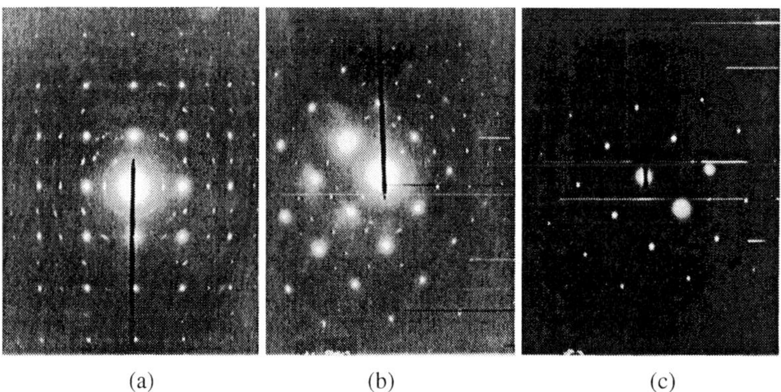

(a) (b) (c)

Figure 7: Selected Area TEM diffraction patterns from in-situ doped silicon films: (a) $N=1x10^{21}$ cm^{-3}, 1000 °C, 10 sec RTA, (b) $N=3x10^{21}$ cm^{-3}, 1000 °C, 10 sec RTA, (c) $N=3x10^{21}$ cm^{-3}, 1050 °C, 10 sec RTA.

As discussed previously, surface morphology of the in-situ doped films change significantly as the boron concentration is increased. 3-D film growth begins to take place at boron concentrations around $4x10^{20}$ cm^{-3}. Since a high boron concentration at the silicide/silicon interface is essential for a low resistivity contacts, and that diffusion behavior improves at concentrations above 10^{21} cm^{-3}, the changes observed in the surface morphology of these films is of interest. The surface roughness figures obtained from AFM measurements indicate that these films undergo structural changes during annealing and the RMS roughness is reduced appreciably (Fig. 6). Selected area diffraction patterns of films with 4 atomic percent boron failed to identify a Si-B phase such as SiB_4which has been previously proposed as the reason for diffusion enhancement above 6 % (5). We have found that heavily boron-doped films ($\geq 4x10^{20}$ cm^{-3}) are amorphous as-deposited and crystallize upon annealing. However, crystallization after a lower temperature RTA anneal (1000 °C, 10 sec) is incomplete leaving distinct signatures of (111), (220), and (311) oriented crystallites (Fig. 7(a) and 7(b)). After a higher temperature anneal, only the Si(100) signature is observed which indicates either larger crystallites or complete crystallization (Fig.7(c)). It was also found that the annealing kinetics is faster for films doped to higher than $1x10^{21}$ cm^{-3} (compare diffraction patterns in Fig. 7(a) and 7(b)). Our results indicate that enhanced diffusion in heavily boron-doped films could still take place without SiB_4 formation. We believe that this enhancement, is a result of self-interstitial injection during recrystallization. Furthermore, our results suggest that boron concentrations higher than ~$2x10^{21}$ cm^{-3} may be necessary since self-interstitial injection is less severe in that regime based on our findings.

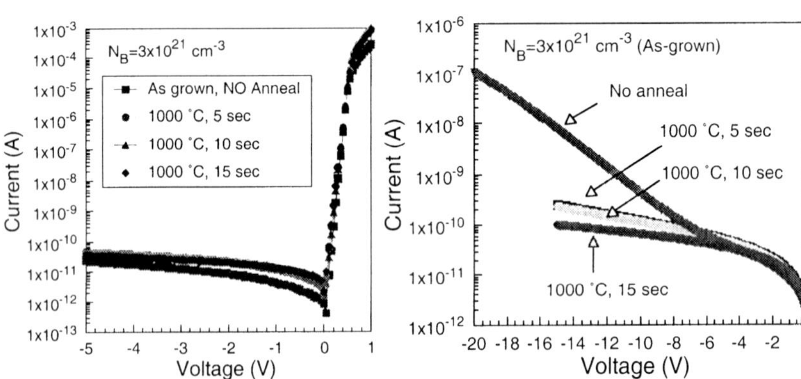

Figure 8: Current-voltage characteristics of P^+-n junctions fabricated by diffusion from a heavily doped ($N=3x10^{21}$ cm^{-3}) silicon films.

Figure 9: Reverse biased junction leakage of P^+-n junctions fabricated by diffusion from a heavily doped ($N=3x10^{21}$ cm^{-3}) silicon films.

Using these films, we have formed ultra-shallow junctions ($X_j \leq 300$ Å) (Fig. 4(a)). Diodes fabricated using this process exhibit excellent low reverse leakage current even when the dopant drive-in anneal is not carried out (Figure 8). Forward characteristics of the junctions formed without a drive-in anneal are also identical to those of deeper junctions, which indicates that these junctions are also highly ideal. This result can be particularly useful in forming elevated source/drain junctions for CMOS technology nodes below 100 nm. During deposition, little or no diffusion into the substrate occurs.

Boron profiles obtained from as-deposited samples indicate a very abrupt interface considering the fact that profiles are not corrected for the SIMS cascading effect. Therefore, these junctions can be referred to as *zero-depth* junctions. The low leakage obtained from the as-deposited sample suggests that the depletion region on the heavily boron doped side can not extend into the heavily defective deposited layer which exhibits a 3-D structure. This is possible only if the junction is indeed heavily doped and it is super abrupt. Reverse bias junction leakage currents of these junctions are given in Figure 9. The *zero depth* junctions exhibit higher generation current only when reverse bias becomes more than 5 V. Experiments are currently in progress to fully characterize these junctions including those that received anneals to drive-in the dopant into the substrate.

CONCLUSIONS

Ultra-shallow junction formation using selectively deposited in-situ boron doped Si was investigated. Very smooth epitaxial silicon films with no visible defects were grown up to a boron concentration of ~ 10^{20} cm^{-3}. Further increase in boron concentration resulted in defective layers and eventually 3-D film growth. Boron profiles obtained from these films revealed that boron diffusion was significantly enhanced for 950 °C and 1000 °C anneals for films doped to boron levels between 10^{20} cm^{-3} and ~ 10^{21} cm^{-3}. This enhancement occurred within approximately 5 sec of the annealing cycle. Another severe diffusion enhancement was observed for the 1050 °C/10 s anneals. We attributed this enhancement to structural changes occurring in the film during the anneal resulting in injection of self-interstitials. This idea is supported by the AFM measurements showing reduced RMS roughness after annealing and by the selected area diffraction patterns confirming the change in the crystalline structure. We have demonstrated very abrupt, ultra-shallow junctions using this process. *Zero-depth* junctions with no junction drive-in anneal exhibit excellent forward and reverse bias characteristics.

ACKNOWLEDGEMENTS

This research was partially supported by NSF Engineering Research Centers Program through the Center for Advanced Electronic Materials Processing (CDR-8721545) and by the NSF Presidential Faculty Fellow Award.

REFERENCES

1. "National Technology Roadmap for Semiconductors" (Semiconductor Industry Association, 1997).
2. A. Agarwal, et al., *Appl. Phys. Lett.* **73**, 2015-2017 (1998).
3. D. F. Downey, J. W. Chow, W. Lerch, J. Niess, S. D. Marcus, The Effects of Small Concentration of Oxygen in RTP Annealing of Low Energy Boron, BF_2 and Arsenic Ion Implants, M. C. Öztürk, F. Roozeboom, P. J. Timans, S. H. Pas, Eds., Mat. Res. Soc. Symp. Proc., San Fransisco, CA (1998).
4. A. Agarwal, et al., Boron-Enhanced-Diffusion Of Boron: The Limiting Factor For Ultra-Shallow Junctions, IEDM, Washington D.C. (1997).
5. A. Agarwal, H.-J. Gossmann, D. J. Eaglesham, *Appl. Phys. Lett.* **74**, 2331-2333 (1999).

6. I. Ban, et al., *J. Electrochem. Soc.* **146**, 1189-1196 (1999).
7. M. K. Sanganeria, K. E. Violette, M. C. Öztürk, G. Harris, D. M. Maher, *J. Electrochem. Soc.* **142**, 285-289 (1995).
8. J. M. R. Sardela, H. H. Radamson, J. O. Ekberg, J.-E. Sundgren, G. V. Hansson, *Journal of Crystal Growth* **143**, 184-193 (1994).

SHALLOW JUNCTION FABRICATION BY RAPID THERMAL OUTDIFFUSION FROM IMPLANTED OXIDE

J. Schmitz[1], M. van Gestel[2], P. A. Stolk[1], Y. V. Ponomarev[1], F. Roozeboom[1], F. N. Cubaynes[3], J. G. M. van Berkum[4], W. M. van de Wijgert[4], P. C. Zalm[4] and P. H. Woerlee[1,5]

[1] Philips Research Laboratories, Prof. Holstlaan 4, 5656 AA Eindhoven, The Netherlands
[2] Eindhoven University of Technology, Eindhoven, The Netherlands
[3] INSA, Toulouse, France
[4] Philips CFT, Eindhoven, The Netherlands
[5] University of Twente, Enschede, The Netherlands

ABSTRACT

Outdiffusion experiments were performed from ion implanted thermally grown SiO_2 into silicon. The purpose of these experiments is to fabricate ultra-shallow junctions in a manner compatible to standard CMOS processing. Shallow p^+ junctions were formed in the 30-100 nm range under various implant and Rapid Thermal Processing conditions. In this paper, a comparison is made between B^+ and BF_2^+ implanted oxides at the same equivalent B energy. Furthermore, we present results from rapid thermal outdiffusion in various ambients (N_2, N_2/H_2 and O_2). The implantations of B^+ and anneals in N_2 give the best results in terms of the tradeoff between sheet resistance and junction depth.

INTRODUCTION

The formation of ultra-shallow p^+ and n^+ regions in opposite-type doped silicon is a crucial step in the fabrication of MOS transistors. The ever-decreasing size of MOS transistors requires the downscaling of all lateral and vertical dimensions of the transistor. In conventional scaling scenarios, the depth of the junctions that form the sources and drains of MOS transistors scales linearly with the gate length. This is illustrated by the goals set by the Semiconductor Industry Association in its 1997 Roadmap [1]. The junction depth targets for the coming CMOS generations from this document are depicted in Figure 1.

The figure shows that within 10 years from now, the fabrication of junctions as shallow as 15-30 nm must be feasible on an industrial scale. It is expected that in this range, other methods than the conventional fabrication scheme are necessary to fulfill this need [1].

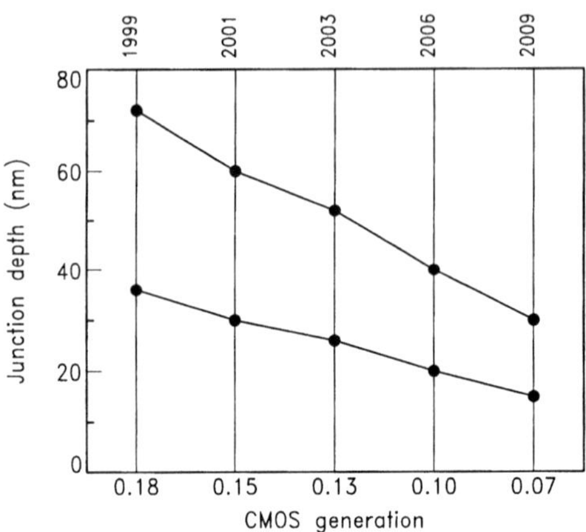

Figure 1: Junction depth range of source and drain extensions in future CMOS generations (indicated by their nominal gate length in μm), according to the NTRS Roadmap [1]. The year in the upper axis is the expected year of the first shipment of products in this generation. Process and application requirements will determine the exact junction depth between the upper and lower limits depicted in the graph.

In modern CMOS processes, the shallow junctions are formed by ion implantation followed by a (rapid thermal) anneal. Several authors have shown the feasibility of this technique down to at least 30-40 nm junction depth (see e.g. [2,3]), which is particularly difficult for p-type shallow regions due to the implant and diffusion properties of boron. Crucial issues are control of dopant channeling, reduction of thermal diffusion, and suppression of transient enhanced diffusion in the case of boron and phosphorus. Moreover, good device performance is only obtained with a low sheet resistance of the shallow regions, i.e. a high impurity concentration. The scaling tendency has been to reduce the ion implant energy while the total dopant level is kept more or less constant; and to reduce the thermal budget, without deteriorating the dopant activation level too much (by the introduction of rapid thermal anneals, and spike anneals).

This conventional scaling is expected to become difficult below 40 nm junction depth, particularly for p^+ junctions. The technical difficulty in making a high-current, low-energy ion implantation beam may be alleviated by the use of Plasma Doping (also called Plasma Immersion Ion Implantation, see e.g. [4]) or decaborane ($B_{10}H_{14}$) implantation [5]. Many alternative processes that avoid implantation altogether are also studied, such as (Rapid Thermal) Vapor Phase Doping (see e.g. [6]), Gas Immersion Laser Doping (GILD, see e.g. [7]), and the method used most commonly in early semiconductor manufacturing processes, Solid State Outdiffusion (e.g. from BSG [8] or PSG [9]). Of these techniques, GILD gives the best depth/sheet resistance combination. However, all these methods face one or more problems with industrialization, particularly the poor reproducibility of the outdiffusion processes. A thorough discussion of shallow junction formation methods for silicon technology is presented in [10].

OUTDIFFUSION FROM IMPLANTED OXIDE

Recently we have reported of a new method to fabricate shallow junctions: outdiffusion from implanted oxide [11]. A thin oxide on top of the silicon is implanted with a high dose of boron or phosphorus, and in a subsequent RTA-step this dopant is driven into the silicon. The principle is sketched in Figure 2. The method relies on a high concentration of B or P in the oxide to give sufficient dopant diffusion through the Si-SiO_2 interface. Therefore, given the capabilities of modern ion implantation equipment, this method is only practical if a screening oxide thinner than 10 nm is used.

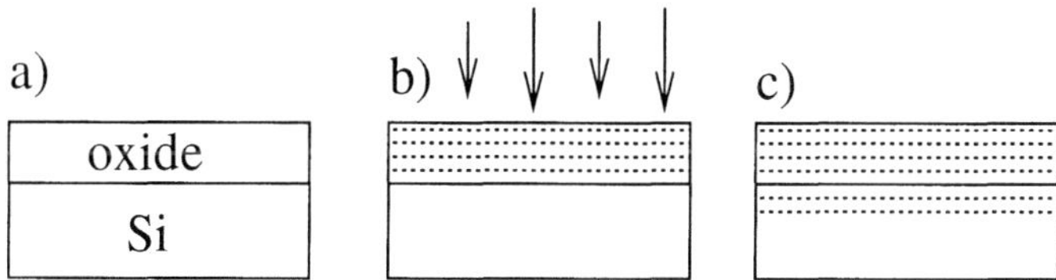

Figure 2: Principle of outdiffusion from implanted SiO_2 (a). A low-energy implant into a thin screening oxide (b) is followed by a rapid thermal anneal, driving the impurities from the oxide into the silicon (c).

We have shown [11] that shallow junctions with competitive sheet resistance and junction depth were obtained with boron. Moreover, the outdiffusion effect was also observed with phosphorus, and full process integration of this method was shown to be feasible in deep-submicron CMOS. The ultra-shallow junctions, used as Medium Doped Drain extensions, significantly improved the short channel behaviour of NMOS transistors fabricated in a 0.13 μm process.

The purpose of the present paper is to extend the range of outdiffusion experiments for p-type junctions, by comparing outdiffusion from B^+ implanted and BF_2^+ implanted oxides, and by investigating the outdiffusion in RTP equipment under various gas ambients.

EXPERIMENTAL

Silicon substrates were oxidised in a conventional furnace and subsequently implanted with an Applied XR80-LEAP low-energy implanter. The wafers were diced into 2.6 cm^2 squares. Rapid Thermal Anneal steps were performed in a STEAG-AST SHS1000 system. Carrier concentration and sheet resistance of the obtained shallow junctions were determined using a Hall-Van der Pauw measurement. A Cameca/ION-TOF time-of-flight and a Cameca ims4f SIMS analyzer were used to determine the boron and fluorine concentrations as a function of depth, employing $\leq$ 1 keV primary O_2 beams, using oxygen flooding in the TOF measurements. The depth scale was calibrated assuming a constant erosion rate; this is probably more valid for the Cameca ims4f than for the TOF analyzer.

RESULTS

A) COMPARISON OF B^+ AND BF_2^+ IMPLANTED OXIDES

An oxide with 7 nm thickness was thermally grown on (100) wafers. Subsequently we implanted the oxide with either 1 keV B^+ (dose $3x10^{14}$ cm^{-2}, $1x10^{15}$ cm^{-2} and $3x10^{15}$ cm^{-2}) or 4.5 keV BF_2^+ (same dosage). The fractional energy of the B ions in a 4.5 keV BF_2^+ beam is 1.01 keV (only isotope ^{11}B is used in these experiments), so that the same as-implanted distribution of boron can be expected[1]. Still, differences may be expected in the outdiffusion process because:

- The BF_2^+ implant causes more damage in the SiO_2-layer than the B^+ implant;
- The fluorine may influence the B diffusivity, both in the SiO_2 and in the silicon; it may also influence the segregation coefficient;
- The fluorine that remains in the silicon after annealing may influence the electrical properties of the shallow p-type layer.

SIMS analysis has shown that the fluorine implanted with the BF_2^+ beam is not entirely removed after a 1 minute, 1025°C anneal. About 10% of the original fluorine dose is still present, and is located in the first 20 nm of the sample. We conclude that the anneal is not sufficiently long or hot to remove the fluorine entirely, and the F present can influence the electrical properties of the shallow junction.

Indeed, we observe clear differences between B^+ and BF_2^+ implanted samples after annealing. Figure 3a shows the sheet resistances of these samples, for various implant dosages and two different anneal periods. The BF_2^+ implanted samples systematically have a higher sheet resistance. The resistance differences originate from a difference in hole concentration, as illustrated by Figure 3b. The difference in resistance can be corrected for by a slightly longer anneal, as the 60 second data in the same figures indicate. The hole mobility is comparable in the B^+ and BF_2^+ implanted samples, which suggests that even though fluorine is still present in the shallow junctions after anneal, it does not influence the carrier mobility.

In Figure 4, the B depth profile is shown for B^+ and BF_2^+ implanted samples after outdiffusion. This figure shows that the higher sheet resistance of the BF_2^+ samples is the result of a retarded outdiffusion. It appears that the fluorine does act in the outdiffusion process: it seems to reduce the boron diffusion coefficient at 1025°C. Given the fact that advanced CMOS fabrication processes must have a minimum thermal budget, the B^+ result can be considered superior (based on the data in Figure 4), because a lower outdiffusion temperature (or time) will suffice to achieve a shallow junction with the same sheet resistance and junction depth.

---

[1] An instrumental effect may cause different implantation results. The B^+ beam is extracted at a higher energy and decelerated, which introduces the risk of energy contamination in the beam. The BF_2^+ beam on the other hand has been extracted directly at the appropriate energy. With SIMS analysis, we determined the energy contamination in the B^+ implanted samples to be less than 0.1%, being low enough to have a negligible effect on the results presented in this paper.

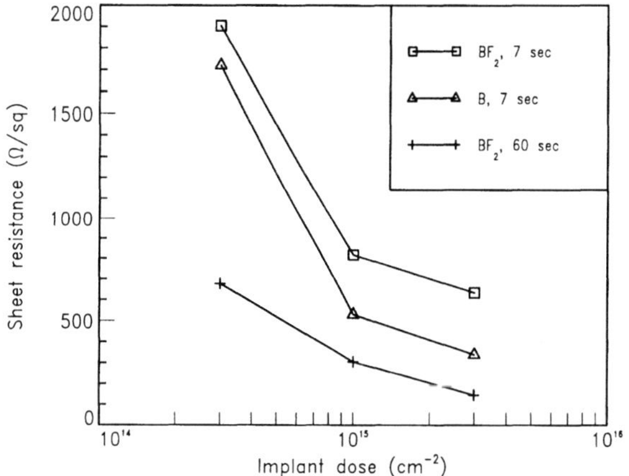

Figure 3a: Sheet resistances of B^+ and BF_2^+ implanted samples as a function of dose and anneal time at 1025°C; the boron (fractional) energy is 1 keV.

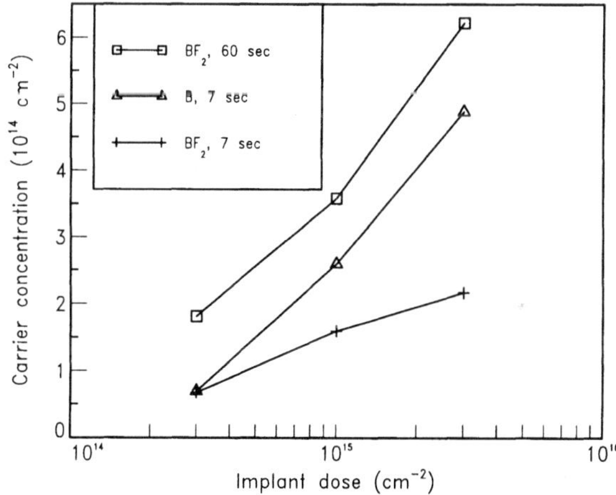

Figure 3b: Carrier concentrations of the samples as in Figure 3a.

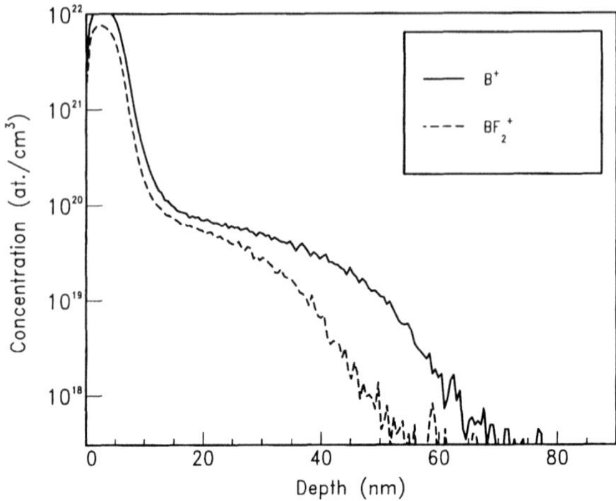

Figure 4: Depth profiles of B^+ and BF_2^+ implanted samples with an implantation dose of 10^{15} cm^{-2}. The rapid thermal anneal was 7 seconds 1025°C.

B) COMPARISON OF ANNEALING AMBIENTS

To study the possible participation of ambient gases to the outdiffusion process, samples were annealed during 20 seconds at 1025°C in N_2, N_2/H_2 (90/10) and O_2 ambients. Figure 5 shows SIMS results, indicating that with respect to (supposedly inert) N_2 ambient, the N_2/H_2 ambient retards the B outdiffusion. This results in a marginally lower junction depth and slightly higher sheet resistance in N_2/H_2 at a fixed thermal budget. This is quantitatively shown in Figure 6, where sheet resistance is plotted against the junction depth for these samples. It is unclear what causes the difference between N_2 and N_2/H_2. We expect that the hydrogen penetrates the SiO_2 and the Si, passivates dangling bonds, physically occupies room in the SiO_2 matrix, and may bind to B atoms. These three mechanisms can influence the B diffusion process and might also affect the B solubility.

Figure 5 and 6 also show the result of the oxygen anneal. In this ambient, two competing mechanisms are present during the anneal. The oxidation process enhances boron diffusivity (in the silicon) by the injection of interstitials; meanwhile the thickness of the screening oxide increases during the thermal step, which is detrimental to the boron outdiffusion. The net result is a lower boron peak concentration in the silicon, yielding a relatively high sheet resistance for a given junction depth (as illustrated in Figure 6).

The described effects of the anneal ambient have been observed both with B^+ and BF_2^+ implanted samples in the same manner. It should be noted that the shallowest junction in Figure 6, fabricated with a BF_2^+ implant, has a junction depth of 33 nm at a sheet resistance of 800 Ω, which is a very competitive result.

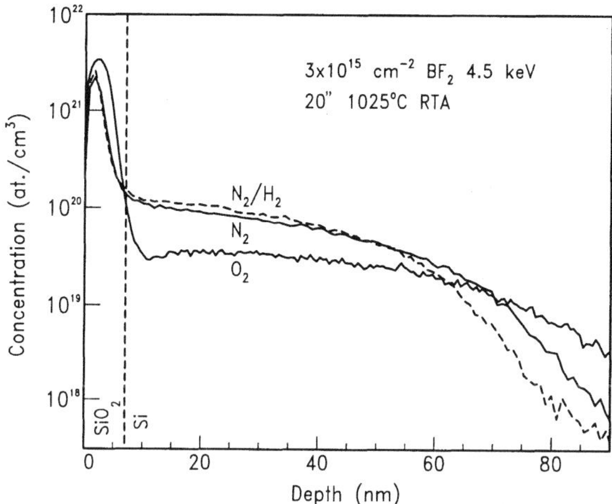

Figure 5: Depth profiles of B in silicon after outdiffusion of B in various ambients.

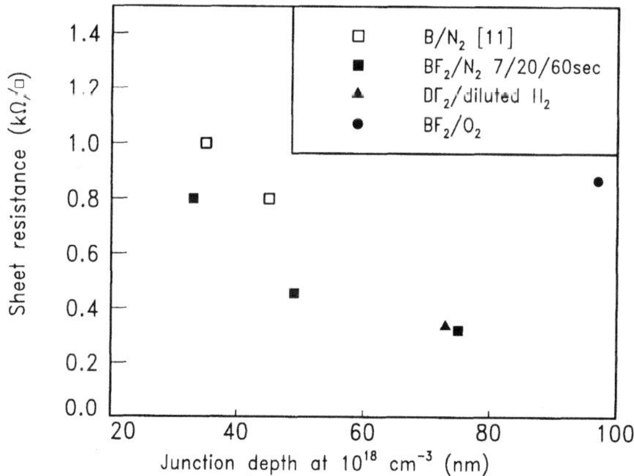

Figure 6: Relation between junction depth and sheet resistance for samples implanted with B^+ and BF_2^+ and annealed in various ambients and for various anneal times.

Concluding the results in Figure 5 and 6, there is no advantage of outdiffusion in N_2/H_2; in view of safety considerations, the N_2 anneal is therefore preferred. Samples annealed in O_2 show a decreased peak concentration of boron in the silicon. As a result, a considerably larger sheet resistance is obtained at a given junction depth. Oxidizing anneals are therefore not attractive for application in the source and drain regions of MOS transistors.

CONCLUSIONS

We have fabricated shallow p^+ junctions in silicon using outdiffusion from implanted oxide. The junctions were compared on the basis of Hall carrier concentration, sheet resistance and SIMS depth profiling. In this paper it is demonstrated that the outdiffusion process can be initiated from both B^+ and BF_2^+ implanted oxides. Outdiffusion is more pronounced when a B^+ implantation is applied; the fluorine remaining after BF_2^+ implantation has no effect on the carrier mobility. Rapid thermal outdiffusion has been carried out in N_2, N_2/H_2 and O_2 ambients. The inert (N_2) ambient is considered to be the most straightforward and effective way to achieve shallow p^+ junctions with competitive sheet resistance.

ACKNOWLEDGEMENTS

We would like to thank Operations WAG for the processing of the wafers, and F. P. Widdershoven for his contributions to the Hall-Van der Pauw measurements. This work was partly supported by the EC within the ULTRA project (Esprit E23806).

REFERENCES

[1] Semiconductor Industry Association, National Technology Roadmap for Semiconductors 1997.
[2] S. Shishiguchi et al., VLSI 1998 Technical Digest p. 134.
[3] E. J. H. Collart et al., J. Vac. Sci. Technol. B16(1) p. 280 (1998).
[4] M. Takase et al., IEDM Technical Digest 1997, p. 475.
[5] K. Goto et al., IEDM Technical Digest 1997 p. 471.
[6] Y. Kiyota et al., IEICE Trans. Electron. E77C, p. 362 (1994).
[7] K. H. Weiner et al., Microelectron. Eng. 20 (1993) p. 107.
[8] M. Bianco et al., Microelectron. Eng. 15 (1991) p. 525;
M. Miyake, J. Electrochem. Soc. Vol. 138, No. 10 p. 3031 (1991).
[9] M. Ono et al., IEDM Technical Digest 1993, p. 119.
[10] E. C. Jones and E. Ishida, Mater. Sci. Eng. R24 (1998) p. 1.
[11] J. Schmitz et al., IEDM Technical Digest 1998, p. 1009.

Section III

MOSFET Source/Drain Engineering: Metals and Silicides

RTP FOR ADVANCED DEVICE FABRICATION USING SHALLOW, ELEVATED, AND SILICIDED JUNCTIONS

C.M. Osburn
Center for Advanced Electronic Materials Processing
North Carolina State University
Raleigh, NC 27695-7911

Rapid thermal processing is needed to minimize motion of dopants in CMOS device channels and in extension junctions. The thermal cycle associated with growth or deposition of the gate dielectric is shown to impact the final channel doping profile, and a variety of deposited gate stack dielectrics have been used successfully to minimize the thermal budget. The use of silicides as diffusion sources reduces transient enhanced dopant diffusion and junction leakage; more importantly, it maximizes the doping concentration at the metal (silicide) - semiconductor contact and thereby minimizes the interfacial contact resistivity. However, junctions beneath the silicide must have sufficiently low sheet resistance, i.e. they must be sufficiently deep, so as to minimize the overall contact resistance. This consideration requires that the SADS junction anneal be around 10 sec at 950°C or a 1050°C spike anneal; thus the silicide technology must be thermally stable under these conditions. To be useful, elevated source/drain processes must employ RTP in order to minimize the impact on the extension junction depth. Epi faceting is an issue for ESD devices. On the one hand, faceting is needed to confine the epi along the edges of trench isolation; on the other hand, facets can lead to locally deeper junctions and silicides. When facets are present, special care must be taken to avoid junction penetration by the silicide. Using SADS technology in ESD devices is a good way to combine the advantages of each, and good device characteristics have been observed.

INTRODUCTION

One of the key challenges in scaling the channel lengths in CMOS devices for future generations of technology is the need to control the channel doping profile and to reduce the depth of both the extension junction and the contacting junction. The need to minimize the parasitic series resistance drives the quest for low resistivity and abrupt extension junctions. Although it is not as deep as the extension junction is, the contacting junction provides two additional challenges: first, it must readily accommodate silicidation, and second, it must provide a low contact resistance. Both of these requirements become crucial as the technology is scaled to below 100 nm.

Figure 1 shows the evolution of junction structures. The technology initially employed a simple, single junction. The LDD structure was invented to reduce hot-electron problems (1). This drain extension structure employs two junctions, one extending to the channel and one at the contact, which are separated by a dielectric spacer. In today's technology the doping in the extension junction is high, and contacting junctions are typically clad with self-aligned silicides. Because the silicide consumes part of the junction, further scaling of junction depths in the substrate is likely to require that the contact junction be elevated. The use of selective deposition processes in the elevated source/drain (ESD) structure allows many process options. For example, the silicide can be used as a diffusion source, Si-Ge can be selectively deposited to reduce contact resistance, and selective silicide or metal might be deposited to totally eliminate the need for a selective epi growth (SEG) step.

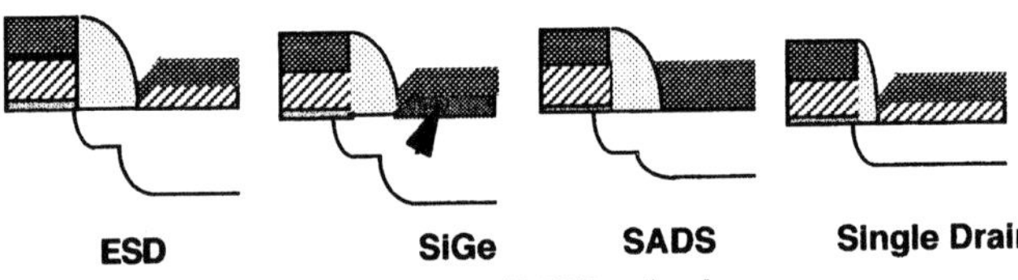

Figure 1. Evolution of junction structure in CMOS technology

A key concern for any junction formation process is its thermal budget. Rapid thermal processing is needed to minimize the depth of the extension junction and the channel doping profiles. A representative set of profiles for 100 nm devices is portrayed in Fig. 2, where it can be seen that control of the dopants to within 30 nm is necessary. Groups at NCSU have successfully employed rapid thermal processing at several points in the process to fabricate 100 nm devices. We have used RTP for: gate-stack formation (2-3), doping of Poly-Si (3), extension junction formation in conjunction with low-energy ion implantation, formation of elevated source/drain junctions (4), and contact junction formation using silicides as diffusion sources (SADS)(5). For maximum scalability, the SADS process has been used in conjunction with ESD technology.

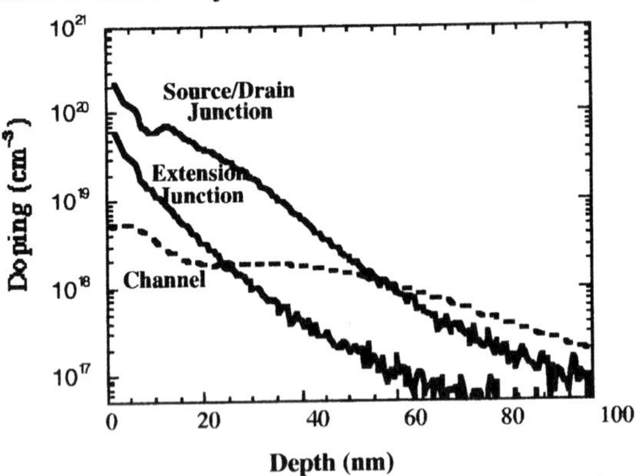

Figure 2. Doping profiles in the channel, extension junction, and contacting (source/drain) junctions for a 100 nm PMOS technology

EXPERIMENTAL RESULTS AND DISCUSSION

Low Thermal Budget Gate Dielectrics to Control Channel Doping

NMOS and CMOS devices were fabricated, starting with lightly doped wafers into which channel dopants were implanted. As can be seen from Fig. 3, the gate dielectric formation step plays a large role in redistribution of channel dopants--more so than subsequent junction formation steps. Thus the thermal cycle used to form the gate is important. For work within NC State Center for Advanced Electronic Materials Processing, alternate gate-stack dielectrics were successfully formed using rapid thermal CVD or remote plasma enhanced CVD of oxide or of oxide/nitride. These processes not only reduced the overall thermal budget, but they also allowed the controlled incorporation of nitrogen at either the Si-SiO_2 interface to enhance reliability or within the bulk of the dielectric to retard boron penetration. For today's generation of 250 nm devices, the lower thermal budget associated with these deposited dielectrics was seen to provide sharper channel doping profiles such that the use of buried-channel PMOS devices was viable—in contrast to the case when thermally grown dielectrics were employed (6). As the gate dielectric becomes thinner, the primary benefit of the rapid thermal dielectrics lies in the ability to control nitrogen incorporation.

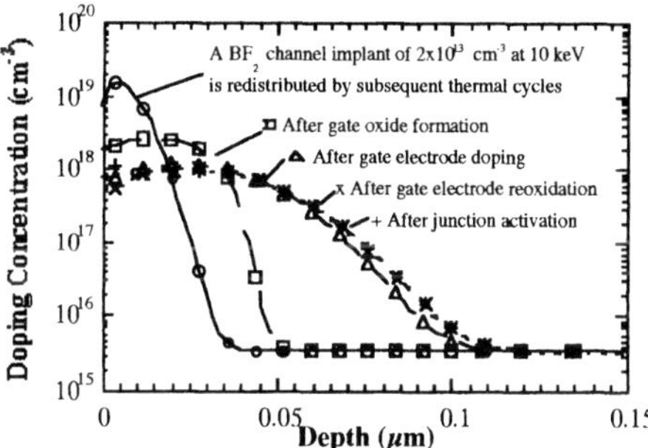

Figure 3. Effect of process steps on channel doping profile. From Yee et al. (2)

Extension Junctions

Through the use of RTA, 20 to 30 nm deep extension junctions, similar to the one portrayed in Fig. 2, having a moderate sheet resistance, have been demonstrated, even without the need to employ ultra-low implant energies. However, once voltages are scaled such that hot carrier degradation is no longer of concern, higher-dose, lower energy extension implants along with fast ramp RTA in a controlled annealing atmosphere are required to minimize the parasitic series resistance. A considerable amount of current research work is being performed in this area (7).

Silicide as a Diffusion Source to Form Contact Junctions

Alternative approaches to the formation of contacting junctions were evaluated. Since this junction is deeper than the extension junction is, the biggest challenge is the formation of the silicided contact. Figure 4 illustrates some of the issues associated with silicided contacts. Because the silicide consumes silicon, the dopant at the silicide/Si contact is less than that at the surface, and the interfacial contact resistivity is higher. Segregation of dopants from the silicon into the silicide, along with evaporation of dopants from the silicide surface lead to further reductions in dopant at the contact and higher resistance. To overcome the silicon-consumption problems associated with traditional self-aligned silicide processes, we examined the use of SADS processes to minimize dopant excursion beyond the silicide.

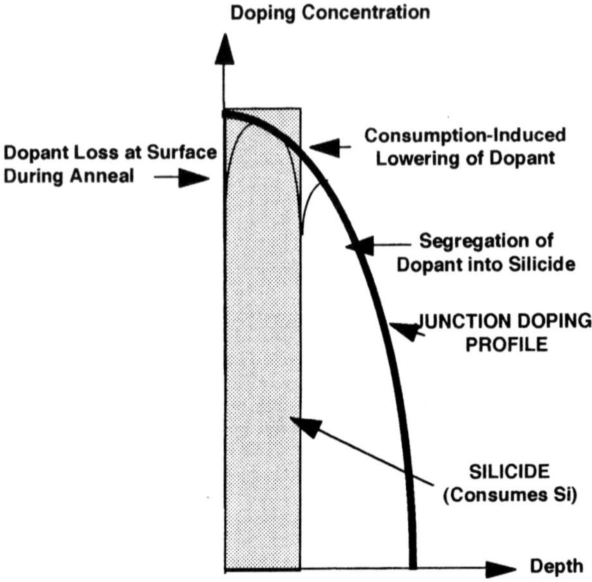

Figure 4. Contact resistance issues associated with silicided contacts

The key potential advantage of the SADS process is that it avoids consumption-induced degradation of the interfacial contact resistivity, along with dopant segregation into the silicide. It also eliminates implant-induced damage in the substrate, thereby permitting use of a low thermal budget anneal. The contact resistance, R_{co}, depends on the interfacial contact resistivity, ρ_c, the sheet resistance *beneath the contact* of the source/drain contacting junction, R_{sd}, the width, W, and the length, L_c, of the contact hole, which for a self-aligned silicide contact amounts to the length of the diffusion region between the edge of the spacer on the Poly-Si gate and the onset of the isolation (LOCOS or trench) region. (8,9).

$$R_{co} = \frac{\sqrt{\rho_c R_{sd}}}{W \bullet \tanh(L_c/L_t)} \quad (1)$$

Where L_t, the transfer length, is the average distance that carriers travel in the diffusion before going into the contact, is given by:

$$L_t = \sqrt{\frac{\rho_c}{R_{sd}}} \quad (2)$$

Ideally the transfer length should be greater than the contact length, so that current flows into the entire length of the contact, and the contact resistance is minimized. Thus care must be exercised when designing SADS junctions, to ensure that the underlying junction is sufficiently deep, i.e., its sheet resistance is sufficiently low, so that the transfer length is comparable to the diffusion width. Otherwise, the SADS process can result in higher resistance than a conventionally silicided junction, having a thinner silicide or a deeper junction. As equation 2 indicates, the lower the interfacial resistivity, the lower the junction sheet resistance must be to maintain an acceptably long transfer length. To achieve a low resistance of the junction beneath the silicide, the doping must be high and the junction must be deep. Assuming that the SADS junction is doped to the electrical activity limit, Fig. 5 shows how the contact resistance varies with junction depth. For a value of ρ_c of 10^{-7} ohm-cm^2, typical for a mid-gap silicide/Si barrier height and electrical activity limit doping, the junction needs to be about 10 nm beyond the silicide before the contact resistance starts to reach a plateau for a contact length (diffusion width) of 100 nm, and it should be over 20 nm deep for a 200 nm long contact. For deeper junctions, increasing the contact length (diffusion width for self-aligned silicided contacts) can be used to lower the contact resistance; however, this requires additional circuit layout area. As a tradeoff, diffusion widths are typically two to three times the lithographic feature size.

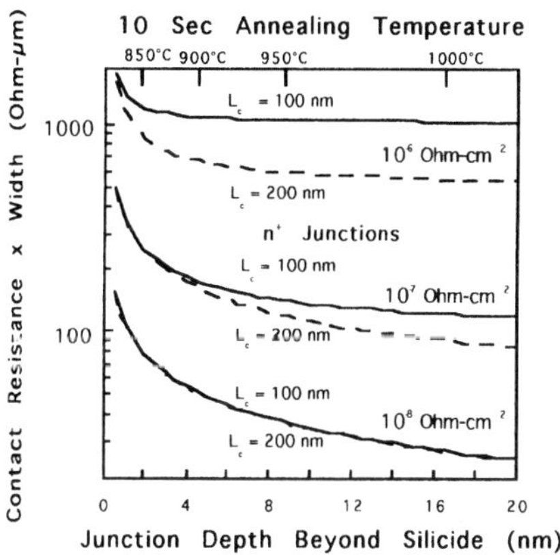

Figure 5. The effect of junction depth beyond the silicide on the contact resistance for different interfacial contact resistivities and contact lengths. The 10 sec RTA temperatures corresponding to the junction depths were calculated using the analysis in (10).

In the absence of transient-enhanced diffusion in SADS processes, a junction depth of 10 nm beyond the silicide requires an RTA cycle of at least 10 sec at 950°C or a spike anneal to 1050°C. To withstand those annealing cycles, the silicide stability must be good. Figure 6 shows that the use of a quasi-epitaxial $CoSi_2$ is preferred over a polycrystalline silicide. A discernible increase in sheet resistance is observed for lines narrower than 0.5 μm when a polycrystalline diffusion source is used; the epitaxial material, made using Ti/Co bilayers, is needed to preserve low sheet resistance in narrow lines. Thus, the proper design of SADS junctions requires the use of a stable silicide and a junction sufficiently beyond the silicide to have a low enough sheet resistance to give a low contact resistance. Under those conditions, devices with SADS junctions exhibit higher drive currents than conventional ones. On the other hand, if these conditions are not met, the contact resistance can be high and the drive current degraded.

Scaling of SADS contacts to ever decreasing dimensions will be a continuing challenge. The minimum contact resistance is achieved when the transfer length is long. In that case the contact resistance varies as the contact area:

$$R_{co} = \frac{\rho_c}{W \bullet L_c} \tag{3}$$

Then the contact resistance varies as the reciprocal square of the dimensional scaling factor. Since the resistance of scaled devices stays roughly constant, ultimately contact resistance will dominate the overall resistance, unless ways, such as metastable dopant activation or lowered contacting barrier height, can be found to lower the interfacial contact resistivity. To maintain a long transfer length, as new ways are found to lower the contact resistivity, concurrently requires a comparable scaling of the sheet resistance of the junction residing beneath the contact--an especially difficult challenge if the total junction depth is also scaled down.

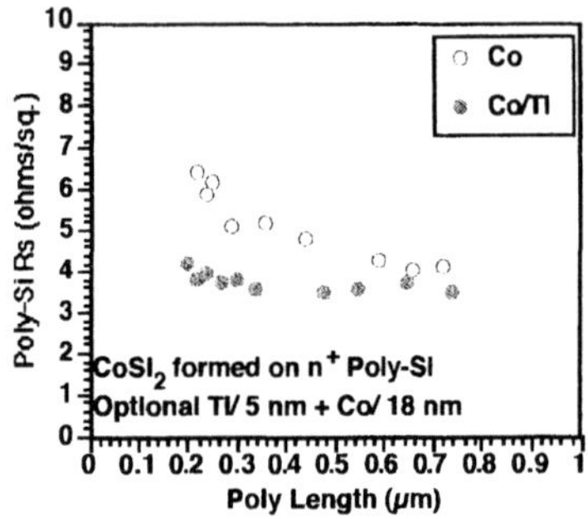

Figure 6. Linewidth dependence of the sheet resistance of silicided Poly-Si in a SADS process. From J.Y. Tsai (5)

Elevated Source/Drain Junctions

Elevation of the contacting junction provides a sacrificial layer of silicon to enable the formation of a thick silicide contact, while maintaining a shallow junction in the substrate. And it allows for a thick, low-sheet-resistance junction beneath the contact. Thus, when the epi layer is heavily doped, the contact resistance is low, and the drive current is higher than in conventional devices. The use of low thermal budget processing for the pre-cleaning and the selective epi steps is critical to minimize the extension junction depth. For the 100-180 nm generations of devices, the selective epi deposition must be constrained to below about 850°C; otherwise diffusion of the extension junctions degrades the device short-channel behavior. Figure 7 shows that extension junction depths less than 40 nm are needed to keep the off-state leakage below 100 pA/μm, for a given substrate doping; as a result, the thermal cycle associated with the selective epi growth (SEG) process must be restricted to less than about 5 min at 850°C (11). Device design constraints for the channel doping and the extension junction depth require that RTP be used for the SEG process.

Faceting of the selectively deposited epi has been found to be a potential issue in ESD devices (11, 12). When junction formation occurs after epi, slightly deeper junctions occur beneath the facet, closest to the device channel where they can aggravate the short-channel threshold voltage rolloff. When junction formation occurs before epi, the silicide penetrates deeper into the substrate along the facets and can result in high leakage and even shorting of the junctions, as can be seen in Fig 8.

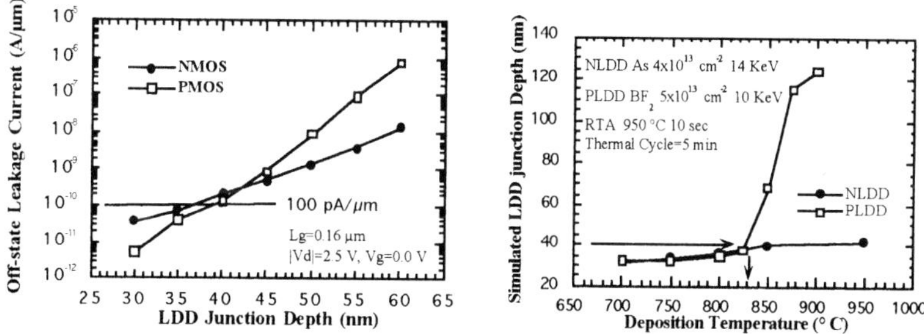

Figure 7. Effect of extension junction depth on off-state leakage of ESD devices (left). Effect of SEG deposition temperature on extension junction depth (right). After Sun *et al.* (11)

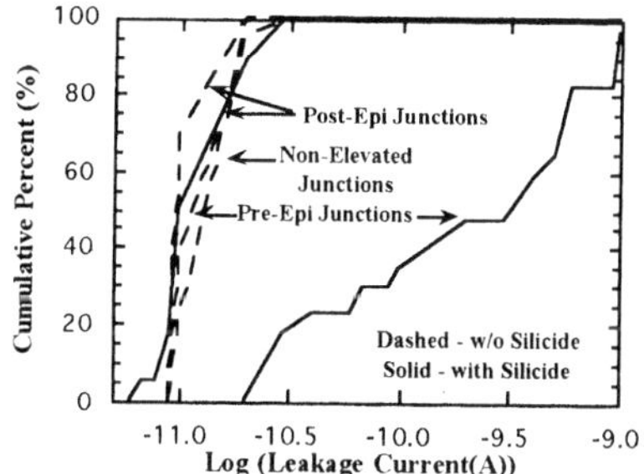

Figure 8. Junction leakage of elevated junctions before and after silicidation for the cases where junction formation precedes and where if follows SEG deposition. Data from J.J. Sun al. (11)

Selective deposition processes which eliminate faceting might seem to eliminate both problems, namely the deeper, local junction that occurs when junction formation is done after epi or the junction leakage (shorting) that occurs when junctions are formed before epi deposition. However, the SEG interaction with the device isolation must also be considered, especially when trench isolation is employed. Ideally one would like a SEG process that a) meets the thermal budget constraint discussed earlier and b) does not facet along the spacer dielectric at the gate edge but does not appreciably extend laterally over a planar dielectric at the (trench) isolation edge. Some of the results of our study of the facet formation sequence are shown in Fig. 9 (13). Using conditions that give highly selective deposition (14, 15), Fig. 9 shows the facet evolution along an essentially co-planar oxide, as would be present at an oxide-filled trench. For thin epi, {113} and {111} facets are present; as the epi gets thicker, {110} and then retrograde $\{11\bar{1}\}$ planes appear. A similar sequence, except for the $\{11\bar{1}\}$ planes, occurs for deposition along a tapered nitride sidewall. A model developed from this data (13), along with other observations (16) suggest that only the {113} and {111} facets will be formed when moderately selective deposition is employed.

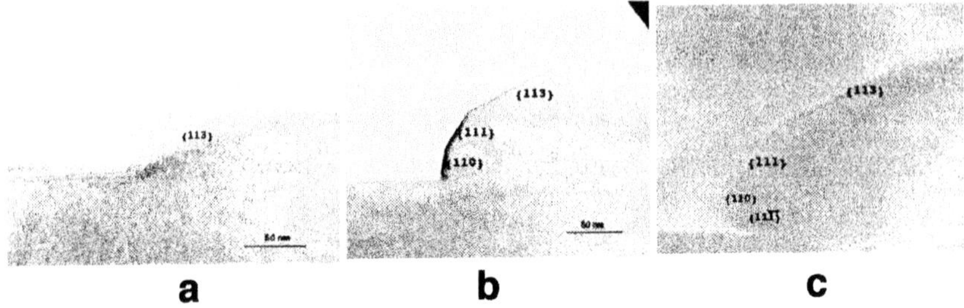

Figure 9. Facet profiles during selective epi deposition of a) 45 nm, b) 90 nm and c) 225 nm. From Srivastava (13) The SEG mask was 5 nm of oxide oriented along the <110> direction, and the deposition conditions are described in (14, 15)

Elevated SADS Junctions

The viability of using SADS for junction formation in conjunction with ESD technology has also been demonstrated (17). As observed previously with planar SADS junctions (4), elevated SADS junctions generally exhibit lower diode leakage than conventional junctions (see Fig 10); the lower leakage is attributed to the fact that the ion implantation damage is confined to the silicide layer in the SADS process. Low leakage is observed even under conditions where a significant doping tail is implanted through the silicide (115 keV As in Fig. 10). A comparison of ESD device characteristics using conventional junction formation and using SADS is shown in Fig. 11, where it is seen that elevated SADS provides a viable, alternative junction formation technology. Elevated SADS is seen to provide the advantages of both ESD and SADS technologies. The elevated layer provides a silicon source for consumption during silicidation, thereby allowing the formation of a thicker, lower sheet resistance, silicide. The SADS process minimizes the junction depth beneath the silicide and minimizes the interfacial contact resistivity by maximizing the interfacial dopant concentration.

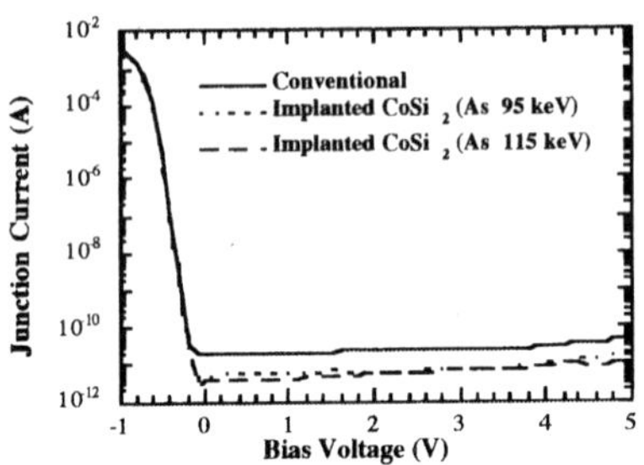

Figure 10. Leakage of elevated n+ SADS junctions. These junctions were formed on 100 nm elevated epi by 4e15 As implants into $CoSi_2$ (25 nm Co deposition) followed by 10 sec 900°C RTA. After J.J. Sun et al. (17)

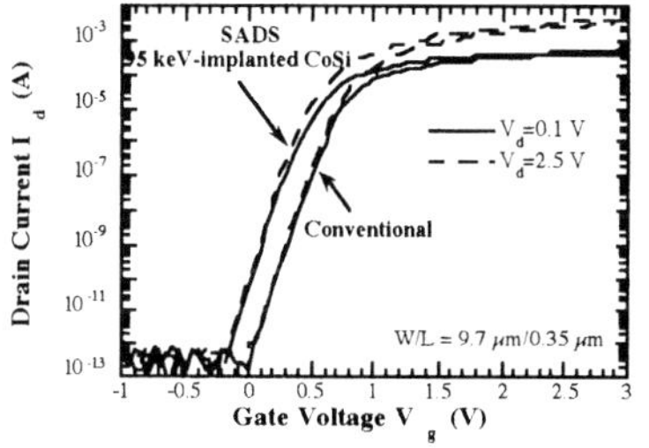

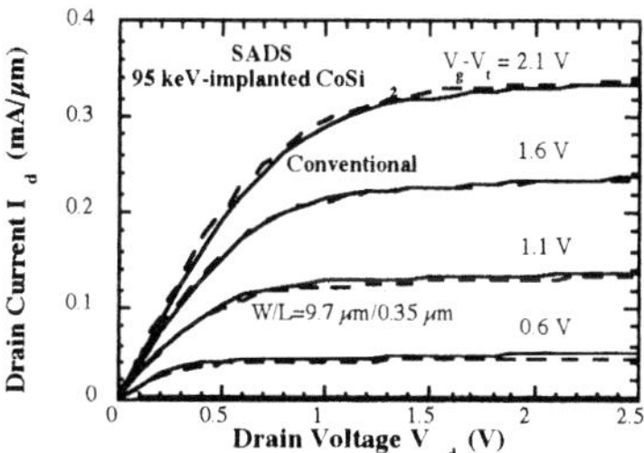

Figure 11. Characteristics of SADS junctions: left) subthreshold region; right) linear and saturation regions. After J.J. Sun et al. (17)

SUMMARY AND CONCLUSIONS

Rapid thermal processing plays an important role in the integration of advanced CMOS technologies. RTP is especially needed to minimize motion of dopants in the device channel and in the extension junction. The thermal cycle associated with growth or deposition of the gate dielectric was shown to impact the final channel doping profile, even though the majority of thermal steps follow gate formation. A variety of deposited gate stack dielectrics have been successfully used to minimize the thermal budget.

Several candidate advanced technologies for junction formation have been evaluated. These include the use of silicides as diffusion sources. A key advantage of this technology is that it maximizes the doping concentration at the metal (silicide) - semiconductor contact and thereby minimizes the interfacial contact resistivity. Although it seems possible to use SADS to make vanishingly shallow junctions beneath the silicide, in part because transient-enhanced diffusion is reduced or eliminated, such a configuration is not desirable from a contact resistance point of view. Junctions beneath the silicide must have sufficiently low sheet resistance, i.e. they must be sufficiently deep, so as to minimize the overall contact resistance. This consideration requires that the SADS junction anneal be at least about 10 sec at 950°C or a 1050°C spike anneal; thus the silicide technology must be thermally stable under these conditions.

Elevation of junctions using SEG provides a good way to make a shallow junction in the substrate while providing a thick sacrificial silicon layer for silicidation. To be useful, the SEG process must employ RTP in order to minimize the impact on the extension junction depth. Epi faceting is an issue for ESD devices. On the one hand, faceting is needed to confine the epi along the edges of trench isolation; on the other hand, facets can lead to locally deeper junctions and silicides. When facets are present, special care must be taken to avoid junction penetration by the silicide. Using SADS technology in ESD devices is a good way to combine the advantages of each. Good characteristics have been observed in elevated SADS devices.

ACKNOWLEDGEMENTS

A team of faculty and students within the Center for Advanced Electronic Processing at NC State University performed the work described here, including: J. Hauser, G. Lucovsky, D. Maher, N. Masnari, M.C. Öztürk, J.J Wortman, R. Bartholomew, K. Bellur, H. Jiang, A. Srivastava, J. Sun, J. -Y. Tsai, Q. -F. Wang, K.F. Yee, X. Zhang, I. Ban, M. Celik, H. Heinisch, W. Henson, P. O'Neil, C. Parker, and E. Vogel. Other collaborations with D.F. Downey (Varian), B. Fowler (Motorola-

SEMATECH), E. Ishida (AMD), and R. Westhoff (Lawrence Semiconductor) on shallow, silicided, and elevated junction devices are gratefully appreciated.

REFERENCES

1. S. Ogura, P.J. Tsang, W. Walker, D. Critchlow, and J. Shepard, *IEEE Trans, Electron Devices*, **ED-27**, 1359 (1980).

2. K.F. Yee, C.M. Osburn, N.A. Masnari, J.R. Hauser, C.G. Parker, G. Lucovsky, W.K. Henson, and J.J. Wortman, *Proceedings of the Rapid Thermal and Integrated Processing VII Conference*, Vol 525, p.157, Materials Research Society, San Francisco, April (1998).

3. A. Srivastava, H.H. Heinisch, E. Vogel, C. Parker, C.M. Osburn, N.A. Masnari, J.J. Wortman, and J.R. Hauser, *Proceedings of the Rapid Thermal and Integrated Processing VII Conference*, Vol 525, p.163, Materials Research Society, San Francisco, April (1998).

4. A. Srivastava, J. Sun, R. Bartholomew, R. O'Neil, M. Celik, C.M. Osburn, N.A. Masnari, M.C. Öztürk, R. Westhoff, and B. Fowler, in *ULSI Science and Technology/* 1997, H.Z. Massoud, H. Iwai, C. Claeys, and R.B. Fair, Editors, PV 97-3, p.571, The Electrochemical Society Proceedings Series, Pennington, NJ (1997).

5. C.M. Osburn, J.Y. Tsai, and J. Sun, *J. Electronic Mater.*, **25**, 1725, (1996).

6. X. Zhang, Ph.D. Thesis, Duke University (1994).

7. For a representative sampling, see the Proceedings of the Ion Implant Technology Conferences.

8. D.B. Scott, W.R. Hunter, and H. Shichijo, *IEEE Trans. Electron Devices*, **ED-29**, 651, (1982).

9. D.B. Scott, R.A. Chapman, C.C. Wei, S.S. Mahant-Shette, R.A. Haken, and T.C. Holloway, *IEEE Trans. Electron Devices*, **ED-34**, 562, (1987).

10. H. Jiang, C.M. Osburn. Z. -G. Xiao, G. McGuire, and G.A. Rozgonyi, *J. Electrochem. Soc.*, **139**, 211, (1992).

11. J. Sun, R.F. Bartholomew, K. Bellur, P.A. O'Neil, A. Srivastiva, K.E. Violette, M.C. Öztürk, and C.M. Osburn, *Proceedings of the Rapid Thermal and Integrated Processing V Conference*, Vol. **429**, p.343, Materials Research Society, (1996).

12. J.J. Sun and C.M. Osburn, *IEEE Trans. Electron Devices*, 45, 1377, (1998).

13. A. Srivastava. Ph.D. Thesis, NC State University (1998).

14. K.E. Violette, M.K. Sanghaneria, M.C. Ozturk, G. Harris, and D. M. Maher, *J. Electrochem. Soc.*, **141**, 3269, (1994).

15. P.A. O'Neil, M.C. Ozturk, K.E. Violette, D. Batchelor, K. Christensen, and D.M. Maher, *J. Electrochem. Soc.*, **144**, 3042, (1997).

16. C.I. Dowley, G.A. Reid, and R. Hull, *Appl. Phys. Lett.*, **52**, 546, (1988).

17. J.J. Sun, J. -Y. Tsai, and C.M. Osburn, *IEEE Trans. Electron Devices*, **45**, 1946, (1988).

Pre-deposition Treatments for Selective Rapid Thermal Chemical Vapor Deposition of $TiSi_2$ on Arsenic-Implanted Silicon Substrates

Hua Fang, Mehmet C. Öztürk and Patricia A. O'Neil
Department of Electrical and Computer Engineering, North Carolina State University
EGRC Building, 1010 Main Campus Drive, Box 7920, Raleigh, NC 27695-7920

Ed Seebauer
Department of Chemical Engineering ,University of Illinois at Urbana Champaign,
Urbana-Champaign, Illinois 61801

During the chemical vapor deposition of $TiSi_2$ on As implanted Si using $TiCl_4$ and SiH_4, As introduces a barrier to nucleation and a mechanism that results in enhanced substrate consumption. These undesirable effects have been attributed to surface passivation by As. In this work, two pre-deposition treatments have been examined to alleviate these problems: 1. insertion of an epitaxial silicon buffer layer and 2. annealing in vacuum or hydrogen. The experimental results indicate that either annealing in vacuum or the insertion of a silicon buffer layer may potentially suppress the undesirable effects of As.

INTRODUCTION

Selective chemical vapor deposition (CVD) of titanium disilicide ($TiSi_2$) has gained increasing interest because of the possibility of consumption-free silicide formation and process simplicity (1,2). However, recent studies have shown that depositions on arsenic-implanted substrates exhibit anomalous substrate consumption - even on substrates with doses as low as 3×10^{14} cm^{-2}. On heavily doped substrates (5×10^{15} cm^{-2}), nucleation of continuous $TiSi_2$ films is challenging even at temperatures as high as 850°C (3). Arsenic surface passivation is believed to cause these two effects. These effects were not observed for $TiSi_2$ deposition on boron doped silicon substrates (4). In order to use this process in CMOS device integration, the effects caused by As must be eliminated. In this work we have examined two pre-deposition treatments as potential solutions to the problem: the use of an epitaxial silicon buffer layer or annealing in vacuum or H_2 to remove the As passivated surface at a low temperature. The effects of these treatments on $TiSi_2$ nucleation as well as substrate consumption have been investigated.

EXPERIMENTAL

In this study, 100mm, p-type, 5-8 Ω/square (100) silicon wafers were used. Following a standard RCA clean, a 1000Å SiO_2 layer was thermally grown on the substrates. Silicon windows were opened in this oxide by photolithography and etching in a buffered HF solution. Arsenic was then implanted at different doses ranging from

$3x10^{14}$ cm^{-2} to $5x10^{15}$ cm^{-2} at a fixed energy of 50 keV. Annealing was performed in a Heatpulse 210T^{Tm} rapid thermal annealing system at 1000°C for 10 seconds in Ar.

Titanium silicide depositions were carried out in a rapid thermal chemical vapor deposition (RTCVD) reactor designed at North Carolina State University. The reactor consists of a water cooled stainless steel deposition chamber and a load-lock. The base pressure of the system is maintained below $1x10^{-6}$ Torr. Heating is achieved by an array of tungsten halogen lamps. A more detailed description of the reactor can be found elsewhere (5). $TiSi_2$ depositions in this system were achieved using pure $TiCl_4$ and SiH_4 diluted in He (10 %) as the precursors.

Selective epitaxial silicon films were grown in an ultra-high vacuum (UHV) RTCVD system, which consists of three chambers: a sample entry chamber, an intermediate chamber, and a main process chamber. Cryopumps are used to maintain an ultrahigh vacuum base pressure at 10^{-9} Torr. A magnetically levitated oil-free turbormolecular/molecular drag combination pump is used during epitaxial growth. Epitaxial growth is achieved at 800°C using Si_2H_6 and Cl_2. In-Situ surface preparation consists of a pre-bake in vacuum at 830°C for 10 s. A detailed description of the system and the growth process can be found in earlier publications from this laboratory (6).

The thickness of the selectively deposited $TiSi_2$ layers and the amount of silicon substrate consumption were determined from step-height measurements using stylus profilometry. Atomic force microscopy (AFM) were used to study the surface morphology. Arsenic profiles were obtained using secondary ion mass spectroscopy (SIMS).

RESULTS AND DISCUSSION

1. Using An Epitaxial Silicon Buffer Layer

The purpose of using a silicon buffer layer is to cap the heavily arsenic passivated surface. Regolini et al. have also considered the use of selectively grown epitaxial silicon layers prior to the CVD of $TiSi_2$ (7,8). In their work, however, no attempts were made to reach an in-depth understanding of the impact of this approach on nucleation and substrate consumption. In this study, selective silicon epitaxial films were utilized. The lack of dopants in the buffer layer will cause increased series resistance since the dopant concentration at the $TiSi_2$/Si interface is critical in achieving a low resistivity junction contact. This problem can be solved if the buffer layer is fully consumed during $TiSi_2$ deposition and/or As diffuses readily into the layer to maintain a high As concentration at the interface. In this work, we have considered selective silicon buffer layers with different thicknesses ranging from ultra-thin (~ 20 nm) to quite thick (~ 130 nm) to examine the effectiveness of the buffer layer on $TiSi_2$ nucleation and substrate silicon consumption. The arsenic dose used in these experiments was $5x10^{15}cm^{-2}$ which is high enough to completely suppress $TiSi_2$ nucleation. $TiSi_2$ films were deposited onto these Si revealed consumption free $TiSi_2$ deposition at this temperature. The deposition time was kept constant at 36 seconds. The flow rates of SiH_4 and $TiCl_4$ were 30 sccm and 0.5 sccm, respectively.

The thickness of the deposited $TiSi_2$ layers and the silicon consumption are shown in Figure 1 as a function of the Si buffer layer thickness. The consumption figures shown consider the buffer layer as part of the substrate. Thus, if the silicon consumed is less

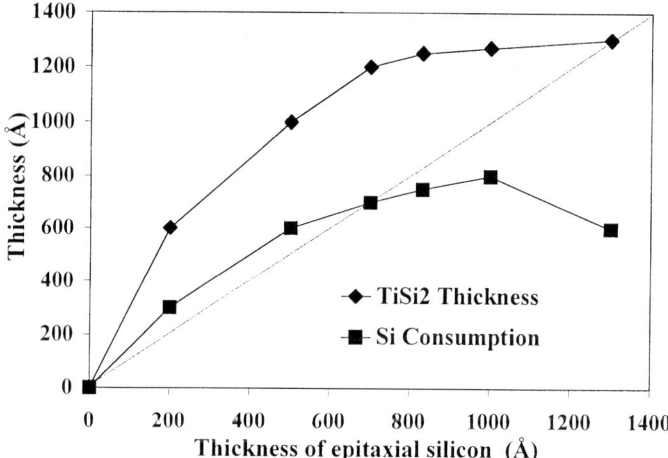

Figure 1: Dependence of $TiSi_2$ thickness and substrate silicon on the thickness of epitaxial silicon buffer layer.

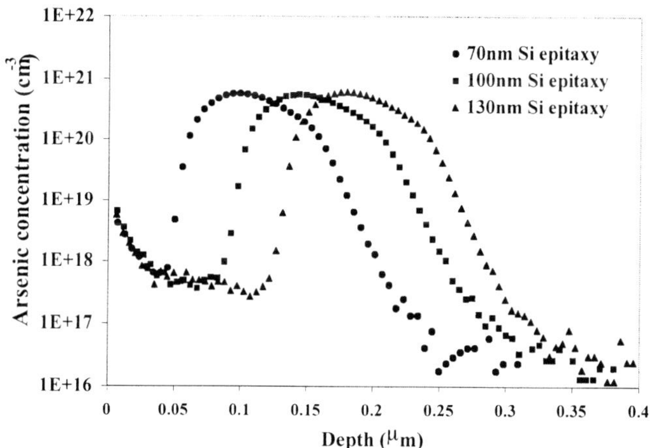

Figure 2 Arsenic profiles after depositions of epitaxial Si with various thickness obtained by SIMS (Arsenic-implant dose: $5 \times 10^{15} cm^{-2}$).

than the thickness of the buffer layer, the original junction is expected to remain untouched. Without a buffer layer, previous results have shown that $TiSi_2$ nucleation can not be achieved even at 850°C for 48 seconds (3). Figure 1 shows that with a 20nm silicon buffer layer, nucleation can be significantly accelerated and $TiSi_2$ film can be formed at 750°C. However, a buffer layer with a thickness of at least 70nm is required to form $TiSi_2$ with an average growth rate comparable to that on undoped substrates (~200 nm/min). Figure 1 also reveals that when the thickness of the buffer layer is less than 70nm, the layer is fully consumed and consumption stops almost right at the interface of the epitaxial silicon layer and the original silicon substrate.

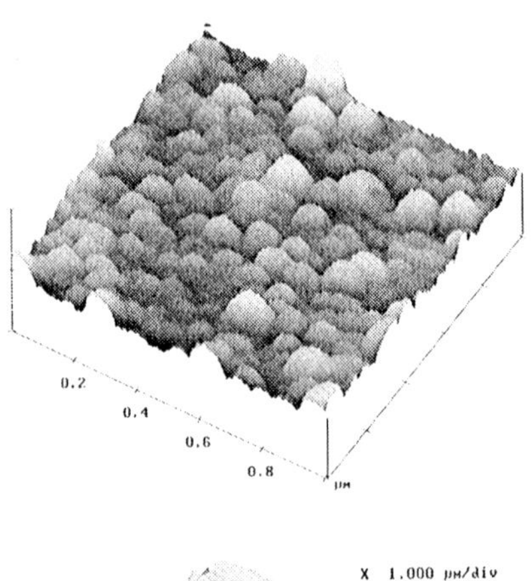

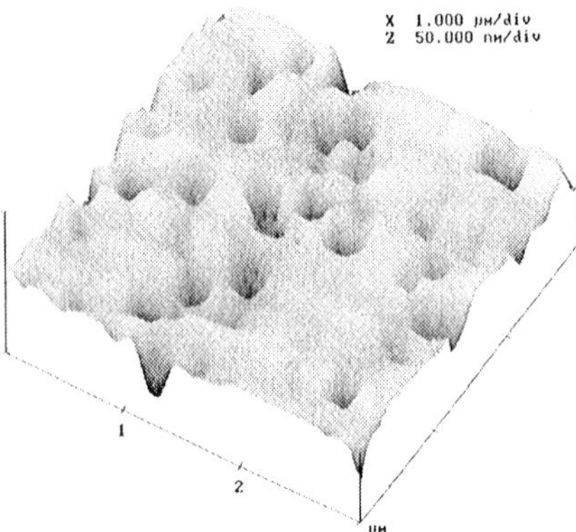

Figure 3 AFM images of 20 nm and 100 nm thick silicon epitaxial layers grown on heavily As implanted Si.

Figure 2 shows arsenic profiles after epitaxial Si growth. The profiles indicate a relatively low As concentration in the bulk (~$5x10^{17}$ cm^{-3}) and an As peak at the surface of the buffer layer. The profiles indicate a sharp As transition between the substrate and the buffer layer. The presence of arsenic on the surface and in the bulk of the epitaxial layer could result from arsenic out-diffusion as well as auto-doping. The existence of As at the surface and in the bulk of epi-Si buffer layers pushes the following $TiSi_2$ deposition into the consumption mode. Moreover, the epitaxial silicon layers are far from perfect, which is also a result of the high As concentration in the substrate. Figure 3 shows the AFM surface scans of 20nm and 100nm thick epitaxial layers. As shown, the thinner layer is in the form of discrete nucleation islands and a continuous silicon epitaxial film has not been formed yet. This suggests that arsenic introduces a nucleation barrier to Si deposition as well. On the thicker sample, the surface still reveals pyramidal voids with high density. Fortunately, these voids can be smeared out after $TiSi_2$ deposition (Figure 5). As it can be seen, even though the Si buffer layers are defective, continuous and uniform $TiSi_2$ films can be formed.

Figure 1 reveals that by adding a silicon buffer layer, the nucleation of $TiSi_2$ can be significantly improved. $TiSi_2$ films can be deposited at 750°C on substrates with an

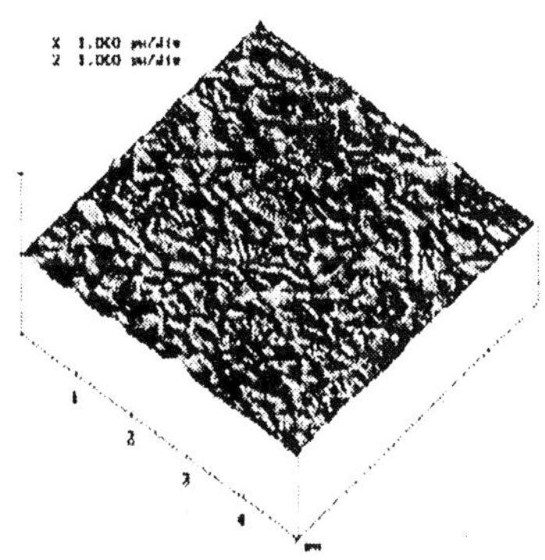

Figure 4 AFM image of $TiSi_2$ deposited on 100nm Si buffer layer.

arsenic dose of $5x10^{15}$ cm^{-2}. The effect is visible even for buffer layers as thin as 20 nm. When the thickness of the buffer layer is approximately 60 nm, the entire buffer layer is consumed resulting in an As concentration of approximately $2x10^{20}cm^{-3}$ at the $TiSi_2/Si$ interface (SIMS profiles after $TiSi_2$ deposition not shown here). While the interface concentration is still somewhat lower than that desired for contacts for the sub 0.1 μm CMOS technology nodes, the result is promising. By optimizing the Si thickness a higher As concentration may be achieved at the interface. Figure 1 also shows that if Si buffer layer is thinner than 70nm, consumption will automatically stop at the surface of the original n^+ junction. This property can be used to achieve symmetrical deposition of $TiSi_2$ on n^+ and p^+ areas because $TiCl_4$ to SiH_4 flow ratio can be adjusted to consume the entire buffer layer on p^+ areas. Unfortunately, one problem still exists with CMOS integration caused by epitaxial facets formed during Si buffer layer deposition. Facets formed on arsenic implanted substrates are not problematic because consumption will stop at the original junction. However, facets formed on p^+ areas could result in uneven consumption. Therefore, one must either consider growing buffer layers only on As doped regions or find an optimum thickness that results in acceptable contact resistance on both junctions.

2.In-Situ Annealing in Vacuum or Hydrogen

Arsenic surface passivation is proposed as the reason for the undesirable effects caused by arsenic (3). It is therefore conceivable that at least the nucleation problem can be alleviated if the As layer on the surface is somehow removed prior to deposition. It may also be possible to modify the surface such that in spite of the surface As, the Si surface lends itself to deposition more readily (e.g. surface roughening). In this study, we have considered in-situ bakes in vacuum as well as in hydrogen. Our hope was to supply sufficient energy to desorb As from the Si surface. However, in doing this, the baking temperature can not be very high, otherwise, arsenic atoms in the substrate can diffuse to the surface and replenish the As lost from the surface. In this experiment, all in-situ anneals were performed at 800°C.

Figure 5(a) shows the thickness of deposited $TiSi_2$ as a function of in-situ baking time in vacuum and in hydrogen. Samples were implanted with an arsenic dose of 5×10^{15} cm^{-2}. $TiSi_2$ depositions were preformed at 830°C for 36 seconds. As shown in Figure

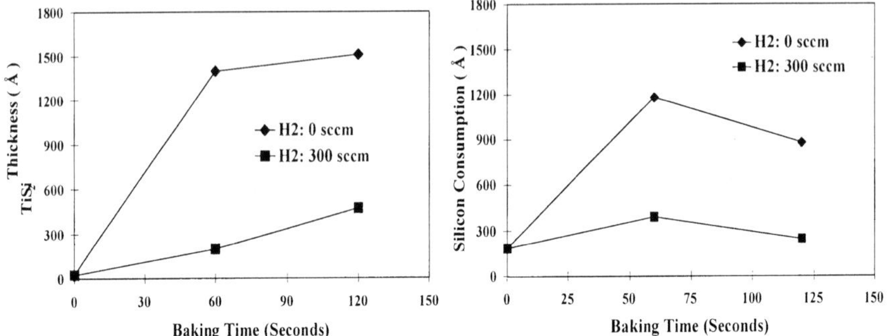

Figure 5: Deposited TiSi2 thickness (a) and substrate silicon consumption (b) as a function of in-situ baking time. Arsenic implant dose: 5×10^{15} cm^{-2}.

5(a), $TiSi_2$ nucleation could not be achieved without in-situ baking. However, by annealing the sample in vacuum for 60 s, the nucleation barrier was completely removed yielding a $TiSi_2$ thickness comparable to that deposited on undoped substrate with the conditions used. Interestingly, Figure 5(a) also shows that adding H_2 into the in-situ baking environment is not as effective as baking in vacuum. In both cases, increasing the baking time improves nucleation. This behavior can be explained by considering two potential mechanisms. One possibility is that with a hydrogen background in the annealing environment, As can not desorb readily from the Si surface. The second possibility is that annealing in vacuum modifies the surface either by roughening or by introducing a contamination layer (e.g. a carbon rich surface) which provides a better surface for nucleation. Experiments are currently underway to isolate the reasons for the improved nucleation.

The corresponding substrate silicon consumption is shown in Figure 5(b). As it can be seen, although nucleation is no longer a significant issue, substrate consumption is still present. As profiles obtained from SIMS measurements indicate large concentrations of As in the deposited $TiSi_2$ layers. It is clear that As readily moves into the $TiSi_2$ layer during deposition and maintains a high concentration on the $TiSi_2$ surface. Therefore, As atoms residing on the $TiSi_2$ surface may conceivably suppress Si adsorption from the gas phase and leave the surface with an ample amount of adsorbed Ti atoms. We propose that these Ti atoms are then available to react with the silicon substrate thereby accounting for the observed Si consumption. Provided that this explanation is valid, the only feasible way to suppress the observed substrate consumption would be to suppress As diffusion into the $TiSi_2$ layer.

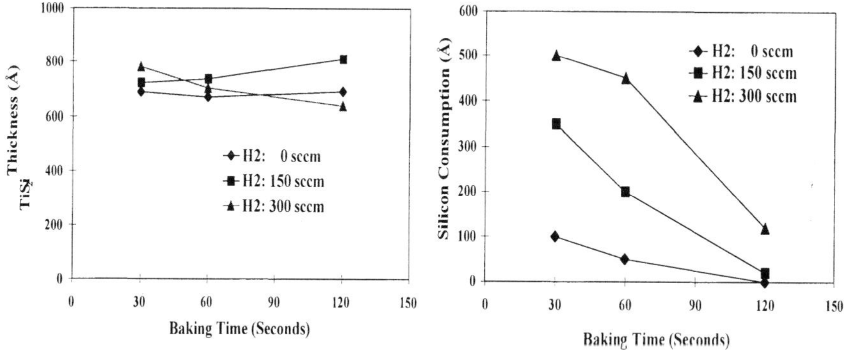

Figure 6: $TiSi_2$ thickness (a) and substrate silicon consumption (b) as a function of in-situ baking time. Arsenic dose is 1×10^{15} cm^{-2}

To further study the effect of in-situ baking on substrate consumption, depositions were performed on substrates with a relatively low arsenic dose ($1\times10^{15}cm^{-2}$). The conditions for in-situ baking and $TiSi_2$ deposition were the same as above. The results are presented in Figure 6, which shows the effect of in-situ baking on $TiSi_2$ thickness and substrate consumption for different H_2 flows. It can be seen that with this lower As dose, $TiSi_2$ can be deposited on Si with no apparent incubation period without the need for in-situ baking. However, without in-situ annealing appreciable consumption still persists. As it can be seen, the $TiSi_2$ thickness is virtually not affected by the in-situ baking time or the H_2 flow rate. On the other hand, both parameters have a strong impact on the substrate consumption. Figure 6(b) shows that longer in-situ baking times and lower hydrogen flows result in reduced substrate consumption. With a 60 s in-situ bake in vacuum, $TiSi_2$ deposition can be achieved without substrate consumption. These results suggest that the initial stages of $TiSi_2$ deposition can influence consumption as well. It appears that a link exists between nucleation and consumption. Figure 6 is also supportive of our previous argument regarding the impact of hydrogen. As shown, increasing the hydrogen background results in enhanced substrate consumption. This could be due to one or more of the following reasons: 1. H_2 itself passivates the Si surface (9). H_2 desorption from Si is known to be affected by dopants in Si. It has been shown that hydrogen desorption is faster with boron but slower with phosphorus (10,11). Because such a phenomenon is related to the electron configuration of atoms, it is expected that arsenic will retard H_2 desorption as well. 2. The As surface concentration increases with H_2 baking as observed with a microwave hydrogen plasma (12). The dissociation of As clusters by atomic hydrogen is believed to be responsible for the observation. 3. The surface defects that help adsorption of reactants are annihilated by baking in H_2. Annealing in H_2 has been used to decorate damage because broken bonds formed during implantation preferentially adsorb hydrogen (13).

CONCLUSIONS

Experimental results indicate that an undoped epitaxial Si buffer layer can significantly enhance $TiSi_2$ nucleation on heavily As implanted substrates. To achieve a low contact resistance, this approach requires consumption of the buffer layer yielding a sufficiently high As concentration at the $TiSi_2$/Si interface. By using an optimum silicon buffer layer thickness (50-70nm in this study), normal growth of $TiSi_2$ can be achieved with the buffer layer fully consumed and substrate consumption fully suppressed. However, facet and defect free epitaxial silicon growth is critical in order for successful CMOS integration.

Annealing in vacuum prior to $TiSi_2$ deposition appears to be a promising approach to suppress the nucleation barrier as well as enhanced consumption. We have deposited consumption free $TiSi_2$ layers on heavily As implanted substrates by annealing the wafers at 800°C for 60 s. On the other hand if hydrogen is used in the annealing ambient, these advantages are lost. Further studies are underway to fully understand the mechanisms involved during these in-situ bakes.

(This work is supported by NSF Engineering Research Centers Program through the center of Advanced Electronic Materials Processing (Grant CDR-8721545) and SEMATECH.)

REFERENCES

1. P. S. Southwell and E. G. Seebauer, *J. Electrochem. Soc.*, **143**, 1726 (1996).
2. J. Mercier, J. L. Regolini, S. Bondar, D. Maury and C. Morin, *Appl. Surf. Sci.*, **100/101**, 566 (1996).
3. H. Fang and M. C. Öztürk, in *Advanced Interconnects and Contact Materials and Processes for Future Integrated Circuits/1998*, Shyam P. Murarka, Moshe Eizenberg, David B. Fraser, Roland Madar and Raymond Tung, 514, p.231, The Materials Research Society Proceedings Series, Warrendale, PA (1998).
4. M. C. Öztürk and C. E. Weintraub, to be published.
5. G.C. Xing, and M.C. Öztürk, Mater. Res. Soc. Proc. ULSI-VIII, p.309 (1993).
6. K. E. Violette, M. K. Sanganeria, M. C. Öztürk, G. Harris, and D. M. Maher, *J. of Electrochem. Soc.*, **141,** 3269 (1994).
7. J. L. Regolini, J. Margail, C. Morin, and P. Gouy-Pailler, *Mater. Res. Soc. Proc.*, **342**, p.249 (1994).
8. J. L. Regolini, J. Margail, S. Bondar, D. Maury, and C. Morin, *Appl. Surf. Sci.*, **100/101**, 566 (1996)
9. H. Bender, S. Verhaverbeke and M. M. Heyns, *J. Electrochem. Soc.*, **141**, 3128 (1994)
10. D.S. Yoo, M. Suemitsu and N. Miyamoto, *J. Appl. Phys.*, **78**, 4988 (1995)
11. B. Doris, J. Fretwell, J. L. Erskine and S. K. Banerjee, *Appl. Phys. Lett.*, **70**, 2819 (1997)
12. K. Yokota, K. Hosokawa, K. Terada, K. Hirai, H. Takano, M. Kumagai, Y. Ando and K. Matsuda, *J. Electrochem. Soc.*, **145**, 1208 (1998)
13. K. Saito, Y. Sato, N. Yabumoto and Y. Homma, *J. Electrochem. Soc.*, **143**, 4101 (1996)

Section IV

MOSFET Source/Drain Engineering: Contacts

JUNCTION PERIMETER LEAKAGE CONSIDERATIONS FOR THE INTEGRATION OF $CoSi_2$ AND DAMASCENE W LOCAL INTERCONNECT IN DYNAMIC LOGIC COMPATIBLE, SUB 0.25μm CMOS TECHNOLOGIES

Paul D. Agnello
IBM Semiconductor Research and Development Center, Hopewell Junction, NY 12533

The integration of low leakage $CoSi_2$ silicided junctions and damascene W local interconnect is demonstrated for advanced 0.25μm and sub-0.25μm technologies. Distributed and localized leakage can be controlled by modifying silicide morphology and optimizing junctions, resulting in microprocessors with low standby current and active power. These manufacturable technologies are suitable for low power, high performance microprocessors employing dynamic logic.

INTRODUCTION

$CoSi_2$ does not display narrow line resistance problems like $TiSi_2$ but has been reported to have higher junction leakage[1]. Some high speed circuit designs employ dynamic logic whose function can be degraded with excessive node leakage. Most reports on integration of $CoSi_2$ with sub 0.25μm technologies have little data on junction leakage[2,3,4] or focus on the area component of leakage[5]. Here we report the integration of W local interconnect with $CoSi_2$ for 0.25μm and sub-0.25μm technologies. The addition of W local interconnect and the use of aggressive lateral doping profiles to control short-channel effects makes maintaining low junction leakage associated with gate and isolation perimeter problematic.

TECHNOLOGY FEATURES

The 0.25μm technology is a 1.8V dual workfunction high performance logic process with 4.0nm gate oxide and shallow extensions combined with halos to control chort channel effects. Hierarchical wiring with optional fat-thick levels at M5 and M6 are offered for large, high-performance processors[6]. Additionally, a 0.3 Ω/□ local interconnect formed after the completion of the salicide process, with borderless contact to isolation, enables a 10% improvement in random logic density and a 6 transistor embedded SRAM cell of 8.6μm^2 in a 0.25μm process and 6.8 μm^2 in a sub-0.25 μm process[7]. Borderless local interconnect to isolation edge (Fig. 1) is allowed making the control of leakage at the isolation edge difficult.

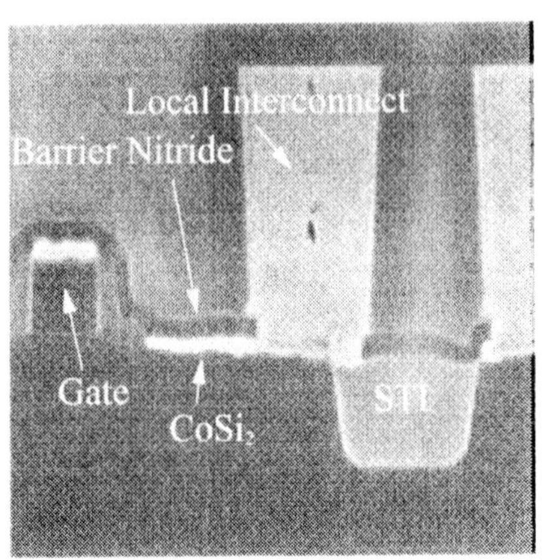

Fig. 1. SEM X-Section of W Interconnect borderless to STI

$CoSi_2$ FORMATION

$CoSi_2$ forms at lower temperatures as Co diffuses into the Si substrate. At higher temperatures, Si diffusion into the metal plays an increasingly important role. The resulting $CoSi_2$ film has an embedded interfacial layer as seen in the TEM cross section Fig. 2. The layer has been identified as SiO_x by microspot EDAX analysis. The location of the film is determined by the formation and transformation temperatures. The $CoSi_2$ below the interface is that formed by Co diffusion while that above the interface formed by Si diffusion. The embedded film was also seen in as deposited Co samples at the Co/Si interface, and thus could not be attributed to RTA ambient purity or capping layer effectiveness.

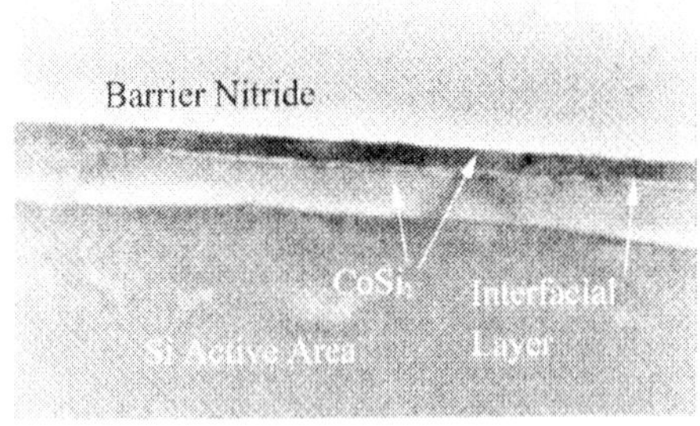

Fig. 2. X-TEM Image of Completed $CoSi_2$ Film showing Embedded SiO_x Interfacial Layer

Morphology at the isolation and gate edges has a dramatic impact on the perimeter components of junction leakage. To control the morphology at the isolation and gate edges, balancing the fluxes of the two species is important. The morphology of the silicide at the edge of the isolation and gate (Figs. 3 & 4) drives the perimeter components of junction leakage as well as the leakage associated with the W local interconnect overlapping the isolation. Note that the silicide thickness is well controlled except in the vicinity of silicon/dielectric discontinuities. Excess Co is transported laterally, presumably from over the dielectric regions.

Fig. 3. X-TEM Image of $CoSi_2$ Spike at STI Edge for a Non-Optimized Process

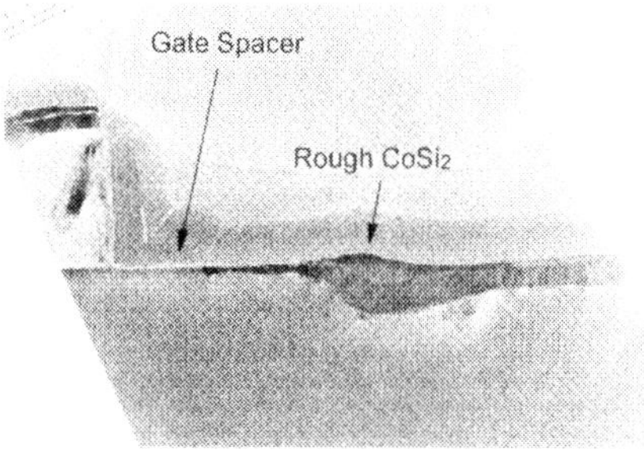

Fig. 4. X-TEM Image of Rough $CoSi_2$ at Gate Edge for a Non-Optimized Process

For silicide formed by an optimized process (Figs. 5 & 6), the silicide thickness is uniform right up to the silicon/dielectric boundaries at gate and STI edges.

Fig. 5. X-TEM Image of Smooth $CoSi_2$ at Gate Edge for an Optimized Process

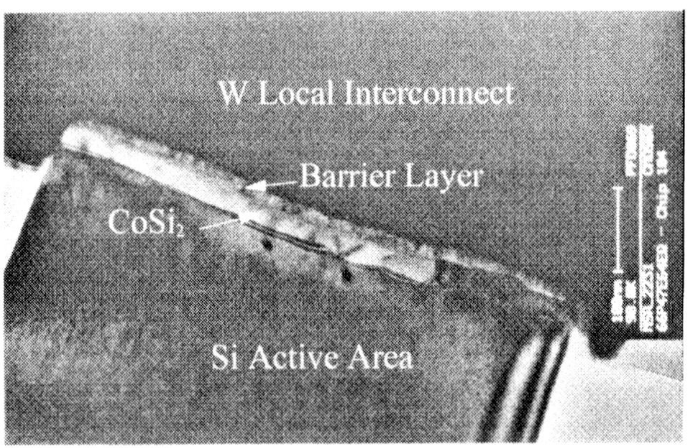

Fig. 6. X-TEM Image of Smooth $CoSi_2$ strapped by W local interconnect at STI Edge for an Optimized Process

If, in order to eliminate the interfacial SiO_x layer, an Ar ion sputter clean or hydrogen terminated surface is used prior to Co deposition the resulting $CoSi_2$ layer will exhibit epitaxial faceting as seen in the cross-section TEM image, Fig. 7. The silicide/Si boundary displays numerous features oriented along the Si <111> plane for these processes, both near the Si/dielectric boundaries and away from the perimeter of structures. At as much as twice the thickness of the surrounding film, these regions are expected to cause junction leakage for shallow junction technologies, and will be shown later to cause increased area leakage.

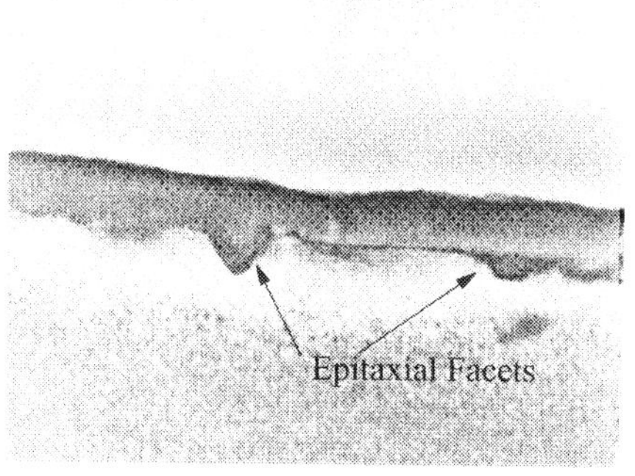

Fig. 7. X-TEM Image of $CoSi_2$ film formed on Hydrogen Terminated Surface

ELECTRICAL RESULTS

The silicide morphology affects both the silicide sheet resistance and the junction leakage. Figure 8 is a plot of sheet resistance of n+ polysilicon lines as a function of the linewidth for the non-optimized and optimized silicide processes The non-optimized process (diamonds) shows some reduction in the sheet resistance as a function of linewidth due to the enhanced thickness of the rough silicide near the perimeter (Fig. 4). At the smallest linewidths the silicide sheet resistance displays high flyers due to the thinner silicide that is formed immediately adjacent to the perimeter. The optimized process (squares) displays nearly linewidth independent sheet resistance due to the uniform thickness of the film up to the edge of the gate (Fig. 5). The morphological differences seen in TEM cross-section and affecting the narrow line sheet resistance are also expected to have an effect on the junction leakage performance, as will be shown in the next section.

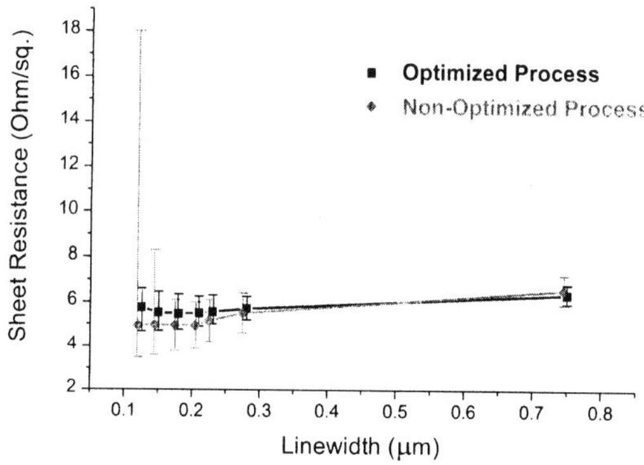

Fig. 8 Sheet Resistance of $CoSi_2$ as a Function of n^+ Polysilicon Linewidth

The cumulative distributions for the n^+ to substrate reverse-biased leakages are shown in Figs. 9 and 10. Figure 9 is the leakage associated with the gate polysilicon perimeter and Fig. 10 is the leakage associated with the local interconnect at the isolation edge. The rough or non-optimized process represented by Figs. 3 & 4 (which displays a distinct bulge in the silicide thickness associated with the perimeter of structures) results in the distributions plotted as the solid triangles, while the smoother, optimized $CoSi_2$ process shown in cross section in Figs. 5 & 6 results in the distributions plotted as the solid squares.

The choice of dopant can also play a significant role in the reduction of junction leakage. Improvement is observed for the leakage tail of the smooth $CoSi_2$ morphology by switching the n^+ junction dopant from As to phosphorus. The resulting leakage distributions are plotted with the open square symbols in Figs. 9 & 10.

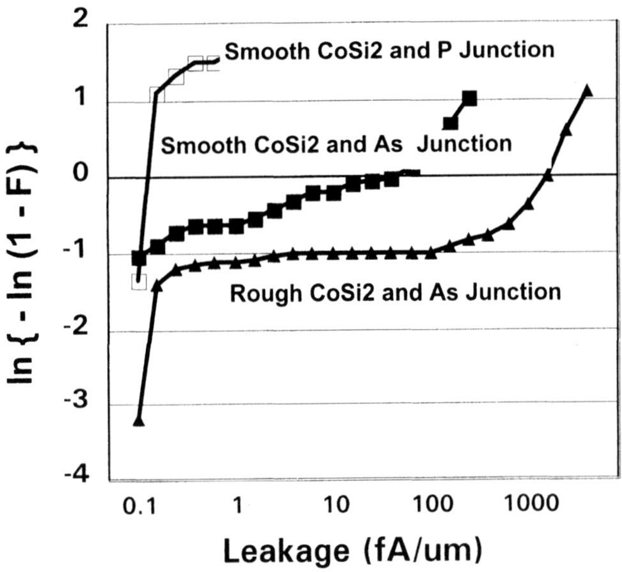

Fig 9. Weibull Plot of 1.8V Reverse-Biased Junction Leakage Associated with the Gate Perimeter

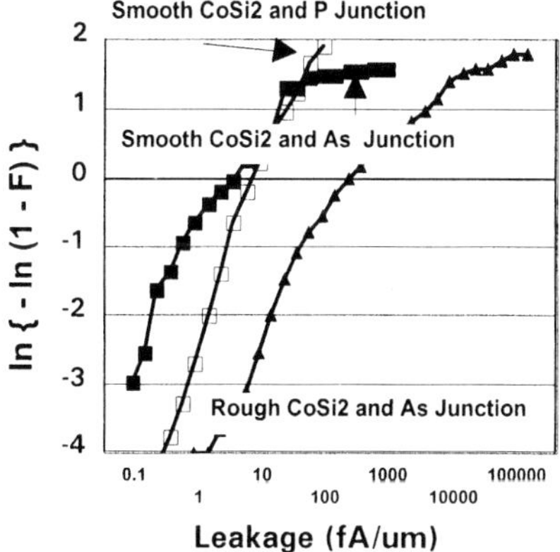

Fig 10. Weibull Plot of 1.8V Reverse-Biased Junction Leakage Associated with the Local Interconnect at the STI Perimeter

Hydrogen termination of the Si surface prior to Co sputtering or Ar ion sputter precleaning of the Si surface (which resulted in extensive faceting as shown in Fig. 7) results in substantially worse perimeter leakage than for the cases plotted in Figs. 9 & 10 and is not plotted. In addition this faceted morphology also leads to an elevation in the reverse-biased leakage associated with the junction area, Fig. 11, due to localized leakage at epitaxial spike locations.

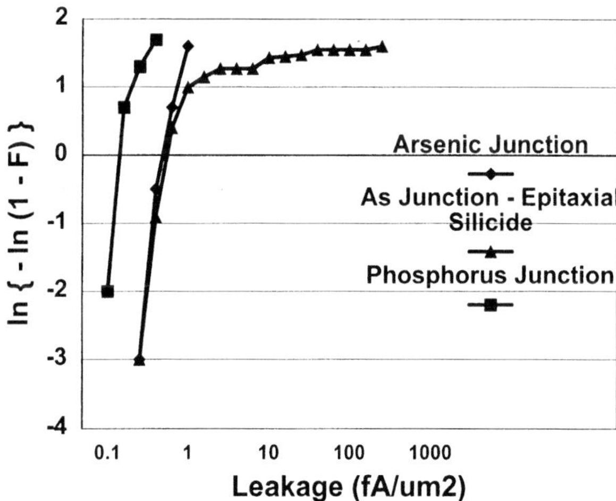

Fig. 11. Weibull Plot of the Area Component of 1.8V Reverse-Biased Junction Leakage

Note that additional improvement in area and perimeter junction leakage behavior as shown in Figs. 9-11 can be obtained without degrading the device characteristics, Table 1, by using P rather than As for the n^+ dopant. The P junction provides better junction perimeter leakage behavior, reduced external resistance and lower junction capacitance, while eliminating the need to predope the gate polysilicon prior to reactive ion etching. This is achieved without sacrificing on-current or short-channel control. Similarly p^+/n junction leakage was also dependent on the silicide morphology and junction optimization.

n+ Junction		As+ Predope	Phos + Predope	Phos
Cinv/Cox	%	85.4	87.5	83.2
Rext	O-um	200	160	176
Cj	fF/um^2	1.32	1.17	1.17
Ion at Lnom	uA/um	595	604	595
Ioff at Lnom	nA/um	< 1	<1	<1

Table 1. NFET Parametrics at 1.8V Vdd

Representative I_d-V_{ds} characteristics for NFET and PFET from the 0.25μm process, Fig. 12, show high current drive at V_{ds} of 1.8V and greater than 10 year DC lifetime at worst-case supply voltage, Fig. 13, leaving margin for process tolerance. The 1.8V unloaded inverter delay is 27ps at 25C for nominal channel length of 150nm, demonstrating that a competitive 0.25μm technology can be fabricated with a phosphorus n^+ junction.

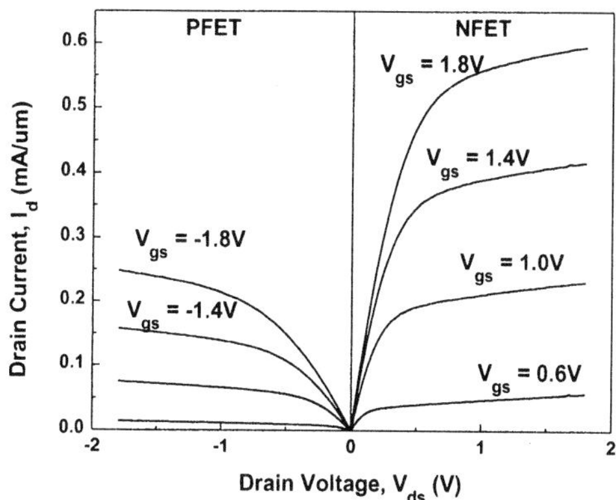

Fig 12. I_d-V_{ds} characteristics of nominal NFETand PFET devices.

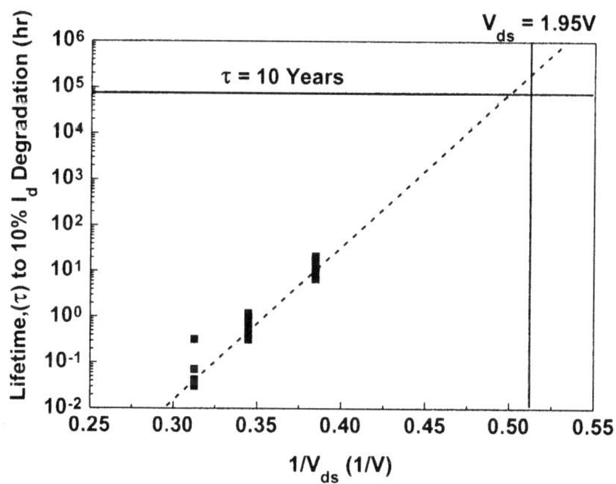

Fig 13. DC lifetime for 10% I_{dsat} degradation in reverse saturation at nominal NFET L_{eff}

The junction leakage has been investigated in detail through the use of optical emission microscopy. The tail of the leakage distribution is due to single point leakage sources with current values from 100nA to 1μA. In these cases it is no longer appropriate to normalize the junction leakage to the test structure's perimeter (as in Figs. 9 & 10). Contrary to previous work[8], we have not found localized light emission on area junction structures (except in the case where $CoSi_2$ formed with epitaxial faceting caused by the absence of the interfacial SiO_x film). Only the structures with significant poly-bound perimeter or local interconnect/STI perimeter showed localized emission of light.

The highly localized leakage can have a significant impact on the yield of high-speed logic employing dynamic nodes. The calculated dynamic node limited yield of a hypothetical 6M transistor chip with 10% dynamic nodes assuming a Poisson distribution of defects is plotted in Fig. 14 assuming a defect threshold current of 100nA or 1μA. The limited yield is plotted for three cases i) the rough silicide in combination with the As n^+ junction, ii) the smoother silicide with the As n^+ junction and iii) the smoother silicide with the phosphorus n^+ junction. The impact of the leakage tail for the non-optimized processes is now more apparent.

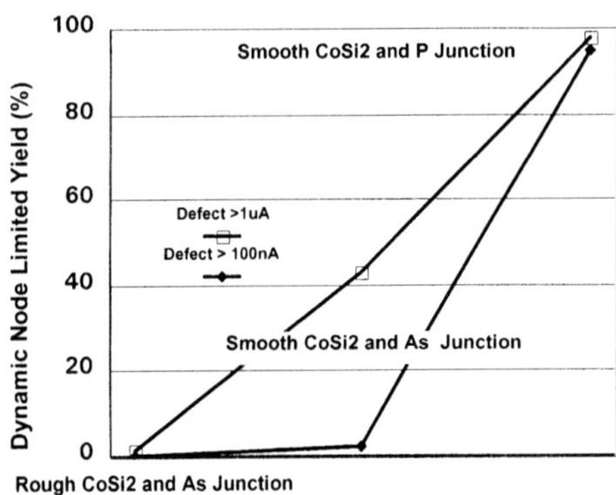

Fig 14. Limited Yield Due to Point Junction Leakage Impact on Dynamic Nodes.

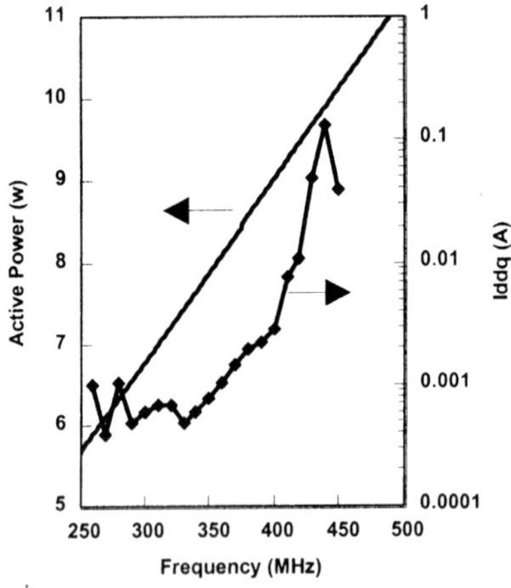

Fig 15. Active Power and Quiescent Current as a Function of Maximum Processor Speed

A 3rd generation 604e+***PowerPC***tm microprocessor was fabricated in this 0.25 μm generation technology at speeds in excess of 450MHz. The chip is 47 mm^2 and contains over 6M transistors. The technology supports 3.3V I/O interface while maintaining 1.8V operation for low power dissipation. The low leakage characteristics of the optimized $CoSi_2$ process permits the use of dynamic circuits to enhance system performance. The low active power, Fig. 15, benefits from the reduced capacitance of the phosphorus n^+ junction as well as the reduced junction area due to local interconnect borderless to isolation. Low quiescent current which increases as a function of clock frequency is an indication that the quiescent current of the microprocessor is dominated by offstate current for shorter channel lengths, not junction leakage.

The Extendibility of this optimized process for creating smooth $CoSi_2$ has been investigated by carrying out the formation sequence on 25nm physical gate length polysilicon. A cross sectional TEM image of a sub-25 nm gate with this process is seen in Fig. 16. The sheet resistance is linewidth independent and under 10 Ω/□ illustrating extendibility well beyond the 0.25μm generation.

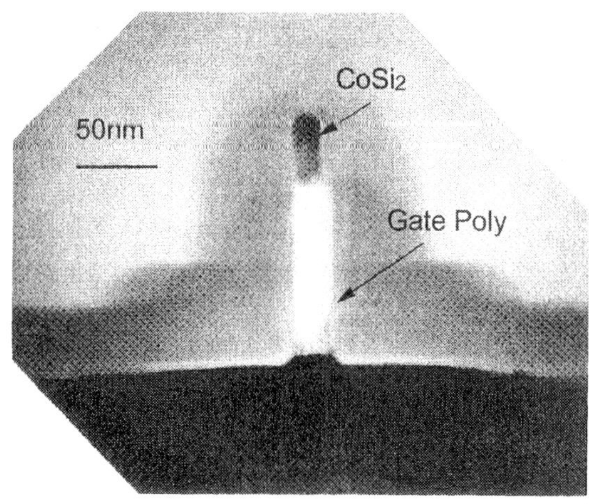

Fig 16. Sub 25nm Physical Gate Length NFET Salicided with Optimized Process

CONCLUSIONS

In conclusion, we have demonstrated the integration of low leakage $CoSi_2$ junctions and W local interconnect for high performance 0.25μm logic with dynamic nodes. Controlling localized leakage is important to dynamic node circuit operation and is a strong function of the Co/Si surface preparation and the choice of junction dopant. Lack of an interfacial film, as can result from hydrogen termination of the Si surface or Ar ion sputter cleaning, results in faceting of the $CoSi_2$ along the Si <111> planes and elevated area leakages. It also results in enhanced formation of silicide on narrow lines due to lateral transport of Co from over the insulator regions. Inadequate surface cleaning can result in a restriction in the silicide formed near the Si/dielectric boundaries and "bulged" or rough silicide adjacent to it, which results in elevated perimeter leakage. An optimal $CoSi_2$ process can be achieved which yields a uniform film thickness up to the edge of Si/dielectric discontinuities. This process has the desirable properties of a nearly linewidth independent sheet resistance and, when combined with a phosphorus n^+ junction, perimeter leakage characteristics that are compatible with dynamic node circuit requirements. This process is utilized on our sub-0.25μm technology and is extendible to 0.18μm and beyond.

ACKNOWLEDGMENTS

The author would like to acknowledge the contributions of his many colleagues in the Advanced Silicon Technology Center, the CMOS Logic Development organization, and the support of the Power PC Product Engineering Team. In particular I would like to recognize the contributions of S. Brodsky, E. Crabbé, E. Nowak, J. Lasky, B. Davari, R. Schulz, S. Crowder, A. Domenicucci, J. Benedict, W. Natzle, L. Gignac and P. Flaitz.

REFERENCES

1. T. Yamazaki, K. Goto, T. Fukano, Y. Nara, T. Sugii and T. Ito, IEDM Tech. Dig., pp.906-909, (1993).

2. S. Shimizu, T. Kuroi, Y. Kawasaki, T. Tsutsumi, H. Oda, M. Inuishi and H. Miyoshi, in *1996 Symposium on VLSI Technology Digest of Technical Papers*, pp.64-65, (1996).

3. T. Yoshitomi, T. Ohguro, M. Saito, M. Ono, E. Morifuji, H. Momose and H. Iwai, in *1996 Symposium on VLSI Technology Digest of Technical Papers*, pp.34-35, (1996).

4. A. Hori, H. Umimoto, H. Nakaoka, M. Sekiguchi, M. Segawa, M. Arai, M. Takase and A. Kanda, IEDM Tech. Dig., pp.575-578, (1996).

5. K. Goto, A. Fushida, J. Watanabe, T. Sukegawa, K. Kawamura, T. Yamazaki, and T. Sugii, IEDM Tech. Dig., pp.449-452, (1995).

6. G. Sai-Halasz, Proc. IEEE, V83, pp. 20-36, (1995).

7. S. Subbanna, P. Agnello, E. Crabbe, R. Schulz, S. Wu, K. Tallman, M. Saccamango, S. Greco, V. McGahay, A. Allen, B. Chen, T. Cotler, E. Eld, J. Lasky, H. Ng, A. Ray, J. Snare, L. Su, D. Sunderland, J. Sun, and B. Davari, IEDM Tech. Dig., pp.275-278, (1996).

8. T. Sukegawa, H. Tomita, A. Fushida, K. Goto, S. Komiya and T. Nakamura, Jap. J. of Appl. Phys., V36 pp. 6244-6249, (1997).

Influence of Rapid Thermal Ramp Rate on Phase Transformation of Titanium Silicides

Yao Zhi Hu, Sing Pin Tay, Jiting Yang
Steag RTP Systems Inc., 4425 Fortran Dr., San Jose, CA 95134
Randhir Thakur
Steag Electronic Systems Inc., 4425 Fortran Dr., San Jose, CA 95134
Paul Martin Smith, Glenn Bailey
Sandia National Laboratories, Albuquerque, NM 87185

ULSI technology requires low resistance, stable silicides formed on small geometry lines. Titanium disilicide ($TiSi_2$), which is the most widely used silicide for ULSI applications, exists in two crystallographic phases: the high resistance, metastable C49 phase and the low resistance, stable C54 phase. The major issue with $TiSi_2$ is the increasing thermal budget required to transform the C49 phase into the low resistance C54 phase as linewiths decrease below 0.25 μm. Annealing above 900°C to obtain this transformation often results in thermal degradation, so it is desirable to reduce the transformation temperature. The transformation temperature has been shown to be a function of many factors including microstructure, grain size, and impurities. In this paper we report an investigation of rapid thermal silicidation of titanium films (250, 400, and 600 Å) on single crystalline silicon at temperatures from 300 to 1000°C. The ramp rates for these experiments are 5, 30, 70, and 200°C/s. The transformation temperature decreases as the ramp rate increases and as the initial film thickness increases. Scanning electron microscopy (SEM) is used to analyze the resultant film microstructure. The ramp rate influence on Ti silicidation is also investigated on polycrystalline Si lines with widths ranging from 0.27 to 3.0 μm.

Introduction

The coming generation of submicron MOS ULSI circuits is expected to require more advanced self-aligned silicide (SALICIDE) technology. Because of the stability of their contacts with silicon and because of their self-passivating property in an oxygen-rich environment, silicides have been preferred over pure metals (1). To date, $TiSi_2$ SALICIDATION has been the process of choice in the semiconductor industry due to its temperature stability and low resistivity. Unfortunately, submicron lines of $TiSi_2$ made from very thin initial Ti films are predominantly the high resistivity C49 phase (2). The troublesome but necessary transformation of the high resistivity C49 phase into the low resistivity C54 phase, an intrinsic material limitation, complicates further scaling of the $TiSi_2$ process. Recently, efforts have been made to overcome these problems. Silicidation was enhanced by pre-amorphization, using ion-implantation with ions of B, Sb, Ge, or others on narrow gate and source/drain regions (3-6). By introducing small quantities of a refractory metal, such as molybdenum, tantalulm, niobium, or tungsten at or near the titanium/silicon interface, the temperature required to form the C54 phase

$TiSi_2$ can be reduced by as much as 100°C (7). Furthermore, the resulting C54-$TiSi_2$ film exhibits small grain size and improved thermal stability.

Nickel mono-silicide (NiSi) has certain advantages over $TiSi_2$ and has been considered as a potential candidate for next generation SALICIDE applications. However, NiSi is limited since it transforms to the higher resistivity $NiSi_2$ phase and agglomerates above 700°C (8). $CoSi_2$ has also been studied extensively (9,10), especially because of the ease of formation on submicron lines without the phase transformation problem. $CoSi_2$ has a resistivity of 16-18 μΩ-cm, which is comparable to $TiSi_2$. In production, cobalt technology is also easily implemented on deep submicron lines; has better stability in the presence of dopants; less stress; better plasma etch resistance; lower contact resistance, and a larger process window. As device dimensions are reduced, some advantages are even more pronounced (11). Therefore $CoSi_2$ is a serious candidate to replace $TiSi_2$ for future technologies. However, the fact that $CoSi_2$ consumes more Si than either $TiSi_2$ or NiSi is a significant shortcoming.

In order to extend Ti SALICIDE manufacturing to deep submicron applications there have been detailed studies of the C49 to C54 transformation. The transformation temperature has been shown to be a function of many factors including the silicon microstructure (12), film thickness (13,14,15), C49 grain size (15,16), and impurities present in the film (17). Earlier work has shown (15,16) that the grain size in C49 $TiSi_2$ specimens can be controlled by varying the heating ramp rate (dT/dt). The largest grain sizes were obtained after annealing in a furnace with a ramp rate of 0.4°C/s. Smaller grained $TiSi_2$ films were made by rapid thermal processing (RTP) with ramp rates up to 220°C/s. More recent work (18) also showed that higher ramp rates result in lower sheet resistance for deep submicron polysilicon lines. In this paper we report on RTP of thin titanium films (250, 400, and 600 Å) on single crystalline silicon substrates at temperatures from 300 to 1000°C. The ramp rates for these experiments are 1, 30, 70, and 200°C/s. In order to study the phase transformation, scanning electron microscopy (SEM) is used to characterize the resulting microstructure. The ramp rate influence on Ti SALICIDATION was also investigated on patterned polycrystalline Si lines with widths ranging from 0.27 to 3.0 μm.

Experiment

An RTP system was used to perform the anneals in a nitrogen ambient. One reason for the success of RTP for Ti SALICIDATION is its excellent ambient and temperature control (19) as this process is extremely sensitive to trace amounts of O_2 and H_2O. With our RTP system it is easy to reach O_2 concentrations below 1 ppm and to control wafer temperatures down to 300°C. Samples were prepared by sputtering Ti (250, 400, or 600 Å) on single crystal silicon substrates. RTP of these samples was carried out at temperatures from 300 to 1000°C. The target ramp rates used were 5, 30, 70, and 200 °C/s. The sheet resistance of the as-deposited films and the silicided films were measured with a four point probe. In order to study the ramp rate effects on the resulting film morphology, SEM was used to characterize the thin silicide microstructure.

Patterned samples of electrically isolated, thin salicided lines with Si_3N_4 spacers were produced on 150 mm Si wafers. The samples were produced by first growing a 125 Å thermal oxide, then depositing 3000 Å of poly-Si. This poly-Si layer was implanted with

8×10^{15} As/cm^2 at 120 keV, capped with 5000 Å of TEOS oxide, and annealed at 1100 °C in N_2 for 180 minutes. The oxide was stripped and the poly-Si lines were patterned to varying linewidths of 0.27 μm and greater, as measured by SEM. An 1800 Å Si_3N_4 film was then deposited on the samples and etched back to form Si_3N_4 sidewall spacers. The 250 Å or 400 Å Ti films were deposited after cleaning the samples in 5:1 H_2SO_4:H_2O_2 for 5 minutes, SC-1 for 10 minutes, SC-2 for 5 minutes, and 30:1 BOE for 1 minute. The as-deposited sheet resistances of the Ti films were measured to be approximately 36 Ω/sq. for the 250 Å film and 19 Ω /sq. for the 400 Å film. The patterned samples were annealed in the RTP system at 700°C for 20 seconds with three ramp rates of 5, 30, and 200°C/s (RTA1). After RTA1 the patterned samples were etched in SC-1 to remove any TiN and unreacted Ti. Finally, the samples were annealed at 850°C for 30 seconds with a ramp rate of 30°C/s (RTA2).

Results and Discussion

1. Titanium Silicidation for Various Ti Film Thicknesses

Figure 1(a) shows sheet resistances vs. RTP temperature for 250, 400, and 600 Å Ti films on single crystal silicon <100> substrates. For all three Ti thicknesses, the plots of sheet resistance versus anneal temperature reveal three distinct transitions. The first transition is a drastic resistivity increase that occurs above 300 C and corresponds to the formation of metal rich silicides (a mixture of TiSi and Ti_5Si_3). Referring to Figure 1(b) the second transition, which corresponds to the formation of the C49 $TiSi_2$ phase, begins to occur above 600°C and is completed by 620°C in all three cases. The third transition, which corresponds to the transformation from C49 to C54 $TiSi_2$, occurs at increasing temperature, T_2, as the initial film thickness decreases. Once this third transition is complete the resistivity remains stable until agglomeration begins. It is interesting to note that the lower temperature extreme, the C49 to C54 transformation temperature, is highest for the 250 Å film and the upper temperature extreme, where agglomeration begins, is lowest for the 250 Å film.

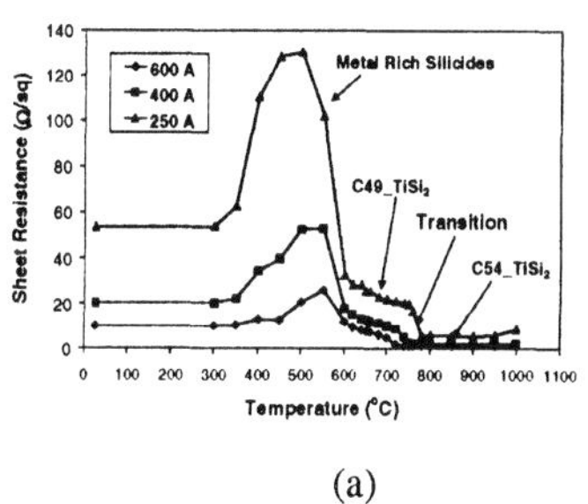

(a)

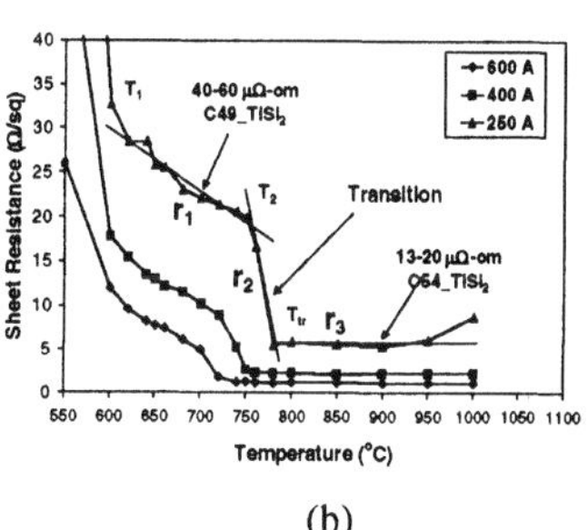

(b)

Fig.1 (a) Sheet resistance as a function of RTA temperature for 250, 400, and 600 Å Ti films on Si; (b) Enlarged section of Fig.1(a)

The $TiSi_2$ film is in the high resistivity C49 phase between the second and third transitions. The resistivity of the C49 $TiSi_2$ decreases with increasing temperature, which is probably caused by grain coarsening. The slope of this change, denoted by r_1 in Figure

1(b) and Table 1 is known as the sensitivity of RTA1, the first salicide RTP process. It increases with decreasing film thickness. The slope of the third transition, where the low resistivity C54 $TiSi_2$ phase (13-20 μΩ-cm) is formed, is denoted by r_2 in Figure 1(b) and Table 1. The beginning of this last transition, T_2, is defined as the temperature at which the linear fit in the C49 region and the linear fit in the transition region intersect. It is interesting to note that the slope in this third transition region, or the rate of C54 formation, increases as the film thickness decreases. Table 1 also indicates that there is no resistivity change ($r_3 = 0$) after the third transition occurs (until agglomeration begins). This implies that the transformation to the C54 phase produces stable material and that annealing to higher temperatures serves no purpose. T_{tr} is defined as the temperature at which the linear fits to the C49 to C54 transition region and the linear fit to the C54 resistivity intersect. As shown in Table 1, the transition temperature, T_{tr}, rises considerably with decreasing film thickness.

Table 1: Ti Silicidation Characteristics for Various Thicknesses

Ti Film Thickness (Å)	RTA1 (T_1-T_2) (°C)	r_1 (RTA1) Ω/sq/°C	r_2 (TRAN.) Ω/sq/°C	r_3 (RTA2) Ω/sq/°C	T_{tr} (C-54) (°C)
600	620-700	0.056	0.143	0	724
400	620-720	0.061	0.217	0	753
250	620-740	0.072	0.512	0	780

2. Ramp Rate Effect on C49 to C54 Phase Transformation

$TiSi_2$ is a polymorphic material and may exist as an orthorhombic base-centered (C49) phase with 12 atoms per unit cell, or as the thermodynamically favored orthorhombic face-centered (C54) phase with 24 atoms per unit cell (20). When a Ti coated Si wafer is heated at temperatures above 500°C, the higher-resistivity (40 – 60 μΩ-cm) metastable C49 phase forms first. Once the C49 phase has formed, a sufficient amount of additional energy must be supplied to overcome the nucleation barrier and enable the phase transformation to the lower resistivity (13 – 20 μΩ-cm) C54 phase. The activation energy required to convert a thin film of C49-$TiSi_2$ on c-Si (100) to C54-$TiSi_2$ is ≥5.6 eV (21). Previous investigations have indicated that the driving force for the C49 to C54 transition is determined by the geometrical component of the free energy (surfaces, grain size), stoichiometry, and impurities (15). It was known from the previous publications (16,22) that the phase transformation temperature decreases with the RTA1 ramp rate. This is due to the smaller grain size of the C49 phase formed with higher RTA1 ramp rate.

Measurements of the temperature at which the C49 to C54 transformation occurs depend on the analysis method (16,23-25). The transformation temperature can be determined from sheet resistance measurements, X-ray diffraction (XRD) analysis, and other techniques. It has been shown (16) that the transformation temperatures derived from XRD and sheet resistance measurements are the same to within experimental error (±5°C). In this paper we define the transformation temperature, T_{tr}, as the minimum temperature at which the C54-$TiSi_2$ film has reached its lowest resistivity, as shown in Figure 1(b).

In order to study the effect of ramp rate on transformation temperature, samples with different Ti thicknesses were annealed for 20 seconds with ramp rates from about 5°C/s to about 200°C/s. The annealing temperatures ranged from 620 to 840°C. The RTP results are shown in Figure 2(a) for 600 Å Ti films, Figure 2(b) for 400 Å Ti films, and Figure 2(c) for 250 Å Ti films

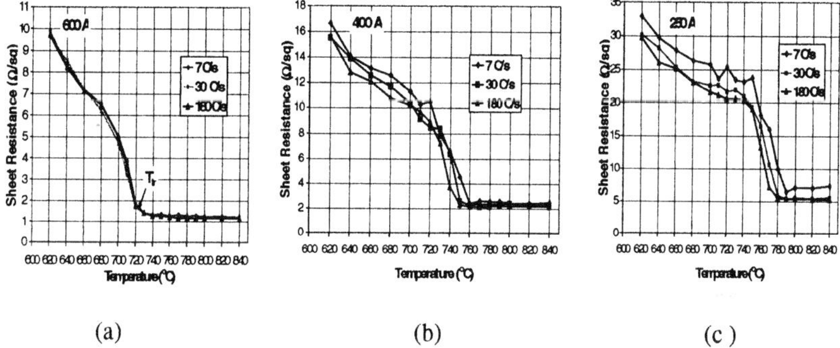

Figure 2. Ramp rate effect on the transformation temperature for various Ti thicknesses. (a). 600 A; (b) 400 A; and (c) 250 A.

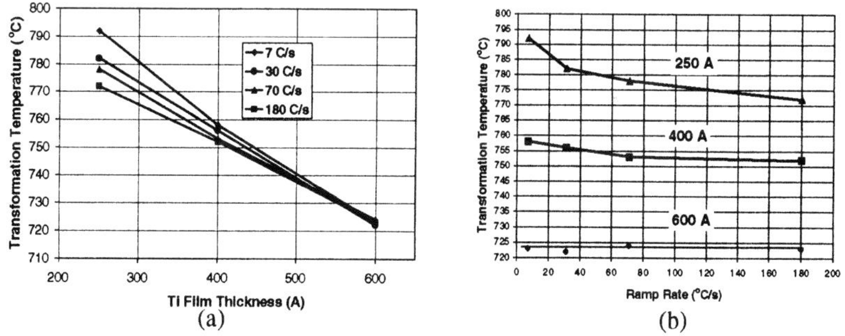

Figure 3. Transformation temperature as a function of Ti film thickness (a) and as a function of RTA ramp rate (b).

It was found that both thickness and ramp rate affect the transformation temperature. Thicker Ti films cause the transformation to occur at lower temperatures. The ramp rate effects are different for different Ti film thickness. For the 600 Å Ti films, ramp rates from 7 to 180°C/s do not affect the transformation temperature. However, for the 250 Å Ti film increasing the ramp rate significantly reduces the transformation temperature. The transformation temperature as a function of Ti film thickness and as a function of

RTP ramp rate is shown in Figures 3(a) and 3(b), respectively. The transformation temperature increases by more than 50°C as the thickness decreases from 600 Å to 250 Å in all cases. On the other hand, the transformation temperature of the 250 Å Ti film decreases by 20°C as the ramp rate increases from 7°C /s to 180°C /s. A summary of transformation temperature dependence on both ramp rate and thickness is shown in Table 2.

Table 2. Transformation Temperatures for Various Thicknesses and Ramp Rates

Ramp Rate °C/s	T_{tr}(°C) 600 Å Ti Film	T_{tr}(°C) 400 Å Ti Film	T_{tr}(°C) 250 Å Ti Film
180	723	752	772
70	724	753	779
30	722	756	782
7	723	758	792

3. SEM Analysis of Ramp Rate Effects

The transformation temperature, T_{tr}, is dependent on the microstructure of the C49 $TiSi_2$ film (15, 16, 22). It has been shown that by varying the ramp rate it is possible to control the grain size in the C49 $TiSi_2$ film (15,16). The largest C49 grain sizes are produced with the lowest ramp rates, while smaller grains are produced by higher ramp rates. In order to study the microstructure of the C49 $TiSi_2$ films, 400 Å Ti films were annealed at 700°C for 20 seconds with ramp rates of 3, 30, and 200°C/s. After annealing, the samples were etched in an SC-1 solution (NH_4OH:H_2O_2:H_2O at 1:1:5) at 75°C for 120 seconds, and the resulting surface was decorated with a 1% HF solution. The samples were then inspected by SEM and the resulting micrographs are shown in Figures 4(a), 4(b), and 4(c). It is clear that the higher ramp rates create smaller grains, which is in agreement with the previous work (16).

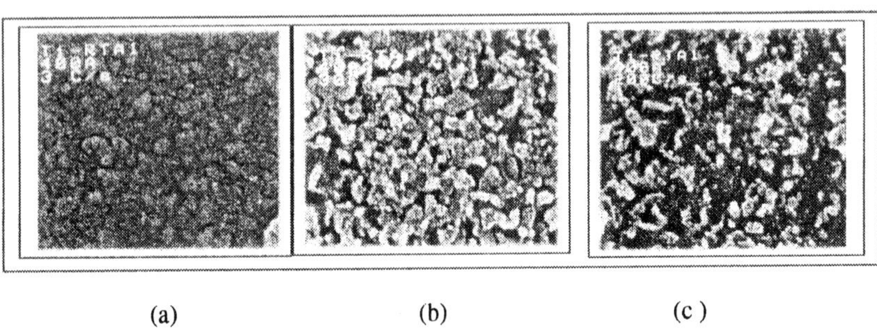

(a) (b) (c)

Fig.4 SEM micrographs of 400 Å Ti film on Si after RTA1 at different ramp rates: (a) 3 °C /s; (b) 30 °C /s ; (c) 200 °C /s

4. Ramp Rate Effect on Patterned Samples with Narrow Linewidths

Figure 5 shows the sheet resistance as a function of linewidth and ramp rate for the patterned samples with 400 Å Ti films. As previously shown, the transformation is inhibited by reduced volume (i.e. thinner lines) and reduced ramp rates. The linewidth effect is much more pronounced for the samples ramped at 5°C/s and a significant resistivity increase is already seen for lines of 0.6 μm. It is interesting to note that the sheet resistance for ramp rates of 30 and 200°C/s remains low (<4.2 Ω/sq) for patterned lines as narrow as 0.4 μm. The effect of the very high ramp rate only becomes apparent at very narrow line widths, for example, the 0.27 μm patterned lines. This decrease in resistivity with increasing ramp rate may be of significant technological importance as the critical dimensions for semiconductor devices continues to decrease.

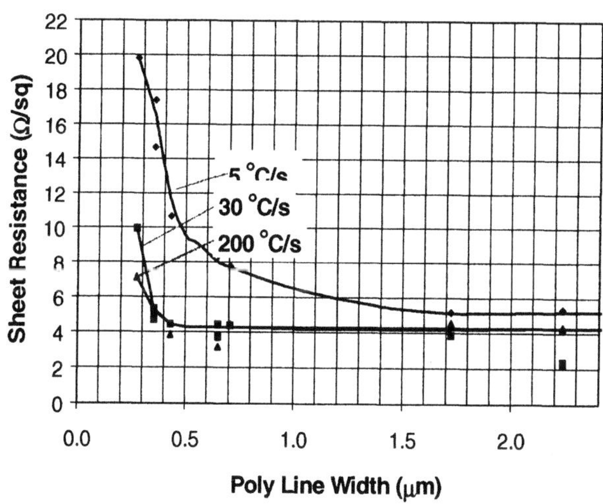

Figure5. Ramp rate effect on sheet resistance of poly lines with different line widths.

Conclusions

It has been found that the C49 $TiSi_2$ to C54 $TiSi_2$ transformation temperature is dependent on both the initial Ti thickness and temperature ramp rate. The transformation temperatures increase by more than 50°C when the thickness decreases from 600 Å to 250 Å. The effect of the ramp rate on the transformation temperature is stronger for thinner films. For the 250 Å Ti films the transformation temperature decreased about 20°C when the ramp rate was increased from 7 to 180°C/s. SEM analysis showed that the ramp rate controls the C49 grain size, with higher ramp rates producing a smaller grain structure. This is of critical importance because the C49 grain structure significantly affects the C49 $TiSi_2$ to C54 $TiSi_2$ transformation. The studies of thin patterned lines indicate that the sheet resistance produced by ramp rates of 200°C/s is significantly lower than the sheet resistance produced by ramp rates of 5°C/s. This ramp rate effect becomes most pronounced for lines with widths below 0.4 μm.

Acknowledgements

The authors would like to thank the personnel at Sandia's Microelectronics Development Laboratory (MDL) for wafer processing. Sandia is a multiprogram laboratory operated by Sandia Corporation, a Lockheed Martin Company, for the United States Department of Energy under Contract DE-AC04-94AL85000.

References

1. M.E. Alperin, T.C. Hollaway, R.A. Haken, C.D. Gosmeryer, R.V. Karnaugh, and W.D. Parmantie, IEEE Trans. Electron Devices, **32**, 141 (1985).
2. E. Ganin, S. Wind, P. Ronsheim, A. Yapsir, K. Barmak, J. Bucchignano, and R. Assenza, Mat. Res. Soc. Symp. Proc. **303,** 109 (1993).
3. Q. Xu and C. Hu, IEEE Transactions on Electron Devices , **45,** 2002 (1998).
4. K. Fujii, et al., Mat. Res. Soc.Symp. Proc. **402,** P83, (1996).
5. Z. Ma, et al., 1995 4th International Conference on Solid-State and Integrated Circuit Technology. Proceedings, P.35, (1995) .
6. K. K. Larsen, et al., Appl. Phys. Lett., **67**, 2931 (1995).
7. R.W. Mann, et al., Mat. Res. Soc. Symp. Proc. **402,** 95 (1996).
8. Y.Z. Hu, S.P. Tay and Y. Wasserman, 5th Int. Conf. On Advanced Thermal Processing of Semiconductors, **RTP'97,** 429 (1997).
9. Q.F. Wang, A. Daniel, S.P. Tay and Y. Wasserman, 4th Int. Conf. On Advanced Thermal Processing of Semiconductors, **RTP'96**, 395 (1996).
10. K. Maex, Semiconductor International, P75, March 1995.
11. Q.F. Wang, A. Lauwers, F. Jonckx, M.de Potter, C.C. Chen, and K. Maex, Mat. Res. Soc. Symp. Proc. **402,** 221 (1996).
12. H.Jeon, C.A.Sukow, J.W.Honeycutt, G.A. Rozgonyi and R.J. Nemanich, J. Appl. Phys., **71,** 4269 (1992).
13. R.V.Nagabushnam, S.Sharan, G.Sandhu, V.R.Rakesh, R.K.Singh and P.Tiwari, Mat. Res. Soc. Symp. Proc. **402,** 113 (1996).
14. Z.Ma and L.H.Allen, Phys. Rev. **B 49**, 13502 (1994).
15. H.J.W. van Houtum and I.J.M.M.Raaijmakers, Mater. Res. Soc. Symp. Proc. **54,** 37 (1986).
16. H.J.W. van Houtum, I.J.M.M. Raaijmakers, and T.J.M. Menting, J. Appl. Phys. **61,** 3116 (1987).
17. R. Beyers and R. Sinclair, J. Appl. Phys. **57,** 5240 (1985)
18. C.C.D. Tan, L.Lu, A.See, S.Y. Chen, L.H. Chua, K.L.T. Chan and L.Chan, 1st International Conference on Advanced materials and Processes for Microelectronics, San Jose, CA, March15-19, 1999.
19. K.S. Balakrishnan, and T.F. Edgar, Special Edition of the Journal of Thin Solid Films, accepted for publication (1998).
20. R.W. Mann and L.A. Clevenger, J. Electrochem. Soc. **141,** 1347 (1994).
21. R.W. Mann, L.A. Clevenger and Q.Z.Hong, J. Appl. Phys., **73,** 3566 (1993).
22. R. Sikora and W. Lundy, J. Appl. Phys. **72,** 1160 (1992).
23. X.W. Lin and D. Promanik, Mat. Res. Soc. Symp. Proc., **429,** 181 (1996).
24. K.L. Saenger, C. Cabral,JR., L.A. Clevenger and R.A. Roy, Mat. Res. Soc. Symp. Proc., **402,** 275 (1996).
25. R.T. Tung, Mat. Res. Soc. Symp. Proc., **402,** 101 (1996).

LOW RESISTIVITY CONTACTS TO ULTRA-SHALLOW JUNCTIONS IN ULSI DEVICES

L. J. Chen, S.L. Cheng, and L.W. Cheng
Department of Materials Science and Engineering, National Tsing Hua University, Hsinchu, 30013 Taiwan, Republic of China

Abstract

Low resistivity $TiSi_2$, $CoSi_2$ and NiSi are the three primary candidates for metal contacts in sub-0.25 μm devices. For ultrashallow junction formation, nitrogen ion implantation has been succesfully introduced to suppress both channeling and transient enhaneduion of dopant. In the present paper, we review recent progresses in the investigations of low-resistivity contacts on ultrashallow junctions, which include 1. improved thermal stability of C54-$TiSi_2$, $CoSi_2$ and NiSi by nitrogen ion implantation in (001)Si, 2. enhanced formation of C54-$TiSi_2$ on nitrogen ion implanted (001)Si by an interposing Mo layer.

INTRODUCTION

As the microelectronics device dimensions scale down to the deep submicron size, silicides in the source/drain area are needed to reduce the contact resistance to an acceptable level. For self-aligned silicidation process, silicides are formed on source, drain and gate regions simultaneously to ease the lithography requirements and to lower the contact resistance. Low-resistivity $TiSi_2$ is currently the most common silicide for this application (1). On the other hand, $CoSi_2$ and NiSi are also being considered as the metal contacts on shallow junctions for sub-0.25 μm devices (2,3). For the deep submicron devices, it is essential to understand the interactions of metal contacts with highly doped shallow junctions (4).

Shallow junction formation has long been one of the major challenges in the fabrication of ultralarge scale integrated circuits (ULSI) devices (5). Recently, nitrogen ion implantation had been used in deep submicron devices to suppress the B and As diffusion as well as hot-carrier degradation (6,7). Preamorphization of silicon was reported to be effective in suppressing channeling in Si (8,9). However, it is not effective for suppressing transient enhanced diffusion (TED). Nitrogen ion implantation was introduced in an attempt to suppress both channeling and TED (10).

As the device dimensions scale down to deep submicron region, it has been found that the high-resistivity C49- to C54-$TiSi_2$ conversion becomes increasingly difficult, the fact attributed to the low density of nucleation sites for the C54-$TiSi_2$ phase (11). The desire to form C54-$TiSi_2$ completely in a submicron structure has led to extensive research in increasing the C54-$TiSi_2$ nucleation density by using refractory metal (Mo or W) ion implantation (12), Sb additions (13), preamorphization of the silicon substrate by ion implantation of Xe (14), or by interposing a thin layer of molybdenum (15). In the present paper, we review recent progresses in the investigations of low-resistivity contacts on ultrashallow junctions, which include 1. improved thermal stability of C54-$TiSi_2$, $CoSi_2$ and

NiSi by nitrogen ion implantation in (001)Si, 2. enhanced formation of C54-$TiSi_2$ on nitrogen ion implanted (001)Si by an interposing Mo layer.

EXPERIMENTAL

Single crystal, 3-5 Ω-cm, (001)Si wafers were used in the present study. Following a standard cleaning procedure, (001)Si wafers were implanted by 30 keV N^+ to doses of 5 x 10^{14} to 1 x 10^{16} /cm^2 and 20 keV N_2^+ to doses of 2.5 x 10^{14} or 1 x 10^{15} /cm^2. Some of the N^+- or N_2^+- implanted samples were then implanted by 30 keV BF_2^+ to a dose of 2 x 10^{15} /cm^2 or 30 keV As^+ to a dose of 3 x 10^{15} /cm^2. The wafers were oriented 7° off the incident beam direction to alleviate the channeling effects. The implantation of N_2^+ was carried out to achieve an effective N^+ implantation of half energy and double dose. The projected ranges (R_p) and stragglings (ΔR_p) for 20 keV N_2^+, 30 keV N^+ and As^+ were calculated to be about (28.1 nm, 76 nm, 27.1 nm) and (11.9 nm, 27.1 nm, 9.7 nm), respectively, using TRIM program. For 30 keV BF_2^+ implant, the R_p and ΔR_p of B and F ions are (28.3 nm, 11.9 nm) and (27.3 nm, 11.9 nm), respectively (16). Table I lists the implantation conditions and corresponding R_p and ΔR_p of different ions in silicon.

Table I Implantation conditions and corresponding projected ranges and stragglings.

	N^+, 30KeV	N_2^+, 20KeV	As^+, 30KeV	BF_2^+, 30KeV	
Dose	5x10^{14}- 1x10^{16}/cm^2	2.5x10^{14}- 1x10^{15}/cm^2	3x10^{15}/cm^2	2x10^{15}/cm^2	
				B^+, 6.645KeV	F^+, 11.678KeV
R_p (nm)	76	28.1	27.1	28.3	27.3
ΔR_p (nm)	27.1	11.9	9.7	11.9	11.9

30-nm-thick Ti, Co or Ni thin films were deposited onto the samples. For other samples, a 0.5-nm-thick Mo layer was deposited onto the blank or implanted samples first, then 30-nm-thick Ti thin films were deposited onto the interposing Mo layers. Heat treatments were carried out in a three-zone diffusion furnace at 600-1000 °C for 1 h or in a rapid thermal annealing (RTA) apparatus at 600-1000 °C for 30 s in N_2 ambient unless otherwise mentioned. The samples are characterized by both planview and cross-sectional transmission electron microscopy (XTEM), scanning electron microscopy (SEM), grazing incidence X-ray diffractometry (GIXRD), scanning Auger electron spectroscope (AES), secondary ion mass spectroscopy (SIMS), electron spectroscopy for chemical analysis (ESCA) and Fourier transform infrared spectroscopy (FTIR).

ELECTRICAL ACTIVATION AND SUPPRESSION OF DOPANT DIFFUSION

Figures 1 (a) and (b) show the sheet resistance data of 30 keV N^+ + 30 keV BF_2^+ and 30 keV N^+ + 30 keV As^+ implanted samples after different heat treatments. For BF_2^+-N^+ samples with the implantation of N^+ to doses of and higher than 5 x 10^{15} /cm^2, electrical activation of B appeared to be rather poor after annealing at a temperature as high as 1000 °C. On the other hand, for As^+-N^+ samples, difficulty in electrical activation was encountered only for samples with the implantation of N^+ to a dose as high as 1 x 10^{16} /cm^2. In other words, for BF_2^+-N^+ and As^+-N^+ implanted samples, from standpoint of electrical activation, the N^+ doses of 5 x 10^{15} /cm^2 and 1 x 10^{16} /cm^2, respectively, are too high to be useful in forming the shallow junctions. In addition, to activate the BF_2^+-2 x 10^{15} /cm^2 N^+ samples, annealing at a temperature of 800 ^{0}C was not sufficient. A recent study indeed showed the formation of B-N complex by FTIR analysis in BF_2^+ + N^+ implanted samples (17). The formation of N-containing complex is also likely to be the cause of the retardation of the diffusion of dopants.

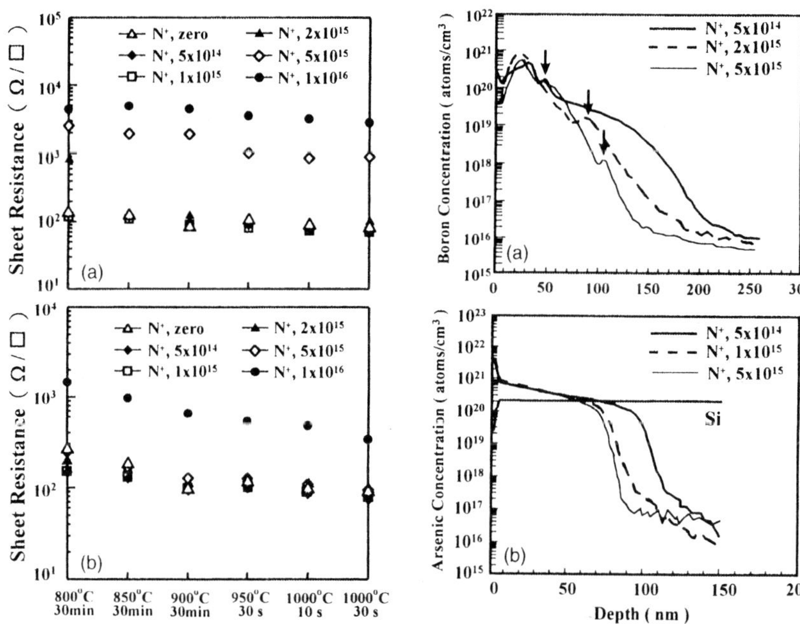

Fig. 1. Sheet resistance vs. heat treatment curves for (a) N^+ + BF_2^+ and (b) N^+ + As^+ samples.

Fig. 2. SIMS concentration-depth profiles for (a) N^+ + BF_2^+ and (b) N^+ + As^+ samples after annealing at 900 ^{0}C for 0.5 h.

Figures 2 (a) and (b) show the SIMS profiles of 30 keV BF_2^+ + 30 keV N^+ and 30 keV As^+ + 30 keV N^+ implanted samples after annealing at 900 ^{0}C for 30 min, respectively. The suppression of boron and As diffusion is seen to increase with the N^+ dose. It is seen that shallower junctions can be formed with the N^+ implantation. From XTEM images, 110-, 90- and 45-nm-thick amorphous layers were formed in 1 x 10^{16} /cm^2 N^+, 2 x 10^{15} /cm^2 N^+ and 5 x 10^{14} /cm^2 N^+ + BF_2^+ implanted samples, respectively. As a result, it can be concluded that the nitrogen and boron atoms tended to accumulate at the original amorphous/crystalline (a/c) interfaces. SIMS concentration-depth profiles of B and N were further indicated that the distribution of B atoms in silicon did not change significantly with the formation of titanium silicide in samples. Significant nitrogen outdiffusion was previously found in N^+ implanted samples without boron implantation (18). On the other hand, most of the nitrogen atoms remained in the implanted layer after annealing in N^+-

BF_2^+ implanted samples. The results indicated that boron and nitrogen interact with each other. The interaction is thought to retard the boron diffusion during annealing.

Figure 3 shows the sheet resistance data of titanium on blank, 30 keV BF_2^+- + 20 keV N_2^+- and 30 keV As^+- + 20 keV N_2^+ implanted samples after annealing by RTA at different temperatures for 30 s. A sharp drop in sheet resistance was found for all implanted samples at 700 ^{0}C. The sheet resistances maintained the same low level in implanted samples annealed at 800-900 ^{0}C. The data indicated that complete transformation from high-resistivity C49-$TiSi_2$ to low-resistivity C54-$TiSi_2$ was largely achieved for all implanted samples at 700 ^{0}C. The results are consistent with those obtained by the GIXRD and XTEM analysis. The sheet resistance was found to be largely independent of N_2^+ dose in the dose range of 2.5 x10^{14} to 1 x 10^{15} /cm^2. The results suggested that the electrical activation levels are about the same in samples implanted by N_2^+ to doses in this range.

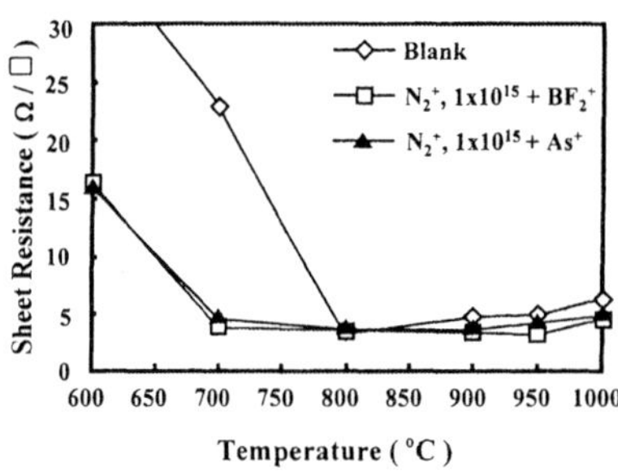

Fig. 3. Sheet resistance vs. annealing temperature curves for titanium on blank, N_2^+ + BF_2^+, and N_2^+ + As^+ implanted (001)Si.

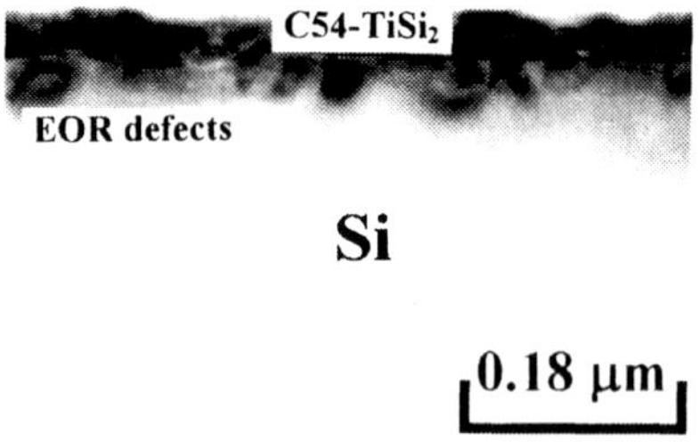

C54-$TiSi_2$
EOR defects
Si
0.18 µm

Fig. 4. Ti on 1 x 10^{15} /cm^2 N_2^+ + BF_2^+ implanted (001)Si, 900 ^{0}C, 30 s, c-s.

In 1 x 10^{15} /cm^2 N_2^+ implanted samples without Ti overlayer and annealed at a temperature as high as 900 ^{0}C for 1 h, a high density of EOR defects were found at the original a/c interface. In contrast, for Ti on 1 x 10^{15} /cm^2 N_2^+ implanted samples, no EOR defects were evident in all samples annealed at 600-950 ^{0}C. From XTEM micrographs, the thicknesses of $TiSi_2$ layer ranged from 38 to 45 nm in this temperature range. A simple calculation indicated that the formation of the $TiSi_2$ layer consumed the regions with EOR defects in silicon. For 2.5 x 10^{14} N_2^+ implanted samples, the dose was below the critical dose for amorphization of the surface layer. As a result, no EOR defects were observed in both samples with and without Ti overlayer. For 1 x 10^{15} /cm^2 N_2^+ + BF_2^+ implanted samples, EOR defects were found to be prevalent in 600-950 ^{0}C annealed samples. In 2.5 x 10^{14} /cm^2 N_2^+ + BF_2^+ implanted samples without Ti overlayer, a high density of fluorine bubbles were observed to distribute in the surface layer after annealing at 900 ^{0}C for 1 h. On the other hand, no EOR defects were observed in N_2^+ + As^+ implanted samples annealed at 700-950 ^{0}C. In 2.5 x 10^{14} /cm^2 N_2^+ + As^+ implanted samples without Ti overlayer, the rodlike defects located at the surface region can be observed. However, no EOR defects were evident after annealing at 900 ^{0}C for 1 h.

The elimination of EOR defects were correlated to the low leakage currents for shallow junctions (19). The projected range of 20 keV N_2^+ (28.1 nm) is close to those of As^+ (27.1 nm), B (28.3 nm) and F ions (27.3 nm) so that the presence of doping species may significantly influence the removal of the EOR defects. In the present study, with the appropriate selection of energy and dose, the EOR defects were indeed eliminated in N_2^+ + As^+ implanted samples. However, EOR defects remained in N_2^+ + BF_2^+ samples even after annealing at 900 ^{0}C as seen in Fig. 4. The presence of EOR defects may induce too high a leakage current for memory devices. On the other hand, the requirement for limiting the leakage current is much less stringent for logic devices (20,21). The present scheme is expected to be applicable to form low-resistivity $TiSi_2$ contacts on ultrashallow junctions in logic devices.

Incorporation of nitrogen in the processing of silicon devices has been shown to improve the reliability of gate dielectrics, suppress boron penetration from the p^+-polysilicon gate, improve the peak transconductance and reliability of polycrystalline thin film transistor devices and stabilize the $TiSi_2$ (22-24). In addition, nitrogen ion implantation also finds applications in silicon on insulator (SOI) devices and nitrogen ion implanted layers can serve as an etch stop. On the other hand, heavy nitridation and nitridation induced surface donor layers in silicon might be the reason for carrier mobility degradation. Slight decreases in the peak value of transconductance for both n- and p-MOSFET's and nitrogen related n-type doping for SOI have also been reported. Further works are needed for structural and electrical characterization for exploiting the nitrogen ion implantation in device applications. (25)

IMPROVEMENT OF THERMAL STABILITY OF $CoSi_2$ BY NITROGEN ION IMPLANTATION

Formation of $CoSi_2$ on 30 keV BF_2^+-30 keV N^+ and 30 keV As^+-30 keV N^+ implanted samples has been investigated. $CoSi_2$ was found to be the only silicide phase formed in all samples annealed at 700-1100 ^{o}C. Polycrystalline $CoSi_2$ structure was observed in all samples. No island formation was observed in N^+ implanted samples. The average grain size was found to decrease with the N^+ dose. Nitrogen ion implantation has been proved to be beneficial in improving the thermal stability of $CoSi_2$ thin films and suppressing the formation of pinholes. The thermal stability of all N^+ implanted samples is better than that of samples without implant. The thermal stability and morphology of pinholes in samples implanted with N^+ and As^+-N^+ are similar. Pinholes were nearly eliminated in BF_2^+-N^+-implanted samples. The thermal stability of BF_2^+-N^+ samples was the best among all samples. (26).

IMPROVEMENT OF THERMAL STABILITY OF NiSi BY NITROGEN ION IMPLANTATION

The phase formation of nickel silicides on nitrogen implanted (001)Si was found to be suppressed and shifted to a higher temperature compared to that of blank samples. The process window of NiSi was extended to a higher temperature compared to that of blank sample. The sheet resistance maintained the same low level in a wide temperature range. It indicates that NiSi is the dominant phase for the 1 x 10^{15} /cm^2 N_2^+ implanted samples annealed at 400-800 ^{0}C. An example is shown in Fig. 5. The SIMS concentration profiles reveal that nitrogen and boron atoms are confined in nickel silicide. In contrast, arsenic atoms were found to diffuse out of silicide during annealing at 800 ^{0}C. An example is shown in Fig. 6. The diffusion of nickel atoms is apparently retarded by the presence of nitrogen atoms. The presence of nitrogen ion was also found to improve the thermal stability of NiSi. The effects of nitrogen on NiSi formation become more pronounced with the increasing N^+ dose (27).

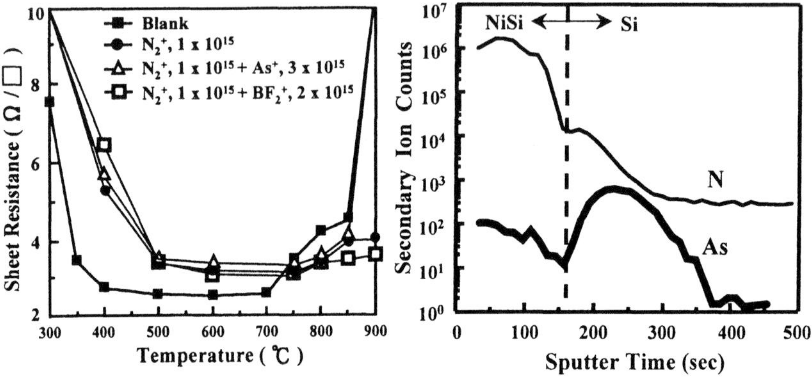

Fig. 5. Sheet resistance of nickel silicides as a function of annealing temperature.

Fig. 6. SIMS profiles in a 1 x 10^{15} /cm^2, N_2^+ + As^+ implanted sample after annealing at 800 °C.

FORMATION OF C54-$TiSi_2$ IN TITANIUM ON NITROGEN ION IMPLANTED (001)Si WITH A THIN INTERPOSING Mo LAYER

Ti (30 nm) / (001) Si and Ti (30 nm) / Mo (0.5 nm) / (001) Si Samples

Figure 7 shows the sheet resistance data of Ti (30 nm) / (001) Si and Ti (30 nm) / Mo (0.5 nm) / (001) Si samples (A and B samples) after different heat treatments. The data clearly demonstrate the impact of the Mo layer on the C49- to C54-$TiSi_2$ transformation. A sharp drop in sheet resistance was found for Ti/Mo bilayer on (001)Si samples at 650 ℃. On the other hand, for titanium on (001)Si samples, as the annealing temperature was increased to 750 ℃, a sharp drop in sheet resistance was found indicating that appreciable C49- to C54-$TiSi_2$ transformation occurred. The increases in sheet resistances for A and B samples after annealing at 900 ℃ were correlated to the formation of island structure of C54-$TiSi_2$.

From the GIXRD analysis, the low-resistivity C54-$TiSi_2$ was observed to be the only silicide phase formed in 750 ℃ annealed A samples and 650 ℃ annealed B samples, respectively. On the other hand, even after annealing at 700 ℃, only diffraction peaks corresponding to the C49-$TiSi_2$ were found for A samples. Therefore, with an interposing Mo thin layer, the formation temperature of C54-$TiSi_2$ was lowered by about 100 ℃ compared to the what is usually needed for the C49- to C54-$TiSi_2$ transformation. In the XRD spectra of the samples containing Mo, a weak peak at about $2\theta=41.5^o$ was observed for samples annealed at 650 ℃. This peak can be associated with Mo and Ti, and assigned to a ternary (Ti, Mo)Si_2 silicide phase (15). $(Ti_{0.4}Mo_{0.6})Si_2$ and $(Ti_{0.8}Mo_{0.2})Si_2$ are two possible candidates (28,29). Both are of a hexagonal structure.

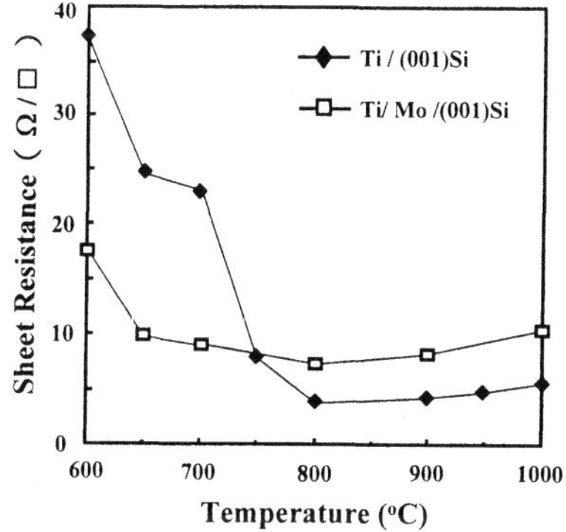

Fig. 7. Sheet Resistance of titanium silicides versus annealing temperature curves for A and B samples.

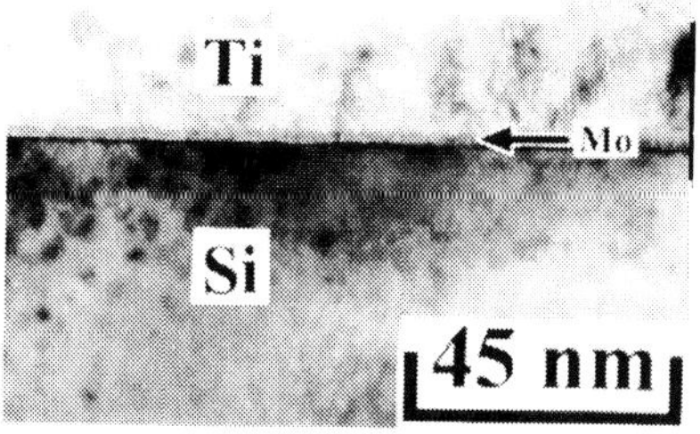

Fig. 8. XTEM micrograph, BF, of an as-deposited Ti (30nm) / Mo (0.5nm) / (001)Si sample.

XTEM images revealed the presence of an as-deposited B sample with a distinct Mo layer seen between Ti and Si. An example is shown in Fig. 8. After annealing, the presence of the interposing Mo layer was found to slow down the interaction between Ti and Si. XTEM micrographs of A and B samples after annealing at 800 ℃ for 30 s showed that the thicknesses of C54-$TiSi_2$ in B samples (with the Mo layer) are thinner than those in A samples (without the Mo layer). In addition, the ternary (Ti, Mo)Si_2 silicide phase was formed in B sample after annealing. These results are correlated with the difference in sheet resistance of C54-$TiSi_2$ between A and B samples in Fig. 7. The island formation of C54-$TiSi_2$ and formation of pinholes were found for 900 ℃ annealed samples. The result is correlated with the slight increase in sheet resistance at 900 ℃.

Following the silicide formation, the distinct Mo layer was no longer present. SIMS concentration-depth profiles revealed that as the annealing temperature increases, Ti and Si reacted to form $TiSi_2$ and the Mo atoms diffused from the interface to disperse in $TiSi_2$. Based on the TEM and SIMS data, it is conjectured that the redistribution of Mo atoms, in the form of a ternary Ti-Mo-Si phase, leads to the enhancement of the formation of C54-$TiSi_2$ by providing more heterogeneous nucleation sites needed for the transformation from C49- to C54-$TiSi_2$ phase.

Ti / Mo on Nitrogen, Nitrogen and Dopant Implanted Si Samples

Figure 9 shows the sheet resistances of Ti on 20 keV, 2.5 x 10^{14} or 1 x 10^{15} /cm^2 N_2^+, 30 keV, 2 x 10^{15} /cm^2 BF_2^+ + 20 keV N_2^+, and 30 keV, 3 x 10^{15} /cm^2 As^+ + 20 keV N_2^+ implanted samples with an interposing Mo thin layer, after annealing by RTA at different temperatures for 30 s. A sharp drop in sheet resistance was found for all implanted samples at 650 ℃. The data indicated that complete transformation from high-resistivity C49-$TiSi_2$ to low-resistivity C54-$TiSi_2$ was largely achieved for all implanted samples at 650 ℃. The corresponding XRD spectra are shown in Fig. 10. It is obvious that the substantial amount of C54 phase already forms at 650 ℃, the same as that of the Ti (30nm)/Mo (0.5nm)/blank (001)Si samples. In the XRD spectra of the 650 ℃ annealed implanted samples containing Mo, a weak peak at about $2\theta=41.5^{\circ}$ was also found. This peak corresponds to a ternary phase containing Ti, Mo and Si atoms as described earlier. The increase in sheet resistances for Ti on 20 keV N_2^+, 30 keV BF_2^+ + 20 keV N_2^+, and 30 keV As^+ + 20 keV N_2^+ implanted samples with an interposing Mo layer after annealing at 1000 ℃, 1000 ℃ and 950 ℃, respectively, were correlated to the formation of pinhole structure of C54-$TiSi_2$.

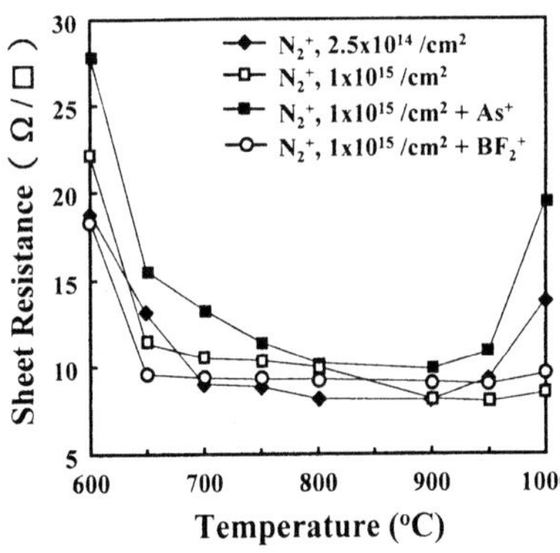

Fig. 9. Sheet Resistance of titanium silicides versus annealing temperature curves for different implanted samples with a thin Mo interposing layer.

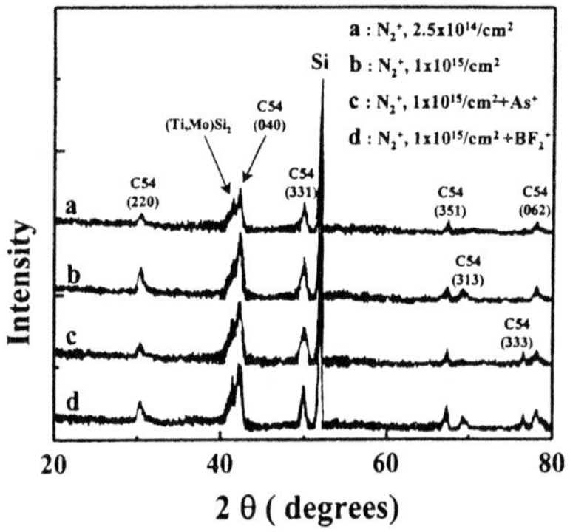

Fig. 10. Glancing angle XRD spectra of Ti (30nm) / Mo (0.5nm) on (a) 2.5x 10^{14} /cm^2 N_2^+-, (b) 1 x 10^{15} /cm^2 N_2^+-, (c) 1 x 10^{15} /cm^2 N_2^+ + As^+-, and (d) 1 x 10^{15} /cm^2 N_2^+ + BF_2^+-implanted samples after 650 ℃ annealing.

No amorphous surface layer was observed in samples implanted by N_2^+ to a dose of 2.5 x 10^{14} /cm^2. On the other hand, as the N_2^+ dose was increased to 1 x 10^{15} /cm^2, a 33-nm-thick amorphous layer was found to form in as-N_2^+ implanted samples. For 650 ℃ annealed samples, C54-$TiSi_2$ was the only silicide phase. In addition, after annealing at 1000℃, formation of pinholes in C54-$TiSi_2$ was found. The slight increase in sheet resistance is correlated to the initiation of the formation of pinholes.

For Ti/Mo on 1 x 10^{15} /cm^2 N_2^+ implanted samples, no EOR defects were evident in all samples annealed at 600-950 ℃. From XTEM observation, the formation of the $TiSi_2$ layer was found to consume the regions with EOR defects in silicon. For 2.5 x 10^{14} N_2^+ implanted samples, the dose was below the critical dose for amorphization of the surface layer. As a result, no EOR defects were observed.

For nitrogen implanted samples, the presence of N was found to be effective in retarding the degradation of surface morphology of the C54-$TiSi_2$ thin film. From the SIMS analysis, the N atoms were found to diffuse out of Si substrate to disperse in $TiSi_2$ during annealing. In addition, the presence of N atoms in $TiSi_2$ is thought to lower the silicide/silicon interface energy and/or the silicide surface energy to stabilize the integrity of the C54-$TiSi_2$ layer at high temperatures. The alleviation of agglomeration of C54-$TiSi_2$ in nitrogen implanted samples may also be attributed to the accumulation of N atoms at the grain boundaries.

In order to study further the interactions between N atoms and Ti during silicide formation, ESCA was used to investigate the chemical binding of silicide thin film. If any interaction occurred between N atoms and Ti during annealing, the binding energy of Ti is expected to shift to the higher energy position in ESCA spectra. Figs. 11 (a) and (b) show the ECSA spectra of Ti on blank (001)Si and 20KeV, 1 x $10^{15}/cm^2$ N_2^+ implanted samples after annealing at 800°C, respectively. It clearly shows that the peak positions of binding energy of Ti in ESCA spectra are essentially the same. The results indicated that N atoms did not bond strongly with Ti during silicide formation.

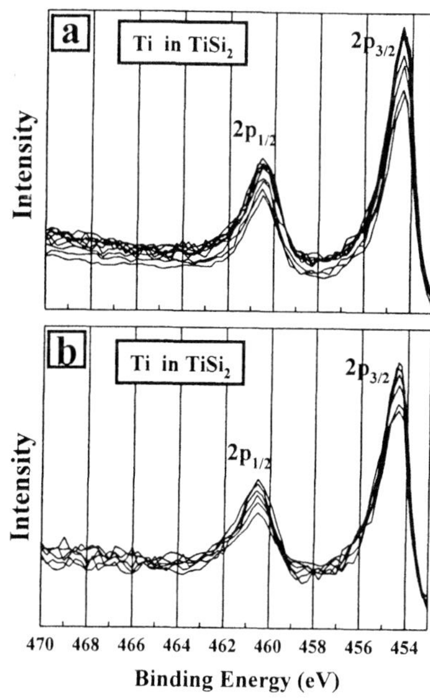

Fig. 11. ESCA spectra of Ti(30nm) on (a) blank (001)Si, (b) 20keV, 1 x $10^{15}/cm^2$ N_2^+ implanted samples, 800 °C, 30 s.

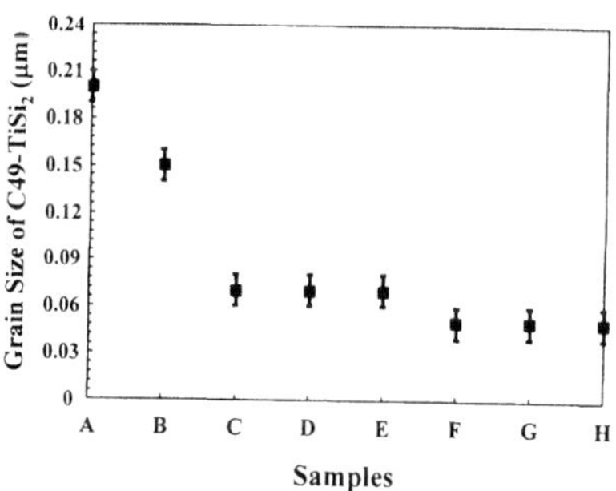

Fig. 12. The grain size of different samples after 600 °C annealing.

A : Ti (30nm) / (001) Si
B : Ti (30nm) / Mo (0.5nm) / (001) Si
C : Ti (30nm) / N_2^+, $1x10^{15}/cm^2$
D : Ti (30nm) / N_2^+, $1x10^{15}/cm^2$ + As^+
E : Ti (30nm) / N_2^+, $1x10^{15}/cm^2$ + BF_2^+
F : Ti (30nm) / Mo (0.5nm) / N_2^+, $1x10^{15}/cm^2$
G : Ti (30nm) / Mo (0.5nm) / N_2^+, $1x10^{15}/cm^2$ + As^+
H : Ti (30nm) / Mo (0.5nm) / N_2^+, $1x10^{15}/cm^2$ + BF_2^+

For Ti on 1 x 10^{15} /cm^2 N_2^+ + BF_2^+ and 1 x 10^{15} /cm^2 N_2^+ + As^+ implanted samples, 45-nm-thick and 51-nm-thick amorphous layer were observed, respectively. In all sets of samples annealed at 600 ℃, C49-$TiSi_2$ was found to be the only silicide phase present. For 650 ℃ annealed samples, C54-$TiSi_2$ was the only silicide phase. The results are correlated with the difference in sheet resistance of $TiSi_2$ in Fig. 9 and the XRD spectra in Fig. 10. A continuous C54-$TiSi_2$ layer was found to form in N_2^+ + BF_2^+ implanted samples annealed at 650-950 ℃ and N_2^+ + As^+ implanted samples annealed at 650-900 ℃. In addition, for Ti / Mo on 20 keV N_2^+ + BF_2^+ and 20 keV N_2^+ + As^+ implanted samples after annealing at 1000℃ and 950℃, respectively, formation of pinholes of C54-$TiSi_2$ were found. The formation of pinholes of C54-$TiSi_2$ is correlated to the slight increase in sheet resistance in Fig. 9.

For 1 x 10^{15} /cm^2 N_2^+ + BF_2^+ implanted samples, EOR defects were found to be prevalent in 600-950 ℃ annealed samples. On the other hand, no EOR defects were observed in 1 x 10^{15} /cm^2 N_2^+ + As^+ implanted samples annealed at 700-950 ℃.

The Effects of Grain Size on the Transformation of C49- to C54-$TiSi_2$

In order to study the effects of grain size on the transformation of C49- to C54-$TiSi_2$, 30-nm-thick Ti films were also deposited on 20 keV N_2^+, 20 keV N_2^+ + BF_2^+ and 20 keV N_2^+ + As^+ implanted Si samples without a Mo thin interposing layer. The average grain sizes of Ti films were measured to be about the same in all as-deposited samples. However, after annealing at 600 ℃, prior to the transformation of C49- to C54-$TiSi_2$, the grain sizes of C49-$TiSi_2$ on ion implanted samples were observed to be much smaller than those on samples without implant. Fig. 12 shows the grain size data for different samples after annealing at 600 ℃. The decrease in grain size in implanted samples can be attributed to the increase in the nucleation density of C49-$TiSi_2$ by the formation of an implantation amorphous layer. Table II lists the grain size and phase formation data for various samples.

In a polycrystalline structure, grain boundary nucleation for anew phase is generally the dominant mode in phase transition. However, in Ti on implanted Si samples without a thin interposing Mo layer, both C49- and C54-$TiSi_2$ were found to form after annealing at 650 ℃. On the other hand, for Ti on blank (001)Si samples with a thin interposing Mo layer, C54-$TiSi_2$ was found to be the only silicide phase in 650 ℃ annealed samples although the average grain size of C49-$TiSi_2$ was significantly larger as seen in Fig. 12. The effect of grain boundary on the lowering of transformation temperature was therefore found to be less crucial than the presence of Mo layer. Instead, the presence of a ternary (Ti, Mo)Si_2 phase is likely to provide more heterogeneous nucleation sites to enhance the formation of C54-$TiSi_2$ at a temperature as low as 650 ℃. The results indicated that the dominant effect for enhancing formation of C54-$TiSi_2$ is from the presence of Mo atoms and not directly from the increase in grain boundaries.

In the present study, it is shown that with appropriate control, an interposing Mo layer and N_2^+-implantation can be successfully implemented in forming low-resistivity $TiSi_2$ contacts and enhancing the thermal stability of $TiSi_2$ layer on shallow junctions in deep submicron devices (30). A recent report demonstrated successful implementation of a novel one-step RTP Ti SALICIDE process combining low dose Mo and preamorphization implant into a 0.12 μm gate length CMOS structure (31).

CONCLUSIONS

Nitrogen ion implantation was found to be advantageous to the thermal stability of $TiSi_2$ significantly. Thermal stability of $CoSi_2$ and NiSi was also improved by nitrogen ion implantation. Enhanced formation of low-resistivity $TiSi_2$ in titanium on nitrogen implanted (001)Si with a thin interposing Mo layer has been found.

The presence of Mo thin interposing layer was found to decrease the formation temperature of C54-$TiSi_2$ by about 100 ℃. A ternary (Ti, Mo)Si_2 phase was found to distribute in the silicide layer. The ternary compound is conjectured to provide more heterogeneous nucleation sites to enhance the formation of C54-$TiSi_2$. On the other hand, the effect of grain boundary for decreasing transformation temperature was found to be less crucial. For Ti/Mo bilayer on 30 keV BF_2^+ or As^+ + 20 keV, 1 x 10^{15} /cm^2 N_2^+ implanted samples, a continuous C54-$TiSi_2$ layer was found to form in all samples annealed at 650-950 ℃. The presence of nitrogen atoms in $TiSi_2$ is thought to lower the silicide/silicon interface energy and/or the silicide surface energy to maintain the integrity of the C54-$TiSi_2$ layer at high temperatures.

For future devices, metal contacts on shallow junctions would require elevated source/drain by SEG of Si and suppression of TED of dopants. Rapid thermal annealing has already become an essential processing step. For practical implementations, integration issues pertinent to other schemes such as application of nitrogen ion implantation to the device fabrication need to be further explored.

ACKNOWLEDGMENT

This research was supported by the Republic of China National Science Council through a grant No. NSC87-2215-E-007-029.

REFERENCES

1. C.Y. Wong, L.K. Wang, P.A. McFarland, and C.Y. Ting, J. Appl. Phys., 60, 243 (1983).
2. J. P. Gambino and E. G. Colgan, Mater. Chem. Phys., 52, 99 (1998).
3. S.P. Murarka, Mater. Res. Soc. Symp. Proc., 320, 3 (1994).
4. W. Lur and L.J. Chen, J. Appl. Phys., 66, 3604 (1989).
5. M. Ono, M. Saito, T. Yoshitomi, C. Fiegna, T. Ohguro, and H. Iwai, IEEE IEDM 1995 Tech. Digest, 119 (1995).
6. T. Kuroi, S. Shimizu, A. Furukawa, S. Komori, Y. Kawasaki, S. Kusunoki, Y. Okumura, M. Inuishi, N. Tsubouchi, and K. Horie, Digest of 1995 Symposium on VLSI Technology, 19 (1995).
7. S. Shimizu, T. Kuroi, Y. Kawasaki, S. Kusunoki, Y. Okumura, M. Inuishi, and H. Miyoshi, IEEE IEDM 1995 Tech. Digest, 859 (1995).
8. T.O. Sedgwick, A.E. Michel, V.R. Deline, S.A. Cohen, and J.B. Lasky, J. Appl. Phys., 63, 1452 (1988).

9. A.C. Ajimera and G.A. Rozgonyi, Appl. Phys. Lett., 49, 1269 (1986).
10. T. Murakami, T. Kuroi, Y. Kawasaki, M. Inuishi, and Y. Matsui, Nucl. Instrim. Methods, 121, 257 (1997).
11. J.B. Lasky, J.S. Nakos, O.J. Cain, and P.J. Geiss, IEEE Trans. Electron Devices, ED-38 (1991) 262.
12. R. W. Mann, G. L. Miles, T. A. Knotts, and D. W. Rakowski, L. A. Clevenger, J. M. E. Harper, F. M. D'Heurle, and C. Cabral, Jr., Appl. Phys. Lett., 67 (1995) 18.
13. X. -H. Li, R. A. Carlsson, S. F. Gong, and H. T. G. Hentzell, J. Appl. Phys., 72 (1992) 514.
14. H. Kuwano, J. R. Phillips, and J. W. Mayer, Appl. Phys. Lett., 56 (1990) 440.
15. A. Mouroux, S. -L. Zhang, W. Kaplan, S. Nygren, M. Ostling, and C. S. Petersson, Appl. Phys. Lett., 69 (1996) 975.
16. J. Ziegler, J.P. Biersack, and U. Littmark, The Stopping and Range of Ions in Matter (Pergamon, New York, 1985).
17. T.S. Chao, M.C. Liaw, C.H. Chu, C.Y. Chang, C.H. Chien, C.P. Hao, and T.F. Lei, Appl. Phys. Lett., 69, 1981 (1996).
18. T. Murakami, T. Kuroi, Y. Kawasaki, M. Inuishi, and Y. Matsui, Nucl. Instrim. Methods, B121, 257 (1997).
19. T. Nitta, T. Ohmi, Y. Ishihara, A. Okita, T. Shibata, J. Suigura, and N. Ohwada, J. Appl. Phys., 67, 7404 (1990).
20. A.F. Tasch, Jr. And L.H. Parker, Proc., IEEE 77, 374 (1989).
21. M. Rodder, S. Aur, and I.C. Chen, IEEE 1995 IEDM Digest, 415 (1995).
22. T. Kuroi, Y. Yamaguchi, M. Shirahata, Y. Okumura, Y. Kawasaki, M. Inuishi, and T. Tsubouchi, IEEE IEDM 1993 Digest, 325 (1993).
23. C.Y. Yang, T.F. Lei, and C.L. Lee, IEEE IEDM 1994 Digest, 505 (1994).
24. A. Nishiyama, Y. Akaska, Y. Ushiku, K. Hishioka, Y. Suizu, and M. Shiozaki, Proc. IEEE VLSI Multilevel Interconnection Conference, 310 (1990).
25. S.L. Cheng, L.J. Chen, and B.Y. Tsui, J. Mater. Research, 14, 213 (1999).
26. K.M. Chen, S.L. Cheng, L.J. Chen, and B.Y. Tsui, Mater. Chem. Phys. 54, 71 (1998).
27. L. W. Cheng, S. L. Cheng, J. Y. Chen, L. J. Chen, and B. Y. Tsui, Thin Solid Film, (in press, 1999).
28. Standard JCPDS diffraction pattern 6-607 [hexagonal $(Ti_{0.4}Mo_{0.6})Si_2$], JCPDS-International Center for Diffraction Data, PDF-2 Database, 12 Campus Blvd., Newton Square, PA 19073-3273.
29. Standard JCPDS diffraction pattern 7-331 [hexagonal $(Ti_{0.8}Mo_{0.2})Si_2$], JCPDS-International Center for Diffraction Data, PDF-2 Database, 12 Campus Blvd., Newton Square, PA 19073-3273.
30. S.L. Cheng, J.J. Jou, L.J. Chen, and B.Y. Tsui, J. Mater. Research, (in press, 1999).
31. J.A. Kittl, Q.Z. Hong, M. Rodder, and T. Breedijk, IEEE IEDM97 Digest, 111 (1997).

Attainment of Low Resistivity Polycide Films Using Rapid Thermal Annealing

H. A. Yoon, C. Chen, A. Singhal, D. Lopes, G. Miner, S. Hong
Applied Materials, Thermal Processing Organization, TPI Product Business Group
2727 Augustine Drive, Santa Clara, CA 95054

M. Yamazaki, Y. Maeda
Applied Materials, Japan
14-3 Shinizaumi, Narita-Shi, Chiba-ken 286-8516, Japan

Polycide films with lower resistivity were developed by combining Dichlorosilane-SiH_2Cl_2 (DCS) based tungsten silicide (WSi_x) films and Rapid Thermal Annealing (RTA). The lowest resistivity was achieved with high temperature WSi_x deposition and Rapid Thermal Nitration (RTN). The polycide film shows as-deposited resistivity of ~800 μΩ·cm and ~ 65 μΩ·cm resistivity after annealed in NH_3 at 1000°C for 30 seconds. The low resistivity polycide films also contains relatively low levels of fluorine and chlorine compared to current DCS based WSi_x films.

Introduction

Tungsten silicide (WSi_x) on polycrystalline silicon films are used widely for gate level interconnect and bit lines in advanced dynamic random access memory (DRAM) devices. As the DRAM technology approaches 0.18 μm and beyond, there is a strong need for polycide with lower resistivity for gate and bit line applications since a large resistance results in a delay in the operating speed of a device. There has been a development in metal gate area [1,2], however, it is still faces many integration issues. As an alternative to metal gate, low resistivity polycide with DCS based WSi_x offers many advantages - there is little issue for the process integration since polycide with WSi_x has been widely used. This combination of W-rich interfaced WSi_x and Rapid Thermal Annealing (RTA) has been demonstrated [3]. Here we have developed a low resistivity polycide by combining high temperature DCS based WSi_x with Rapid Thermal Nitration (RTN) to achieve ~65 μΩ·cm resistivity. In this paper, the process development and film characterization results are presented.

Experiments

The deposition of polysilicon and WSi_x films were carried out in a single wafer, integrated polycide system. Polysilicon deposition is carried out in a high temperature quartz chamber. The wafer is transferred to silicon carbide (SiC) coated susceptor on which wafer is heated by banks of lamp arrays both top and bottom of the wafer. Process and dopant gases (SiH_4 and PH_3) enter from one side of the chamber, then flows over the wafer before exit through the opposite side. Wafer is rotated during deposition in order to achieve better thickness and sheet resistance uniformity. After polysilicon films deposition, wafers are transferred to DCS-WSi_x chamber through transfer chamber which is kept in 6 Torr of N_2. This insures no contamination of the Si surface during transfer. WSi_x deposition chamber is equipped with resistively heated heater. Process gases flow to chamber from two separate lines for WF_6 and DCS then mixed at mixer block assembly at the top of the chamber. Then mixed gases pass through shower head, flow over wafer then exit through pumping plate. Subsequent to polycide deposition, wafers are annealed in a single wafer Rapid Thermal Process (RTP) chamber in various conditions. RTP temperature were varied from 950°C to 1050°C while the annealing time varied from 5 to 90 second. Our current DCS based WSi_x process was also tested as a reference.

Sheet resistance (Rs), stress, and thickness were measured after each process steps. Variety of complementary analytical techniques were used to characterized films in the as-deposited stage and after annealed in RTP chamber. Phase identification analysis was determined by x-ray diffraction (XRD). The composition of the films as a function of depth was measured by Rutherford back-scattering spectrometry (RBS) using 2.275 MeV He^{++} ion beam. The depth profile of various impurities of concern such as Cl, F, O and P was carried out by secondary ion mass spectrometry (SIMS).

Results

Effect of RTP Process

The following graph (Figure 1) shows the sheet resistance (Rs) change for low resistivity polycide after RTP process. The thickness of WSi_x and a-Si layers were 1000Å each. As-deposited Rs value was 66.1 Ω/sq. The annealing temperature and the process gases were changed.

As shown below, Rs decreases as the annealing time and temperature increases. Also, one can notice that annealing in NH_3 (referred as Rapid Thermal Nitration, RTN) provides lower Rs than annealing in N_2 (referred as Rapid Thermal Annealing, RTA). Since annealing in NH_3 shows lower Rs, RTN was used as standard annealing process. Figure 2 illustrates the difference between current WSi_x process (with furnace anneal at 850°C for 30 minutes), current WSi_x and low resistivity WSi_x with 1000°C/30seconds RTN process.

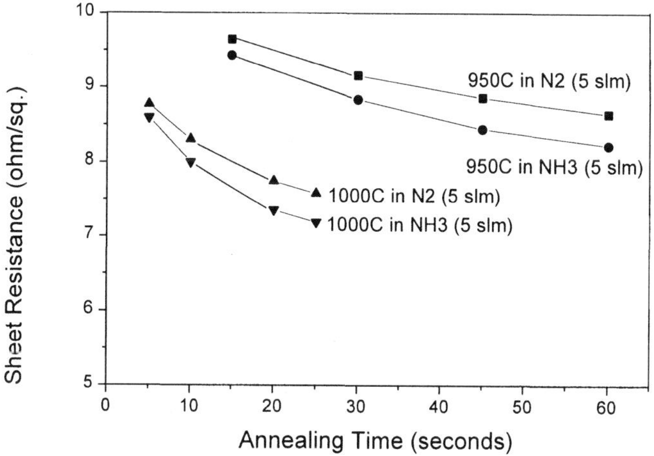

Figure 1. Sheet Resistance of low resistivity polycide films (1000Å WSi_x on 1000Å polysilicon) after various RTP processes. As-deposited Rs value was 66.1 Ω/sq.

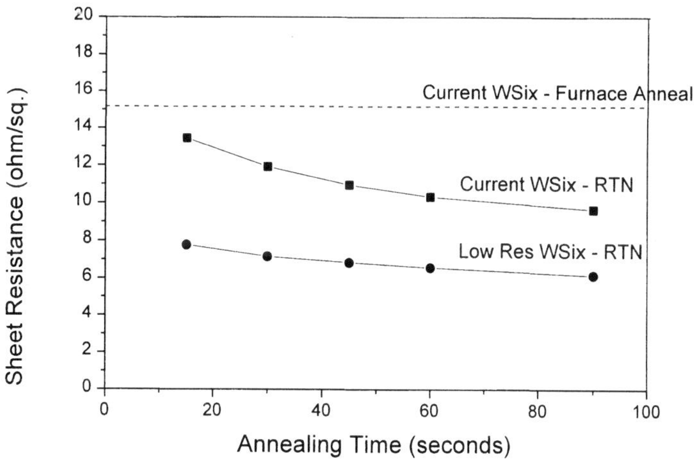

Figure 2. Sheet Resistance of polycide films with current WSi_x process and low resistivity WSi_x process. 1000°C for 30 seconds in 5 slm NH_3 was used for RTN process.

Rs values for polycide with current WSi_x process is 15.2 Ω/sq. after 30 minutes anneal in furnace at 850°C. When compared with current WSi_x process with furnace anneal, Rs for low resistivity polycide film decreases as much as 50 % after 30 second anneal at 1000°C in RTP chamber (with 5 slm flow of NH_3). When compared with current polycide process, low resistivity polycide film shows as much as 42 % reduction in Rs with same RTN conditions. Thickness of WSi_x films were measured from SEM for as-deposited and a few RTN samples to obtain amount of a-Si loss as well as thickness of these films. The following SEM micrographs were taken for as-deposited and after 1000°C 30 second RTN process.

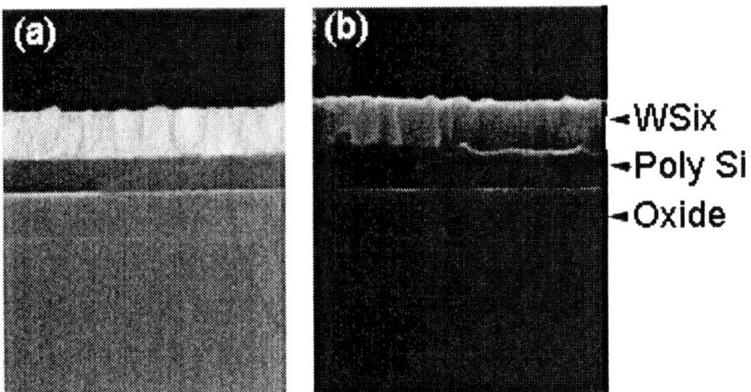

Figure 3. Cross-sectional SEM of low resistivity polycide films (WSi_x / poly Si/ Oxide): (a) As-deposited; (b) after RTN at 1000°C for 30seconds.

From SEM analysis, there is only about 100 Å of a-Si loss after RTN but no apparent trend was observed as RTN conditions changed. With 1000°C 30second RTN, low resistivity WSi_x shows ~ 65 μΩ·cm resistivity while current WSi_x process with furnace anneal shows 150 μΩ·cm resistivity. There is > 50 % reduction in resistivity for low resistivity polycide compared with current polycide process.

Film Composition

The composition of WSi_x film deposited at 640°C was analyzed with RBS. Figure 4 is RBS spectra for both as-deposited and post RTN (1000°C, 30 seconds) polycide films. For as-deposited films, the Si:W ratio (x) in low resistivity WSi_x was 2.2, compared to 2.4 - 2.5 for our current WSi_x film. Si:W ratio remained at 2.2 even after RTN process.

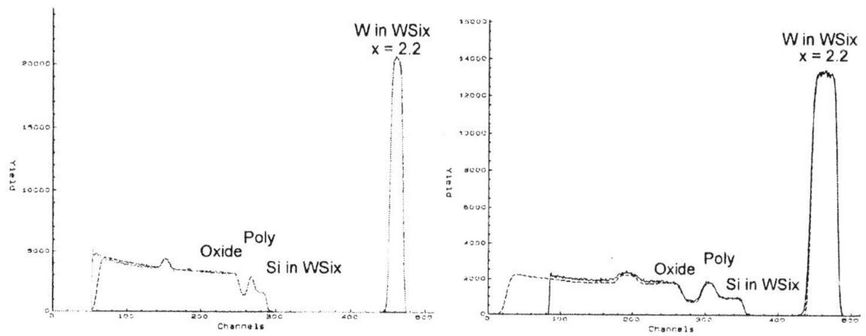

Figure 4 RBS spectra for low resistivity WSi_x (a) as-deposited (160° RBS), (b) post RTN at 1000°C for 30 seconds (107° RBS).

F and Cl concentration in as-deposited WSi_x and a-Si were obtained from SIMS analysis (Figure 5). The concentration of F and Cl in low resistivity WSi_x turned out to be lower than those of current DCS based WSi_x films. [4,5]

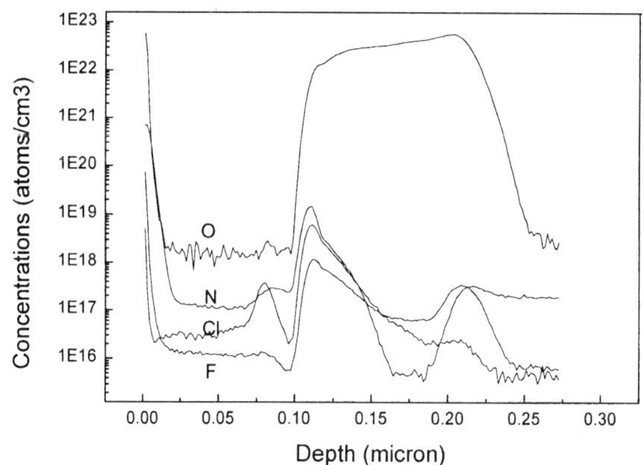

Figure 5. SIMS profiles of impurities in low resistivity polycide films on oxide after 1000°C , 30 seconds RTN.

Concentration	Low Resistivity WSi_x	Current WSi_x
[F] (atoms/cm^3)	1E16	6 E16 - 2 E17
[Cl] (atoms/cm^3)	2E16 - 4E16	5 E 17 - 5 E 18

Table 1. Concentration of F and Cl in low resistivity and current WSi_x by SIMS analysis.

Discussion

Crystalline Structure

X-ray diffraction was taken on as-deposited and post-RTN (1000°C, 30seconds) WSi_x films which were grown at temperatures varying from 550 to 700°C .

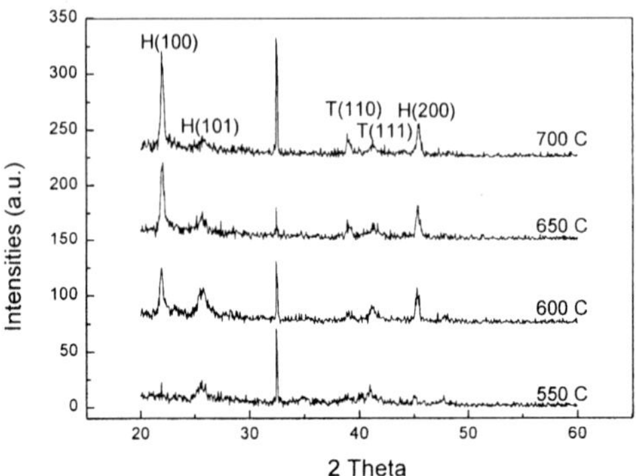

Figure 6. XRD of as-deposited WSi_x at several deposition temperatures.

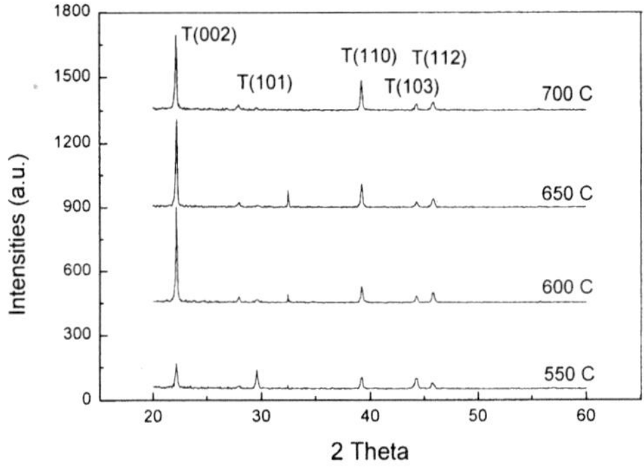

Figure 7. XRD spectrum of WSi_x after RTN at 1000 °C for 30 seconds. The crystalline structure changes from hexagonal to tetragonal and diffraction intensities increases compare to as-deposited film.

XRD spectrum of as-deposited film shows diffraction pattern of mostly hexagonal structure. However, diffraction intensities are very weak, suggesting possible amorphous WSi_x formation.

RTN changes the WSi_x crystalline structure from hexagonal to tetragonal structure. The intensities are also higher than those of as-deposited WSi_x film.

One thing can be noticed here that as the deposition temperature increases, the intensity of hexagonal (100) peak increases in as-deposited films. Also, after RTN, tetragonal (002) intensity hits the maximum at deposition temperature of 650°C. There is a correlation between intensity of tetragonal (002) and resistivity. Therefore suggesting influence of crystalline orientation on the final resistivity of the film.

As mentioned, rapid thermal processing in NH_3 yield much lower Rs than in N_2. It has been suggested [3] that by processing in NH_3 environment, thin nitride is formed at the surface of WSi_x films which blocks the out diffusion of phosphorus atoms. SIMS analysis were carried out in order to examine the phosphorus concentration in WSi_x and polysilicon films after annealed in NH_3 and in N_2.

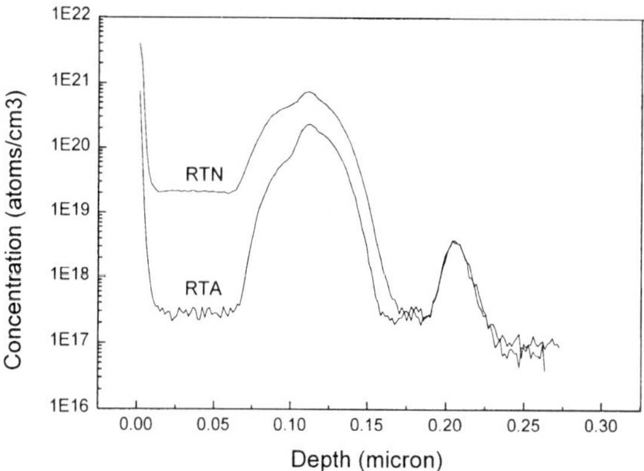

Figure 8. SIMS profile of phosphorus in WSi_x and polysilicon films after RTA and RTN process.

SIMS analysis indicates that concentration of phosphorus is about 2 orders of magnitude higher for RTN case than RTA. This proves the previous claim of nitride formation. This analysis also shows the phosphorus concentration difference in polysilicon layer.

Conclusion

The low resistivity polycide process was developed with HTQ Poly, DCSxZ and RTP chambers and evaluated. Low resistivity WSi_x film (as-deposited) exhibits lower Si:W ratio than that of current WSi_x process (2.2 vs. 2.4-2.5). Upon annealed in RTP chamber, resistivity decreases significantly than current polycide with furnace anneal. There is little consumption of a-Si after RTN, which indicate that the reduction in resistivity seems to be resulted from crystallographic change of WSi_x film, as well as from increase in grain size upon RTN. Also, RTN can lower the resistivity by forming thin nitride which probably blocks the out diffusion of phosphorus from WSi_x and polysilicon films.

References

1. H. Wakabayashi, T. Andoh, K. Sato, K. Yoshida, H. Miyamoto, T. Mogami, T. Kunio, *IEDM Tech. Dig.*, 447 (1996)
2. Y. Akasaka, S. Suehiro, K. Nakajima, T. Nakasugi, K. Miyano, K. Kasai, H. Oyamatsu, M. Kinugawa, M. Takayanagi, K. Agawa, F. Matsuoka, M. Kakumu, K. Suguru, *IEEE Trans. Electron Devices*, **43**, 1864 (1996)
3. J.S. Byun, B. H. Lee, J-S. Park, D-K. Sohn, S.J. Choi, J.J Kim, *J. Electrochem. Soc*, **145,** 3228 (1998)
4. S.G. Telford, M. Eizenberg, M. Chang, A.K Sinha, T.R. Gow, *J. Electrochem. Soc.*, **140**, 3689 (1993)
5. S.G. Telford, M. Eizenberg, M. Chang, A.K Sinha, T.R. Gow, *Appl. Phys. Lett.*, **62**, 1766 (1993)

THERMAL STABILITY IMPROVEMENT OF COBALT DISILICIDE THIN FILMS ON (001)SI BY HIGH TEMPERATURE SPUTTERING DEPOSITION

H. Y. Huang[1], L. J. Chen[1], W. F. Wu[2], and R. P. Yang[2]
[1]Department of Materials Science and Engineering, National Tsing Hua University, Hsinchu, Taiwan, Republic of China
[2]National Nano Device Laboratories, Hsinchu, Taiwan, Republic of China

The thermal stability of $CoSi_2$ thin film was improved by high temperature sputtering deposition (HTSD). There was no clear intermixing between Co and Si on single crystal substrate in samples deposited at room temperature or 300 °C. Polycrystalline CoSi phase was found in as-deposited samples with a sputtering deposition temperature of 450 °C. After the two-step annealing (500 °C, 30s + selective etching + 800 °C, 30s), the average grain sizes of $CoSi_2$ were measured to be 10, 14, and 18 nm in samples deposited at 450, 300 °C and room temperature, respectively. For further annealing treatments, samples deposited at 450 °C were found to possess the best thermal stability. The alleviation of agglomeration of $CoSi_2$ is attributed to the higher nucleation density of $CoSi_2$, which led to the retardation of grain growth in high temperature sputtering deposition samples.

INTRODUCTION

Low resistivity silicides have been widely used in source/drain as contacts in ULSI devices. It reduces both the parasitic source/drain resistance and the contact resistance (1). Among all silicides, NiSi, $TiSi_2$, and $CoSi_2$ are of the lowest resistivity, about 15-20 $\mu\Omega$-cm. $TiSi_2$ is currently the most commonly used silicide in IC industry. However, as devices are scaled down to deep submicron dimension, it is difficult to transform $TiSi_2$ from high resistivity C49 phase (60 $\mu\Omega$-cm) to low resistivity C54 phase (15 $\mu\Omega$-cm) owing to the lack of nucleation sites.

$CoSi_2$, owing to its low resistivity, has been considered as a candidate to substitute $TiSi_2$ (2). For shallow junction applications, the silicide-as-diffusion-source process is a promising approach and $CoSi_2$ is a suitable material in this regard (3). However, as the thickness of silicide scales with the device dimension, the thermal stability of thin silicide films, especially that grown on polysilicon, is of major concern (4). High temperature sputtering deposition (HTSD) was previously found to facilitate the phase transition from C49 to C54-$TiSi_2$. Ti capped Co salicide process was found to improve the thermal stability of $CoSi_2$, and the CoTi phase was formed at the surface (5, 6). The formation of pinholes of $CoSi_2$ on Si can be eliminated as $CoSi_2$ was formed in oxygen-containing ambient (7). A thin oxide layer was formed at the cobalt silicide surface to block fast diffusion paths and reduce rate of kinetic process. In the present study, the Co films were deposited at 450 °C (HTSD), 300 °C (medium temperature sputtering deposition, MTSD), and room temperature (room temperature sputtering deposition, RTSD). The HTSD was found to improve thermal stability of a 65-nm-thick $CoSi_2$ layer.

EXPERIMENTAL PROCEDURES

Single crystal, 1-10 Ω-cm, 6 inches in diameter, p-type (001) oriented silicon wafers were used in the present study. All wafers were initially cleaned by a standard RCA process. Wafers were then loaded in a MRC Primus 2500 PVD cluster tool, equipped with dual cassette module. 20-nm-thick Co thin films were deposited on the wafers with the deposition rate of 1 nm/s. The base presure was 2 x 10^{-9} Torr. The depositions were carried out at room temperature, 300 °C or 450 °C. Two-step rapid thermal annealing (RTA) was employed for the silicidation reaction. During the two-step annealing, samples were first annealed at 480-550 °C for 30-60 s. Unreacted metals were removed in a solution of H_2SO_4:H_2O_2 = 3:1. The second annealing was performed at 800 °C for 30 s. Additional thermal annealing was used to evaluate the thermal stability of $CoSi_2$ thin films at 900-1100 °C. Sheet resistance was measured by the standard four-point probe method. X-ray diffraction (XRD) was used for phase identification. A JEOL-200 CX transmission electron microscope (TEM) operating at 200 kV was used for examination of microstructures. Both planview and cross-sectional TEM (XTEM) were carried out. The elemental distribution was determined by Auger electron spectroscopy (AES) depth profiling.

RESULTS AND DISCUSSION

CoSi was detected to be the only metal phase present in as-deposited HTSD samples, while only Co was found in films deposited at RT and 300 °C. A XRD spectrum is shown in Fig. 1. After annealing at 500 °C for 30 s, CoSi was the only metallic phase in all samples as shown in Fig. 2. The silicides were all converted into $CoSi_2$ after the second RTA at 800 °C.

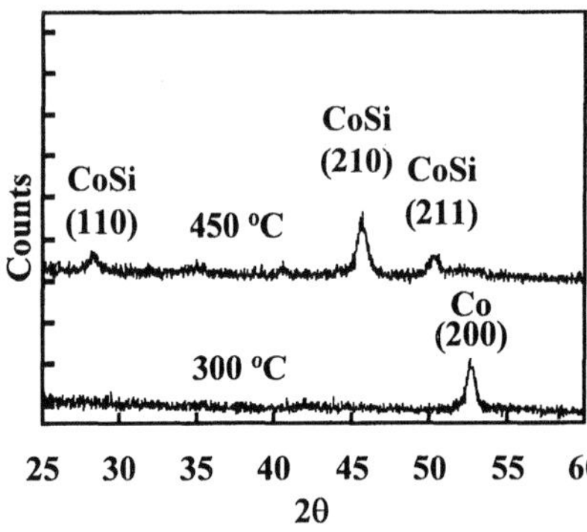

Fig. 1. XRD spectra of the Co films deposited on (001)Si at various deposition temperatures.

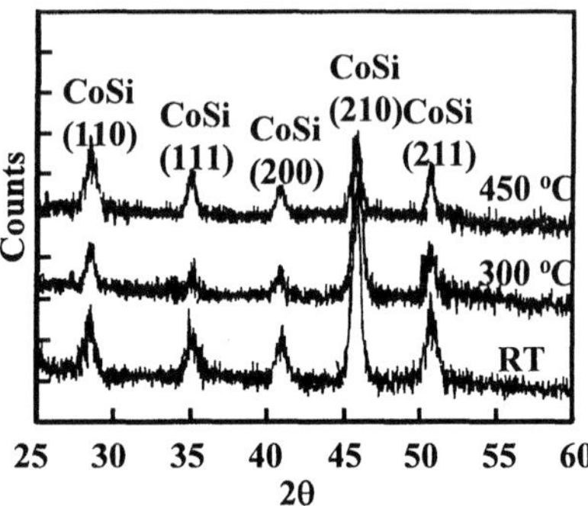

Fig. 2. XRD spectra of the Co films deposited on (001)Si at various temperatures after annealing at 500 °C for 30 sec.

AES depth profiles revealed the intermixing of Co and Si as shown in Fig. 3. In samples deposited at 450 °C, Co metal was completely mixed with Si to form CoSi as determined by XRD. On the other hand, much less interfacial reaction or intermixing between Co metal and Si was found for MTSD and RTSD samples as seen in Fig. 3 (b) and (c).

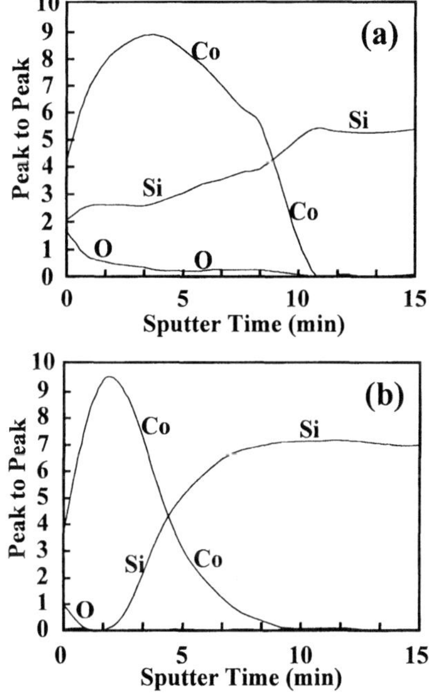

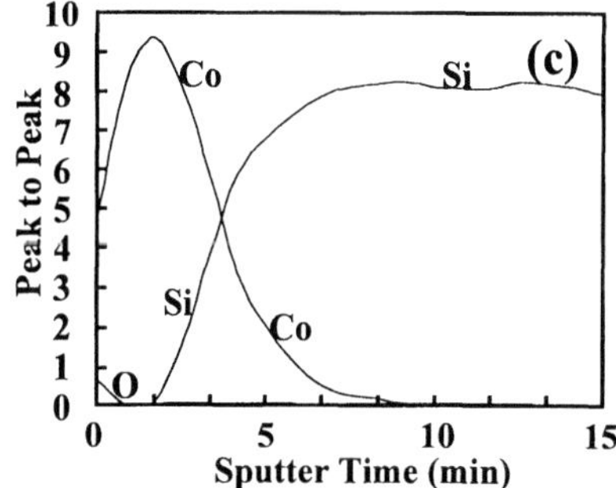

Fig. 3. AES depth profiles of the Co films deposited on Si substrates at (a) 450 °C, (b) 300 °C, and (c) RT.

After stripping the unreacted metal and the second RTA step, only $CoSi_2$ phase was detected in all samples. The average grain sizes of $CoSi_2$ were measured to be 10, 14, and 18 nm for the Co films deposited at 450, 300 °C and RT, respectively. Planview TEM images are shown in Fig. 4. The average grain size is smaller in samples deposited at higher temperatures. A previous study showed that thicker amorphous $TiSi_x$ interlayer was found in samples deposited at higher temperature (8). $TiSi_2$ phase was nucleated from embedded crystallites in the amorphous layer. In the Co/Si system, no amorphous interlayer between Co and Si was found. However, high temperature deposition apparently also facilitates the nucleation of CoSi phase. The nucleation of $CoSi_2$ phase was at the triple points of the CoSi grains and the silicon surface (9). Smaller CoSi grains provide more nucleation sites for $CoSi_2$ phase. As a result, the HTSD samples were of smaller grain than that of MTSD or RTSD samples. Figure 5 shows the XTEM images of $CoSi_2$ after 2-step annealing. Smaller $CoSi_2$ grain was also observed for the samples deposited at 450 °C. However, HTSD did not seem to improve the interface roughness.

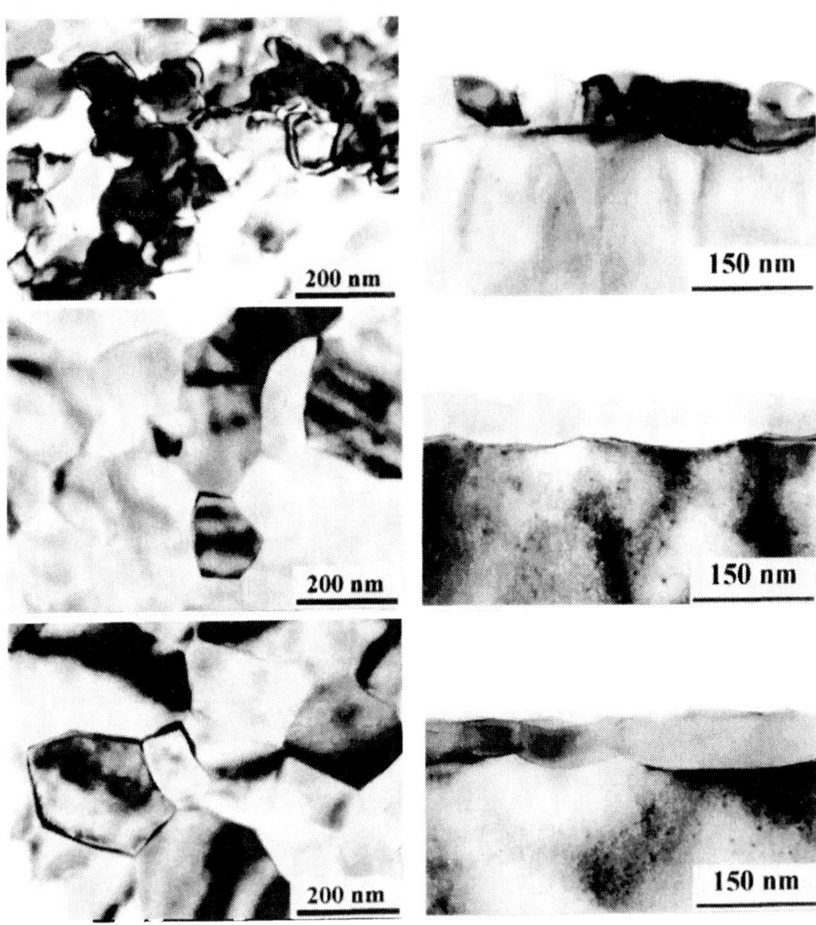

Fig. 4. Planview TEM images of $CoSi_2$ after the two-step annealing. The deposition temperatures of films were (a) 450 °C, (b) 300 °C, and (c) RT.

Fig. 5. XTEM images of $CoSi_2$ after the two-step annealing. The deposition temperatures of films were (a) 450 °C, (b) 300 °C, and (c) RT.

Additional RTA treatments were performed to evaluate the thermal stability of $CoSi_2$ thin films after the two-step annealing. The sheet resistance data revealed that better thermal stability was obtained for the films deposited at 450 °C, as shown in Fig. 6. The thermal stability of the samples deposited at 300 °C was very similar to that deposited at RT. The sheet resistance of the HTSD samples was higher than that of the MTSD or RTSD samples after annealing at 800 °C for 30 s. Several layers of $CoSi_2$ grains were randomly grown on Si substrate in HTSD samples, while only a monolayer of $CoSi_2$ grains was grown in MTSD or RTSD samples. The additional thermal annealing reduced the sheet resistance of the HTSD samples due to the grain growth of $CoSi_2$ into monolayer grains. Then the formation of pinholes and the agglomeration of $CoSi_2$ phase increase the sheet resistance in all samples. The fine grain of $CoSi_2$ alleviates the agglomeration. Adopting a criterion of 30 % increase in sheet resistance as unacceptable degradation, $CoSi_2$ was found to be stable after annealing at 1050 °C for 10 s.

SUMMARY AND CONCLUSIONS

The thermal stability of $CoSi_2$ thin film was improved by HTSD. No significant intermixing was found between Co and Si in samples deposited at RT and 300 °C. Polycrystalline CoSi phase was formed in as-deposited HTDS samples. After the same two-step annealing, the average grain sizes of $CoSi_2$ were found to be 10, 14, and 18 nm for samples with the Co films deposited at 450, 300 °C and RT, respectively. For additional thermal annealing, samples with films deposited at 450 °C were found to possess the best thermal stability. The alleviation of agglomeration of $CoSi_2$ is attributed to the higher nucleation density of $CoSi_2$, which led to the retardation of grain growth in high temperature sputtering deposition samples.

REFERENCES

1. K. K. Ng and W. T. Lynch, IEEE Trans. Electron. Device, **ED-34**, 503 (1987).
2. J. B. Lasky, J. S. Nakos, D. J. Cain, and P. J. Geiss, IEEE Trans. Electron. Device, **ED-38**, 262 (1991).
3. H. Jiang, C. M. Osburn, P. Smith, Z. G. Xiao, D. Griffis, G. Mcguire, and G. A. Rozgonyi, J. Electrochem. Soc., **139**, 196 (1992).
4. K. Maex, Mater. Sci. Eng. Rep., **R11**, 53 (1993).
5. R. T. Tung and F. Schrey, Appl. Phys. Lett., **67**, 2164 (1995).
6. D. K. Sohn, J. S. Park, B. H. Lee, J. S. Byun, and J. J. Kim, Appl. Phys. Lett., **73**, 2302 (1998).
7. R. T. Tung, Appl. Phys. Lett., **72**, 2538 (1998).
8. S. M. Chang, H. Y. Huang, H. Y. Yang, and L. J. Chen, Appl. Phys. Lett., **74**, 224 (1999)
9. A Appelbaum, R. V. Knoell and S. P. Murarka, J. Appl. Phys., **57**, 1880 (1985)

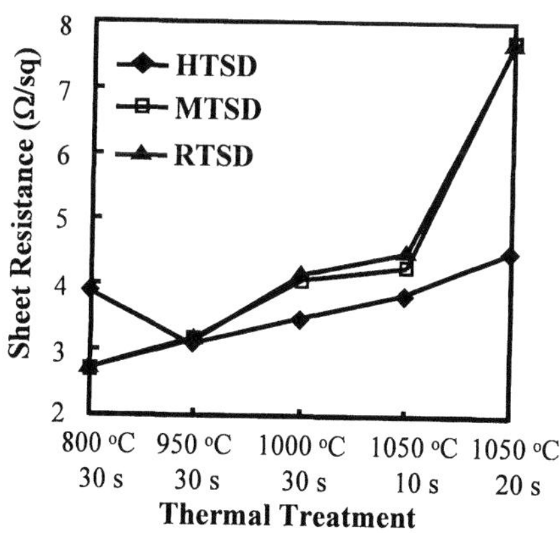

Fig. 6. Sheet resistance of $CoSi_2$ versus annealing conditions.

Section V

Novel Applications

RAPID THERMAL PROCESSING OF HIGH-PERFORMANCE DIELECTRICS AND SILICON SOLAR CELLS[1]

A. Rohatgi

University Center of Excellence for Photovoltaics Research and Education,
School of Electrical and Computer Engineering, Georgia Institute of Technology,
Atlanta, GA. 30332, USA

Rapid and potentially low-cost process techniques are analyzed and successfully applied toward the fabrication of high-efficiency monocrystalline Si solar cells. First, a novel dielectric passivation scheme (formed by stacking a plasma silicon nitride film on top of a rapid thermal oxide layer) is developed that serves as antireflection coating and reduces the surface recombination velocity (S) of the 1.3 Ω-cm p-Si surface to approximately 10 cm/s. The essential feature of the stack passivation scheme is its ability to withstand short 700-850 ^{o}C anneal treatments used to fire screen printed (SP) contacts, without degradation in S. The stack also lowers the emitter saturation current density (J_{oe}) of 40 and 90 Ω/□ emitters by a factor of 3 and 10, respectively, compared to no passivation. Next, rapid emitter formation is accomplished by diffusion under tungsten halogen lamps in both beltline and rapid thermal processing (RTP) systems (instead of in a conventional infrared furnace). Third, a combination of SP aluminum and RTP is used to form an excellent back surface field (BSF) in 2 minutes to achieve an effective back surface recombination velocity (S_{eff}) of 200 cm/s on 2.3 Ω-cm Si. Finally, the above individual processes are integrated to achieve: 1) >19% efficient solar cells with emitter and Al-BSF formed by RTP and contacts formed by vacuum evaporation and photolithography, 2) 17 % efficient manufacturable cells with emitter and Al-BSF formed in a beltline furnace and contacts formed by SP, and 3) 17 % efficient bifacial cells with surface passivation accomplished by the stack and grid front and back contacts formed by SP and co-firing.

INTRODUCTION

Mankind's dependence on fossil fuels has negatively impacted the environment and is also believed to contribute to global warming. Photovoltaics (PV) offers a unique opportunity to solve the energy and the environmental problems simultaneously because solar energy is essentially unlimited and solar cells can covert it into electrical energy without any undesirable impact on the environment. The real challenge is to reduce the cost of PV generated electricity, which is about a factor of four higher than the traditional energy sources. Crystalline silicon accounts for more than 80 % of the PV modules shipped today. The cost of current crystalline Si PV modules is about $4/watt, approximately 45 % of that is attributed to Si material, 25 % to cell processing, and 30 % to module assembly [1]. For widespread implementation of PV, module cost should be reduced to $1-2/watt. This can be accomplished by lowering the cost of solar cell materials and processing without sacrificing cell efficiency. In addition, to reach a 50 MW/yr production capacity, close to 50,000 silicon cells with 15% efficiency (100 cm^2

[1] Dielectric Science and Technology Callinan Award Address

cell producing 1.5 W) must be processed every day. This translates into approximately one cell per second [2]. Therefore, processes must be invented that can significantly reduce the cell processing time and number of processing steps and corresponding equipment must be developed that can handle more wafers per unit time. Rapid thermal processing (RTP) has the potential to accomplish high throughput, lower processing cost, and high performance solar cells simultaneously. Reduced number of processing steps and shorter processing times also reduce material handling. This may allow the use of thinner material in the future (100 μm thick Si as opposed to 300-400 μm being used today) which can also lead to significant cost reduction. RTP, which utilizes banks of tungsten-halogen lamps to radiatively heat a semiconductor rapidly, can reduce the time for diffusion and oxidation because of extremely high ramp-up rates (30-300 ^{o}C/s) and the presence of high energy visible and ultraviolet (UV) photons [3-5]. Due to the thermal mass of a boatload of wafers, conventional furnaces cannot be used for short time processing. In the case of rapid isothermal processing via lamps, sample is thermally isolated and heating and cooling is dominated by radiation. Heating times are limited by thermal response time, which for most materials is less than 1 second [3]. Unlike the conventional furnace processing (CFP), RTP is a cold wall process, which allows for simultaneous diffusion of junctions on both sides or diffusion on one side and oxidation on the other, without any cross contamination [6]. This feature can reduce the number of processing steps. The challenges in RTP involve preserving minority carrier lifetime, achieving optimum doping profiles, and providing high quality surface passivation. These difficulties limited the efficiencies of RTP cells with screen printed contact to less than 13 % prior to 1992 [7-10]. Since then several investigators have successfully implemented RTP to make higher efficiency silicon solar cells [6, 11-15].

Most manufactures, in an attempt to reduce manufacturing cost, use a minimal number of low-cost processing steps. However, the resulting cell efficiencies are in the 10-14 % range. In contrast to industrial cells, laboratory cells have reached efficiencies over 24 % [16] by incorporating numerous processing steps and high efficiency features such as selective emitter, advanced light trapping and surface texturing, low front and back surface recombination velocity, and high quality contacts formed by vacuum evaporation and photolithography. Because of this, current laboratory cells are too expensive and industrial cells are not efficient enough to meet the cost and efficiency targets simultaneously. Therefore objective of this study is to develop rapid thermal processes that can reduce manufacturing cost without sacrificing efficiency. This paper reports on the rapid and improved formation of 1) dielectrics for surface passivation and antireflection coating, 2) phosphorus diffused n^{+}- emitters and 3) Al alloyed back surface field. Rapid thermal technologies, such as RTP diffusion and oxidation, screen printed contacts, and plasma enhanced chemical vapor deposition (PECVD) of silicon nitride (SiN), are optimized and integrated to achieve high efficiency cells on monocrystalline silicon. Solar cells with full Al-BSF and back metal coverage as well as grid back bifacial devices are fabricated to demonstrate the throughput and efficiency potential of RTP and beltline processing.

RTP OF DIELECTRICS FOR FRONT AND BACK SURFACE PASSIVATION AND ANTIREFLECTION COATING FOR SILICON SOLAR CELLS

Minimizing minority carrier recombination at the surfaces of silicon is crucial for performance of many semiconductor devices including solar cells [17]. The objective of this section is to provide a quantitative comparison of surface passivation quality of several promising dielectrics for diffused and non-diffused silicon surfaces. The emphasis is placed on rapid technologies like rapid thermal oxidation (RTO) under tungsten-halogen lamps, PECVD SiN, and TiO_2. These technologies have much lower thermal budget than oxides grown in conventional furnace. Films like TiO_2 and SiN are investigated because they also provide good antireflection coating. Since cell fabrication often involves 400 °C forming gas anneal (FGA) and 700-800 °C beltline firing of screen printed contacts, the impact of such heat treatments on dielectric passivation quality is also investigated.

Samples for the emitter passivation experiment were diffused on both sides in an RTP system using spin-on dopant sources. Emitters with sheet resistances of 40 and 90 Ω/□, which correspond to emitters that can accommodate screen-printed and evaporated contacts, respectively, were investigated. After removal of the residual phosphorus silicate glass, part of the diffused and non-diffused p-type samples were oxidized in the same RTP system used for the diffusions. This rapid thermal oxidation at 900 °C for 150 s resulted in an oxide thickness of approximately 8-10 nm on diffused surfaces and about 6 nm on non-diffused 1 Ω-cm p-Si. The oxidized low-resistivity samples were then annealed in forming gas at 400 °C for 15 min. After this, ~ 600 nm SiN was deposited at 300 °C in a 13.6 MHz parallel plate PECVD reactor to obtain a good SiN AR coating [18]. Although its passivation is known to be poor, TiO_2 films are also compared for completeness because it is by far the dielectric most commonly employed by the PV industry as an AR coating.

The minority carrier lifetime was measured after each step (deposition and anneal) using a commercially available inductively coupled PCD tester. From these data, the emitter saturation current J_{0e} (for diffused samples) and the surface recombination velocity S_{eff} were (for undiffused samples) calculated. The PCD measurement of J_{0e} is discussed by Kane and Swanson [19] and S_{eff} was calculated from the measured τ_{eff} value according to equations and methodology described in Ref [20]. Thus dielectric passivation is measured in terms of S and J_{0e} which are not only influenced by D_{it} but also by the fixed charged density in the dielectric [21,22].

A. Dielectric Passivation of Diffused Si Surfaces

The passivation of diffused front surfaces was investigated for both 40 Ω/□ and 90 Ω/□ emitters. On a relatively opaque 40 Ω/□ emitter (which is generally needed to accommodate screen-printed contacts), the surface is largely decoupled from the bulk, because of the high surface doping concentration and depth of the doping profile. Thus, the introduction of RTO or SiN passivation resulted in a moderate decrease in J_{0e} of about a factor of two to three, as can be seen from Fig. 1. While TiO_2 showed hardly any passivation, SiN was clearly inferior to RTO. The combination of RTO and SiN

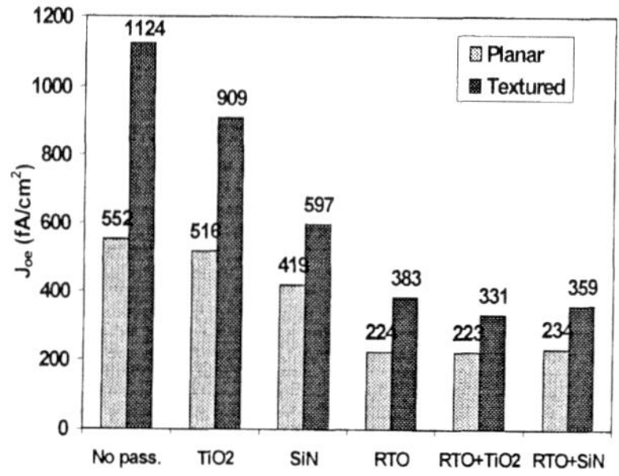

Fig. 1. Emitter saturation current densities for different passivation schemes on 40 Ω/□ RTP emitters.

resulted in the best passivation. Note, that the high-temperature treatment during RTO growth changed the doping profile and leads to a lower surface doping concentration, which allowed for better surface passivation. The J_{0e} values for textured samples were about 1.5 to 2 times higher than those for planar surfaces, which resembles the 1.73 times increase in surface area resulting from regular pyramidal texturing.

On the relatively transparent 90 Ω/□ emitter, (which is generally used for evaporated or photolithographically-defined contacts) the difference in the degree of passivation for various dielectrics was more apparent, as shown in Fig. 2. Again, TiO_2 does not provide any appreciable reduction in J_{0e}. For the planar surface, RTO growth reduced J_{0e} by more than a factor of ten to below 100 fA/cm^2, as did the deposition of SiN. However, on the textured surface, RTO is not as effective, resulting in a moderate J_{0e} value of 400 fA/cm^2. Again, RTO+SiN double layer was clearly superior resulting in lowest J_{0e} values.

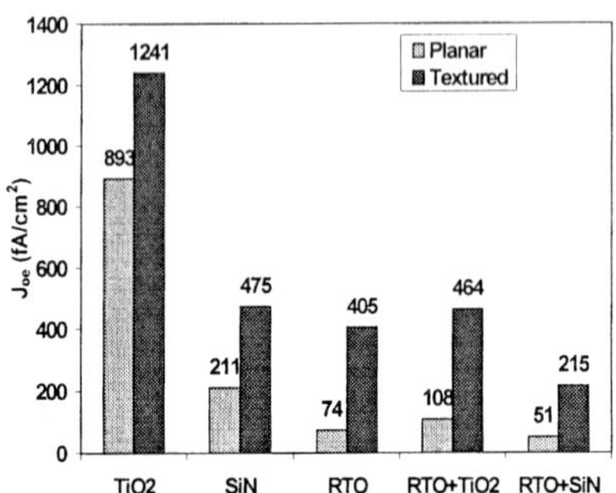

Fig. 2. Emitter saturation current densities for 90 Ω/□ RTP emitter.

B. Dielectric Passivation of Undiffused Si Surfaces

Dielectric passivation of undiffused surface is crucial for bifacial solar cells. Fig. 3 shows the effectiveness of several *individual*-passivating dielectrics. The dielectrics studied include 100 Å RTO alone, 100 Å thick conventional furnace oxide (CFO) alone, and 650 Å SiN alone. The individual layers result in S values in excess of 10,000 cm/s on 1.3 Ω-cm Si immediately after growth or deposition. This extremely poor S is reduced to lower levels (20-200 cm/s) after an additional FGA at 400 °C. However, a subsequent 730 °C beltline anneal (simulating SP contact firing) degrades each passivation, increasing S above 1000 cm/s for the RTO and SiN films and above 175 cm/s for CFO.

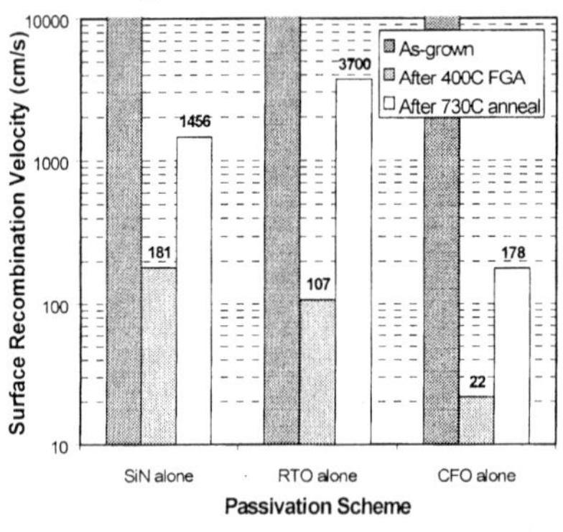

Fig. 3. The effect of beltline annealing on the S value of individual dielectric passivation schemes.

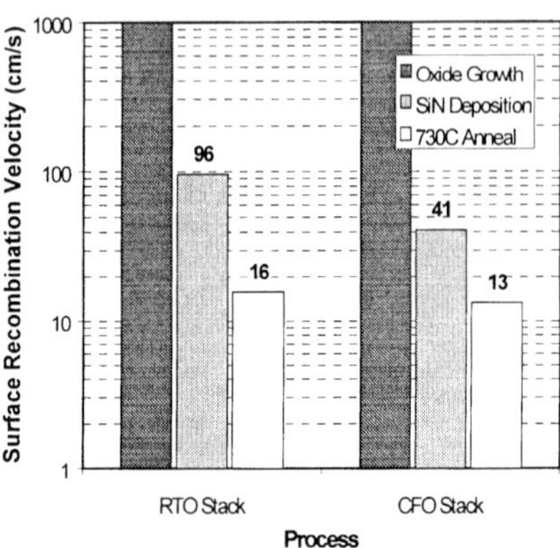

Fig. 4. Improvement in S of the stack passivation scheme after 730 °C / 30 s annealing.

Contrary to the response of the individual films, annealing the RTO/SiN and CFO/SiN stacks clearly *enhances* the passivation quality. The stepwise effect of stacking SiN on top of the RTO or CFO layer and then annealing at 730 °C is shown in Fig. 4. An S value of nearly 10 cm/s is attained at the 1.3 Ω-cm Si surface after the final anneal. This anneal is believed to enhance the release and delivery of atomic hydrogen from the SiN film to the Si-SiO_2 interface, thereby reducing the interface state density (D_{it}). This hypothesis was supported by FTIR measurements of the SiN film, which revealed the evolution of hydrogen from Si-H and N-H bonds after annealing and corresponding high-frequency CV measurements of MNOS capacitors, which showed a reduced distortion in the C-V response after annealing. Applying the modified Terman method to these C-V curves revealed a mid-gap D_{it} reduction by a factor of 20 after the anneal treatment.

RAPID THERMAL PROCESSING OF PHOSPHORUS DIFFUSED EMITTER FOR SILICON SOLAR CELLS

The total process time for phosphorus diffusion in a conventional furnace requires on the order of one hour at 850-900 °C to achieve emitters with appropriate junction depth for SP contact formation. This severely limits the throughput of a manufacturing line. In this section, emitter formation in a beltline furnace and RTP unit is described and compared with conventional furnace diffusion. Both beltline and RTP systems utilize tungsten halogen lamps to optically heat the sample. At normal diffusion temperatures, these lamps produce high-energy photons that increase the effective diffusion of phosphorous in Si when utilized in conjunction with spin-on dopants (SOD) [23,24].

In order to support this hypothesis, three different drive-in treatments were performed at 880 °C after the application of SOD phosphorus. These were: 1) conventional furnace diffusion where the sample temperature is raised by standard resistance heating, 2) beltline furnace diffusion where the sample is heated by a combination of radiation from tungsten halogen lamps, conduction from the belt, and convection, and 3) single wafer RTP diffusion where the sample is heated primarily by radiation from the tungsten-halogen lamps. In the conventional furnace, only infrared photons are present. In the RTP system the sample sees large number of higher-energy photons. In the lamp heated three-zone beltline furnace, the sample sees fewer high-energy photons because the lamps in the first two zones are not as hot as the third zone and there is additional heat via conduction and convection. Fig. 5 shows that after the SOD application, an 880 °C / 6 min. RTP drive-in produces significantly more diffusion compared to conventional furnace processing (CFP). The diffusion profile attained with beltline processing (BLP) generally falls between the two.

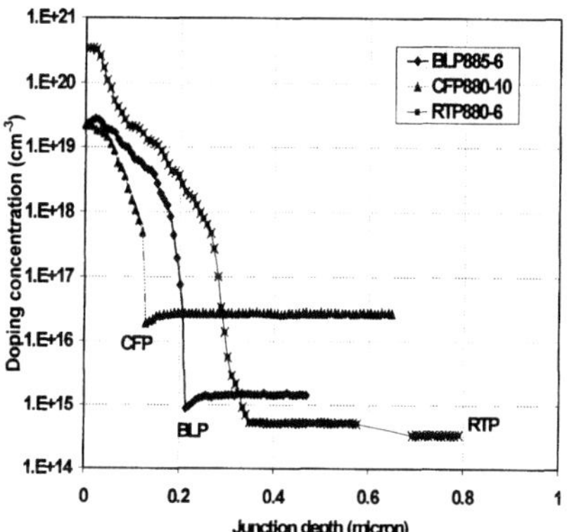

Fig. 5. Comparison of CFP, BLP, and RTP phosphorus diffusion (using SOD sources) at 880 °C for 6 minutes.

This is consistent with the high-energy photon content of the three spectra. As a result, it required 30 minutes, 6 minutes, and 3 minutes at 880 °C to obtain a 40 Ω/□ emitter by CFP, BLP and RTP, respectively.

RAPID THERMAL PROCESSING OF ALUMINUM BACK SURFACE FIELD FOR SILICON SOLAR CELLS

The Al-BSF is widely used for reducing the effective back surface recombination velocity (S_{eff}) in solar cells with full metal rear contacts. The advantage of the Al-BSF (over a boron) is that the p^+ region can be formed in a short alloying step at lower temperature instead of a lengthy boron diffusion process at much higher temperatures. However, the electrical quality of an Al-BSF is extremely sensitive to uniformity and process conditions. If the p^+ is not formed appropriately, full performance gains will not be realized. This section demonstrates that RTP of Al BSF not only reduces thermal budget but also produces much higher quality back surface field. The quality of back surface field is assessed in terms of effective surface recombination velocity at the p^+ -p interface. S_{eff} value was determined from the long wavelength response by matching the measured and calculated internal quantum efficiencies.

An important factor that affects the electrical quality of an Al-BSF is junction uniformity. The uniformity of an Al-BSF, or any metal-Si junction in general, is controlled to a large extent by the ramp-up rate used to reach the alloying temperature. This effect was noted early on in Refs.[25-26]. The same effect was observed later, in qualitative terms, for the Al-Si system [27]. This study analyzes quantitatively the effect of RTP on junction uniformity of Al-BSF and the resulting S_{eff} value and solar cell efficiency.

Due to the slow ramp rates typical for conventional furnace processing, the alloying between Al and Si occurs at certain sites before others (a form of *local wetting*). This results in a non-uniform p^+-region. The non-uniformity can include variations in BSF junction depth, loss of surface planarity, Al spiking, and possibly non-formation of the p^+ region. Under fast ramp conditions present in RTP, a sample goes through the Al-Si eutectic point and reaches the process temperature very quickly. At typical process temperatures (≈800-900 °C), the Al layer melts and readily wets the entire Si surface. This promotes more uniform alloying, which in turn improves Al-BSF uniformity.

In order to correlate the degree of BSF uniformity to device performance, two sample BSFs were formed by thermally evaporating 10 μm of Al onto Si and then alloying at 850 °C. The sample undergoing the slow ramp process was pushed into a conventional furnace below the Al-Si eutectic temperature and ramped-up at a rate of 5 °C/min. The sample undergoing the fast ramp procedure was alloyed in an RTP unit and ramped-up at a rate of 1200 °C/min. After processing, the p^+ regions were delineated by etching in an acid solution [28]. Fig. 6 shows cross-sectional SEM micrographs of the Al-BSF regions formed under slow ramp and fast

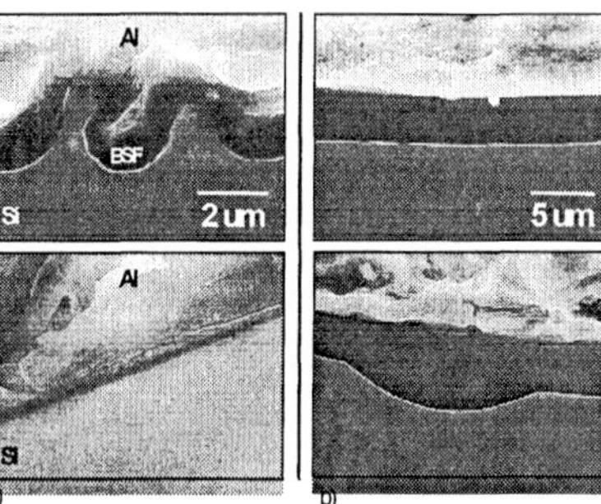

Fig. 6. BSF regions formed at a) slow ramp rates (top shows severe junction non-uniformity, bottom shows non-formation and b) fast ramp rates (both top and bottom show clean formation and improved uniformity.

ramp conditions. Clearly, the slow ramp process results in significant non-uniformity in the p^+ junction (pitting and non-formation across much of the wafer surface). On the contrary, the sample alloyed under fast ramp conditions shows a higher degree of junction uniformity and planarity.

Fig. 7 shows the synergistic effect of ramp-up rate and deposited Al thickness on solar cell performance. The open circuit voltage (V_{oc}) and long wavelength internal quantum efficiency (IQE) measurements were used to assess the Al-BSF quality. The results show that for the slow ramp-up condition (5 °C/min in a conventional furnace), the BSF effectiveness is not proportional to the deposited Al thickness in the range of 1-10 μm. This behavior is not consistent with theoretical predictions based on Al-Si phase diagram considerations [29]. Based on theory alone, BSF thickness is proportional to the deposited Al thickness, and thicker BSF should produce lower S_b and higher V_{oc}. In contrast to the slow ramp-up, the fast ramp-up in the RTP system shows a monotonic rise in V_{oc} with increasing Al deposition thickness. RTP-alloyed Al-BSF cells formed on 2.3 Ω-cm Si with 1 μm evaporated Al resulted in V_{oc} of 606 mV, whereas cells with 10 μm evaporated Al resulted in V_{oc} of 632 mV.

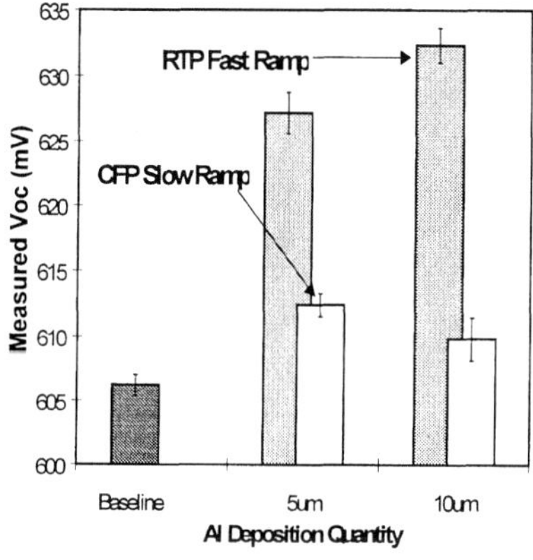

Fig. 7. The effect of Al layer thickness and ramp-rate on solar cell V_{oc}

The thick film (> 10 μm) Al deposition can be rapidly achieved by the SP technique. Therefore, this technique was also investigated for BSF formation in conjunction with RTP. SP/RTP Al-BSFs (alloyed at 850 °C in 2 minutes) gave slightly higher V_{oc} than even the 10 μm evaporated Al case Fig. 8. This shows that the lengthy and impractical 10 μm Al evaporation step can be completely replaced by SP without sacrificing cell performance. Modeling and analysis of the measured long wavelength IQE for these cells indicated an S_{eff} values ≈ 1000 cm/s and of 200 cm/s for the evaporation/CFP and SP/RTP Al-BSF, respectively, on 2.3 Ω-cm Si. Thus RTP of Al BSF is not only faster but also superior.

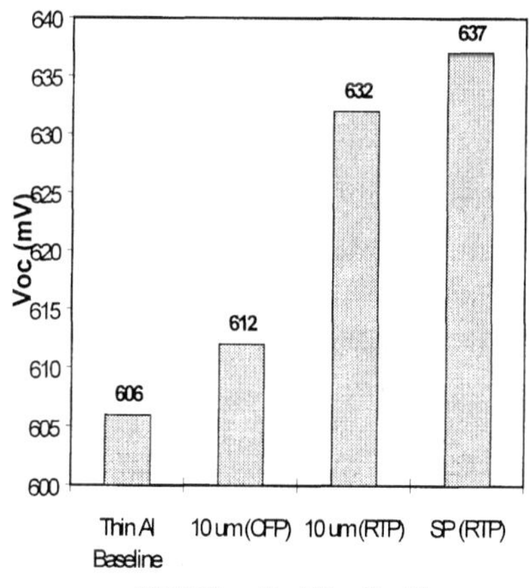

Fig. 8. Effect of different Al-BSF processes on solar cell V_{oc}.

INTEGRATION OF RAPID PROCESS TECHNOLOGIES FOR HIGH EFFICIENCY SILICON SOLAR CELLS

Silicon solar cell fabrication involves processing of emitter, back surface field, dielectric passivation, antireflection coating, and front and back contact formation. Previous sections demonstrated the use of beltline and RTP systems for rapid and improved formation of emitter and back surface field, and deposition of RTO/SiN stack that can provide not only an excellent passivation for front and back surfaces and but also withstand 700-800 °C firing of low-cost screen-printed contacts. In this chapter we show the integration of these rapid technologies for achieving high efficiency cells on mono-crystalline silicon.

A. Conventional Furnace Processing of Si Solar Cells with Photolithography Contacts

Figure 9a shows the fabrication sequence of a baseline cell using conventional furnace processing (CFP) and photolithography contacts. In this process 80 Ω/□ phosphorus diffusion, 1 μm Al back surface field formation, and front oxide passivation was done in conventional furnace. Process sequence involved about 6 hours of high temperature processing, 2 ½ hours of metal evaporation, and 5 hours of photolithography resulting in a total cell processing time of about 16 hours with cell efficiencies of about 18 % (Fig. 9a) without any surface texturing.

B. Rapid Thermal Processing of Si Solar Cells with Photolithography Contacts

The above process was modified and shortened significantly by replacing furnace processing by rapid thermal processing in which phosphorus diffusion, screen printed Al BSF formation, and oxide passivation was done in a single wafer RTP system from AG Associates. Front and back contacts were formed by evaporation and photolithography. Figure 9b shows the detailed process sequence and the corresponding cell performance. The 80 Ω/□ phosphorus diffusion was performed in about 6 minutes by application of spin-on-dopant, followed by a short drive in the RTP unit under the tungsten halogen lamps. Al back surface field was formed by screen printing 15 μm thick Al paste in less than a minute followed by RTP in an oxygen ambient for less than 5 minutes. Besides forming a very effective and deep BSF, this step also produced a high quality rapid thermal oxide on the front simultaneously. Thus, this RTP process sequence reduces the total high temperature processing time from 6 hours (conventional furnace processing in Fig. 9a) to less than 15 minutes (Fig. 9b). In addition to reducing the total processing time from 16 hours to 8½ hours, this RTP process produced higher efficiency cells compared to conventional furnace processing. The RTP cell efficiencies of 19.1% were achieved compared to 17.8% for the CFP cells. This is primarily due to the thicker and more uniform RTP SP Al back surface field. As indicated earlier, 1 μm evaporated Al BSF formed in conventional furnace, using typical slow ramp up rate, is neither uniform nor deep enough to be effective.

C. Rapid Thermal Processing of Screen Printed Si Solar Cells

In an attempt to reduce the cell processing time further, the evaporation and photolithography contacts on the front were replaced by screen-printed Ag contacts. Due to the high contact resistance and junction shunting, the emitter sheet resistance of the SP cells was decreased from 80 to 40Ω/□ to achieve good contacts and high fill factors (FF). This increased the emitter diffusion time from 3 to 7 min. An 800 Å thick PECVD SiN coating was deposited on top of RTO for stack passivation and single layer AR coating. Ag contact grid was screen printed on the front and fired through the SiN AR coating in less than 5 min. This reduced the contact formation time from 330 to 8 min. Figure 9c shows the modified process sequence along with the cell performance. This RTP/SP process reduced the cell processing time from 8.5 hours to less than 2 hours and produced a planar cell efficiency of 17 % with a fill factor of 0.798. The 2 % reduction in absolute efficiency (19 % to 17 %) is largely attributed to heavy doping effects in the emitter, increased shading and reflectance, and somewhat inferior front surface passivation due to higher surface doping concentration [30]. Formation of selective emitter ($\leq$ 40 Ω/□ underneath the grid and $\geq$ 80 Ω/□ between the grid lines) for SP cells should be able to recover majority of the 2% loss in the efficiency.

D. Grid-Back Contact (Bifacial) Screen-Printed Si Solar Cells

Most PV manufacturers fabricate cells on 300-400 μm thick Si substrates. The material cost can be lowered significantly by reducing the thickness to <200 μm while maintaining or increasing cell efficiency. This can be accomplished only if the S_b remains low. Even though the SP/RTP Al-BSF is extremely effective in reducing S_b, the stress imparted to the substrate during formation can lead to warpage or breakage of thin wafers. Process-induced stress can be virtually eliminated by employing a passivated rear structure in which metal covers only a small fraction of the surface area. In addition, this structure is suited for bifacial operation and has the potential to produce increased power production per unit area. However, to take advantage of the bifacial capability, the rear surface passivation must be excellent. The RTO/SiN stack described earlier is an ideal passivation scheme for this structure. It simplifies the cell processing by permitting co-firing of the SP contacts on both sides. This process sequence, shown in Fig. 9c, resulted in a front-illuminated efficiency of 17 % (V_{oc} of 641 mV, J_{sc} of 33.4 mA.cm^2, and FF of 0.792), and a rear-illuminated efficiency of 11.6%. Detailed analysis of rear-illuminated IQE gave in a spatially averaged S_b (with ohmic contact coverage) of 340 cm/s.

E. Rapid Beltline Processing of Screen Printed Si Solar Cells

Since there is no continuous RTP system available today, continuous beltline processing (BLP) has been modified to bridge the gap between the single wafer RTP and BLP cells. Recall that P diffusion in the belt furnace is slower than in an RTP system. In an effort to keep the BLP emitter diffusion time to 6 minutes, the diffusion temperature was raised from 880 to 925 °C to achieve a $\approx$ 40 Ω/□ emitter. Instead of the stack passivation scheme, SiN alone was utilized for emitter surface passivation and as a single layer AR coating. The SP Al-BSF was alloyed in a beltline furnace at a set point temperature of 850 °C for 2 minutes. Finally, SP front contacts were fired through the SiN layer in 30 seconds at 730 °C. The total beltline processing time was less than 2 hours, and this resulted in a planar cell efficiency of 17 % (Fig. 9d).

Next generation cells will involve fabrication of beltline bifacial cells on thin PV grade Si. Model calculations (Fig. 10) show that 100 μm thick cells with a bulk lifetime of 20 μs can produce 17 % efficient SP cells, even without surface texturing, if the spatially averaged back surface recombination velocity can be reduced to $\approx$ 100 cm/s.

Investigations on growing high-quality oxide layers in a beltline furnace are currently underway to realize the stack passivation with this equipment. The approach outlined in this paper can transform today's 12-14 % efficient industrial cells on 300-400 μm thick Si to greater than 17 % SP cells on 100-200 μm thick Si in the future.

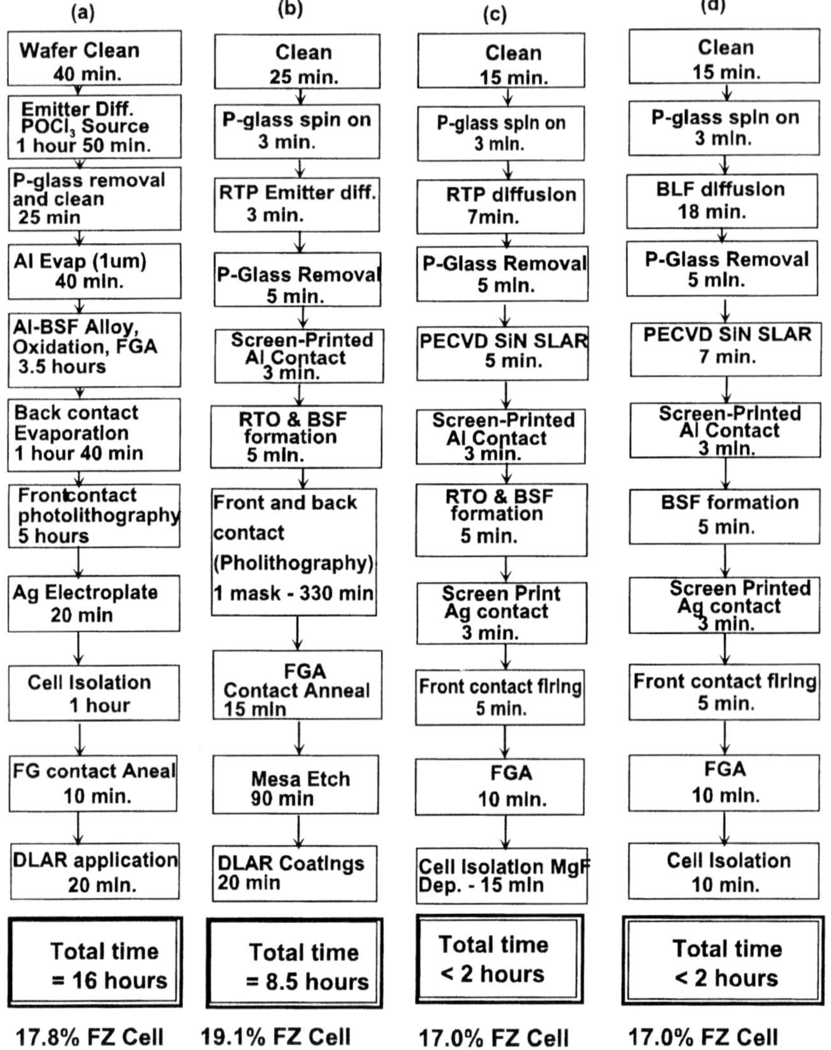

Fig. 9: Process sequence and cell performance for (a) baseline CFP cells with photolithography contacts (b) RTP cells with photolithography contacts (c) RTP cells with screen-printed contacts, and (d) BLP cells with screen-printed contacts.

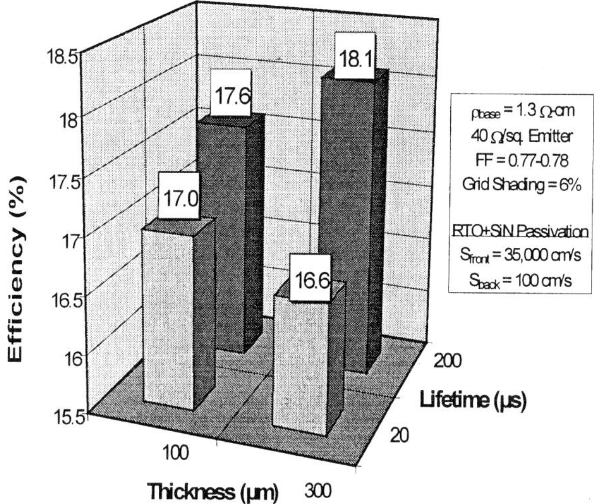

Fig. 10 Model calculations showing the efficiency resulting from high-quality RTO + SiN passivation of the front and back of SP cells on thick and thin materials.

CONCLUSIONS

Rapid Thermal Processing of each layer of a silicon solar cell was investigated and optimized. It was shown that an RTP alloyed, SP Al-BSF, formed in 2 minutes yields low S_{eff} of 200 cm/s on 2.3 Ω-cm Si. In the same thermal cycle, an effective RTO layer can be grown for front surface passivation. A novel stack passivation scheme (consisting of plasma SiN stacked on top of an RTO layer) was developed that can attain S values approaching 10 cm/s on 1.3 Ω-cm Si and J_{oe} values as low as 230 fA/cm^2 and 50 fA/cm^2 on 40 Ω/□ and 90 Ω/□ phosphorus diffused emitters, respectively. These emitters were formed in an RTP unit or a beltline furnace in 3 to 7 minutes by tungsten halogen lamp heating. Formation of the n^+ emitter, front surface oxide passivation, and Al-BSF by RTP resulted in a substantial reduction in cell processing time. Cell efficiencies of 19.1 % and 17 % were achieved on FZ Si using photolithography and SP contact techniques, respectively. SP bifacial cells with stack passivated surfaces, as well as SP Al-BSF devices (full back metal coverage) formed rapidly in a commercial beltline machine also resulted in 17% efficiency. These results endorse the potential of rapid thermal technologies for cost-effective Si PV in the future.

ACKNOWLEDGMENT

The author would like to thank A. Ebong, S. Narasimha, P. Doshi, S. Kamra, J. Moschner and A. Ristow for the help in the experimental work modeling and analysis. The author would also like to thank Professor Rajendra Singh of Clemson University for numerous technical discussions on RTP. This work was supported by Sandia National Labs and Subcontract No. AO-6162 and NREL under Subcontract No. XD-2-11004-2.

REFERENCES

1. J. Szlufcik, S. Sivoththaman, J. F. Nijs, R. P. Mertens and R.Van Overstraeten, Proc. IEEE, **85**, 709 (1997)

2. K Mitchell, R. King, T. Jester and M. McGraw, in Conf. Proc., 24th IEEE Photovoltaic Specialist Conference (Piscataway: IEEE) 1266-1269, (1994).
3. R. Singh, J. Appl. Phys., **63**(8), R59-R114, (1988)
4. R. Singh, in Conf. Proc., TMS International Conf. On Beam Processing of Advanced Materials, Warrendale, PA, 619-688 (1993).
5. R. Singh, F. Radtour and T. Chou, J. Vac. Sc. Technol. A, **A7**, 1456-1460 (1989).
6. A. Rohatgi, Z. Chen, P. Doshi, T. Pham, D. Ruby, Appl. Phys. Lett., **65**, 2087-2089 (1994)
7. A. Usami, M. Ando, M. Tsunekane, K. Yamamoto, T. Wada and Y. Inoue, in Conf. Proc., 18th ECPVSEC, 797-803, (198 5).
8. A. Usami, M. Tsunekane, T. Wada, Y. Inoue, S. Shimada, N. Nakazawa, Y. Meada, in Conf. Proc., 18th ECPVSEC, 1078-1083, (1985).
9. R. Campbell and D. Meier, J. Electrochem. Soc., **133**, 2210-2211 (1986)
10. B. Hartiti, A. Slaoui, J. C. Muller, P. Siffert, B. Wagner, R. Schindler, A. Eyer, A. Rauber, in Conf. Record, 11th ECPVSEC, 420-422 (1992)
11. B. Hartiti, A. Slaoui, J. C. Muller, P. Siffert, in Conf. Record, 23rd IEEE PVSC, 224-228, (1993).
12. R. Schindler, I. Reis, B. Wagner, A. Eyer, H. Lautenschlager, C. Schetter, W. Warta, in Conf. Record, 23rd IEEE PVSC, 162-166, (1993).
13. B. Hartiti, R. Schindler, A. Slaoui, B. Wagner, J. C. Muller, I. Reis, A. Eyer, P. Siffert, Progress in Photovoltaics, **2**, 129-142, (1994).
14. S. Sivoththaman, W. Laureys, P. De Schepper, J. Nijs, T. Mertens, in Conf. Record, 25th IEEE PVSC, 621-624, (1996).
15. S. Sivoththaman, J. Horzel, W. Laureys, F. Duerinck, P. De Schepper, J. Szlufzcik, J. Nijs, R. Mertens, in Conf. Record, 14th ECPVSEC, 420-422 (1997).
16. J. Zhao, A. Wing, P. Altermatt and M. A. Green, Appl. Phys. Lett., **66**, 3636-3638, (1995)
17. A. Goetzberger, J. Knobloch and B. Voss, Crystalline Silicon Solar Cells, 90-98, John Willey & Sons, New York (1990)
18. P. Doshi, G. E. Jellison and A. Rohatgi, Appl. Optics, **36**, 7826-7837 (1997).
19. D. E. Kane and R. M. Swanson, in Conf. Proc., 18th ECPVSEC, 578-583, (1985).
20. D.K. Schroder, Semiconductor Material and Device Characterization. Wiley, New York, 367-374 (1990).
21. R. Hezel, in Conf. Proc., 16th IEEE PVSC, 1237, (1982).
22. J. Schmidt, T. Lauinger, A. G. Aberle and R. Hezel, in Conf. Record, 25th IEEE PVSC, 162-166, (1996).
23. S. Noel, L. Ventura, A. Slaoui, J. C. Muller, B. Groh, R. Schindler, B. Froeschle and T. Theiler, in Conf. Record, 14th ECPVSEC, 104-107, (1997).
24. R. Singh, R. Sharangpani, K. C. Cherukuri, Y. Chen, D. M. Dawson, K. F. Poole, A. Rohatgi, S. Narayanan and R. P. S. Thakur, in MRS Symp. Proc., **429**, 81-94, (1996).
25. F. M. Roberts and L. G. Wilkinson, J. Mat. Sci., **3**, 110-119, (1968).
26. F. M. Roberts and L. G. Wilkinson, J. Mat. Sci., **6**, 189-199, (1971).
27. L. L. Chalfoun, Master Thesis, Massachusetts Institute of Technology, (1996).
28. W. R. Runyan, Semiconductor Measurements and Instrumentation (McGraw Hill: New York), (1975).
29. S. Narasimha, A. Rohatgi and A. W. Weeber, to be published in IEEE Trans. Electron. Dev. (1999).
30. P. Doshi, J. Mejia, K. Tate, S. Kamra, A. Rohatgi, M. Narayanan and R. Singh, in Conf. Record, 25th IEEE PVSC, 421-424, (1996).

POLY-$SI_{1-x}GE_x$ PROCESS INTEGRATION FOR LOW SHEET RESISTANCE GATE CMOS TECHNOLOGY

Hideki Takeuchi* and Tsu-Jae King
Department of Electrical Engineering and Computer Sciences
University of California at Berkeley, Berkeley, CA 94720
* on leave from Nippon Steel Corporation

Three different gate stack structures have been investigated to lower the sheet resistance of poly-$Si_{1-x}Ge_x$ gate lines for CMOS application. Directly formed Ti and Ni germanosilicides are not suitable because of poor thermal stability. A poly-metal structure(W/barrier metal/a-Si/ poly-$Si_{1-x}Ge_x$) can tolerate the thermal budget of a conventional CMOS process. For a poly-$Si_{1-x}Ge_x$/polycide stack structure, significant improvement against Ge upward diffusion from the bottom poly-$Si_{1-x}Ge_x$ to the top polycide layer was achieved by allowing an intermediary native oxide layer to form before polycide formation.

INTRODUCTION

In deep-submicron MOSFETs, the increase in effective gate-oxide thickness imposed by the gate poly-depletion effect is one of the biggest issues to be solved. Poly-$Si_{1-x}Ge_x$ is an attractive alternative to conventional poly-Si, because it provides reduced gate-depletion effect[1] as well as boron penetration[2] for devices with thin gate oxides. In addition to this advantage, its adjustable workfunction[3] provides an added degree of flexibility in device structure design. Because its workfunction is close to the midgap of Si, B-doped poly-$Si_{1-x}Ge_x$ with high Ge content ($x>0.6$) allows us a low threshold voltage to be achieved without increasing surface concentration of dopant, for a super-steep-retrograde(SSR) channel profile[4].

For practical applications, it is necessary to lower the sheet resistance of the gate down to 5ohm/sq[5]. Although the resistivity of poly-$Si_{1-x}Ge_x$ is lower than that of poly-Si[6], this low sheet resistance can not be obtained without increasing the gate thickness beyond a practical value(~200nm). Therefore, it is necessary to stack a lower-resistance silicide or a metal on top of the poly-$Si_{1-x}Ge_x$.

In conventional poly-Si gate technology, SALICIDE(self-aligned silicide) is widely used to reduce the sheet resistance of the gate and the source/drain regions simultaneously. After formation of the gate, side wall spacers and source/drain junctions, metal is deposited over all the regions and silicide is selectively formed on the gate and the source/drain regions during a thermal anneal. Unreacted metal is then stripped away. Ti is commonly used to form the silicide. Co is an attractive alternative, because sheet resistance of CoSi does not increase for narrow gate lines, whereas the sheet resistance of $TiSi_2$ increases for gate lines narrower than 0.2μm [7]. The sheet resistance of NiSi is also independent of gate length[8].

Recently, poly-metal structure composed of W/ barrier metal/ poly-Si stack has been investigated for DRAM applications[9]. WN_x, TiN or WSiN is used as a barrier metal. In this structure, W layer lowers the total sheet resistance, while the barrier metal prevents a silicidation reaction between the top W and bottom poly-Si layers. The formation of Si-N bonds at the interface between the barrier and poly-Si suppresses the diffusion of Si into the W layer[10].

Solid phase reactions between $Si_{1-x}Ge_x$ and Ti or Co were investigated by Aldrich et al.[11][12]. Because the heats of formation of silicides($TiSi_2$, CoSi) are lower than those of germanides($TiGe_2$, CoGe), both $Ti(Si_{1-x}Ge_x)_2/Si_{1-x}Ge_x$ and $Co(Si_{1-x}Ge_x)/Si_{1-x}Ge_x$ systems are thermally unstable. After $Ti(Si_{1-x}Ge_x)_2/Si_{1-x}Ge_x$ or $Co(Si_{1-x}Ge_x)/Si_{1-x}Ge_x$ is initially formed, Ge diffuses through grain boundaries to form Ge-enriched clusters at surface. This Ge precipitation results in increased resistance.

To avoid the thermal instability problems of germanosilicide systems and realize a SALICIDE structure, Ponomarev et al. adopted a poly- $Si_{1-x}Ge_x$ /polycide stack structure consisting of two layers[13]: bottom poly- $Si_{1-x}Ge_x$ to control workfunction and top poly-Si to retain the standard SALICIDE process. Although Ti SALICIDE CMOS devices were successfully fabricated, the following issue was raised. During post-gate thermal processing, Ge diffuses upwards from the bottom layer into the top poly-Si layer, so that the Ge content in the bottom layer is reduced. Because the change in Ge content leads to the change of the gate workfunction, this phenomenon is not desirable for practical use.

The purpose of this study is to develop a thermally stable, low resistance poly-$Si_{1-x}Ge_x$ gate stack. Three different structures were examined; 1) Ti or Ni SALICIDE, 2) a poly-metal structure, and 3) a stacked poly-$Si_{1-x}Ge_x$/polycide structure.

EXPERIMENTAL

Poly-$Si_{1-x}Ge_x$ films were deposited in a conventional LP-CVD system, onto 100nm thermal oxides grown on P-type 4" (100) Si wafers. Phosphorus(30keV, $4 \times 10^{15} cm^2$) and boron(15keV, $4 \times 10^{15} cm^2$) were implanted, followed by 900°C 1hr furnace annealing. Metals(Ti, TiN, W, WN_x and Ni) were deposited in a DC sputtering system. Wafers were then cut into $1 cm^2$ dies and annealed in a rapid thermal processing system or a furnace. An H_2SO_4:H_2O_2 mixture was used to remove any remaining unreacted metal. Sheet resistance was measured by a four point probe.

RESULTS AND DISCUSSION

Direct reaction with Ti / Ni

Direct Ti germanosilicidation was firstly examined. 50nm Ti was used, and a 2-step annealing process was applied similarly to a conventional Ti SALICIDE process. The first anneal was carried out in nitrogen ambient to form $Ti(Si_{1-x}Ge_x)_2$ at the Ti/poly-$Si_{1-x}Ge_x$ interface and TiN on the metal surface simultaneously. After stripping the formed TiN and the unreacted Ti, a second anneal was carried out in argon ambient to transform the silicide structure from the high-resistance C49 phase to the low-resistance C54 phase. **Fig.1** and **Fig.2** show the sheet resistance after the 1st and 2nd anneals, respectively. For relatively low temperatures, the sheet resistance of poly-$Si_{1-x}Ge_x$ germanosilicide can be lower than those of single crystal and polycrystalline Si. However, for high temperatures required to achieve low resistance in the single crystal Si source/drain regions, the sheet resistance of $Ti(Si_{1-x}Ge_x)_2$ goes up. For the samples with high sheet resistance, precipitation on the film surface was observed through a microscope. This resistance rise occurs because germanides are thermodynamically less stable than silicides and thus Ge atoms agglomerate at grain boundaries during high-temperature annealing[12]. Although the use of a thicker Ti film can lower the sheet resistance, it increases the junction leakage and thus a SALICIDE process is not possible.

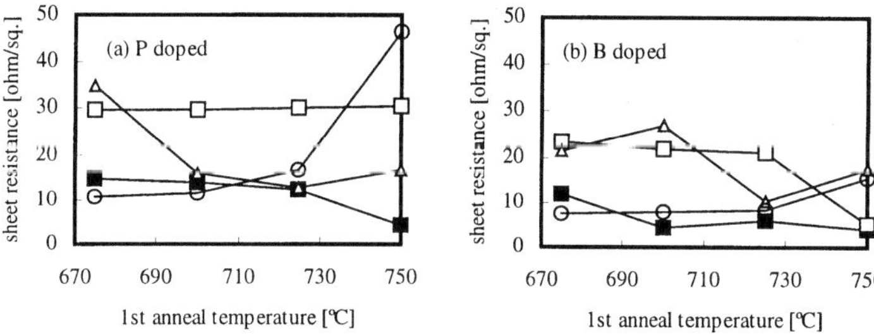

Fig.1 Sheet resistance of Ti germanosilicide after the 1st anneal (a)P and (b) B doped ■ single-Si □ poly-Si △ poly-$Si_{0.7}Ge_{0.3}$ ○ poly-$Si_{0.5}Ge_{0.5}$(Ti 50nm, poly-$Si_{1-x}Ge_x$ 200nm, 60sec in N_2)

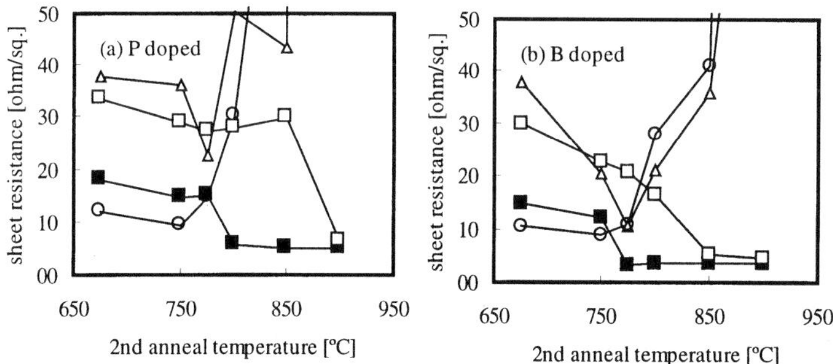

Fig.2 Sheet resistance of Ti germanosilicide after the 2nd anneal (a)P and (b) B doped ■ single-Si □ poly-Si △ poly-$Si_{0.7}Ge_{0.3}$ ○ poly-$Si_{0.5}Ge_{0.5}$(1st anneal 675°C in N_2, 2nd 10sec in Ar)

As an alternative to Ti, Ni SALICIDE was also investigated. 20nm-thick Ni was used to form germanosilicide. Because Ni is the diffusing species instead of Si in this reaction system, a single-step anneal in Ar was used. **Fig.3** shows the sheet resistance after anneals. Low sheet resistance was simultaneously obtained for both single crystal Si and poly-$Si_{1-x}Ge_x$ by 30s annealing at 500°C. However, as **Fig.4** shows, an additional 400°C 1hr annealing, which is necessary as a backend process thermal budget, increases the sheet resistance up to more than three times higher value. This phenomenon is also considered to be caused by the difference in thermal stability of silicides and germanides.

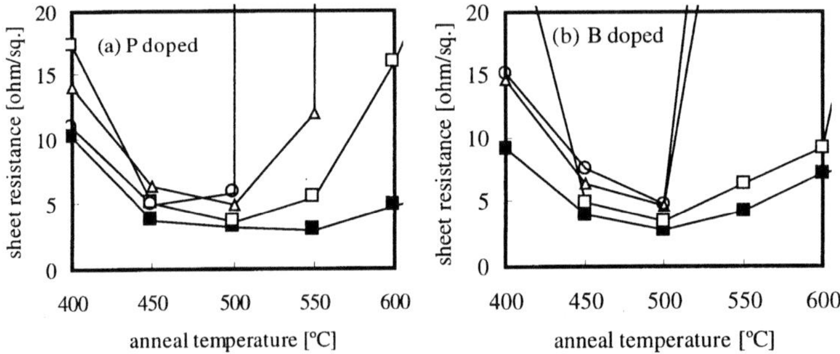

Fig.3 Sheet resistance of Ni germanosilicide (a)P and (b) B doped
■ single-Si □ poly-Si △ poly-$Si_{0.7}Ge_{0.3}$○poly-$Si_{0.5}Ge_{0.5}$(30sec in Ar)

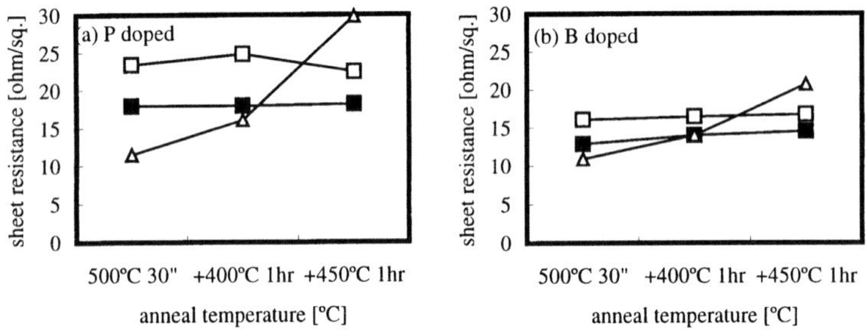

Fig.4 Sheet resistance of Ni germanosilicide after additional annealing (a)P and (b) B doped ■ single-Si □ poly-Si △ poly-$Si_{0.7}Ge_{0.3}$ (silicidation 500°C 30sec in Ar)

As these directly formed silicides are thermally unstable, suppression of Ge diffusion during high-temperature processing is necessary to realize low sheet resistance poly-$Si_{1-x}Ge_x$ gate lines.

Poly-metal structure

The W(50nm)/ barrier metal(5nm)/poly-$Si_{1-x}Ge_x$ structure was investigated with x=0.8,0.9,1.0. TiN or WN was used as a barrier metal. For each case, the sheet resistance after a 950°C 30sec anneal (which corresponds to a typical source/drain activation anneal) was too high to measure. Considering that Ge-N bonds are unstable during high temperature anneals[14] and more difficult to form than Si-N bonds[15], a diffusion barrier was not formed in this sample. However, by inserting a 5nm a-Si layer beneath the barrier layer, the sheet resistance can be lowered to below 5ohm/sq (**Fig.5**). This improvement is attributed to the formation of Si-N bonds at the interface between WN_x(or TiN) and the a-Si layers[10], which prevents the underlying Ge from diffusing into the W layer.

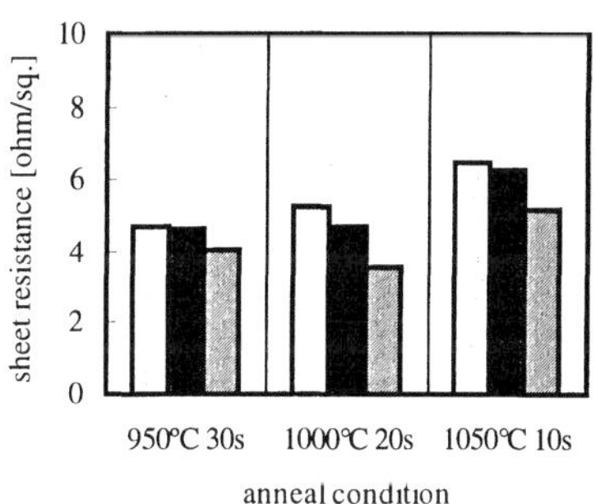

Fig.5 Sheet resistance of W(50nm)/TiN(5nm)/a-Si(5nm)/poly-$Si_{1-x}Ge_x$ stacked structure after annealing in Ar (X=0.8,0.9,1.0, from left to right column)

The stability during furnace annealing was also examined. To prevent the oxidation of W during the furnace anneal, CVD SiO_2 was deposited onto W(100nm)/WN_x(5nm)/a-Si(5nm)/poly-$Si_{0.2}Ge_{0.8}$(50nm, B-doped) at 450°C and then the sample was annealed at 950°C for 30min. After stripping the SiO_2 by dry etching, the sheet resistance was measured to be 1.2ohm/sq. This result suggests that the structure is stable to withstand the thermal budget of a backend process.

Poly-$Si_{1-x}Ge_x$/polycide stack structure

Although the poly-metal structure was found to be able to withstand the backend thermal budget of CMOS process, SALICIDE is still attractive from the view point of process cleanliness and compatibility with current standard processes. Therefore, a poly-$Si_{1-x}Ge_x$/polycide stack structure was also examined. The problem of this structure is the Ge diffusion from the bottom poly-$Si_{1-x}Ge_x$ layer to the top poly-Si layer during post-gate processing[13]. Because no Ge diffusion was observed in MOS capacitors with 1.8nm thin gate oxide and poly-$Si_{1-x}Ge_x$ gate[2], we tried to use a native oxide as a barrier layer against Ge diffusion.

Two different stack structures were used to examine the effect of native oxide. 50nm undoped poly-$Si_{0.2}Ge_{0.8}$ and then 10nm poly-Si to form native oxide on top were continuously deposited on a 100nm-thick thermal oxide. 2E15cm^{-2} B was implanted at 20keV and then the samples were furnace annealed at 800C for 30min. in N_2. Before the deposition of a 100nm-thick top poly-Si layer, native oxide formed on the bottom poly-Si was removed in HF for one sample and not removed for samples were annealed at 950C for 30sec. Ti was then deposited and a conventional 2-step anneal was applied. Samples were additionally annealed at 800°C for 2hrs.

Fig.6 shows the sheet resistance comparison between the two samples. Although a slight increase was still observed, the sheet resistance of the sample with native oxide was one order of magnitude lower than that without native oxide. **Fig.7** shows the SIMS profile of the sample with native oxide. The Ge content in the top layer is suppressed down to 10ppm. In this profile, the Ti peak overlaps the Ge peak, which suggests that top Ti diffused into the bottom Ge layer. We believe that this Ti diffusion caused the resistance increase and that this increase can be eliminated by optimizing the thicknesses.

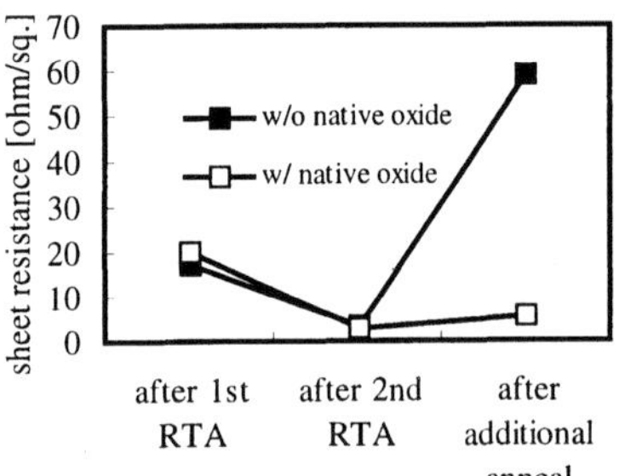

Fig.6 Sheet resistance of Ti polycide(100nm)/ poly-$Si_{0.2}Ge_{0.8}$(50nm) stacked structure
1st RTA: 675°C 60sec in N_2
2nd RTA: 900°C 10sec in Ar
additional anneal: 800°C 2hr in N_2 (FA)

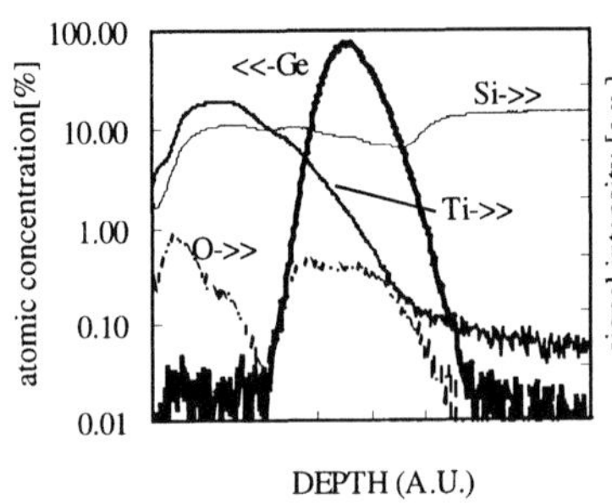

Fig.7 SIMS profile of Ti polycide(100nm)/ poly-Si0.2Ge0.8(50nm) stacked structure

To prevent the reaction between the bottom Ge and metal diffused from the top silicide layer, thin silicide is desirable. Table 1 compares Si consumption needed to obtain 5ohm/sq among Ti, Co and Ni silicides[16]. Ni consumes the least Si for the same sheet resistance. In addition to this advantage, the processing temperature of Ni SALICIDE is the lowest. Therefore, a Ni SALICIDE process should be the best choice for the poly-$Si_{1-x}Ge_x$/ polycide gate stack.

Tabel 1. Si consumption needed to obtain for 5 ohm/sq. silicide[16]

Source metal	Si consumption per unit thickness metal	Si consumption for 5 ohm/sq. silicide formation
Ti	1.83	67nm
Co	3.61	40nm
Ni	2.22	33nm

CONCLUSIONS

Ti SALICIDE and Ni SALICIDE processes have been investigated to lower the sheet resistance of poly-$Si_{1-x}Ge_x$ gate lines. For the Ti SALICIDE process increase in gate sheet resistance was observed after high-temperature annealing required to lower the sheet resistance of source/drain regions. For the Ni SALICIDE process, low sheet resistance can be simultaneously obtained for both gate and source/drain regions, but the sheet resistance of the gate increased after 400°C 1hr annealing(representative of a backend process). As these directly formed germanosilicides are too unstable for practical applications, suppression of Ge diffusion during high-temperature annealing is necessary to realize low sheet resistance poly-$Si_{1-x}Ge_x$ gates.

This can be achieved by using a poly-metal structure(W/barrier metal/a-Si/poly-$Si_{1-x}Ge_x$ stack). Ge diffusion is prevented by the formation of Si-N bonds at the interface between barrier metal and thin a-Si. In addition, for poly-$Si_{1-x}Ge_x$/polycide stack, suppression of Ge was improved by the native oxide formed between top and bottom poly-Si layers. Both of these two structures can be compatible with the thermal budget of a conventional CMOS SALICIDE process.

REFERENCES

1. W.-C. Lee, T.-J. King and C. Hu, *IEEE Electron Device Lett.*, **19**, 247(1998).
2. W.-C. Lee, T.-J. King and C. Hu, *IEEE Electron Device Lett.*, **20**, 9(1999).
3. T.-J. King, J. R. Pfiester and K. C. Saraswat, *IEEE Electron Device Lett.*, **12**, 533(1991).
4. J.Alieu, R. Gwoziecki, M. Paoli, T. Skotnicki, C. Hernandez, F. Martin, C. Mourrain, D. Bensahel, M.-T. Basso, J. Galvier and M. Haond, *Symp. on VLSI Tech*, 192(1998).
5. *The National Technology Roadmap for Semiconductors 1997 Edition*, 74(1997).
6. D. S. Bang, M. Cao, A. Wang, K. C. Saraswat and T.-J. King, *Appl. Phys.s Lett.*, **66**,. 195, (1995).
7. J.B. Lasky, J.S. Nakos, O.J. Cain, adn P.J. Geiss, *IEEE Trans. on Electron Devices* , **38**, 262(1991).
8. K. Goto, T. Yamazaki, A. Fushida, S. Inagaki, and H. Yagi, *Symp. on VLSI Tech*, 192(1998).
9. H. Wakabayashi, T. Yamamoto, K. Yoshida, E. Soda, K.I. Tokunaga, T. Mogami, and T. Kunio, *Intn'l Electron Devices Meeting Tech. Digest*, 393(1998).
10. Y.Akasaka, S. Suehiro, K. Nakajima, T. Nakasugi, K. Miyano, K. Kasai, H. Oyamatsu, M. Kinugawa, M.-T. Takagi, K.Agawa, F. Matsuoka, M. Kakumu, and K. Suguro, *IEEE Trans. on Electron Devices*, **43**,1864(1996).
11. D.B. Aldrich, Z. Wang,. D.E. Sayers, R.J. Nemanich, S.P. Ashburn, and M.C. Öztürk, *J.Appl. Phys.*, **77**,5107(1995).

12. Z. Wang, D.B. Aldrich, Y.L. Chen, D.E. Sayers, adn R.J. Nemanich, *Thin Solid Films*, **270**, p.555(1995).
13. Y.V. Ponomarev, et al., *Intn'l Electron Devices Meeting Tech. Digest*, 829(1997).
14. G.D. Bargatishvili; I.G. Nakhutsrichvili,; , M.R. Katsiashvili; and B.T. Zhorzholianai, *Neorganicheskie Materialy*, **28**, 546(1992).
15. M. Mukhopadhyay,; S.K. Ray, and C.K. Maiti, J.Vac.Sci. & Tech. B, **14,** 1682 (1996).
16. J. Chen, J.-P. Colinge, D. Flandre, R. Gillon, J.P. Raskin, and D. Vanhoenacker, *J. Electrochem. Soc.*,**144**,2437(1997).

SCANNING RAPID THERMAL ANNEALING PROCESS FOR LOW TEMPERATURE POLY-SILICON THIN FILM TRANSISTORS

Tae-Kyung Kim, Gi-Bum Kim, Yeo-Geon Yoon,
Chang-Hoon Kim, Byung-Il Lee, Seung-Ki Joo

Seoul National University, School of Material Science and Engineering
San 56-1, Shillim-Dong, Kwanak-Gu,
Seoul 151-742, Korea

Polycrystalline silicon thin-film transistors are fabricated on transparent glass substrates using lamp-scan annealing.
In order to enhance the annealing effect, we deposited a capping SiO_2 layer and an absorption layer(Mo) over the entire area of the substrate.
The metal-induced lateral crystallization rate of the amorphous silicon under the capping layer or absorption layer was increased by several times.
Thin-film transistors thus fabricated showed considerable electrical characteristics.

INTRODUCTION

There have been continuous research efforts in thin film transistors to grow high quality polycrystalline silicon under the softening temperature (~600 °C) of the glass substrate. The use of rapid thermal annealing (RTA) is advantageous for reducing the thermal budget and manufacturing costs due to the short cycle time of RTA which results in substantially higher throughputs than furnace annealing [1][2][3][4].

However, the fabrication of high-performance poly-Si TFTs typically requires high process temperature (800 °C for anneal times of 1s) if RTA is used[5].
This problem may be circumvented through the use of a new crystallization method, metal-induced lateral crystallization (MILC). It was shown that a few nanometers of nickel-metal film, which is deposited on the amorphous silicon film, can lower the crystallization temperature and the lateral region could be crystallized[6][7].
It has also been reported that high-performance poly-Si TFTs can be successfully fabricated by the Ni-MILC process[8].

In this paper, we report on the scan-RTA apparatus which we have developed (as shown in Fig. 1) to fabricate poly-Si TFTs with metal-induced lateral crystallization.

EXPERIMENTAL

A schematic diagram of the specimens to be crystallized is shown in Fig. 2. Corning 7059 glass was used as a substrate. A 3000 Å thick SiO_2 thin film as a buffer layer was deposited on the glass substrate by Electron Cyclotron Resonance Plasma Enhanced Chemical Vapor Deposition (ECR-PECVD). Next, a 1000 Å-thick amorphous Si thin film was deposited by Plasma Enhanced Chemical Vapor Deposition (PECVD) at 250 °C using SiH_4 and the active island was defined. For the formation of gate dielectric, a 1000 □-thick SiO_2 thin film was formed through two consecutive processes, being ECR plasma oxidation (300 °C) followed by ECR-PECVD (700 °C)[9]. A 3000 □-thick Mo thin film as a gate electrode was sputter deposited. In order to form the gate electrode, the Mo and oxide thin films were etched by plasma and wet chemical methods, respectively. Next, a Ni offset mask pattern was formed on the top of the TFTs using photoresist[10]. The 20 Å-thick Ni thin film was deposited on the entire surface of the TFTs, after which the Ni thin film on the offset mask was removed by lift-off method. Ion doping was performed by an ion mass doping system with 5% PH_3 diluted in H_2 gas.

In order to enhance the MILC rate, a capping oxide and Mo layer were deposited by PECVD and sputtering, respectively, as illustrated in Fig. 4. Crystallization annealing was carried out by scanning RTA with a tungsten-halogen lamp, as shown in Fig. 1.

A line-shaped beam, which was focused with an elliptical reflector, was scanned on the sample that had been preheated by bottom lamps. The annealing temperature was measured in the silicon wafers for reference, using a thermocouple and the real temperature of the glass substrate was likely to be much lower than that.

After crystallization annealing, the aluminum electrode was formed and the device characteristics of TFTs were measured using an HP 4140B.

RESULTS AND DISCUSSION

When the upper lamp scanned, the substrate which was preheated at around 450 °C experienced a steep heating and cooling thermal profile. In the case of solid-phase crystallization of amorphous silicon, the activation energy for crystallization (3 ~ 5 eV) and the incubation time are so high and long that it is not appropriate for a low-temperature process[11]. However, in the case of metal-induced lateral crystallization, the activation energy is about 0.3~1.86 eV [12][4]. Thus, the thermal budget of crystallization can be reduced and it can be applied to a glass-substrate process.

A transparent glass substrate as shown in Fig. 2 is not heated by focused light. However, selectively formed amorphous silicon films absorb most of the energy from the focused light [13]. Thus, an amorphous silicon film can be heated and crystallized without heating the glass substrate[14].

A capping oxide (SiO_2) layer (~3000 Å) transfers the focused light and protects the silicon film from contact with the atmosphere and reduces conductive cooling because thermal conductivity of SiO_2 is lower than that of Si by 2 orders of magnitude. The thermal insulation effect of a capping oxide raises the temperature of silicon; therefore, the crystallization is enhanced.

At first, the Ni film on the a-Si layer crystallizes it by metal-induced crystallization (MIC)[8]. While the crystallization proceeds, the amorphous silicon film is transformed into a polycrystalline silicon film, which is relatively transparent. Also, the temperature of the silicon film decreases since it absorbs little of the light energy[2]. Thus, over heating of the substrate can be minimized.

Fig 3 shows the MILC length versus scanning temperature. The MILC lengths were measured with an optical microscope after the specimen was scanned once at the temperature as labeled on the x-axis.

When silicon films are covered by a capping oxide, they show a 5 times faster crystallization rate. This proves the effect of the capping layer.

In actual devices, there are many silicon islands on the substrate. Therefore, variation of the size and density of islands within a large substrate cause the difficulty of uniform annealing. So in order to achieve uniform annealing in actual poly-Si TFTs, an extra absorption layer must be deposited[2]. We used a Mo film deposited by sputtering between two capping SiO_2 layers as an absorption layer as shown in Fig. 4.

The Mo layer assists the heat absorption and the capping oxide thermally insulates the silicon islands. Then, each silicon island shows further improved MILC rate. The effects of the (Mo) absorption layer are shown in Fig. 5. With a Mo layer that is thicker than 1000 Å, the MILC rate increased several times. Furthermore, a few minutes annealing is enough to fully crystallize micron size devices below 550 °C.

The device characteristics of the TFTs before hydrogenation, which were crystallized by scan-RTA-MILC, are shown in Fig. 6. The scan speed was 1.4 mm/s and maximum temperature was around 700 °C. In the case of scan-RTA-MILC TFTs, dopant activation and crystallization were carried out simultaneously.

However, scan-RTA-MILC TFTs show different electrical properties depending on the annealing conditions. The differences in electrical properties can be attributed to the different thermal profiles that can induce differences in microstructure and concentration of defects in the poly-Si film[15].

Even though, the microstructure of crystalline Si by RTA-MILC and conventional furnace MILC is suggested to be similar[4]. Our experiment shows variation of device parameters depending on their annealing conditions. The experimental details are summarized in Table 1.

There must be another factor that influences crystalline silicon and device characteristics. This point needs further study.

Table. 1 Threshold voltage variation according to scan-RTA condition.
Peak temperatures were measured at reference Si wafers.

Split	*Scan speed (mm/s)*	*Lamp power (kW)*	*Preheat temp. (℃)*	*Peak temp. (℃)*	*Threshold Voltage (V)*
1	0.4	0.8	400	690	2.8
2	1.4	0.3	400	560	1.4
3	1.4	0.5	400	610	4
4	1.4	0.6	400	620	4.8
5	1.4	0.7	400	640	6.6
6	1.4	0.8	400	670	8
7	1.4	1	400	700	9
8	1.4	2	150	900	8

CONCLUSIONS

We have developed a scanning RTA process and fabricated MILC poly-Si TFTs. Annealing for crystallization was carried out by scanning RTA with a tungsten-halogen lamp. The MILC rate could be enhanced buy applying a capping oxide and an absorption layer.

A Mo layer of 1000 Å thickness of and a SiO_2 capping of 3000 Å thickness increased the MILC rate several times. Thus poly-Si TFTs can be fabricated on conventional glass substrate just in a few minutes residence time. This technology enables the crystallization of silicon films using reduced lamp power and short process times.

The threshold voltage of MILC TFTs that were fabricated under different scanning RTA conditions showed different device parameters depending on their annealing conditions.

ACKNOWLEDGEMENT

This work was supported by the G-7 project of Korea through the Research Institute of Engineering Science at the Seoul National University.

REFERENCES

[1] S. Jurichich, T.J. King, K. Saraswat and J. Mehlhaff, Jpn. J. Appl. Phys., **33**, L1139 (1994).
[2] I. Yudasaka and H. Ohshima, Jpn. J. Appl. Phys., **33**, Part 1, No. 3A, 1256-1260 (1994).
[3] Gang Liu and S. J. Fonash, Jpn. J. Appl. Phys., **30,** No. 2B, L269-271 (1991).
[4] L.K. Lam, Szu-ke Chen and D.G. Ast, Appl. Phys. Lett., **74**, No. 13, 1866 (1999).
[5] H. Hovagimian and J. Mehlhaff, Proc. of IDRC, M-52 (1997).
[6] Yue Kuo, Appl. Phys. Lett., **69,** 1092 (1996).
[7] Tae-Kyung Kim, Byung-Il Lee and Seung-Ki Joo, Device Research Conference Digest, p.100 (1998).
[8] Seok-Woon Lee and Seung-Ki Joo, IEEE Electron Device Lett., **17**, 4 (1996).
[9] Tae-Hyung Ihn and Seung-Ki Joo, MRS Proc., **424**, 189-194 (1997).
[10] Tae-Hyung Ihn, Tae-Kyung Kim, Byung-Il Lee and Seung-Ki Joo, Microelectronics Reliability, **39,** 53 (1999).
[11] R. B. Inverson and R. Reif, J. Appl. Phys., **62,** 1675 (1987).
[12] H. Kim, J.G. Couillard and D.G. Ast, Appl. Phys. Lett., **72**, 803(1998).
[13] L. Plevert, S. Mottet, M. Bonnel, N. Duhamel, R. Gy, L. Haji and B. Loisel, Jpn. J. Appl. Phys., **34**, 419 (1995).
[14] Tae-Kyung Kim, Byung-Il Lee and Seung-Ki Joo, J. Korean Phys. Soc., (1999) to be published.
[15] M. Bonnel, N. Duhamel, M. Guendouz, L. Haji, B. Loisel and P. Ruault, Jpn. J. Appl. Phys., **30**B, L 1924 (1991).

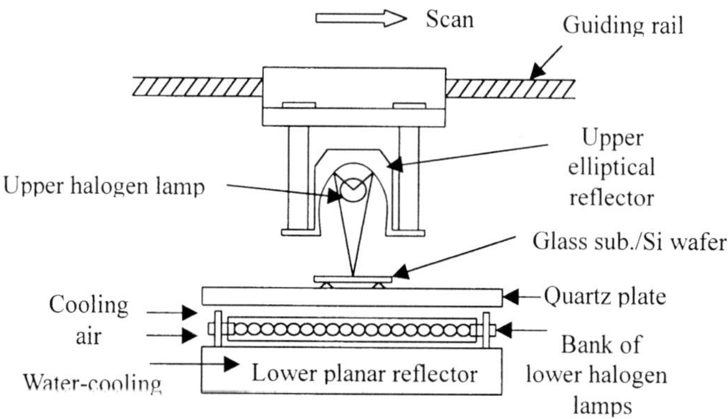

Fig. 1 Scanning rapid thermal annealing apparatus

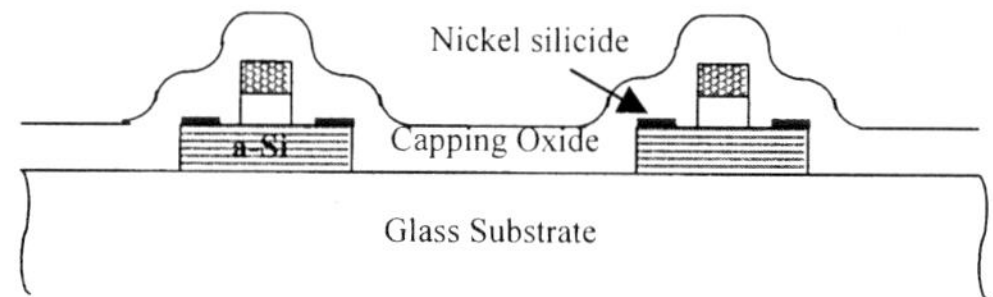

Fig. 2 Schematic diagram of TFT's.
Capping oxide layer was deposited on entire substrate.

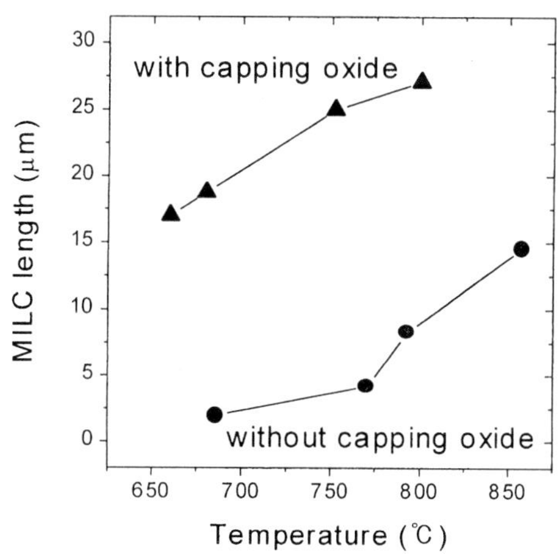

Fig. 3 Metal-induced lateral crystallization length versus scanning temperature.
Capping oxide improved crystallization rate.

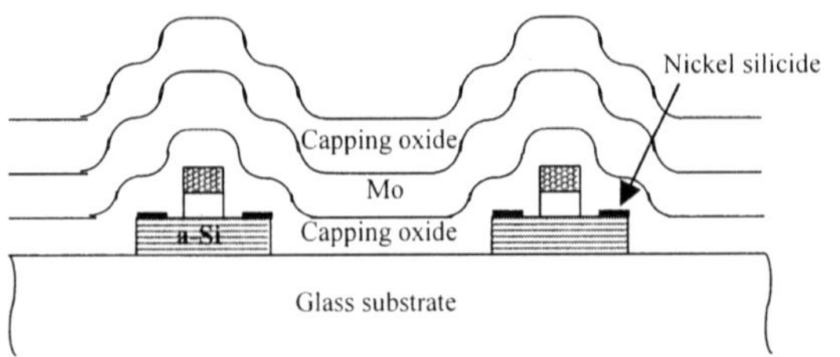

Fig. 4 Schematic diagram of TFT's.
Mo absorption layer deposited between two capping oxide layers.

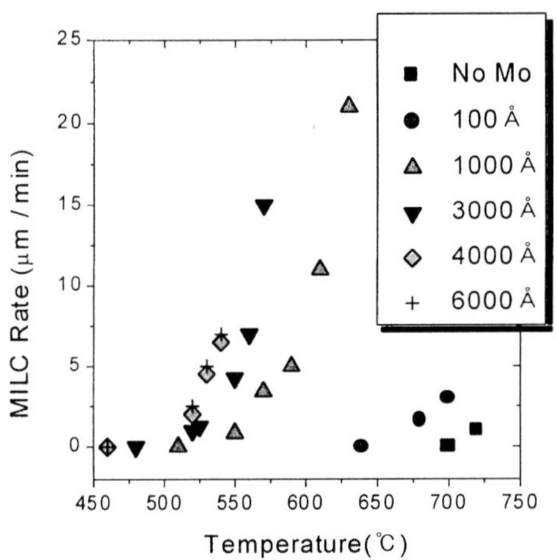

Fig. 5 Metal-induced lateral crystallization rate versus RTA temperature.

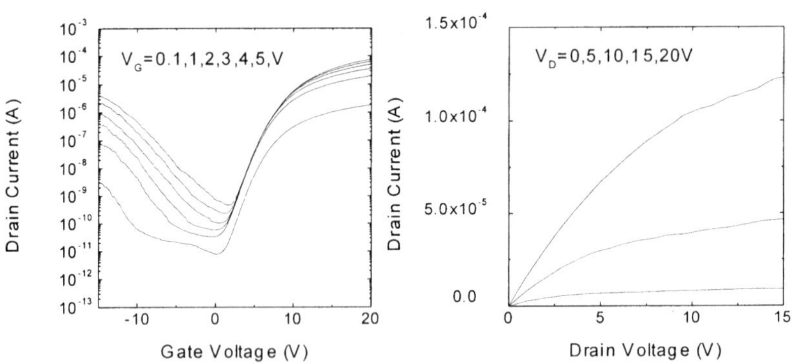

Fig. 6 I-V charateristics of scan-RTA-MILC TFT's.
Scan speed was 1.4mm/sec and maximum temperature was around 700 ℃.

ENHANCED MOBILITY IN BURIED SIGE CHANNEL PMOS FABRICATED USING RAPID THERMAL PROCESSING

Douglas J. Tweet and Sheng Teng Hsu
Sharp Laboratories of America
5700 NW Pacific Rim Blvd. Camas, WA 98607

One design to produce PMOS FET's with enhanced mobility is to use a buried, strained SiGe channel capped by Si. The Si cap is then partially oxidized to form the gate dielectric. We have fabricated long channel devices using rapid thermal processing and studied the dependence of effective hole mobility on Ge content, SiGe channel thickness, and Si cap thickness. With these results a nominally optimized structure is found. In addition, clear evidence for hole confinement in the channel is presented.

INTRODUCTION

During the 1990's a number of different device structures based on SiGe technology have been developed to produce FET's with enhanced mobilities (1-3.) One design for PMOS (1,3-8) is a buried, epitaxially strained SiGe layer capped by Si, with the cap partially oxidized to form the gate dielectric (Fig.1). Due to an offset in the valence band the holes can be confined to the SiGe channel. This enhances the mobility both by the intrinsic properties of strained SiGe and by separating the holes from the SiO_2/Si interface, thereby reducing surface scattering. If the Si cap is too thick, at low gate voltage a parallel inversion layer forms at the SiO_2/Si interface, decreasing mobility. If the cap is too thin, the holes in the SiGe channel begin to scatter from the oxide interface. Consequently, an optimum Si cap thickness is expected.

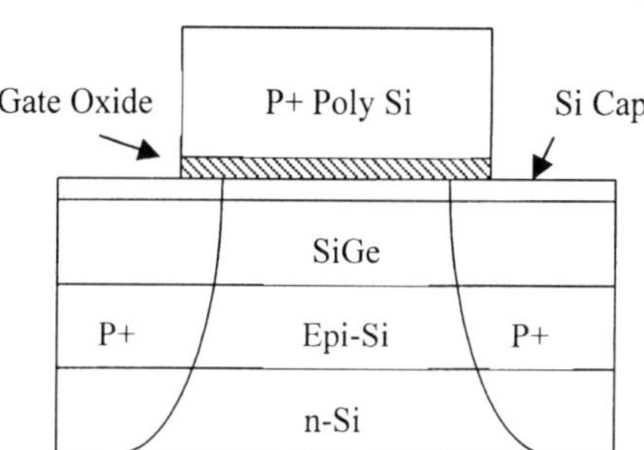

Figure 1. Schematic of transistor structure

Here we describe our initial tests on the dependence of effective hole mobility (μ_{eff}) on film structure. The parameters varied were Ge content (x = 0.2 - 0.34), thickness of the $Si_{1-x}Ge_x$ channel (t_x = 10 – 25 nm) and the thickness of the Si cap (t_{cap} = 7.5 – 25 nm). Due to removal of Si during cleaning and gate oxidation steps the final Si cap thicknesses are

~5 nm thinner than in the as-grown films. Both epi-Si and bulk Si (no epi) controls were also fabricated, which had identical electrical behavior.

EXPERIMENTAL

The samples were grown in an AG IntegraOne rapid thermal chemical vapor deposition (RTCVD) system. Starting with n-type (4-10 Ω-cm) six-inch Si(100) substrates an epi-Si buffer (49 nm) was grown at 950°C using 20 sccm of dichlorosilane (DCS). The SiGe layer was then grown at 660°C with 60 sccm of DCS and 3 to 16 sccm of germane followed by the Si cap at 780°C using 76.5 sccm of DCS. 3 slm of hydrogen flowed continuously throughout film growth. The epi-Si buffer layer was formed in 60 seconds, the SiGe layers took from 13 to 128 s to grow, and the Si cap required from 26 to 75 s. Each layer was undoped and the growth pressure was 0.6 Torr. The background pressure of the deposition chamber was about 5×10^{-8} Torr. X-ray diffraction (XRD) and transmission electron microscopy (TEM) measurements indicate the as-grown layers were free of dislocations with atomically abrupt interfaces.

The 6 nm gate oxide was grown using an atmospheric pressure rapid thermal dry oxidation process at 850°C for 500s. XRD indicates about 1.5-2 nm of Ge diffusion occurs during this step. In comparison, 6 nm dry RTO at 975°C for 100s and 1070°C for 17s resulted in ~3 nm and >4 nm of Ge diffusion, respectively. Prior to formation of the gate oxide RCA cleans (SC1 and SC2) were performed. It was observed that SC1 removed about 2 nm from the Si cap. Together with the consumption of ~3 nm during gate oxide formation, we expect the final Si cap thickness in the active regions to be ~5 nm thinner than in the as-grown films.

The poly-Si gate, source and drain regions were implanted with 5e15/cm^2 BF_2 at 50keV, and activated during an 850°C 20 minute furnace anneal in N_2. Contacts were formed by sputtered Ti/TiN/AlCu and annealed in forming gas at 450°C for 30 minutes. Any materials on the backside of the wafers were removed by plasma etching to ensure good electrical contact with the substrate.

Electrical measurements were made of several die on each wafer, from the center towards the flat. Only large devices, 50x50μm^2 capacitors and 50x100μm^2 (LxW) transistors, were studied to accurately determine the mobility. Electrical measurements were made using a Cascade Microtech Summit station. The x-ray data was obtained with a Philips X'Pert MRD operating at 40 kV and 45 mA using a point focus. An asymmetric Ge(220) Bartels monochromator crystal was used to define the incident x-ray beam and a fixed slit set before the detector collected the diffracted beam. The data was fit with a commercial program based on the Takagi-Taupin formalism (9). XRD data was obtained on several as-grown blanket wafers at positions corresponding to where electrical measurements were made on processed wafers. A small radial variation in film parameters was found, and the results are included in the plots shown in the next section. This turned out to be an efficient way to probe more structures. XRD measurements were also made of several processed wafers.

RESULTS AND DISCUSSION

A wide range of film structures has been examined and some clear trends have been found. Transistors having up to 40% higher mobility than Si have been fabricated with low leakage. The data suggest that further improvements should be attainable with a decrease in the thermal budget. In this section results for some of the best devices are shown first, followed by the dependence of electrical characteristics on film structure. Next, XRD data of wafers before and after processing are presented and discussed. Finally, plots of quasi-static C-V measurements are given, which exhibit clear evidence of hole confinement in the SiGe channel.

Figure 2a compares the I-V curves of large transistors (L=50μm, W=100μm) made on Si and on a wafer with a 29% Ge-15 nm channel capped with 10 nm of Si. The latter is our best device, with 40% higher mobility than the Si control and low leakage. Measurements are shown for gate voltage varying from 1.0V to –3.0V in 0.5V steps. (The threshold voltages are positive because no implant adjustment had been made.) The SiGe device shows a significant increase in drive current of more than 50%. Figure 2b plots the I-V_G curves of the Si and SiGe transistors in the linear region, with V_D=-0.1V. Also shown are the corresponding transconductance values (the slope of I-V_G) for the Si and SiGe devices. At all gate voltages the SiGe transconductance is higher than that of Si, with the maximum transconductance (gm-max) about 42% larger. From the I-V_G curves the threshold voltage (V_T), sub-threshold swing (STS), and off-state drain leakage are determined. The results from these and other structures are listed in Table I.

C-V curves in accumulation were also obtained for 50x50 μm^2 capacitors at 1 MHz. At positive gate voltage electrons accumulate at the SiO_2/Si interface even for the SiGe devices, since the SiGe/Si conduction band offset is very small. The curves are very similar, indicating the gate oxide made from the Si cap layer is equivalent to that made from the control Si wafer. Using gm-max and the measured capacitance per unit area (C/A), the effective hole drift mobility (μ_{eff}) was then calculated for each device. The results are also summarized in Table I.

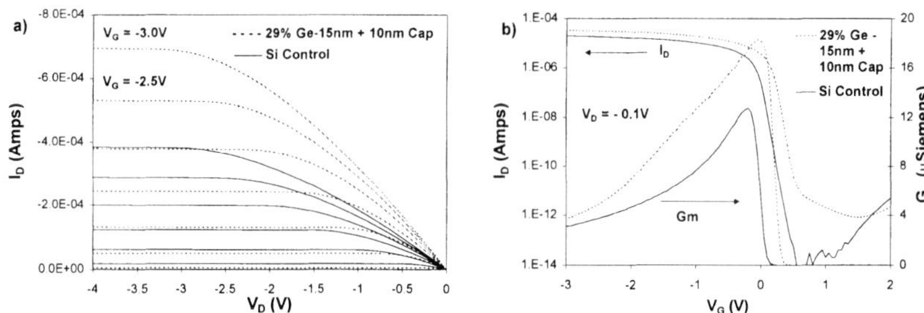

Figure 2. a) I-V_D curves at different V_G and b) I-V_G curves at fixed V_D for a Si control transistor compared with one having a 29% Ge-15 nm channel with 10 nm cap. Also shown in b) are the calculated transconductance values.

x	t_x (nm)	t_{cap} (nm)	gm-max	V_T	STS	Leak	C/A	μ_{eff}	μ/μ_{Si}
0 (epi)	0	0	12.9	0.008	64.8	0.24	4.88	132	0.99
0 (bulk)	0	0	12.8	0.003	64.9	0.25	4.81	133	1.0
0.208	14.5	10.5	16.0	0.142	65.0	0.83	4.96	161	1.21
0.295	9.7	10.5	17.1	0.174	64.5	0.06	4.93	174	1.31
0.295	14.5	8.7	18.1	0.213	70.1	0.22	4.91	184	1.38
0.29	**15**	**10**	**18.3**	**0.202**	**65.3**	**0.82**	**4.93**	**186**	**1.40**
0.295	14.5	15.7	17.2	0.198	68.1	426	5.04	170	1.28
0.345	**10**	**10.5**	**17.9**	**0.200**	**66.5**	**0.08**	**4.84**	**186**	**1.40**
0.345	10	15	16.6	0.160	64.8	74.6	4.93	168	1.26

Table I. Electrical characteristics of best devices. The units are gm-max (μS), V_T (V), STS (mV/dec), off-state drain leakage current measured at 1.5V (pA), capacitance per unit area measured at 4V and 1MHz (10^{-7} F/cm^2), and effective mobility (cm^2/V-s). The last column compares μ_{eff} with that of bulk Si.

Plotting the mobility and other characteristics as functions of the structural parameters show clear trends toward an optimum design. For example, μ_{eff} is found to be independent of t_x for channels 15 nm or thicker, but STS and leakage current get much worse for 20 nm and 25 nm channels. This is expected as dislocations are more easily generated in these thicker strained films. If the channel is too thin, however, diffusion during processing may lower the Ge concentration in the channel, decreasing μ_{eff}.

Figure 3 is a plot of μ_{eff} as a function of the as-grown t_{cap} for devices with 10 nm or 15 nm thick SiGe channels. (The plot for gm-max, not shown, is very similar.) As already mentioned, the actual cap thickness is ~5 nm less. The most striking feature is that μ_{eff} is a maximum at t_{cap} ~ 10 nm regardless of x and t_x. For thicker caps, μ_{eff} decreases gradually. For thinner caps, it decreases rapidly. Also, in general μ_{eff} increases with higher Ge content, but it is apparent that the 29% Ge-15 nm channel devices and the 34% Ge-10 nm channel devices behave similarly. This is because higher strain in the 34% Ge channel has caused enhanced Ge diffusion during processing. XRD of devices of both types shows they have similar structure after processing. Furthermore, μ_{eff} of the 29% Ge-10 nm channel devices is lower than for those with 29% Ge and 15 nm-thick channels. XRD finds that, after processing, the maximum Ge content in the thinner channel device is lower than in the thicker one. Apparently, the thermal budget during processing was high enough to cause the center of the SiGe channel to be partially depleted of Ge.

Figure 4 plots the sub-threshold swing as a function of the as-grown t_{cap}. The STS is close to the ideal value of 60 mV/decade for all devices except those with the thinnest caps. The 34% Ge devices with the thinnest caps tend to have the worst behavior. Since the actual cap thickness is expected to be about 5 nm less than as-grown, and XRD indicates up to ~3 nm of Ge diffusion into the cap, a Si cap that started out at 7.5 nm would be entirely gone. Instead, Ge would be present at the gate oxide interface. This would explain both the precipitous drop in μ_{eff} as well as the increase in STS and leakage.

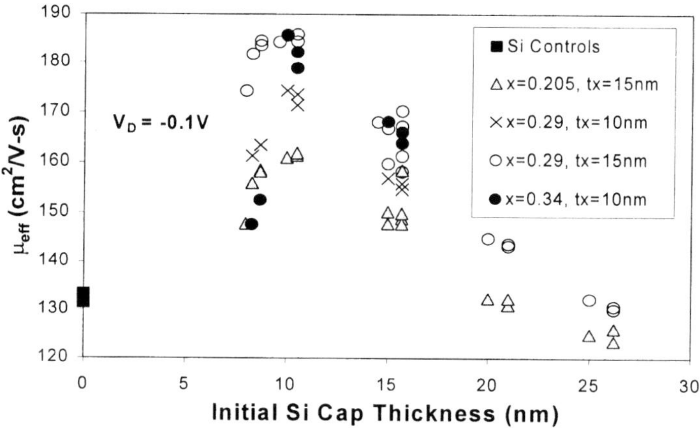

Figure 3. Effective drift mobility versus as-grown Si cap thickness.

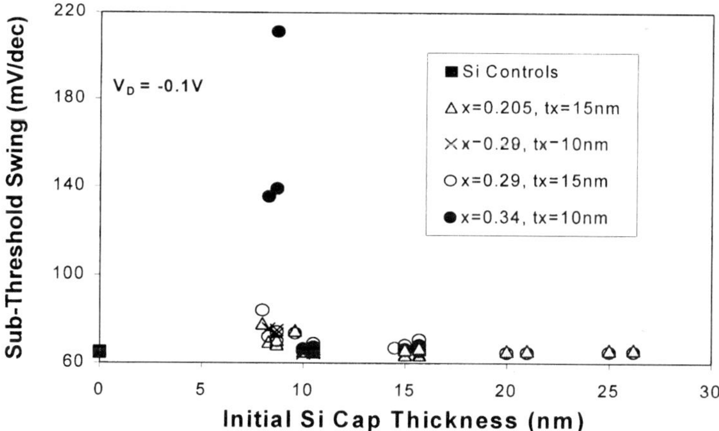

Figure 4. Sub-threshold swing versus as-grown Si cap thickness.

The threshold voltage was found to increase with Ge content due to bandgap narrowing, Figure 5. A positive shift in V_T of ~0.076V per 10% Ge is expected (8). The line shown has a slope of 0.067V per 10% Ge. This dependence of V_T on x needs to be taken into consideration when adjusting V_T. The data that does not fall on the line are from the 29% Ge-10 nm and 34% Ge-10 nm channel transistors. As discussed above, from XRD and mobility measurements both of these have reduced peak Ge concentrations due to diffusion. For example, the devices with an as-grown 29% Ge-10 nm channel have experienced substantial Ge diffusion leaving at most 27% Ge in the channel region (see discussion of Fig. 6 below). Shifting the data in Fig. 5 from x=0.29 to x=0.27 would put it close to the line shown. Similarly, the as-grown 34% Ge devices are found by XRD to have at most 28-29% Ge in the channel after processing.

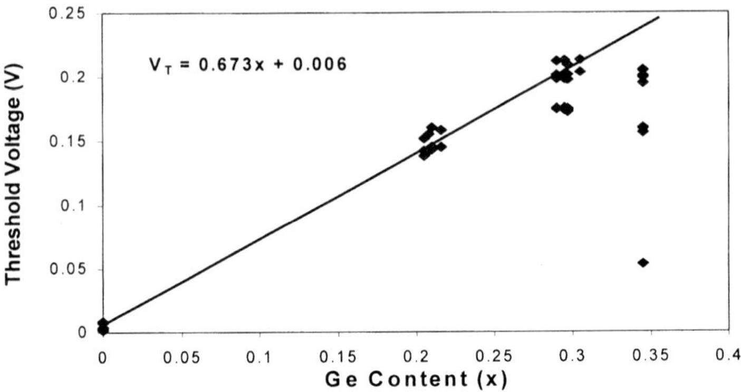

Figure 5. Threshold voltage versus Ge content showing ~67mV/10% Ge shift in V_T.

Figure 6 compares XRD data taken before and after processing of a 29% Ge-10 nm channel with a 10 nm Si cap. After processing, the SiGe(004) peak shifts toward the Si(004) substrate peak, and the extra oscillations seen in the as-grown data disappear. These indicate substantial Ge diffusion out of the channel. Modeling suggests a diffusion length of around 3 nm. However, in spite of this much diffusion, up to 40% improvement in effective mobility has been achieved while maintaining good device properties. These results suggest that if the thermal budget can be reduced, further improvements in device performance should be attainable.

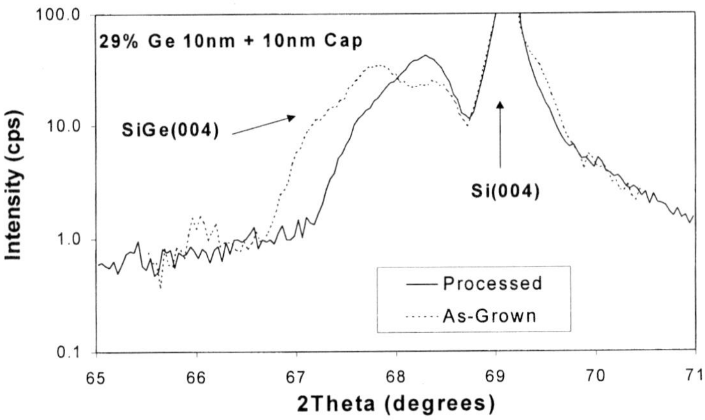

Figure 6. XRD data of a 29% Ge–10 nm channel wafer with 10 nm Si cap before and after processing.

Using quasi-static C-V measurements, we have found clear evidence of hole confinement in inversion (5), as shown in Figure 7. C-V measurements were performed at frequencies ranging from 1 MHz to 100 Hz. Frequency has no effect in accumulation, but in inversion the capacitance increases as the frequency is lowered, stabilizing by 1kHz. It was necessary to have the light on to generate electron-hole pairs; otherwise the data was too noisy. A "plateau" around 0 volts is seen in the capacitance of the SiGe samples that is not observed in the Si control. At low V_G-V_T in inversion, due to the valence band offset between Si and strained SiGe the holes are trapped in the SiGe channel. The observed capacitance at this "plateau" is due to the oxide and Si cap acting as two capacitors in series. As V_G-V_T increases, a parallel channel of holes forms at the Si/oxide interface, increasing the measured capacitance.

Figure 7 compares a Si control with two 29% Ge-15 nm channel capacitors having as-grown Si cap thicknesses of 10 nm and 15 nm. In the device with a 15 nm cap (Fig. 7b), the C-V curve matches that of the Si control very well until a narrow plateau develops. The plateau is at most about 0.5 V wide. In the device with a 10 nm cap (Fig. 7a), the plateau widens to 1.0 V or more, indicating better hole confinement. However, the minimum in the C-V curve also narrows, especially on the inversion side. This indicates that the inversion layer forms at lower gate voltage, which may be due to a narrowing of the band gap caused by Ge diffusion into the cap. Ideally, one wants to have the plateau extend to the gate operating voltage. Under these conditions the majority of holes would be confined to the SiGe channel, optimizing transistor performance.

An estimate of the oxide and Si cap thicknesses can be obtained from the capacitance values using $C=\kappa\varepsilon_0 A/t$, where $\kappa = 3.9$ for SiO_2 and 11.9 for Si, $\varepsilon_0 = 8.85\times10^{-14}$ F/cm, $A = 2.5\times10^{-5}$ cm^2 is the area of the capacitors, and t is the thickness of the dielectric. In inversion, C/A of the gate oxide on the Si control reaches 4.2×10^{-7} F/cm^2, giving an oxide thickness of 8.2 nm. In accumulation we find 4.8×10^{-7} F/cm^2 (see Table I), giving an oxide thickness of ~7.2 nm; this is still higher than the value of 6 nm obtained from spectroscopic ellipsometry of blanket films. In Fig. 7a, the plateau is at about 3.6×10^{-7} F/cm^2. Considering the oxide and Si cap as two capacitors in series and using the inversion value of the oxide capacitance, the Si C/A is 25×10^{-7} F/cm^2, giving an effective Si thickness of about 4.2 nm. This is close to the expected value of 5 nm. In Fig. 7b, the plateau is about 3.0×10^{-7} F/cm^2, leading to a Si capacitance of 10.5×10^{-7} F/cm^2 and a Si cap thickness of 10 nm. This is identical to the expected value.

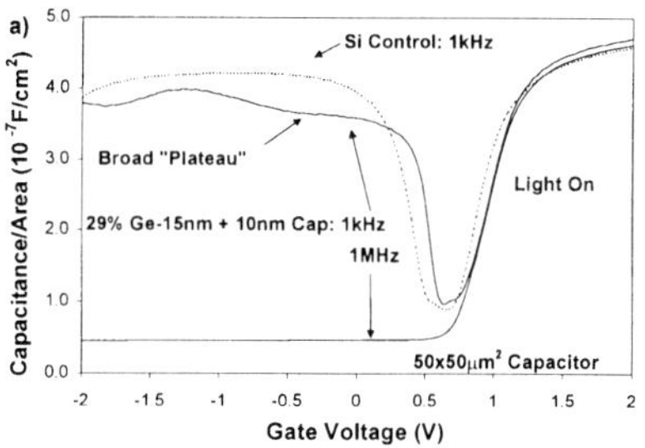

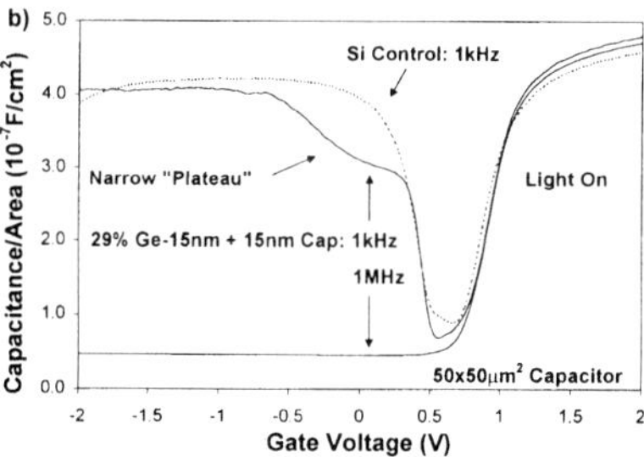

Figure 7. Comparison of C-V curves taken at different frequencies of 50x50μm² capacitors formed on a Si control wafer with those made from 29% Ge-15 nm channel wafers having a) a 10 nm Si cap and b) a 15 nm Si cap.

CONCLUSIONS

We have evaluated the effective hole mobility of buried strained SiGe channel PMOS FET's as a function of device structure. The results indicate that the highest μ_{eff} may be achieved by 1) maximizing the Ge content, x, 2) making t_x thin enough to have a stable channel but not so thin that Ge diffusion reduces x, e.g. 10 nm < t_x <20 nm for x=0.3, and 3) having the Si cap be as thin as possible without having Ge present at the oxide, e.g. ~5 nm. Furthermore, over the range x = 0.2 to 0.34 the optimum cap thickness is independent of x. With the present thermal budget our best results were achieved with an as-grown structure of a 29% Ge-15 nm channel capped by 10 nm of Si, thinned by process steps to ~5 nm. The effective hole mobility for this device is 40% higher than that of the Si controls, and exhibits good STS and low leakage current. Also, clear evidence was obtained for hole trapping in the SiGe channels. Further improvements in performance should be possible with reductions in the thermal budget. These results are in agreement with previously published theoretical and experimental work.

ACKNOWLEDGMENTS

The authors wish to thank David R. Evans, Jer-Shen Maa, and Yoshi Ono for useful discussions.

REFERENCES

1) S.C. Jain, *Germanium-Silicon Strained Layers and Heterostructures*, Academic Press, Boston (1994).
2) M. Arafa, I. Adesida, and K. Ismail, in *Epitaxy and Applications of Si-Based Heterostructures,* E.A. Fitzgerald, D.C. Houghton, and P.M. Mooney, Editors, p. 83, Materials Research Society Symposium Proceedings, vol. 533, Warrendale, PA (1998).
3) D.J. Paul, *Thin Solid Films,* **321**, 172 (1998).
4) D.K. Nayak, J.C.S. Woo, J.S. Park, K.-L. Wang, and K.P. MacWilliams, *IEEE Electron Device Lett.*, **12**, 154 (1991).
5) P.M. Garone, V. Venkataraman, and J.C. Sturm, *IEEE Electron Device Lett.*, **12**, 230 (1991).
6) S.S. Iyer, P.M. Solomon, V.P. Kesan, A.A. Bright, J.L. Freeouf, T.N. Nguyen, and A.C. Warren, *IEEE Electron Device Lett.*, **12**, 246 (1991).
7) S. Verdonckt-Vandebroek, E.F. Crabbe, B.S. Meyerson, D.L. Harame, P.J. Restle, J.M.C. Stork, A.C. Megdanis, C.L. Stanis, A.A. Bright, G.M.W. Kroesen, and A.C. Warren, *IEEE Electron Device Lett.*, **12**, 447 (1991).
8) S. Verdonckt-Vandebroek, E.F. Crabbe, B.S. Meyerson, D.L. Harame, P.J. Restle, J.M.C. Stork, and J.B. Johnson, *IEEE Trans. Electron Devices*, **41**, 90 (1994).
9) P.F. Fewster and C.J. Curling, *J. Appl. Phys.*, **62**, 4154 (1987).

STRAIN RELAXATION OF $Si/Si_{1-x-y}Ge_xC_y/Si$ QUANTUM WELLS GROWN BY RTCVD

M. H. Lee, Y. D. Tseng, C. W. Liu, and M.Y. Chern*
Department of Electrical Engineering, National Taiwan University, Taipei, Taiwan
*Department of Physics, National Taiwan University, Taipei, Taiwan

The $Si/Si_{1-x-y}Ge_xC_y/Si$ quantum wells revealed different relaxation pathways at different temperatures. The relaxation was studied by high-resolution x-ray diffraction, Fourier transform infrared spectroscopy, and defect etching. Adding carbon can release the strain of $Si/Si_{1-x}Ge_x/Si$ quantum wells, and enhanced the Ge outdiffusion in $Si_{1-x-y}Ge_xC_y$ alloys. The lattice structure of $Si_{1-x-y}Ge_xC_y$ layers was as stable as the $Si_{1-x-y}Ge_x$ layers at the annealing temperature of 800°C for two hours. At annealing temperature of 900°C for two hours, the structures of both $Si_{1-x-y}Ge_xC_y$ and $Si_{1-x-y}Ge_x$ started to relax. For the annealing temperatures of 950°C and 1000°C for two hours, the $Si_{1-x}Ge_x$ sample continued to relax with the decrease of strain in the quantum wells, while the $Si_{1-x-y}Ge_xC_y$ sample relaxed with the increase of the strain. This was due to the precipitation of SiC, observed by FTIR measurement. A novel method to determine the in-plane lattice constant is also proposed to study the relaxation of SiGeC layers.

INTRODUCTION

The impressive progress in the growth [1,2] and characterization [3-5] of $Si_{1-x-y}Ge_xC_y$ alloys offer great flexibility to tailor the strain and the electronic properties of Group IV heterostructures [6-8]. Because the lattice constant of diamond (3.56683 Å [9]) is 34 % smaller than that of Si, the substitutional incorporation of C can compensate the compressive strain of $Si_{1-x}Ge_x$ layers grown on Si substrates, where the lattice constant of Ge is 4.17 % larger than that of Si. This increases the critical thickness of pseudomorphic $Si_{1-x-y}Ge_xC_y$ layers on Si. However, the formation of SiC precipitates at high temperature increases the compressive strain and leads to the misfit dislocation formation in the as-grown pseudomorphic $Si_{1-x-y}Ge_xC_y$ layers on Si with the initial thickness bellow its critical thickness [10]. It was also reported that the carbon incorporation into $Si_{1-x}Ge_x$ has the effect of suppressing the boron outdiffusion. To incorporate carbon into the base of $Si/Si_{1-x}Ge_x/Si$ heterojunction bipolar transistors can eliminate the formation of parasitic energy barriers[11], and thus can improve the device performance. Therefore, it is important for further device applications to investigate the thermal stability of $Si_{1-x-y}Ge_xC_y$ alloys.

EXPERIMENTS

The $Si/Si_{1-x-y}Ge_xC_y/Si$ samples were grown on Si<100> substrates by rapid thermal chemical vapor deposition (RTCVD). The $Si_{1-x-y}Ge_xC_y$ layers were grown at 625°C using methylsilane as C source. The growth pressure was 6 torr. The gas flows were 3 slpm for a hydrogen carrier, 26 sccm for dichlorosilane, and 0.8 sccm for germane. The maximum content of carbon incorporation is around 1.2%. The Si cap layer was grown at 700°C, using a 26 sccm dichlorosilane flow and a 3 slpm hydrogen flow. They were annealed in furnace from 800°C to 1000°C in a nitrogen flow. After different annealing temperature, the vertical and in-plane lattice constants were obtained from high-resolution x-ray diffractometry (HR-XRD), using the (400) and (422) diffraction. The Fourier transform infrared spectroscopy (FTIR) was used to study the bonding change of the annealed samples. . The defect etching (4 parts of 49% HF and 5 parts of 0.3M CrO_3) was used to decorate the dislocation network of these samples.

RESULTS AND DISCUSSION

The Ge fraction, the C fraction and thickness were extracted by fitting (400) x-ray rocking curves. Three quantum well structures were investigated in this study. The well compositions were $Si_{0.77}Ge_{0.23}$, $Si_{0.762}Ge_{0.23}C_{0.008}$ and $Si_{0.758}Ge_{0.23}C_{0.012}$ with the thickness of 20nm, 18nm, 18nm, respectively. The nominal thickness of the Si cap was about 60nm. All as-grown samples were pseudomorphic and fully strained due to the low temperature growth (625°C). No defect was observed in these as-grown films after defect etching. For reference, the critical thickness of the $Si_{0.77}Ge_{0.23}$, $Si_{0.762}Ge_{0.23}C_{0.008}$ and $Si_{0.758}Ge_{0.23}C_{0.012}$ samples were 10, 17, 23 nm, respectively, estimated from Matthews and Blakeslee's theory (Fig.1).

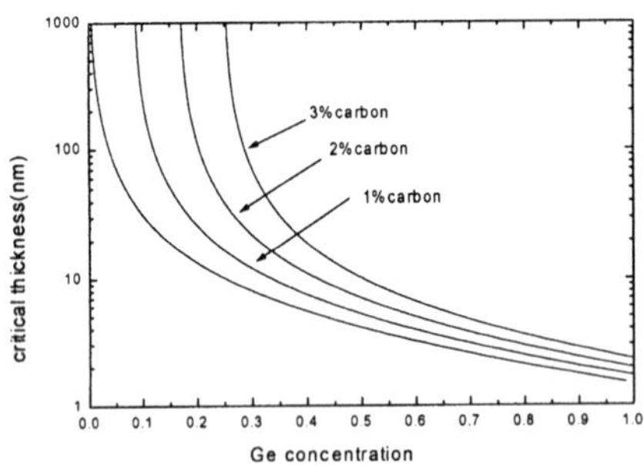

Fig.1 Critical thickness of SiGeC alloys grown on Si substrate. The addition of carbon increases the critical thickness of SiGeC.

There is no change of the x-ray rocking curves of $Si_{1-x-y}Ge_xC_y$ and $Si_{1-x}Ge_x$ alloys at the annealing temperature of 800°C for two hours. The structures of $Si_{1-x-y}Ge_xC_y$ and $Si_{1-x}Ge_x$ alloys are both stable at 800°C. At annealing temperature of 900°C for two hours, the (400) diffraction peaks of both $Si_{1-x-y}Ge_xC_y$ and $Si_{1-x}Ge_x$ samples started to shift towards Si substrate peak (Fig.2, 3, 4), indicating the relaxation of both $Si_{1-x-y}Ge_xC_y$ and $Si_{1-x}Ge_x$ samples. For the annealing temperatures of 950°C and 1000°C for two hours (Fig.2, 3, 4), the diffraction peak of $Si_{1-x}Ge_x$ sample continues to shift towards the Si substrate peak. This indicates that the strain and the vertical lattice constants in the $Si_{1-x}Ge_x$ quantum wells decrease. (Fig.5)

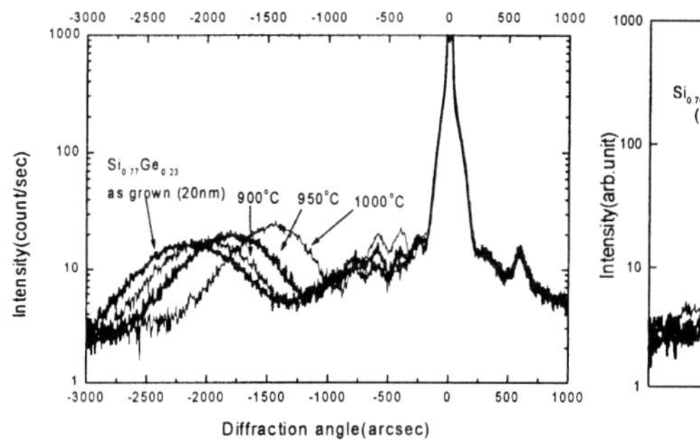

Fig.2 The (400) diffraction peak of $Si_{0.77}Ge_{0.23}$ layer continued to shift towards (400) Si peak as the annealing temperatures increased.

Fig.3 The high-resolution x-ray diffraction spectra of $Si/Si_{0.762}Ge_{0.23}C_{0.008}/Si$ quantum well.

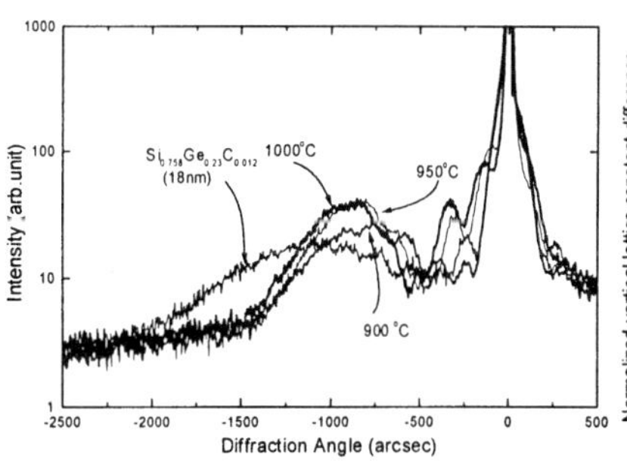

Fig.4 The high-resolution x-ray diffraction spectra of $Si/Si_{0.758}Ge_{0.23}C_{0.012}/Si$ quantum wells.

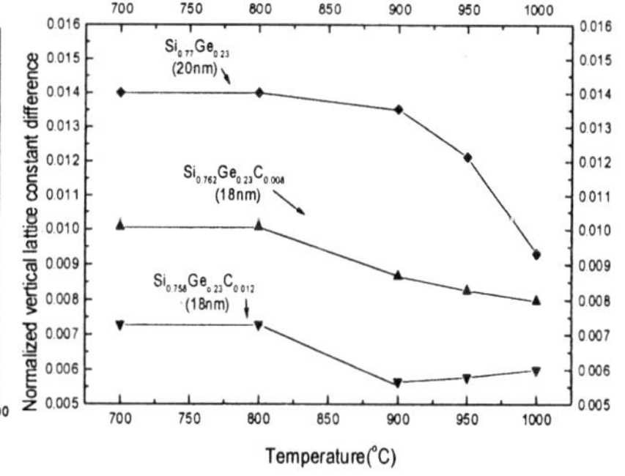

Fig.5 The difference between vertical lattice constant of $Si_{1-x-y}Ge_xC_y$ and Si substrate.

For 2 hr annealing at 950°C and 1000°C, the diffraction peak of $Si_{0.762}Ge_{0.23}C_{0.008}$ samples shifts slightly towards the Si substrate peak, and the diffraction peak of $Si_{0.758}Ge_{0.23}C_{0.012}$ samples even moves to the opposite direction. The vertical lattice constant in the $Si_{0.758}Ge_{0.23}C_{0.012}$ sample increased after 950°C and 1000°C annealing, indicating the increase of strain in the quantum wells. The SiC precipitate can reduce the substitutional C content in the alloys and can be responsible for the strain increase [12]. To confirm this conjecture, the FTIR spectra of an as-grown 40nm $Si_{0.698}Ge_{0.28}C_{0.022}$ sample and a 1000°C annealed sample were measured. The FTIR spectrum of the as-grown sample shows a substitutional C vibration peak at 600cm^{-1}(Fig.6). After annealing at 1000 °C for 2 hr, this substitutional C vibration peak vanishes, and a broad peak from 670 cm^{-1} to 900 cm^{-1} was observed, very similar to amorphous silicon carbide absorption[13]. The increase of vertical lattice constant is due to the precipitation of SiC.

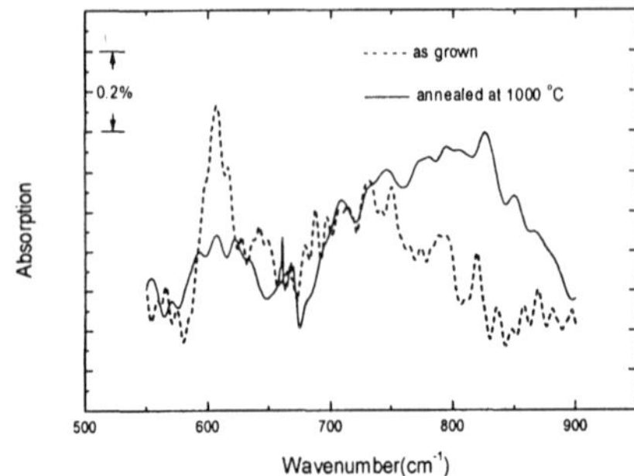

Fig.6 The FTIR spectra of a 40nm $Si_{0.698}Ge_{0.28}C_{0.022}$ sample.

To study the interdiffussion of SiGeC alloys, the diffusion length of Ge was extracted from fitting the x-ray rocking curves. The Ge profile [14] used in the simulation is

$$X_{Ge}(z)= X_{Ge}^{o}/2\{-erf[(-W/2+z)/2L]+erf[(W/2+z)/2L]\} \tag{1}$$

where X_{Ge}^{o} is the initial Ge concentration, W is the quantum well thickness, z is the position along the growth direction (z=0 at the well center), erf is the error function, and L is the diffusion length. The width of carbon profile is relatively unchanged, as compared with that of Ge profile [15]. From the simulation, Ge diffusion length can be obtained. The diffusion length increased as the carbon content increased at all annealing temperatures (Fig.7). This indicates that C can enhance Ge diffusion.

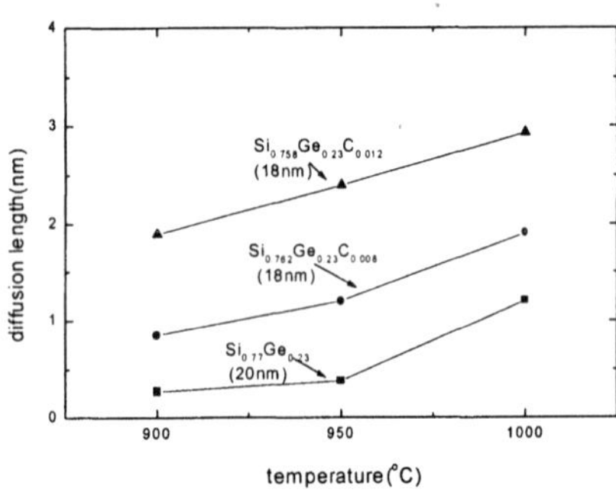

Fig.7 Diffusion length at different annealing temperature. The addition of C increases the diffusion length.

To further resolve the relaxation due to misfit dislocation and the residual carbon content, we measured the asymmetrical (422) diffraction of all three 1000°C annealed samples. The in-plane lattice constant is a useful parameter to determine the relaxation of $Si_{1-x-y}Ge_xC_y$ layers on (100) Si. The in-plane lattice constant of pseudomorphic layers is the same as that of the Si substrate, while the in-plane lattice constant of a fully relaxed layer maintain its own lattice constant. For partially relaxed layers, a relaxation parameter can be defined as $R=(a_{\parallel}-a_{si})/(a_r-a_{si})$, where $a_{\parallel}$ is the in-plane lattice constant of $Si_{1-x-y}Ge_xC_y$ alloys, a_r is the lattice constant of fully relaxed $Si_{1-x-y}Ge_xC_y$ layers, and a_{si} is the lattice constant of Si substrates. Therefore, it is important to measure the both in-plane lattice constant and vertical lattice constant to determine the relaxation of the strained layers. The vertical lattice constant can be obtained from symmetrical (400) x-ray diffraction for the $Si_{1-x-y}Ge_xC_y$ layer grown on (100) Si. The in-plane lattice constant was conventionally determined by x-ray rocking curves using two sets of asymmetrical diffractions [16]. However, for some high temperature annealed samples, only one set of asymmetrical diffraction (glancing incident geometry) can be measured. We, therefore, propose a new method to determine the in-plane lattice constant using only one set of asymmetrical diffraction.

In the glancing incident geometry (Fig.8(a)), the planes of the epilayer and the planes of the substrate would be parallel, if there were no strain in the epilayer. The peak separation between the epilayer and the substrate in the rocking curve would be solely due to the Bragg angle difference (Fig.8(b)). For thin $Si_{1-x-y}Ge_xC_y$ layer, the compressively tetragonal distortion causes the asymmetrical reflection planes of the eiplayer are not parallel with those of the substrates. The misorientation $\Delta\phi$ is defined as the angle between these two planes. The incident angle separation in the rocking curves between the epilayer and the substrate is the sum of the Bragg angle difference and the misorientation ($\Delta\theta_B + \Delta\phi$) as shown in Fig.8(c). Note that for the glancing exit geometry, the x-ray direction in Fig.8 is reversed.

The misorientation $\Delta\phi$, the angle between the (422) planes of epilayers and substrates, is depending on the ratio between vertical lattice constant and in-plane lattice constant, i.e.,

$$\Delta\phi = \arccos\left(4+2\, a_\perp/a_\parallel \right)/ \left(12(a_\perp/a_\parallel)^2+24\right)^{1/2} \quad (2)$$

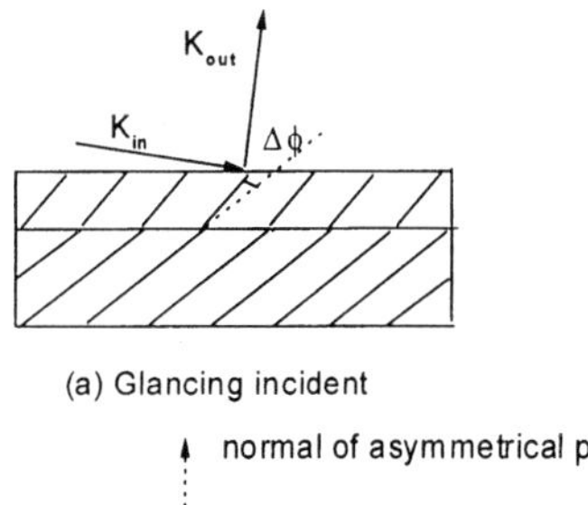

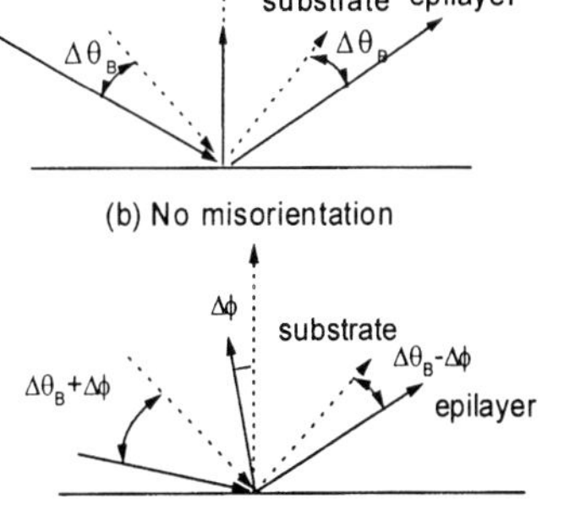

Fig.8 The effect of the misorientation on Bragg diffractions. The misorientation is defined in (a). In (b), the peak separation of the incident angles between the epilayer and the substrate would be the Bragg angle difference, if there were no misorientation. In (c), the peak separation of the incident angles increases for the glancing incident geometry due to the misorientation. Note that for the glancing exit geometry, the x-ray direction is reversed and the peak separation decreases.

In the rocking curve, the separation of the diffraction peaks between substrate and epilayer, Δ, can be expressed as

$$\Delta=\Delta\theta_B \pm \Delta\phi \tag{3}$$

The vertical lattice constant is given by $a_\perp=4d_{400}$, and in-plane lattice constant is

$$a_{\parallel}=[8/(1/d^2_{422}-1/d^2_{400})]^{1/2} \tag{4}$$

where d_{400} and d_{422} are the (400) and (422) plane distance of the epilayers, respectively. Besides (400) diffraction, only one set of diffraction (either glancing incident or glancing exit geometry) is sufficient to solve $a_\perp$ and $a_\parallel$. The vertical lattice constant can be obtained from (400) diffraction. For a range of the initial value of in-plane lattice constant $a_{\parallel,i}$, we can obtain the $\Delta\phi$ and $\Delta\theta_B$ from eq.2 and eq.3, respectively. Finally, a range of final value of in-plane lattice $a_{\parallel,f}$ can be obtained from eq.4. The intersection between the $a_{\parallel,f}$ vs $a_{\parallel,i}$ curve and $a_{\parallel,f} = a_{\parallel,i}$ is the in-plane lattice constant. There are two intersections and the larger one is not physically acceptable (Fig.9). We have compared our method to the previous method for as-grown samples, where both diffraction geometries can be measured. The accuracy of our method is at least the same as that of the previous method [16].

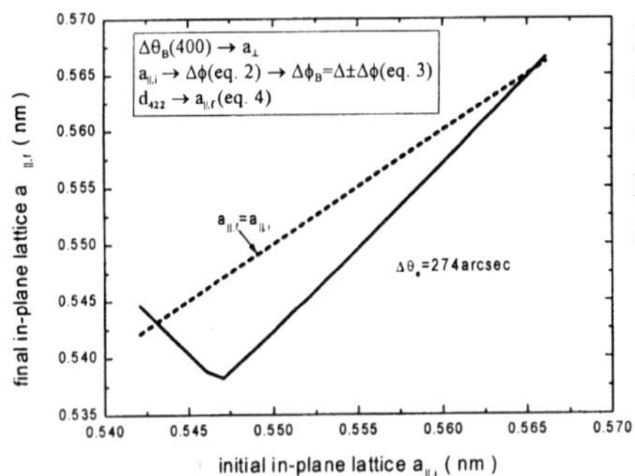

Fig.9 The plot to get the solution of the in-plane lattice constant. The one solution of larger lattice constant is not physically possible. The insert is the procedure to obtain the $a_{\parallel,f}$ vs $a_{\parallel,i}$ curve.

For the three quantum well samples, annealed at high temperature 1000°C for 2 hr, the peak position in the (422) glancing incident geometry can be used to determine the in-plane lattice constant (Fig.10). The results are given in Table I. The in-plane lattice constant increases after annealing, indicating the $Si_{1-x-y}Ge_xC_y$ layers are relaxed. The relaxation parameter increases as the amount of carbon incorporation increases after the 1000°C annealing for 2hr. It is interesting that the 18nm $Si_{0.758}Ge_{0.23}C_{0.012}$ sample is initially below its critical thickness (23 nm), but the critical thickness becomes 15 nm after 1000°C annealing due to the SiC precipitates, and the misfit dislocation network is formed in the sample after the 1000 °C annealing, as confirmed by defect etching.

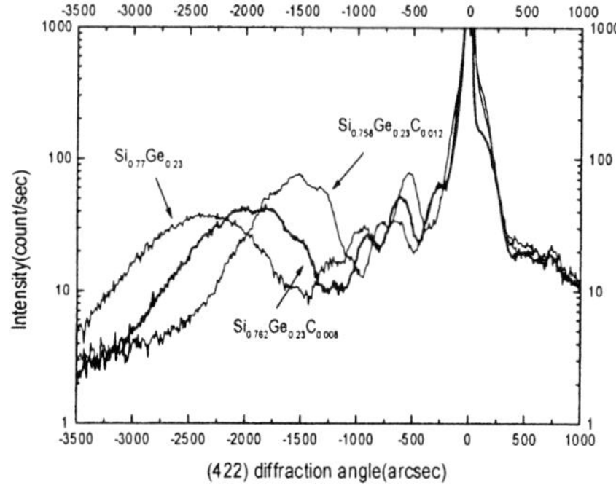

Fig.10 The asymmetrical (422) diffraction spectra of the 1000 °C annealed samples. With the peak position of (400) and (422) diffraction, the in-plane lattice constants can be obtained.

Table. I The measurement results of the glancing incident geometry for the samples annealed 2 hr at 1000°C.

$Si_{1-x-y}Ge_xC_y$ (thickness)	$\Delta\theta_I$	$\Delta\theta_B$	$a_{\parallel}$ from $\Delta\theta_I$	Relaxation parameter
$Si_{0.758}Ge_{0.23}C_{0.012}$ (18nm)	1515	1069	5.44272	26%
$Si_{0.762}Ge_{0.23}C_{0.008}$ (23nm)	1996	1484	5.44155	21%
$Si_{0.77}Ge_{0.23}$ (24nm)	2400	2166	5.43980	17%

SUMMARY

At 800℃ annealing for 2hr, the lattice structure is stable for both $Si_{1-x}Ge_x$ and $Si_{1-x-y}Ge_xC_y$ layers. The thermal budget of device processing would not be degraded due to the C incorporation. At 900℃ annealing for 2hr, we have observed the relaxation of both $Si_{1-x}Ge_x$ and $Si_{1-x-y}Ge_xC_y$ samples. As annealed at 950℃ and 1000℃ for 2hr, SiC precipitation is detected by FTIR, and the defect etching shows the misfit dislocation network of the $Si_{0.758}Ge_{0.23}C_{0.012}$ layer. As annealed at 950℃ and 1000℃ for 2hr, the SiC precipitation was formed, and vertical lattice constant of the SiGeC layer increased due to the consumption of Si and C atoms to form the SiC precipitates. By measuring the asymmetrical(422) diffraction of all three 1000℃ annealed samples, the in-plane lattice constants can be obtained and the relaxation parameter increases with the carbon content. We have proposed a new method to determine in-plane lattice constant of stained layers. Only one of the asymmetrical diffraction conditions is required in this method.

ACKNOWLEGMENTS

We appreciated the sample supply from Prof. Sturm, Princeton University. The support of National Science Council, Taiwan, is highly appreciated (NSC 88-2218-E-002-010).

REFERENCES

[1] J. L. Regolini, F. Gisbert, G. Dolino, and P. Boucaud, *Mat. Lett.*, **18**, 57 (1993).

[2] C. W. Liu, A. St. Amour, J. C. Sturm, Y. R. J. Lacroix, M. L. W. Thewalt, C. W. Magee, and D. Eaglesham, *J. Appl. Phys.*, **80**, 3043 (1996).

[3] P. Boucaud, C. Francis, F. H. Julien, J. M. Lourtioz, D. Bouchier, S. Bodnar, B. Lambert, and J. L. Regolini, *Appl. Phys. Lett.*, **64**, 875 (1994).

[4] A. St. Amour, C. W. Liu, J. C. Sturm, Y. Lacroix, and M. L. W. Thewalt, *Appl. Phys. Lett.*, **67**, 3915 (1995).

[5] C. Y. Lin, and C. W. Liu, *Appl. Phys. Lett.*, **70**, 1441 (1997).

[6] L. D. Lanzerotti, A. St. Amour, C. W. Liu, J. C. Sturm, J. K. Watanabe, and N. D. Theodore, *IEEE Electron Device Lett.*, **17**, 334 (1996).

[7] I. M. Anteney, G. Lippert, P. Ashburn, H. J. Osten, B. Heinemann, G. J. Parker, and D. Knoll, *IEEE Electron Device Lett.*, **20**, 116 (1999).

[8] S. John, S. K. Ray, E. Quinones, S. K. Oswai, and S. K. Banerjee, *Appl. Phys. Lett.*, **74**, 847 (1999).

[9] O. Maselung, Semiconductor – Basic Dara, 2nd Ed. (Spring – Verlag, Berlin, 1996) p.7.

[10] C. W. Liu, Y. D. Tseng, M. Y. Chern, C. L. Chang, and J. C. Sturm, *J. Appl. Phys.*, **85**, 2124 (1999).

[11] I.Manteney, G.Lippert, P.Ashburn, H.J.Osten, B.Heienemann, G.J.Parker, and D.Knoll, *IEEE Electron Device Lett*,vol 20. 116,1999.

[12] M. S. Goorsky, S. S. Iyer, K. Ebert, F. Legoues, J. Angilello, and F. Cardone, *Appl. Phys. Lett.* **60**, 2758 (1992).

[13] C. W. Liu and J. C. Sturm, *J. Appl. Phys.*, **82**, 4558 (1997).

[14] P. Boucaud, L. Wu, C. Guedj, F. H. Julien, I. Sajnes, Y. Campidelli, and L. Garchery, *J. Appl. Phys.*, 80, 1414 (1996).

[15] P. Warren, J. Mi, P. Overney, and M. Dutoit, *J. of Crystal Growth,* **157**, 414 (1995).

[16] M. Fatemi and R. E. Stahlbush, *Appl. Phys. Lett.*, **58**, 825 (1991).

Section VI

RTCVD and Epitaxy of Si and SiGe

EMISSIVITY EFFECTS IN LOW-TEMPERATURE EPITAXIAL GROWTH OF Si AND SiGe

W.B. de Boer and D. Terpstra
Philips Research Laboratories, Prof. Holstlaan 5, 5656 AA Eindhoven, The Netherlands

Control of the susceptor temperature in cold-wall CVD reactors does not necessarily result in a well-defined wafer temperature. Widely varying growth rates during low-temperature epitaxy of Si and SiGe on patterned wafers are not caused by chemical loading effects alone, as is generally assumed, but also by temperature variations. A set of experiments is described separating temperature effects due to varying wafer emissivities from chemical loading effects due to changing silicon/oxide ratios on patterned wafers. The temperature variations are most harmful when blanket Si or SiGe layers are grown on patterned wafers, as the growth rate and the layer composition change during the deposition process.

INTRODUCTION

The introduction of low-temperature epitaxy by means of Chemical Vapor Deposition (CVD) in the second half of the last decade decoupled diffusion of dopants and epitaxial growth of Si layers, allowing the growth of very thin layers with complicated doping profiles and abrupt doping transitions. The low deposition temperature, typically around 700°C, also enabled bandgap engineering through the growth of SiGe alloy layers with varying Ge contents. An important application is the growth of transistors with a SiGe base, so-called Heterojunction Bipolar Transistors (HBTs), with comparatively very high oscillation and cut-off frequencies.

In a low-temperature epi process the base of an HBT is grown and doped simultaneously, in contrast to the conventional way of defining the base width and base dope, which is by implantation and diffusion. The epi-base process not only offers a better control of width and dope content, but the low temperature budget also offers the opportunity to add bipolar circuitry to a CMOS process after the CMOS part has been finished. In such a BiCMOS process the integration of HBTs and MOS requires that the Si and SiGe layers are grown on patterned wafers. This can be done in two fundamentally different ways. Via an appropriate choice of the deposition conditions, a layer can be deposited epitaxially in the windows in the pattern where the substrate silicon is exposed, whereas no growth takes place on the surrounding oxide or nitride fields. This method is called Selective Epitaxial Growth (SEG). The second way is the growth of a blanket silicon layer. The wafer surface is now covered entirely with Si or SiGe. The layer is closed and grows epitaxially in the windows and polycrystalline on the oxide or nitride fields. The method that is chosen has a high impact on the transistor design and the process flow.

From a CVD point of view there are also significant differences between SEG and blanket layer growth. In order to preserve the selectivity and to avoid spurious nucleation on the dielectric, SEG is performed in a Cl-containing chemistry. SiH_2Cl_2 is used as the Si precursor and HCl is added to the gas stream to increase the Cl content. This reduces the growth rate in exchange for a reliable selectivity. Another disadvantage of SEG is a phenomenon called loading effect: the local growth rates depend on the pattern and the lay-out. The windows in a pattern usually cover only a very small fraction of the wafer surface. The deposition reactions proceeding in the windows cause a non-uniform distribution of the active chemical species over the wafer, leading to non-uniform growth. E.g., the growth rates of Si and SiGe vary in opposite ways with the area of exposed Si (1). Blanket layer growth performs better in this respect, as the entire wafer and susceptor area is covered and not just small, isolated areas, albeit that the poly and mono growth rates are different. To be able to grow uniformly on oxide or nitride fields, the Si precursor should not contain Cl. SiH_4 or Si_2H_6 are commonly used for blanket layer growth. Besides the good nucleation properties on oxide and nitride, these Si sources have the additional advantage of a higher growth rate than the chlorinated silanes. A disadvantage of the Cl-free chemistry is that n-type doping cannot be controlled in Si epitaxy. P and As tend to segregate at the surface and long after the dopant supply has been stopped the incorporation of high concentrations of As or P continues. Switching back and forth between SiH_4 and SiH_2Cl_2 so as to combine good nucleation properties and high growth rates with acceptable n-type dopant control may well be the optimum choice for blanket layer growth (2).

The most common epitaxial reactors used in modern Si technology are similar in design. A quartz reaction chamber contains the wafer support, the susceptor, which is rotated to improve the deposition uniformity across the wafer. Only one wafer is processed at a time. Process and carrier gases flow over the wafer in a laminar mode and parallel to the wafer surface. The wafer is heated by tungsten-halogen lamps located underneath and above the the reaction chamber, radiating through the quartz and directly heating the wafer and susceptor. The lamps and the quartz walls of the chamber are air-cooled to protect the lamps and to prevent the risk of Si depositing on the reactor walls. The wafer is loaded and unloaded fully automatically and the reaction chamber is separated from the ambient by load locks and a wafer transfer chamber. It is this last feature, the exclusion of moisture and oxygen from the reaction area, which proved very beneficial for the crystallographic quality of the epi layers and which is the key to low temperature processing. The process temperatures range from very low (500-700°C) for advanced applications to high (1100-1200°C) for conventional epi. The runs can be made at atmospheric pressure and at reduced pressure (10-100 Torr), depending on the process requirements and the personal preference of the epi engineer.

A modern epi reactor resembles Rapid Thermal Processing (RTP) equipment in many aspects. An essential difference is the presence of a susceptor in an epi reactor. This is a carbon disc with a SiC coating to make the surface impervious. In RTP tools the wafers are heated directly and heating rates over 200°C/sec can be reached. The susceptor in an epi reactor acts as a flywheel, slowing down the heating rate to 5-10°C/sec and smoothing local temperature variations. This is necessary in Si epitaxy to prevent the risk of the wafers sustaining crystallographic damage (slip). The temperatures during conventional epi processing are so high that common RTP equipment is not able to process the wafers defect-free. In spite of the fact that the reward for susceptorless epi is a considerable increase in throughput and a decrease in cost per wafer, the additional thermal mass of the susceptor is still indispensable (3). The susceptor also alleviates another well-known RTP problem: the temperature measurement. The pyrometric methods which are applied in RTP to measure the wafer temperature suffer from variations in the wafer emissivity and from varying absorption in the optical path when the measurement is performed through the reactor wall. In epi reactors the susceptor temperature is measured rather than the wafer temperature, e.g. by thermocouples. The susceptor is presumably coupled to the wafer so well that only a negligible temperature difference exists between the wafer and the susceptor. This is assumed to make the epi process insensitive to changing wafer emissivities, in contrast to RTP.

In conventional epi this method has always worked to full satisfaction. The temperatures are high, the wafers are processed at atmospheric pressure or in the 50 - 100 Torr range in reduced-pressure runs. The heat transfer between the wafer and the susceptor under these conditions is good and possible temperature differences between the wafer and the susceptor are so small that they do not affect the process. In this respect two important remarks have to be made: the growth rate in the high temperature regime is relatively insensitive to temperature variations as it is transport-limited and, secondly, the growth is effected on unpatterned wafers. The situation for low-temperature epi is very different. The radiative component of the heat transfer is less efficient and relatively more energy is transferred by conduction through the gas film between the wafer and the susceptor. In order to suppress loading effects, the process pressure is reduced to 20 Torr, which increases the thermal resistance between the wafer and the susceptor (4). In contrast to conventional epi, the growth rate at low temperatures is kinetically controlled and is very temperature-sensitive. The same holds for the incorporation of dopants and Ge in Si. The most important difference is that low-temperature epi growth is often effected on patterned wafers in selective or in blanket growth mode.

We observed growth rate variations during low-temperature epi growth on patterned wafers which could not be explained by chemical loading effects. A strong suspicion that the growth rate variations were related to temperature differences led us to investigate the temperature difference between the wafer and the susceptor in relation to the wafer emissivity.

EXPERIMENTAL SETUP

The Si layers were deposited in an ASM Epsilon One, a commercially available epi reactor for production purposes. The reactor features single-wafer processing and load-locked wafer entry. Processing can be effected at atmospheric or reduced pressure. Fig.1 shows the heart of the system, schematically depicting a simplified lay-out of the process tube, lamps, susceptor and the details of the temperature measurement and control

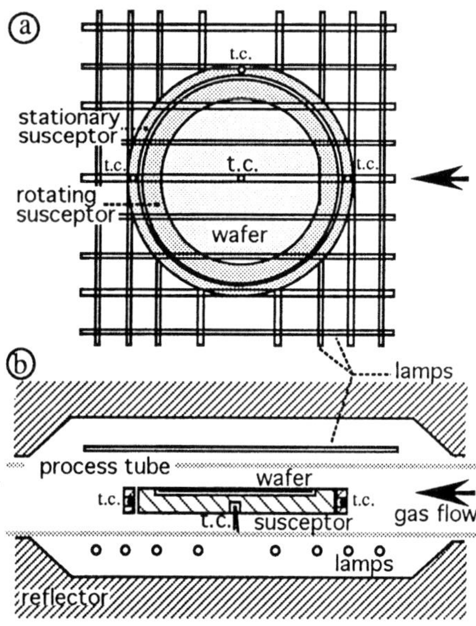

Fig.1 - Top view (a) and cross section (b) of the reactor chamber, showing the geometry and the position of the heating lamps with respect to the wafer, susceptor and thermocouples (t.c.).

system in top view (fig.1a) and in cross section (fig.1b). Longitudinal lamps are positioned on both sides of the quartz process chamber in a reflective cavity. The upper and lower lamp banks are set at an angle of ninety degrees with respect to each other. The susceptor is heated by the lamp radiation and the wafer is in turn heated by the susceptor and by direct radiation from the upper lamps. Process and carrier gases are injected at one end of the process chamber and exhausted at the other end. The susceptor comprises two parts: the center rotating section holding the wafer and a stationary ring at the outside in which three thermocouples are embedded. The fourth thermocouple is located at the center of the susceptor. Each thermocouple controls the temperature in its part of the susceptor via a control system acting on the lamps that have the most influence on that particular susceptor segment. The upstream thermocouple controls the two upstream lamps and the downstream thermocouple controls the downstream lamps in the lower bank. The one thermocouple at the side (fig.1b) controls the four outermost lamps in the upper heater bank and the center thermocouple controls the central lamps in both the upper and the lower heater bank. The balance between the energies supplied from the upper and lower lamp banks can be adjusted by limiting the power in either of the banks with respect to the other. This balance is also affected by the setpoints of the individual thermocouples. In order to avoid thermal stress, which could give rise to wafer bow, loss of contact with the susceptor, more stress and ultimately crystallographic damage to the wafer, it is best to maintain a situation in which the heatflux between the susceptor and the wafer is as small as possible. Although this flux is not a controlled variable, runs performed at high temperatures probably do not deviate too much from the ideal situation, because otherwise the wafer would suffer crystallographic damage. In order to create a maximum unbalance in the energy supply, some of the experiments were carried out with either the upper or the lower heater deliberately switched off .

The epi layers were grown at temperatures around 700°C and at a pressure of 20 Torr. SiH_2Cl_2 was added to the H_2 carrier flow as the Si source and B_2H_6 as the supply of B dope. The susceptor was coated with a Si layer of 0.5μm before the wafer was loaded. The goal of the experiment was to compare growth rates of Si epi on wafers with different emissivities. Any difference in growth rate in the low-temperature regime reflects a difference in wafer temperature as the growth is limited by the reaction kinetics. The problem with patterned wafers is how to separate temperature effects from chemical loading effects. It is also difficult to compare the growth rate of polysilicon layers on the pattern with the growth rate of monocrystalline layers on bare wafers. To circumvent these difficulties and to exclude the extra complexity of patterning, we used bonded wafers in the test. These wafers (Unibond) are characterized by a thin top layer of monocrystalline Si which is separated from the rest of the substrate by a silicon oxide layer. The growth on these wafers from a CVD perspective is equivalent to the growth on our n+ reference wafers (Ø150mm Cz, 6-20mΩcm Sb). In both cases the epi layers grow on (100) Si surfaces, but the Unibond wafer has a different emissivity as a 400nm-thick silicon oxide layer is buried underneath a top layer of 50nm Si . Not only is the intrinsic emissivity different, but the emissivity also changes during the growth, whereas the emissivity of the reference wafer remains unaffected.

The average growth rate is given by the total thickness of the epi layer divided by the deposition time. A refinement, allowing a more detailed determination of the growth rate during the deposition, was made by applying marker layers. At fixed time intervals during the growth of undoped Si-epi layers, B_2H_6 was added to the gas stream to create a thin

B-doped epi layer. This was done without interrupting the deposition process and during no longer than 5% of the deposition time. The effect of the dopant on the growth rate under these conditions is negligible. The layers were analyzed by means of SIMS (Secondary Ion Mass Spectrometry), which showed the B-doped layers as sharp, well-defined B peaks. Each run produced seven peaks and the distance between any two peaks determines the average growth rate in that particular section of the layer. The resolution of this method was more than sufficient to detect growth rate variations during the run.

RESULTS AND DISCUSSION

Phenomenon

Identical Si-epi runs were performed at a temperature setpoint of 700°C on an n+ reference wafer and on a Unibond wafer containing B spikes at fixed time intervals. Fig.2 shows the SIMS B profiles of both layers, measured at the center of the wafer. The epi layer on the Unibond wafer was some 25% thinner than the epi layer on the reference wafer, indicating a lower average temperature during the deposition. A closer look revealed that the B spikes on the reference wafer were equidistant, whereas the Unibond wafer showed an increasing growth rate with an increasing layer thickness. In fig.3 the difference between the wafers has been plotted in terms of growth rates vs. epi layer thickness. The plot suggests that this phenomenon is emissivity-driven. The emissivity of the n+ wafer did not change during the growth and its temperature remained constant, resulting in a constant growth rate and equidistant B spikes. The increasing thickness of the epi layer on top of the buried oxide layer in the Unibond wafer gave rise to attenuation and amplication of varying wavelengths in the absorbed and emitted light spectra, which led to emissivity variations. These emissivity variations are supposed to be oscillatory in nature (5), which is not obvious from fig.3. The absence of these oscillations is probably attributable to the fact that the layer thickness varied considerably over the wafer. The layer thickness at the

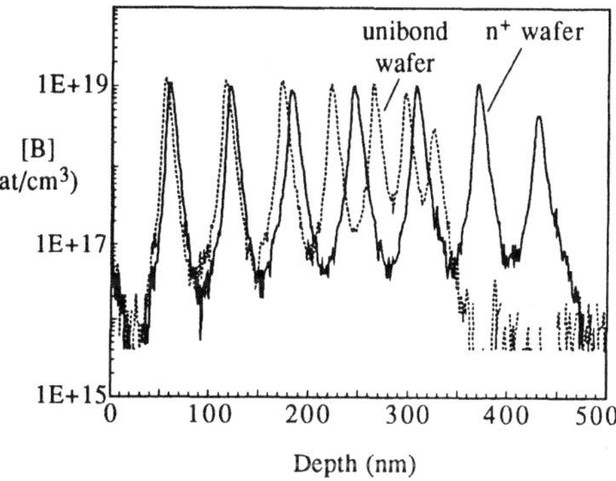

Fig.2 - B spikes grown at fixed time intervals in Si epi layers, measured by means of SIMS. The spikes on the n+ reference wafer are equidistant, whereas the distance between the spikes on the Unibond wafer varies.

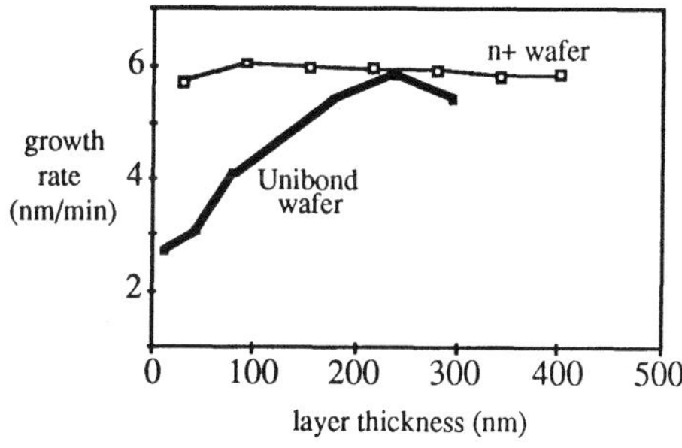

Fig.3 - The epi growth rate on a Unibond wafer compared with an n+ reference wafer. Emissivity variations cause the temperature of the Unibond wafer to change, resulting in growth rate variations. The growth rate on the reference wafer remains constant.

edge was 30% thinner than at the center. The temperature differences coupled to the oscillations which would be apparent if the growth rate across the wafer was uniform, were averaged over the wafer and the susceptor and annihilated each other. What remained is the total absorption of the radiation, which increased with the increasing epi layer thickness and eventually caused the emissivity to approach the emissivity of the reference wafer.

Assuming an activation energy of 2eV for the epi growth (6), the growth rate differences observable in fig.3 translate into stunning temperature differences. The Unibond wafer was more than 30°C colder than the reference wafer at the start of the deposition process and its temperature increased during the process. This not only made it difficult to grow a predetermined layer thickness, but was also detrimental to the incorporation of dopants, which is very temperature-dependent at low temperatures (3). The growth of SiGe layers was also severely affected by the strong temperature dependence of the Ge content.

Explanation

At first sight the cause of the temperature variations may not be clear. The susceptor temperature was measured with thermocouples, a measurement method which is insensitive to variations in the wafer emissivity. An understanding of the cause of the temperature variations of the wafer can be gained in an experiment in which the ratio of the energies supplied by the upper and lower heaters is changed. To this end three epi runs were performed on reference wafers: a run with only the upper lamps switched on, a run with only the lower lamps switched on and a run in normal mode with the upper and lower lamps switched on. The outside lamps on all sides were disabled during all three runs to avoid ratio changes between the center and edge lamps that would be difficult to interprete. Hence, the temperature was controlled by the center thermocouple only, acting on the central lamps. Anticipating the outcome of the experiment, we tried to obtain the same wafer temperature in all three cases by adapting the temperature setpoint from one run to the next. B spikes were again grown to determine the growth rate. The results of the experiment are given in fig.4. The growth rates remained approximately constant during all three runs; no emissivity effects were observed. The different growth rates indicate a maximum difference between the temperatures of the wafers in the three runs of some 15°C. To be able to obtain this, we had to set the thermocouple at temperatures ranging from 625 to 750°C. Obviously there was a substantial temperature difference between the thermocouple and the wafer when the heating was effected from one side. The culprit was the thermal resistance between the wafer and the susceptor and possibly also between the thermocouple and the susceptor. In analogy to a voltage drop over a resistor in an electrical circuit, there is in this case a temperature drop over a thermal resistor. The thermocouple was kept at the set

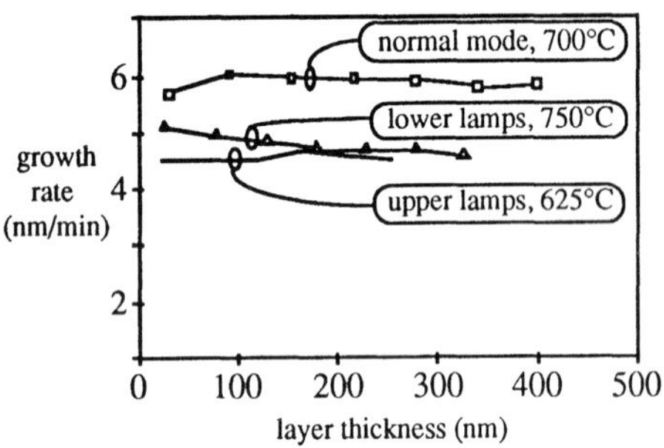

Fig.4 - The temperature setpoints have to lie far apart in order to realize comparable growth rates in different heating modes. This indicates that substantial temperature differences may exist between the wafer and the susceptor.

temperature and the temperature difference between the wafer and the thermocouple was determined by the magnitude and the direction of the heat flux. Heating from only one side is the worst case, causing the maximum heat flux. In normal mode, with two-sided heating, the heat flux is limited and the temperature difference between the wafer and the susceptor is smaller.

The occurrence and prevention of slip in the wafers in high-temperature epi gives some information about the balance of the upper and lower heaters, as mentioned in the experimental section. This limited feedback, together with the better heat transfer at higher temperatures and higher process pressures, reduces the temperature difference to an insignificantly small value. The fact that the reaction rate is tranport-limited in the high-temperature regime and relatively insensitive to temperature variations is also beneficial. At low temperatures the wafers do not develop slip so easily and there is no warning of a large vertical temperature gradient. Consequently, a considerable temperature difference may exist between the wafer and the thermocouple, as can be deduced from fig.4. This is a situation that is not very harmful because it remains constant during the deposition, as long as the balance between the upper and lower heater is not disturbed. It is precisely this condition that is not fulfilled when wafers which change their emissivity during the process are deposited. The varying emissivity causes the absorption and emission of the radiation to vary during the deposition process. This continually changes the balance of the energies supplied by the upper and lower heaters. Assume for instance that the growing layer reaches a thickness at which the absorption of the lamp radiation is at a maximum. More heat is now absorbed from the upper lamps, as the lower lamps radiate at the bottom of the susceptor, which does not change its absorption and emission characteristics. The control system reduces the overall power output to maintain the setpoint and the heat flux from the wafer to the suceptor increases or the flux in the reverse way decreases, depending on the starting condition. It is this changing flux which causes the wafer temperature to change. In this example the wafer temperature increases relative to the thermocouple temperature. Fig.3 indicates that this effect disturbs the deposition process, albeit that the experiment described here represents a worst case. During blanket layer growth on silicon oxide or nitride, polysilicon is deposited on the dielectric rather than epitaxial layers. The emissivity variations are smaller when polysilicon is deposited, as the absorption is higher than in epitaxial silicon. Nevertheless, our experience with the growth of blanket Si and SiGe layers on device wafers proves that the effects described here cannot be neglected.

Possible remedy

The experiment represented in fig.2 was repeated with only the central lamps in the upper heater switched on, controlled by the center thermocouple. The temperature setpoint was fixed at 625°C. We expected a reduced sensitivity to emissivity variations. As the wafer was heating the susceptor in this case, the radiation term of the heat transfer depended on the backside emissivity of the wafer, which was constant. The conduction term depended on the temperature difference between the wafer and the susceptor and the pressure. In an ideal case it is the wafer temperature that determines the susceptor and thermocouple temperature. The temperature difference between the wafer and the thermocouple will be substantial, but constant. Emissivity variations of the top surface of the wafer have no influence, as it is the wafer temperature that drives the thermocouple. The results of the experiment are shown in fig.5. The growth rates on the reference and on the Unibond wafer were measured and compared at the center and at the edge. The results obtained at the centers of the wafers were disappointing. Not only was the temperature of the center of the Unibond wafer lower than the reference wafer, but it also still changed during the growth. However, the growth rates at the edges of the reference and the Unibond wafer were almost the same and the growth rate (temperature) of the Unibond wafer remained constant during the growth. This contrasts with the previous experiment (fig.2), where the growth rate variations at the edge of the Unibond wafer were similar to the variations observed at the center (not shown). We indeed see some emissivity compensation when only the upper heater is switched on, but it is insufficient. The reason for the incomplete compensation of the emissivity variations can be found in fig.1b. The susceptor extended beyond the wafer and the exposed susceptor ring outside the wafer did not change its emissivity during the deposition. It absorbed the heat from the top lamps more effectively and shunted the resistance between the wafer and the susceptor. Long before the center of the wafer was compensated for the varying emissivity, the thermocouple temperature had returned to its setpoint, controlled along a parallel path. A susceptor having the same diameter as the wafer would improve this situation.

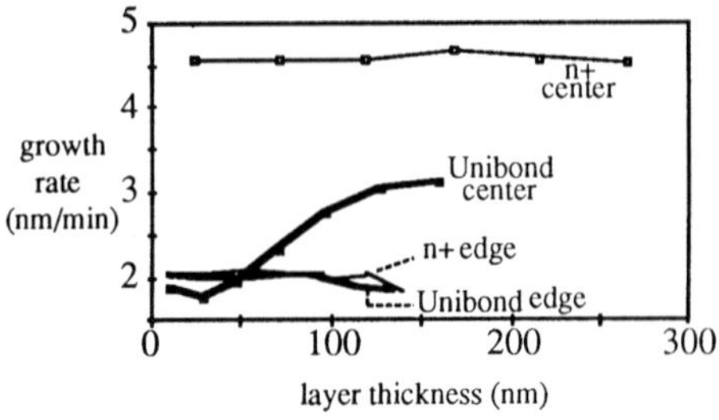

Fig.5 - An attempt to suppress the emissivity dependency by using only the upper lamp bank was only partly successful. The emissivity changes were fully compensated at the edge of the wafer but not at the center.

The last experiment immediately revealed one more problem of the wafer-on-susceptor system. A difference in the emissivity of a wafer will not only give rise to a different temperature and a different growth rate, but the center-to-edge lamp-power ratio will change as well and hence also the thickness distribution across the wafer. This is caused by the different emissivities of the wafer and the susceptor and the fact that the former changes during the growth of blanket layers on patterned wafers and the latter does not. A solution based on the measurement of the true wafer temperature should preferably be a multi-point method, measuring at least the center and the edge of the wafer.

More evidence

Overwhelming evidence showing that varying wafer emissivities have a significant influence on the energy balance was also obtained by simply monitoring the output power of the lamps. In a number of runs only the center thermocouple was used to control the temperature, while the other thermocouples floated. The temperature reading of these thermocouples is a sensitive indication of the output power of the lamps. In the same experimental set up as in the last experiment, with only the upper heater switched on, the temperature reading of the most downstream thermocouple was recorded during the deposition. Three identical deposition runs were performed, one without a wafer on the susceptor, the next one with a reference wafer and the last one with a Unibond wafer. Care was taken to ensure that the starting conditions were the same in every run. Fig.6 shows the temperature readings versus time for the three wafers. The deposition started at time=0. The section from -10 to 0 min demonstrates that the temperature transients had stabilized and that a steady state was reached before the deposition started. The thin line at 625°C represents the temperature of the control thermocouple at the center. In all three cases the control system managed to keep the temperature exactly at the 625°C setpoint. The three temperature plots of the floating thermocouple clearly show that neither growth on the Si-coated susceptor nor growth on the reference wafer changed the emissivity as the lamp power remained constant during the deposition time. This confirms the assumptions made above about the emissivity of the susceptor. The growth on the Unibond wafer differed, showing a continuously varying lamp intensity during the growth of the layer. The intensity increased in the first fifteen minutes of the growth, reaching a maximum, before it decreased more and more towards the level of the n+ reference wafer. This once again demonstrates that the emissivity changes affect the heat balance, thus causing the wafer temperature to vary. Fig.6 also indicates that the coated susceptor had a higher absorption for the lamp radiation than the n+ wafer. The oxide and epi layers on the Unibond wafer made its absorptivity for lamp radiation worse. This does not necessarily mean that the Unibond wafer also emitted less energy. Absorptivity and emissivity are only the same at the same wavelength. The absorbed lamp radiation has a shorter wavelength than the emitted radiation and in this case the absorptivity and emissivity are generally not the same.

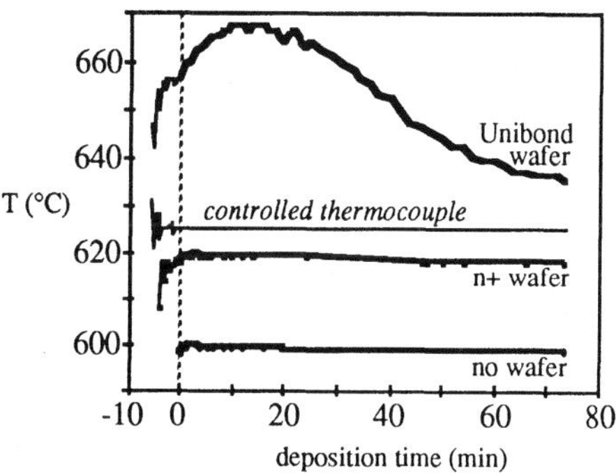

Fig.6 - The temperature reading of a floating thermocouple in the stationary susceptor indicates the intensity of the lamp radiation during deposition at a setpoint of 625°C. The intensity changed only during deposition on a Unibond wafer, confirming that emissivity driven temperature variations play a role in Si epitaxy.

CONSEQUENCES

The consequences of all this for everyday epi practice are severe. The growth rates, the dopant incorporation and the temperature distribution across the wafer change with any emissivity change. The result is that a set of deposition characteristics is pattern-specific. However, during Si SEG and during growth of epitaxial Si on bare substrates the emissivity does not change during the run. After careful calibration of the growth rate, dopant concentration and uniformity across the wafer, the deposition variables valid for a certain pattern remain stable. The situation is a lot worse for blanket growth on patterned wafers. During the growth the emissivity changes due to the increasing thickness of the poly layer on the dielectric. This results in variations in the growth rate and in the dopant and Ge incorporation during the process, which are difficult to control. The magnitude of the effects is determined by the optical thickness of the dielectric, the coverage of the pattern, the thickness of the epi/poly layer and, last but not least, the characteristics of the temperature control system.

Growing blanket SiGe layers on patterned wafers is as difficult as growing Si layers. The growth of SiGe epi layers on bare Si substrates also changes the emissivity, due to the difference in refractive index between Si and SiGe, as opposed to the growth of Si epi. So far we have not performed dedicated experiments to check how serious the temperature deviations are in this case. SIMS plots of the Ge content vs. layer thickness of SiGe layers containing 10-20at% Ge, grown under similar adverse conditions on bare Si substrates, lead us to believe that the emissivity effects are in this case minor, and remain within the the error of measurement of SIMS.

CONCLUSIONS

Emissivity effects during Si epitaxy never have been considered, in spite of the fact that they are significant. The temperature of the susceptor has always been regarded as a reliable measure of the wafer temperature. Indications that temperature variations have an appreciable effect were systematically ignored. We have demonstrated that emissivity effects are important in cold-wall epitaxial reactors. The effects are particularly harmful at low temperatures and reduced pressures.

By using bonded wafers we separated temperature effects from chemical loading effects. When patterned wafers are deposited the two effects mix, which makes it very difficult to conclude that temperature differences are part of the problem. Several authors have in the past tried to explain the growth rate deviations on patterned wafers in terms of chemical loading effects only; this paper reveals that temperature variations should also be taken into account (1,7,8).

ACKNOWLEDGMENTS

This work was partially supported by the European Commission (Esprit project 23229-BETA). The authors would like to thank J.G.M. van Berkum and W.M. van de Wijgert for SIMS support and H. Pijpers and M.A. van den Berg for assistance in processing the wafers.

REFERENCES

1. W.B. de Boer, D. Terpstra and R. Dekker, in *Epitaxy and Applications of Si-based Heterostructures/1998,* E.A. Fitzgerald, D.C. Houghton and P.M. Mooney, Editors, PV 533, p. 315, Materials Research Society, Warrendale, PA (1998).
2. W.B. de Boer, D. Terpstra and J.G.M. van Berkum, *Mat. Sc. and Eng. B*, special issue, in press, (1999).
3. W.B. de Boer, M.J.J. Theunissen and R.H.J. van der Linden, in *Rapid Thermal and Integrated Processing IV/1995*, S.R.J. Brueck, J.C. Gelpey, A. Kermani, J.L. Regolini and J.C. Sturm, Editors, PV 387, p. 287, Materials Research Society, Warrendale, PA (1995).
4. T.I. Kamins, *J. Appl. Phys.*, **74**, 5799 (1993).
5. P.J. Timans, in *Advances in Rapid Thermal and Integrated Processing/1995*, F. Roozeboom Editor, NATO ASI Series Vol.318, p. 35, Kluwer Academic Publishers, Norwell, MA (1996).
6. W.B. de Boer and D.J. Meyer, *Appl. Phys. Lett.*, **58**, 1286 (1991).
7. S. Ito, T. Nakamura and S. Nishikawa, *J. Appl. Phys.*, **78**, 2716 (1995).
8. S. Bodnar, C. Morin, J.L. Regolini, *Thin Solid Films*, **294**, 11 (1997).

Suppressed Phosphorus Autodoping in Silicon Epitaxy for Ultrasharp Phosphorus Profiles by Low Temperature Rapid Thermal Chemical Vapor Deposition

M. Carroll, M. Yang, J. C. Sturm
Department of Electrical Engineering, Princeton University, Princeton, NJ 08540

The production of ultra-sharp phosphorus doping profiles in silicon by low pressure chemical vapor deposition epitaxy is hampered by autodoping effects, which are attributed to phosphorus segregation and redeposition. In this work the phosphorus autodoping effect is completely suppressed by interrupting the growth process to clean the surface ex-situ while the reactor is purged at the same time. A low-temperature in-situ hydrogen bake (10 torr, 800°C, 1-2 min) after reloading the wafer reduces both carbon and oxygen contamination below secondary ion mass spectroscopy detection limits, and yields a high quality n-type/intrinsic silicon interface that is suitable for device application.

INTRODUCTION

Phosphorus and arsenic autodoping effects in low temperature chemical vapor deposition (CVD) of epitaxial silicon have hindered the production of sharp n-type profiles that are desirable for device structures. The autodoping effect has been observed for many n-type silicon dopants sources such as arsenic, phosphorus, and antimony (1), and many different deposition systems such as low pressure chemical vapor deposition (LPCVD), ultra high vacuum chemical vapor deposition (UHVCVD) and molecular beam epitaxy (MBE) (1,2,3). In our laboratory the autodoping effect limits the phosphorus concentration fall off in samples grown by rapid thermal chemical vapor deposition (RTCVD) at 700°C to ~200 nm / decade which is similar to what is observed in other LPCVD systems (3). This is over an order of magnitude slower than the rate of decay of boron profiles, for which 10 nm / decade is achieved using diborane as a dopant source.

Phosphorus autodoping effects are frequently blamed on two potential causes: (i) dopant segregation to the surface (3); and (ii) dopant redeposition from reactor surfaces during growth (4). In this paper we discuss the growth of a sharp phosphorus profile grown in a low pressure, low-temperature rapid thermal chemical vapor deposition (RTCVD) system. Using an interrupted growth process, the sample surface is cleaned in order to remove excess phosphorus with a wet ex-situ etch combined with an in-situ hydrogen bake (5,6), while the reactor is purged with a blank growth in parallel with the ex-situ clean. This concept has been used for a silicon n-channel MODFET structure grown by UHVCVD, but the interrupted interface suffered from oxygen and carbon contamination (2). In this work, an in-situ low-temperature clean, 1-2 minutes at 800°C in 10 torr of hydrogen, reduces oxygen and carbon levels below SIMS detection limits and still yields an ultra-sharp profile. The interrupted interface is found to be nearly indistinguishable to epitaxial silicon grown without interrupting the process characterized by photoluminescence (PL) intensities from strained SiGe layers grown above the cleaned surface and by secondary ion mass spectroscopy (SIMS) of the burried interfaces. Most of this paper will describe the development of this low-temperature (800°C) cleaning process, so that interfaces with low oxygen and carbon can be achieved with out the need of a high temperature bake, which would lead to excessive diffusion of the phosphorus profile.

EXPERIMENTAL PROCEDURE

All structures were grown using a RTCVD cold-wall reactor with a quartz growth chamber heated by halogen lamps located outside the vacuum system (9). The system is pumped by a rotary vane pump and the vacuum integrity relies on o-ring seals. Samples are loaded through a load-lock, also pumped by a rotary vane pump. Source gases used in this work are $Si_2H_2Cl_2$, GeH_4, and PH_3. Hydrogen was used as the carrier gas, which was purified through a Nanochem resin based purifier that insures impurity concentrations to be less than 10 ppb. All silicon and SiGe epitaxy was grown at temperatures of 625°C and 700°C respectively and the reactor pressure was maintained at 6 torr with a dichlorosilane (DCS) partial pressure of approximately 0.052 torr. The growth rates were approximately 100 Å/min and 30 Å/min for the SiGe and Si epitaxy , and oxygen and carbon levels are typically below $2x10^{18}/cm^3$ and $5x10^{17}/cm^3$ respectively.

To examine the quality of various cleaning techniques, silicon substrates were subjected to various wet chemical cleans followed by in-situ hydrogen bakes before epitaxy of $Si_{0.8}Ge_{0.2}$ at 625°C (200Å) followed by a 700°C silicon cap (450Å). All buried Si / $Si_{0.8}Ge_{0.2}$ interfaces were characterized using PL from the pseudomorphically strained SiGe layers as a test of the interface quality (8). Photoluminescence spectra were measured from samples at temperatures of 77 K using a liquid nitrogen cooled Ge detector and an argon laser excitation. The pump power density was approximately ~5 W/cm^2. The luminescence intensity from the strained SiGe layer is extremely sensitive to the carrier lifetime in the SiGe layer and at the hetero-interfaces [7,11]. Interface contamination will lead to increased non-radiative recombination of excited carriers and reduce the overall luminescence intensity emitted from the $Si_{1-x}Ge_x$ layer. The integrated luminescence emitted from the capping $Si_{1-x}Ge_x$ layers were, therefore, compared to the integrated luminescence intensity emitted from the bulk silicon to qualitatively describe the quality of the interface.

Buried interfaces were also characterized using SIMS done at Evans East, in East Windsor, NJ. A 3 keV Cs^+ primary ion beam was used to bombard the surface of the sample and obtain the secondary ions. Sputter rates were between 5-15 Ångstroms/second, producing oxygen and carbon detection limits of approximately $2x10^{18}/cm^3$ and $5x10^{17}/cm^3$ respectively for most samples. Sputter rates were determined using profilometry leading to ~5% uncertainty in depth profiles. Chemical species concentrations were measured to within 15% error.

A standard procedure for the ex-situ wet clean was established beginning with the removal of the native oxide from 5-50 ohm-cm resistivity p-type 4" substrates using a ~1 minute dilute HF dip. The surface is then chemically oxidized to both smooth and clean the surface by consuming the top 5-20Å of silicon surface (10). The oxide is removed using a dilute HF dip, which leaves the surface hydrogen terminated (11). Preparation of the final wet oxidation of the surface was predominately done by emersion of the wafer in H_2SO_4:H_2O_2 3:1 at 70°C. A standard RCA (12) clean was found to be slightly worse as measured by photoluminescence of subsequently grown SiGe/Si capping layer. DI rinses of the wafer surface after the last oxide removal were avoided in all cases except noted, as there exists compelling evidence in the literature that post HF dip rinses reoxidize the silicon surface (13). The DI total organic content was 45 ppb and the resistivity was 18 MOhms-cm.

Following the wet clean the wafer is introduced to the reactor on a single 4 inch wafer quartz stand through a load-lock, which is evacuated to ~50 mtorr by a standard rotary-vane mechanical pump. The wafer is put through a minimum of 4 pump-purge cycles in the load-lock, before the wafer is introduced to the growth chamber. The pump-purge cycle consists of filling the load-lock to ~ 1 torr with dry nitrogen before evacuation. One cycle approximately 5 minutes. After purging the load-lock, the wafer is transferred to the growth chamber which is kept at between 1-10 torr of hydrogen. Immediately following the wafer transfer from the load-lock to the reactor a flow of 1 slpm of hydrogen is passed through the reactor and the reactor pressure is maintained at 1 torr. Immediately before growth the hydrogen flow rate is increased to 3 slpm and the pressure is raised to 6 torr followed by heating the wafer to the growth temperature of 625°C. When the growth temperature is reached, growth is begun by injecting dichlorsilane into the reactor chamber followed by germane injection approximately 5 seconds later, to give a SiGe layer growth rate of 100 Å/min.

Germane is known to react with silicon dioxide to form the volatile species GeO_x. Oxide removal using germane at temperatures of 650-700°C has been reported (16). This report found, however, that for sub-monolayer oxides, germanium adsorbs preferentially to the bare silicon surface rather than forming the volatile germanium-oxide. The sequence of dichlorosilane followed by germane was chosen to allow the SiGe layer to grow as soon as the germane is injected. Because the SiGe layer grows quickly on the silicon surface with no observably long incubation time, and the germane molecule prefers the open silicon surface site over that of the oxide site, it is concluded that the germane induced oxide desorption plays a negligible role in the measured oxygen concentrations found at the buried Si / SiGe interfaces. It is, later, conclusively shown that the low temperature clean technique used to grow the ultra-sharp phosphorus profile does not depend on germane.

EFFECT OF HF CONCENTRATION AND HYDROGEN BAKES

A study of the HF concentration used in the last dilute HF dip of the wet clean on the surface quality was done. The wet cleaning described earlier was reformulated, in which the HF concentration in the final 20 minute dilute HF dip used to strip the chemical oxide was varied from 50-0.1% (pH=0.4-3). After the HF dip, the wafer is moved to the load-lock from the chemical hood. The hydrogen passivated surface is exposed to laboratory atmosphere for 5-20 minutes after removal from the etch solution. High quality interfaces were achieved even after 15-20 minutes of exposure to air, indicated by intense PL from $Si_{1-x}Ge_x$ layers grown above the exposed surface. A monotonic increase in the relative PL intensity from the $Si_{0.8}Ge_{0.2}$ layer was found as the HF concentration was decreased from 0.4-3 (figure 1). Surface morphology roughens as the pH of the HF solution is lowered from 7.8 to 0.4 due to 111 micro-faceting of the etched silicon surface (15). The PL intensity dependence on pH may be linked to the 111 microfaceting of the surface because of the surface roughness at low pH, but chemical contamination of the surface can not be ruled out as an explanation.

Even after using the optimum HF concentration for the ex-situ clean the PL intensity is not as intense as from SiGe layers grown on in-situ grown buffer layers without interrupt. In order to further improve the surface quality after the ex-situ clean we examine in-situ hydrogen baking to further clean the surface. As a first step the condition under which hydrogen does not add more contamination to the interface is examined. To determine whether the hydrogen ambient in the growth chamber contributes any contamination to the wafer surface. After the growth of epitaxial buffer layers the growth was interrupted. The reactor pressure was set between 0.5-250 torr in pure H_2 and the wafers were heated to 700°C for 2-15 minutes. The pressure was then adjusted to 6 torr and DCS was added to resume growth. For pressures equal to and above 6 torr detection of oxygen and carbon were limited by the sensitivity of SIMS. It is believed that at 700°C the desorption rate of oxygen and carbon from the silicon surface is very slow (4), so these measurements only reflect adsorption. Adsorption rates were determined from the integrated carbon and oxygen concentrations found at the interface, determined by SIMS, divided by their respective annealing times (figure 2). The observed increase in detected oxygen and carbon measured at the test interfaces exposed at hydrogen pressures below 6 torr is thought to be due to reduced hydrogen termination of the silicon surface at lower hydrogen pressures (16). The increase in open silicon surface sites, at low hydrogen pressures, will increase the sticking coefficient of impinging oxygen and hydrocarbons from the reactor atmosphere during the hydrogen bake, and therefore leading to increasing oxygen and carbon absorption rates.

To further clean the surface, after the optimum dilute HF dip (0.1% HF), wafers were then subjected to an in-situ bake in hydrogen before growing an epitaxial SiGe/Si cap. A maximum temperature of 800°C was used to minimize the thermal budget of the in-situ clean. A 700°C, 10 minute bake at 6 torr was found to give no improvement of the quality of the Si/SiGe interface as probed by PL from the SiGe layer (figure 2) consistent with ref. 4 and 5. The wafer is heated to 800°C for 1 minute at 0.5-250 torr after which the temperature is reduced to 625°C while stabilizing the pressure to 6 torr. After both pressure and temperature are stable DCS and GeH_4 were injected into the reactor to begin growth of the SiGe / Si capping layer. PL from the thin 200 Å $Si_{.8}Ge_{.2}$ as a fuction of pressure is shown in figure 3, and the corresponding integrated oxygen and carbon concentrations, determined by SIMS, for the same samples are shown in figure 4. The PL intensity is clearly brightest for 1-10 torr bakes, and improved over the ex-situ clean without the hydrogen bake, although the intensity is still slightly dimmer than the PL from an uninterrupted growth of a SiGe/Si layer grown on a silicon buffer layer. Oxygen and carbon concentrations are found to be lower than those without hydrogen bake, representing a clear reduction of oxygen and carbon over only the ex-situ clean. A 2 minute hydrogen bake at 800°C at 10 torr resulted in an interface indistinguishable, by SIMS, from that of an uninterrupted growth.

PHOSPHORUS PROFILE AND DEVICE APPLICATION

The combination of an ex-situ clean and an in-situ clean, have been used to demonstrate sharp phosphorus profiles with low interface contamination. This process has been used to make an ultra-sharp profile for the channel doping in a vertical p-channel MOSFET. Channel lengths as short as 25 nm have been obtained, which rely on the sharp edge of the phosphorus profile (17). The process flow used to grow the ultra-sharp phosphorus profile started first with growth of a thin 300 Å phosphorus doped layer grown at 700°C at a growth rate of 30 Å/min and a phosphorus concentration of $10^{18}/cm^3$ at 6 torr. The growth was halted immediately after the growth of the phosphorus layer was finished (after the phosphine was turned off). The phosphorus doped sample was then removed from the reactor and a blank wafer was loaded into the reactor on the single wafer quartz sample holder. The reactor is then purged by growing an undoped sample, while the phosphorus doped sample surface is oxidized in H_2SO_4:H_2O_2 (3:1) solution at 70°C and then etched in a dilute HF:DI (1% HF in DI) dip. This process was repeated several times in order to insure that the potentially phosphorus rich surface was completely removed. After the last HF dip the wafer was reloaded, as described earlier into the purged reactor. To obtain a contamination free surface with minimal phosphorus diffusion the top surface of the sample was subjected to a short 1 minute, 800°C hydrogen bake at 10 torr, after which the pressure was reduced to 6 torr and the temperature was dropped to 700°C. DCS was injected into the growth chamber as soon as the temperature and pressure stabilized, approximately 30 seconds after the hydrogen bake was finished, initiating the growth of a thin intrinsic silicon layer grown directly above the etched surface to complete the sharp turn-off of the phosphorus profile. In figure 5a, the phosphorus profiles of the uninterrupted process is compared with the interrupted process. The uninterrupted process clearly shows the long phosphorus concentration decay length of 200 nm / decade, whereas the interrupted process completely suppresses the autodoping effect and produces a phosphorus concentration decay length of 17 nm / decade. The oxygen and carbon concentrations at the interface, as seen in figure 5b, are below SIMS detection limits. The best, previously reported, phosphorus concentration decay length for low pressure chemical vapor deposition that is oxygen and carbon free at the top phosphorus edge, that the authors are aware of, is 100 nm / decade of phosphorus concentration (3).

CONCLUSION

In conclusion, phosphorus profiles were grown with a 17 nm / decade fall off at the top edge as measured by SIMS. The phosphorus profile was produced using an interrupt process, which included removing the wafer from the growth chamber to both purge the reactor and to etch the top surface of the sample, before intrinsic silicon was grown above the phosphorus layer. In order to produce high quality silicon epitaxy above the phosphorus doped layer while maintaining a low thermal budget for the ultra-sharp phosphorus profile, a low temperature cleaning technique was developed for the Princeton RTCVD system. A 1-2 minute, 800°C bake of the silicon surface in 10 torr of hydrogen is used to reduce oxygen and carbon concentrations at the interface to below SIMS detection limits, for high quality silicon epitaxy on the phosphorus surface leaving an interrupt interface nearly indistinguishable from an epitaxial layer grown without interrupt. This cleaning result is the lowest thermal budget, that the authors are aware of, for a low pressure chemical vapor system that reduces both carbon and oxygen below SIMS detection limits and is the shortest phosphorus concentration fall off with undetectable oxygen and carbon contamination near the phosphorus profile, known to the authors.

ACKNOWLEDGEMENTS

This work was supported by ONR and DARPA. The authors would like to also thank T. Büyüklimanli at Evans East for his assistance with SIMS analysis.

REFERENCES

1. Friess, Nützel, Abstreiter, APL 60 (18), 4 May 92
2. Ismail, Rishton, Chu, Chan, Meyerson, EDL, vol 14, No 7, July 93.
3. S.K. Lee et al., APL 57 (16), 15 October 1990
4. Greve, Mat. Sci & Eng. B18 (1993) 22-51
5. M. K. Sanganeria, M. C. Öztürk, K. Violette, G. Harris, C. A. Lee and D. Maher, Appl. Phys. Lett. 66 (10), 6 March 1995, 1255-57
6. Wolansky, Tillack, Blum, Bolze, Glowatzki, Köpke, Krüger, Kurps, Rittter, Schley, ECS Meeting, Spring 1998.
7. J. C. Sturm, P. V. Schwartz, E. J. Prinz, H. Manoharan, J. Vac. Sci. Tech B9, 2011 (1991)
8. A. St. Amour, J. C. Sturm, Y. Lacroix, M. L. W. Thewalt, APL 65 (26) 25 Dec 1994 3344-46
9. Z. Matutinovic-Krstelj, E. Chanson, J. C. Sturm, J. of Electronic Materials, vol 24 No6 95
10. W. Kern, RCA Engineer, RCA Review, 31, 187, (1970)
11. C. W. Trucks, K. Raghavachari, G. S. Higashi, Y. J. Chabal, Physical Review Letters, 65 (4), 23 July 1990, 504-507

12. W. Kern, D. Puotinen, RCA Review, 31, 187, (1970)15. T. Takahagi, A. Ishitani, H. Kuroda, Y. Nagasawa, J. Appl. Phys. 69 (2), 15 Jan 1991 (803-7)
13. K. Endo, K. Arima, T. Kataoka, Y. Oshikane, H. Inoue, Y. Mori, APL, 73 (13), 1998
14. C.-L. Wang, S. Unnikrishnan, B.-Y. Kwong, A. F. Tasch, J. Electrochem. Soc., vol. 143, No 7, July 1996.
15. Dumas, Y. J. Chabal, P. Jakob, Surface Science, 269/270 1992 867-78
16. P. V. Scwartz, J. C. Sturm, J. Electrochem. Soc., vol 141, No. 5, May 1994.
17. M. Yang, C-L. Chang, M. Carroll, J. C. Sturm, (to be published), Electron Device Letters, June 1999.

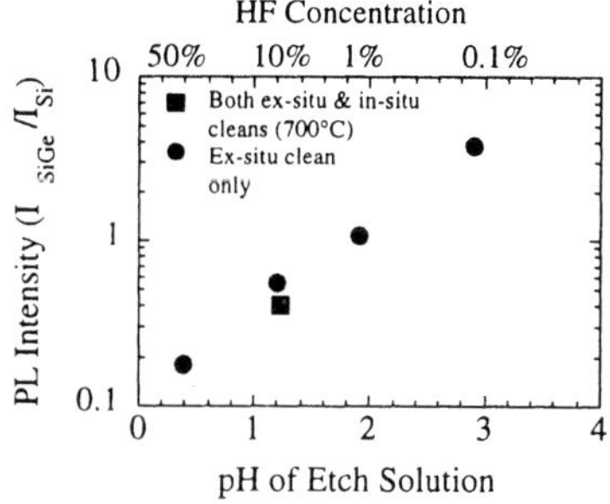

Figure 1. Photoluminescence ratio, which reflects interface quality, as a function of HF dip pH before the growth of the capping SiGe/Si layer. One sample was baked after the ex-situ clean (pH=1.3) at 700°C for 10 minutes in 6 torr of hydrogen before the SiGe/Si cap was grown.

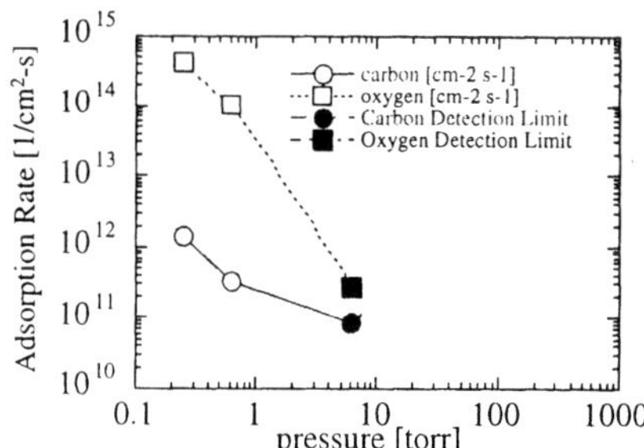

Figure 2. Oxygen and carbon adsorption rates in RTCVD reactor t 700°C as a function of hydrogen pressure. The points at 6 torr were limited by SIMS resolution.

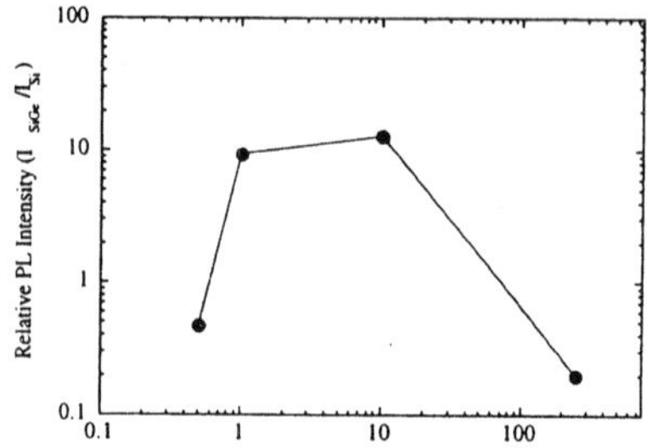

Figure 3. Relative photoluminescence intensity from SiGe layers grown directly above ex-situ cleaned silicon substrates exposed to an in-situ 1 minute bake at 800°C in 0.5-250 torr of hydrogen, before the SiGe/Si epitaxial cap is grown. The relative photoluminescence intensity from SiGe layers grown on in-situ grown buffer layers without interrupt are approximately 30-40.

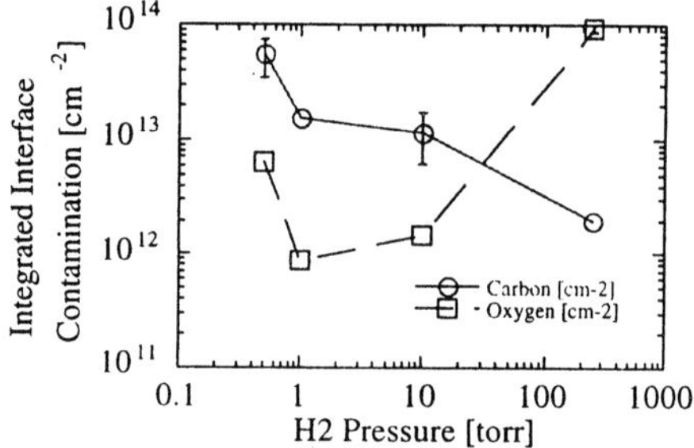

Figure 4. Oxygen and carbon concentrations, determined by SIMS, at the interface between the substrate surface and SiGe/Si cap, in the samples of figure 3. The oxygen and carbon concentrations are plotted as a function of hydrogen pressure during the in-situ 1 minute 800°C bake, before the SiGe/Si cap is grown on the ex-situ cleaned silicon substrate.

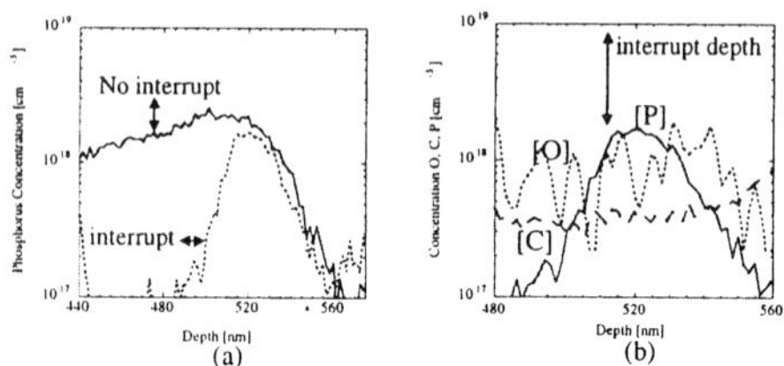

Figure 5. (a) phosphorus profiles grown at 700°C using phosphine with and without an interrupt process to suppress the phosphorus autodoping effect, showing improvement of phosphorus doping fall-off from 200 to 25 nm/decade. (b)Tthe carbon, oxygen and phosphorus profiles from the interrupt process. No carbon or oxygen are detected at the interrupt interface.

COMPARATIVE STUDY OF CRYSTALLINITY AND SURFACE ROUGHNESS OF RTCVD VERSUS LPCVD DEPOSITED POLYSILICON

J.W.H. Maes[1], C. Pomarede[2], M. Mansoori[3], C.W. Werkhoven[3], I.J. Raaijmakers[3]
[1]ASM Europe, Rembrandtlaan 7-9, 3723BG, Bilthoven, The Netherlands
[2]ASM France, 175 Rue Caducee, Parc Euromedecine, 34195 Montpellier Cedex 5, France
[3]ASM America, 3440 E. University Drive, Phoenix, Arizona 85034, USA

A comparison is presented of the crystallinity and surface roughness of undoped and *in-situ* doped RTCVD and LPCVD silicon films. Results show fundamental differences between single-wafer based RTCVD versus batch-type LPCVD deposited films. With RTCVD, excellent film properties in terms of crystallinity, roughness and step coverage are obtained while using substantially higher temperatures and silane pressures than standard for LPCVD. The capability of processing at higher temperatures and silane partial pressures and hence higher deposition rates, enable the use of RTCVD polysilicon for certain device applications.

INTRODUCTION

Since long, low-pressure chemical vapor deposition (LPCVD) is a well-established technique for deposition of undoped polysilicon and *in-situ* phosphorus doped layers. However, batch-type LPCVD of polysilicon has specific limitations, which make single wafer rapid thermal chemical vapor deposition (RTCVD) an interesting alternative (1). With RTCVD *in-situ* doping with arsenic can be done in a straightforward way and recently it was shown that atmospheric pressure silane-hydrogen based RTCVD has exceptionally good gap-filling capabilities for *in-situ* arsenic and phosphorus doped polysilicon films (1). Applications where these characteristics are important are arsenic doped poly-emitters for BiCMOS, DRAM trench capacitor fill and (phosphorus-doped) polysilicon contact fill. The feasibility of this process for poly-emitter and poly-gate applications alrady was demonstrated in Refs. (2) (3). Recently, an ultra-low leakage current was demonstrated using an ultra-thin gate stack technology based on CVD of Si_3N_4, in which the integration of RTCVD Polysilicon is an essential part (4).

In this paper results on the crystallinity and surface roughness of undoped and *in-situ* doped RTCVD silicon films are presented. As a reference, data will be compared with that of LPCVD films. The diffusion of dopants and the electrical characteristics are strongly dependent on the microstructure of the layers. Surface roughness is important in applications where lithography, etching and trench filling are critical. Knowledge of and control over the roughness and crystallinity of silicon layers are essential for successful application of RTCVD films in logic or memory devices. Results suggest that the properties of RTCVD versus LPCVD deposited layers show fundamental differences, which enable application of RTCVD films in logic or memory devices.

EXPERIMENTAL

RTCVD layers were deposited in an ASM Epsilon 2500RTP cold wall, lamp heated, single wafer reactor (5) using H_2-SiH_4-PH_3 or -AsH_3 gas mixtures, at atmospheric pressure conditions. Using this laminar flow reactor, layers were deposited in the temperature range 600°C - 700°C and with different doping levels. Typical flows were 20 slm H_2 and 600 sccm SiH_4. As dopant gasses 1% phosphine and 1% arsine diluted in H_2 were used. Undoped and doped LPCVD layers were deposited in an ASM Advance 400 vertical furnace at 550°C and 623°C. Typical conditions used in this process are 1 slm SiH_4 (no carrier gas) and a pressure of 500 mTorr. All films were deposited on 100 nm thick thermal oxide layers on 150 or 200 mm diameter silicon wafers. Temperatures mentioned for RTCVD are thermocouple temperatures and are within 10°C of the wafer temperature.

Ellipsometry measurements were carried out in the spectral range 200-800 nm. A model consisting of a three-layer structure was fitted to the spectra: 100 nm thick thermal oxide, a silicon layer of unknown degree of crystallinity and thickness and a surface layer (6). Good fits are obtained by considering the silicon layer as a Bruggeman effective medium of amorphous and crystalline silicon (7). The crystalline fraction obtained from the fit was used as a qualitative indicator of the crystalline state (amorphous or polycrystalline) of the layers. Some films also were characterized by transmission electron microscopy (TEM).

The ellipsometric spectra also give an indication of the surface roughness (8) (9). This is because at small wavelength the silicon is nontransparent. Smooth layers could be modeled well using a native oxide surface layer, while for layers showing surface roughness a surface layer consisting of a Bruggeman effective medium of voids and silicon gave a better fit. The thickness of the surface layer is used as qualitative indicator of the roughness, at scanning the process window of RTCVD. To obtain quantitative roughness data, selected number of layers were also analyzed by atomic force microscopy (AFM). Since roughness and crystallinity may depend on layer thickness, a fixed thickness of 300 nm was used.

RESULTS

LPCVD films deposited at 623°C and 550°C are discussed, which are two commonly used temperatures. The first temperature is preferable from a deposition rate standpoint. At 623°C a deposition rate of 12 nm/min is obtained, compared to 2.5 nm/min at 550°C. In Fig. 1, cross-sectional TEM and plan-view scanning electron microscope (SEM) images of LPCVD layers are shown. The 623°C layer is polycrystalline. Close inspection of the TEM images shows cone-shaped, columnar grains. The large number of grain boundaries and the high density of defects inside the grains allow for controlled drive in of dopants from the surface or diffusion of implanted dopants throughout the layer. *In-situ* doping of the layers cannot be done in a straightforward way at 623°C because of difficulties with down-boat and across-wafer uniformities (10).

The LPCVD layer deposited at 550°C (see Fig. 1) is fully amorphous. No crystalline grains are visible. Despite the lower deposition rate, such layers are interesting for several reasons. Their structure is benign for controlled implantation of dopants and the layers are relatively smooth which is critical in many applications. At 550°C *in-situ* doping can be done in a uniform and controlled way, using well-designed injectors for feeding phosphine (diluted in N_2) into the reactor. The transition from amorphous-to-poly (α-poly) takes place in the range 580°C-600°C (10).

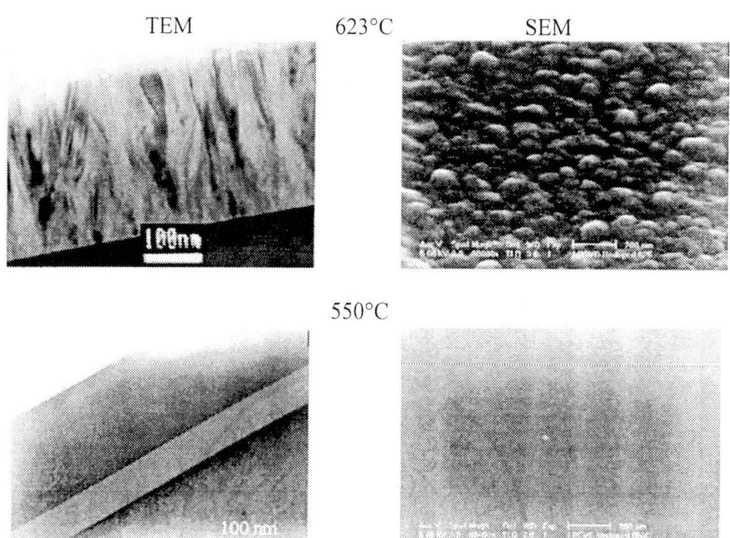

Fig. 1: Cross-sectional TEM and plan-view (at 45° angle) SEM micrographs of 300 nm thick LPCVD deposited silicon films.

Fig 2 shows an Arrhenius plot of the deposition rates of the LPCVD and RTCVD processes. RTCVD gives higher rates than LPCVD because of the high silane partial pressure used, typically 25 Torr versus 0.5 Torr. For equal temperatures, deposition rates are achieved with RTCVD that are at least 5 times higher than those of LPCVD. The maximum allowable silane partial pressure and deposition rate is determined by depositions on the quartz walls, which are unwanted in a lamp-heated reactor. The activation energy found for the atmospheric process is approximately 40kcal/mol, which is a common value for silane based silicon deposition in a surface reaction limited regime (11). Adding dopants to the process reduces the deposition rate due to poisoning effects. For phosphine the effect is rather small (<15%), while for arsenic the decrease still is only a factor 2 for saturated layers (1).

In Fig. 3 crystalline fraction as derived from a Bruggeman effective medium model is shown for various process conditions. A low percentage is indicative of an amorphous film, a high percentage of a polycrystalline film. Fig.3a shows values for undoped films For RTCVD a transition from amorphous to poly-crystalline is found in a narrow temperature window around 650°C. This temperature is much higher than the transition temperature for LPCVD. Fig. 3b shows the effect of adding dopants to the RTCVD process. At 650°C, a transition from an amorphous to a polycrystalline growth mode is found at increasing the dopant flow. Apparently for phosphorus-doped films the amorphous-to-poly (α-poly) transition temperature is lower, and is probably doping concentration dependent. The films deposited at 680°C all are polycrystalline.

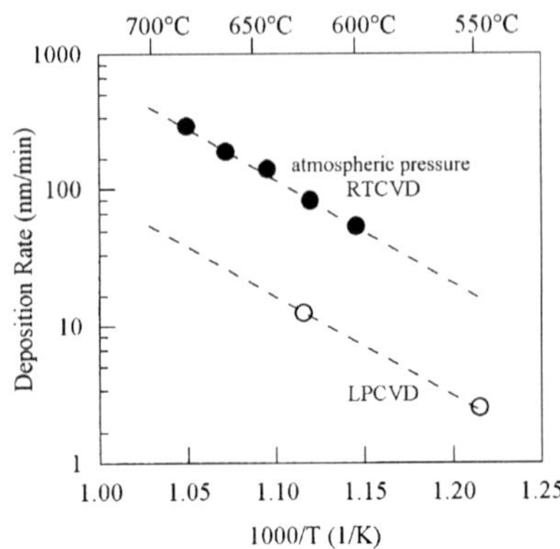

Fig. 2: Arrhenius plot of the deposition rate for RTCVD and LPCVD deposited layers.

In Fig 4, TEM and SEM micrographs of RTCVD layers deposited at three temperatures are shown. For the highest temperature, images for undoped and doped films are shown. The layer deposited at 620°C is amorphous. At the interface with the oxide, only some very small inclusions are visible which probably are small crystallites. In the layer deposited at 650°C, much larger crystallites are visible which confirms that the α-to-poly transition indeed takes place at this temperature. In LPCVD similar layers are obtained at ~580°C. The layer deposited at 680°C is fully poly-crystalline. On the image a narrow grain is visible, extending through the entire height of the film. The photo suggests that the grain size in this layer is larger than that in 623°C LPCVD poly.

The structure of the lightly doped layer is similar to that of the undoped film deposited at 680°C. On some spots in the film grains with a length comparable to the film thickness are visible. Close inspection of the images of the highly P-doped layer show that the grains in this case are very compact. All grains visible have sizes significantly smaller

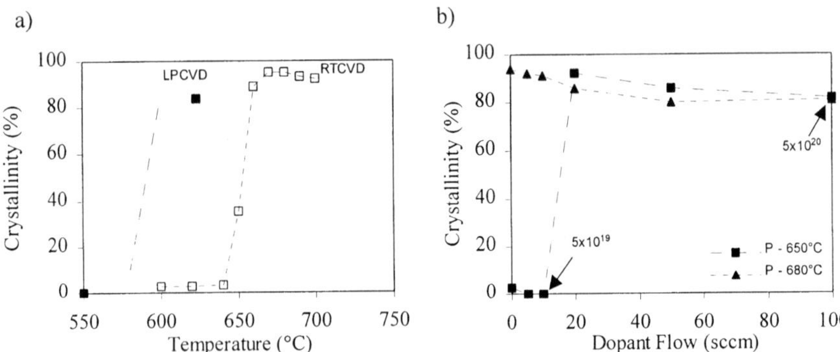

Fig. 3: Crystalline fraction for RTCVD and LPCVD deposited layers. a) Data for different temperatures. b) Effect of adding phosphine for RTCVD. Also shown are two dopant concentrations for the 650°C curve, as measured by SIMS.

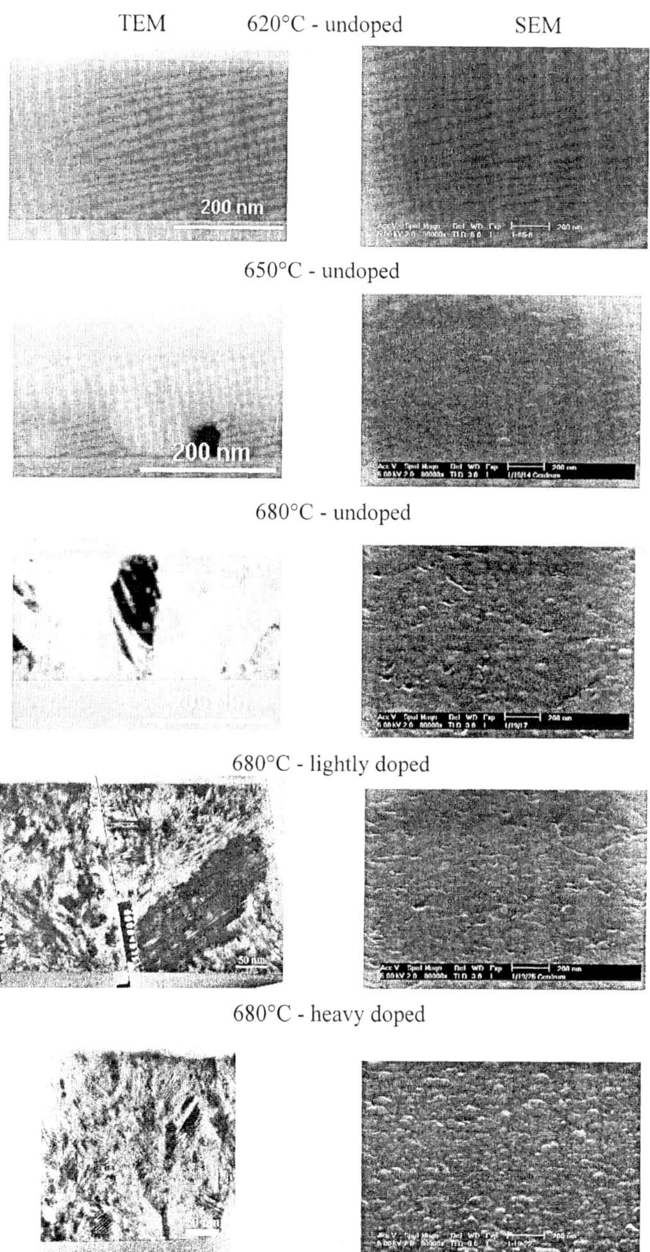

Fig 4: TEM and SEM micrographs of RTCVD layers deposited at three temperatures and doping levels.

than the film thickness. This suggests that grains become smaller at increasing dopant flow.

The TEM and ellipsometry data in Fig. 3 and 4 suggest a correlation between the grains size and the crystalline fraction (range 80-100%) obtained by ellipsometry. Films with a small grain size show a lower crystallinity percentage than films with larger grains. Note that polysilicon films made obtained by annealing of amorphous layers, show a crystallinity value close to 100%. These films have very wide (~1 μm) block-shaped grains. One can imagine that grain boundaries and defects in polysilicon make the optical properties of the film more amorphous like.

In Fig. 5, surface roughness data obtained by ellipsometry are shown. In Fig. 5a data for undoped films are plotted. The open symbols indicate that the surface is described best with a native oxide film, the solid symbols by a roughness layer. The polycrystalline RTCVD films clearly show more surface roughness than amorphous films, likely due to facet formation on grains. The roughness of the poly films increases with temperature. RMS values of the roughness measured by AFM are shown in Table 1. The same trend as in the ellipsometry data is found. An important result is that RTCVD films deposited below 700°C all are less rough than the value found for 623°C LPCVD. Note that the roughness of films is layer thickness dependent. In Table 1 it is shown that the RMS roughness of a 200 nm thick 623°C LPCVD film is only half that of a 300 nm thick film.

Ellipsometry data for *in-situ* doped films are shown in Fig. 5b. The roughness of the films increases with increasing dopant flow. The same trend is visible on the SEM images and is also found by AFM. For 650°C, the roughness increase starts above the α-poly transition. Doped films deposited at 680°C and below all are show less roughness than 623°C LPCVD poly. To achieve roughness values comparable to that of 550°C LPCVD amorphous, a temperature of 650°C or less is required.

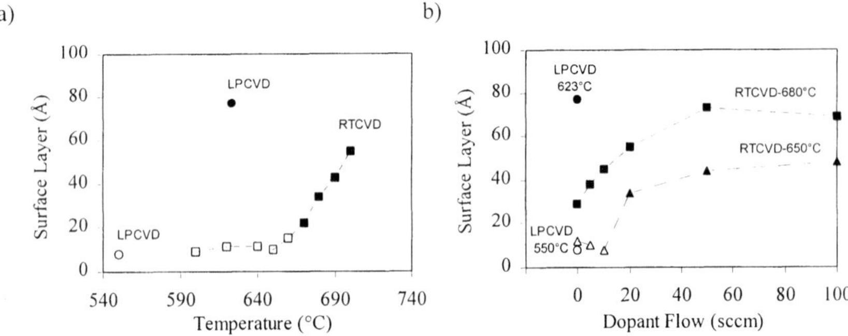

Fig. 5: Ellipsometry data of the surface roughness. a) Data for undoped films. b) Data for doped films.

DISCUSSION

RTCVD versus LPCVD deposited layers show fundamental differences in terms of deposition rate, crystallinity and surface roughness. To be cost effective, much higher deposition rates are required for RTCVD than for batch-type LPCVD. Deposition times typically should be 2 minutes or less for films in the range 100-300 nm. Typically 50-150

Table1: Surface roughness measured by AFM.

	Thickness (nm)	T (°C)	1%PH_3 flow (sccm)	RMS roughness (nm)
LPCVD	300	550	-	NA
LPCVD	200	623	-	3.2
LPCVD	300	623	-	6.8
RTCVD	300	620	-	1.9
RTCVD	300	650	-	1.7
RTCVD	300	680	-	2.6
RTCVD	300	700	-	7.8
RTCVD	300	680	100	5.6

NA: not available

nm/min is required, which sets a lower limit on the allowable temperature. Results indicate that this limit is ~600°C for undoped films. However, thanks to the shift of α-to-poly transition temperature to 650°C, it still is possible to grow amorphous films (e.g. for gate-stack applications) at high enough rates. The crystallinity data suggest that temperatures up to 640°C can be used, where a deposition rate of more than 130 nm/min is obtained. The shift of the transition temperature suggests a reduced surface diffusion length of arriving silicon species in RTCVD compared to LPCVD. Sufficient surface diffusion is critical for obtaining a poly-crystalline growth mode. Reasons for the reduced surface diffusion length in RTCVD (for equal temperatures) are the high deposition rate and a relatively large amount of hydrogen adsorbed on the surface (12). The shift is not related to gas purity because in the same reactor high quality epitaxial growth was achieved with silane–hydrogen mixtures.

Undoped poly-crystalline films are obtained at temperatures above 660°. Undoped poly films deposited at 680°C were found to be relatively smooth compared to LPCVD undoped poly. Temperatures up to 700°C can be used without obtaining roughness values higher than that of 623°C LPCVD poly. The presence of some large grains in these films can be a disadvantage for introducing doping by diffusion of implantation. Further tuning of the process may solve this problem. The decreasing crystalline fraction with increasing temperature suggests a decrease of grain size with increasing temperature.

At adding phosphine to the process, the α-poly transition temperature shifts to lower values and films show more roughness. Also in LPCVD deposited films phosphine is found to enhance crystallization (11). The increase in roughness and decrease of the α-poly transition temperature probably are interrelated. Results for undoped films showed that roughness increases when moving further away from the transition temperature. Roughness values for doped films deposited at 680°C still are lower than those of LPCVD poly. A reduction of the roughness is achieved by reducing the temperature to 650°C. At this temperature, relatively smooth doped poly films are deposited at rates (>100 nm/min) that are still acceptable. Further increasing the silane partial pressure also may reduce the surface roughness. For *in-situ* doped amorphous films RTCVD seems less suitable, unless relatively thin (≤100 nm) films are required.

CONCLUSIONS

The properties of films obtained with RTCVD and LPCVD show fundamental differences. With RTCVD, excellent film properties in terms of crystallinity, roughness and step coverage are obtained, while using substantially higher temperatures and silane pressures than standard for LPCVD. The capability of processing at higher temperatures and silane partial pressures, and hence higher deposition rates, enables the use of RTCVD polysilicon for certain device applications.

ACKNOWLEDGMENTS

We thank M. Schaekers of IMEC and A. Gschwandtner of Siemens for their support in film characterization.

REFERENCES

1. C. Pomarede, J.W.H. Maes, A. Grassl, A. Gschwandtner, I.J. Raaijmakers, *in Proc. 6th Intl. Conf. on Adv. Therm. Proc. of Semicond. – RTP'98*, 120 (1998).
2. S. Niel et. al., in *proceedings IEDM/1997*, 807 (1997).
3. D. Bensahel, Y. Campidelli, C. Hernandez, F. Martin, I. Sagnes, D.J. Meyer, *Solid State Techn.*, March 1998, s5 (1998).
4. C. Pomarede, C. Werkhoven, J. Weidmann, T. Bergman, A. Gschwandtner, M. Houssa, *MRS spring meeting /1999*, to be published, (1999).
5. I.J. Raaijmakers, D.J. Meyer, J. Italiano, in *Proc. 5th Intl. Conf. on Adv. Therm. Proc. of Semicond. - RTP96*, 243 (1996).
6. A. Borghesi, M.E. Giardini, M. Marazzi, A. Sassella, G. De Santi, *Appl. Phys. Lett.*, 70, 892 (1997).
7. G.E. Jellison, Jr. Opt. Mater., 1 ,41 (1992).
8. J.K. Koh, C.R. Wronski, Y. Kuang, R.W. Collins, T.T. Tsong, Y.E. Strausser, *Appl. Phys. Lett.*, 69, 1297 (1996).
9. C. Flueraru, M. Gartner, C. Rotaru, D. Dascalu, G. Andriescu, P. Cosmin, *Microelectron. Eng.*, 31, 309 (1996).
10. B.S. Meyerson, W. Olbricht, *J. Electrochem. Soc.*, 131, 2361 (1984).
11. T. Kamins, *Polycrystalline Silicon for Integrated Circuit Applications*, Kluwer Acedemic Publishers, Massachusetts, (1988).
12. T.I. Kamins, A. Fischer-Colbrie, *Appl. Phys. Lett.*, 71, 2322 (1997).

As PEAKS IN Si (100) FILMS FABRICATED WITH RAPID THERMAL EPITAXY

W.D. van Noort, L.K. Nanver, C.C.G. Visser, A. vd Bogaard, J.W. Slotboom
TU Delft, Dimes/ECTM, Feldmannweg 17, 2628 CT Delft, The Netherlands

A new method is presented for incorperating sharp arsenic n^+ peaks in silicon (100) films that are epitaxially grown in the ASMI Epsilon One reactor. For device fabrication, arsenic is an attractive dopant because it combines low thermal diffusivity with high electrical activation. However, during epitaxial growth very high As concentrations will often adsorb or segregate to the surface and high As autodoping levels may result. This As surface coverage can be completely removed at high temperatures but the corresponding high thermal dopant diffusion is often unacceptable. Optimisation of the growth conditions nevertheless has allowed the fabrication of 500 Å wide As peaks with a maximum doping of 10^{19} atoms/cm^3 and an autodoping level as low as 7 x 10^{16} atoms/cm^3. The latter can moreover be reduced to 3 x 10^{16} atoms/cm^3 by using Ge as a surfactant.

INTRODUCTION

Modern IC processing requires sharp doping gradients. Doping profile engineering during epitaxial growth is a means of combining very sharp gradients with high active doping levels. For example, in SiGe HBT's fabricated by APCVD, phosphorus doped n^+ peaks at the collector-base junction have been successfully grown for the compensation of boron base out-diffusion and for improving the trade-off between cut-off frequency and emitter-collector breakdown voltage [1]. The applicability of phosphorus as a peak dopant is, however, often limited by its high thermal diffusivity and an excessive transient enhanced diffusion at low processing temperatures.

Arsenic would be more attractive than phosphorus as peak dopant. It has a much lower diffusivity and below 900°C both thermal and transient enhanced diffusion are negligible. However, surface segregation and autodoping usually make it impossible to grow steep As doped peaks. In the past, several methods to reduce autodoping have been proposed [4, 5, 6]. These all involve temperatures above 1000°C whereby the resulting doping gradients are determined by thermal diffusion and not the epitaxial peak growth. In this work, methods are presented for reducing the autodoping with a much lower thermal budget, thus allowing the fabrication of narrow highly doped As peaks.

LOW TEMPERATURE ARSENIC DOPED SILICON

All layers are grown in the ASMI Epsilon One epitaxial reactor. The wafers are loaded and unloaded in a nitrogen-purged loadlock, and then placed on a SiC coated graphite susceptor. The quartz reaction chamber is lamp heated. The Si layers are grown from $SiCl_2H_2$ with H_2 as carrier gas and the arsenic dopant source is AsH_3 (4×10^{-4} Torr in atmospheric H_2).

The arsenic doped layers are deposited at 700°C and atmospheric pressure. This low deposition temperature has two advantages. First, the deposition rate is low and a very good control of layer thickness is achieved. Second, the incorporation of arsenic increases considerably with decreasing temperature and a very high doping concentration can be obtained [7]. However, at such low growth temperatures a very high surface

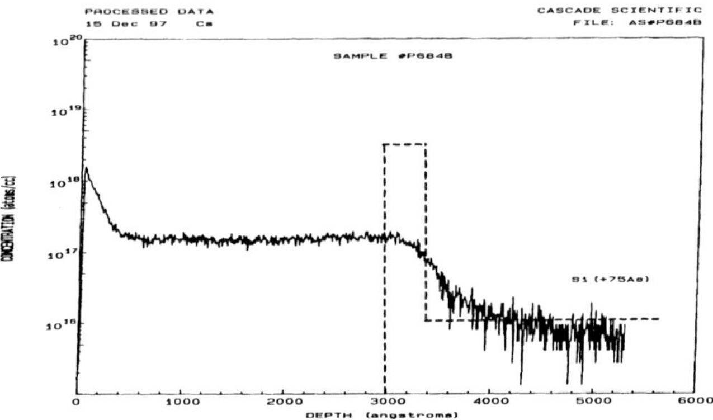

Figure 1: Sims profile and intended profile (dashed) for low temperature As doped Si epitaxy.

concentration of As can adsorb and segregate at the surface [8, 9]. Segregation ratios as high as 10^4 between surface and bulk doping concentrations have been reported [10]. Below 850°C this As (dimer) surface coverage is extremely stable and any desorption at such low temperatures is negligible [8, 9]. If the supply of As is terminated while Si deposition is continued, As from the surface layer will continue to be incorporated. With such a high segregation ratio, several micron of Si must be deposited before the coverage of arsenic on the surface is significantly diminished. This effect prohibits the fabrication of sharp doping profiles at low temperatures and profiles such as shown in Fig. 1 will be the result.

LOW TEMPERATURE ARSENIC DESORPTION

Desorption of segregated As surface layers is commonly performed in the fabrication of arsenic buried layers by growing a cap-layer at high temperature and reduced pressure [5]. A similar approach has been investigated here by performing Si epitaxy after As peak growth at high temperature (1150°C) and reduced pressure (60 Torr). The main disadvantage of this scheme is that the high temperature causes out-diffusion of the original peak. Moreover, the high growth rate at 1150°C also makes it difficult to accurately control the thickness of thin cap-layers. On the other hand, the As autodoping of the cap-layer is low and the processing time is short, only a few minutes.

Examples are given in Fig. 2 where the SIMS profiles show that the autodoping approaches the SIMS detection limit. In this figure the measured profiles are also compared to a SUPREM-3 simulation of a 500 Å wide As peak doped to 2 x $10^{19}/cm^3$ and annealed for 1 min at 1150°C. The resulting profile compares well with the measured profiles, this indicates that thermal diffusion is determining the peak shape.

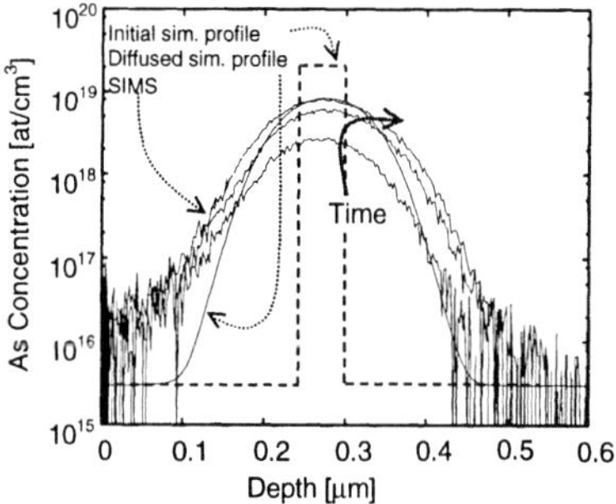

Figure 2: SIMS profiles of As peaks made with the 1150°C cap-layer. Peak growth times 6, 9 and 12 minutes. Smooth line: Simulation (Supem-3) of a 500 Å, box-shaped, $2x10^{19}$ atoms/cm^3 As peak after 1 min at 1150°C. This simulation shows diffusion dominates the peak shape.

The thermal diffusion of As is considerably decreased if the processing temperature is decreased from 1150°C to 1000°C or less. Desorption at 950°C and 1000°C has been investigated by purging the system for 10 minutes at either of these temperatures after peak growth. Reduced pressure (36 - 100 Torr) is also applied and

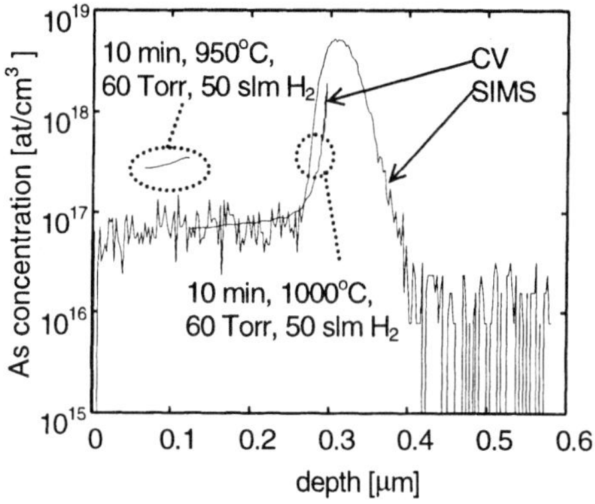

Figure 3: SIMS and CV measurements of doping profiles fabricated with low temperature scheme at 950°C and 1000°C. Peak growth time is 4 min.

subsequently a 300nm Si top-layer is grown at 950°C. An arsenic peak grown with such a low temperature desorption scheme is shown in Fig. 3. The width of the peak is approximately 500 Å with a 1000°C desorption scheme. For a 950°C scheme the peak would be narrower but the autodoping level is increased by about a factor 3. Above 1000°C out-diffusion will start to dominate the profile, so a trade-off must be made between autodoping and out-diffusion.

For temperatures above 950°C, the overall pressure, flow, and temperature are very important for the amount of autodoping [4]. The dependency of autodoping on pressure and flow has been investigated by varying the carrier gas flow and total pressure during the desorption step. The same peak as in Fig. 3 is grown at 700°C and at atmospheric pressure, followed by a desorption step at 950°C during a 10 min H_2 purge. The Si top-layer is 300nm and grown at 950°C. The results are shown in Fig. 4 where it is

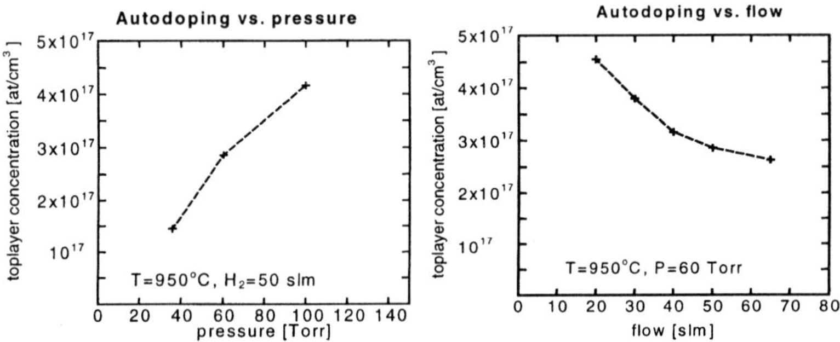

Figure 4: The As concentration in the top-layer (at 150nm) vs. pressure and flow is measured (CV measurements).

clear that the As autodoping of the top-layer depends on both pressure and H_2 flow. When N_2 instead of H_2 was used as a carrier gas (at 900°C), similar results were obtained. This indicates that hydrogen interaction with the surface plays a minor role. The dependency of autodoping on the hydrogen flow shows that the transport of arsenic at 950°C and higher is being hampered by slow diffusion through the stagnant gas layer. Arsenic transport is then enhanced when the flow and temperature are increased or the (overall) pressure is decreased.

ACCURATE DOPING CONTROL USING A TIME-STOP

The arsenic doped peak is grown at 700°C. At such low temperatures the As incorporation is not immediate but increases slowly when arsine has been let into the reaction chamber [3]. The rate of increase is determined by the partial pressure of arsine. To have a maximum rate of increase, the maximum arsine flow is applied. At 700°C and maximum arsine flow the steady state arsenic concentration is 2×10^{19} atoms/cm^3. The experiments show that in takes 4 min and 40 nm of Si deposition to grow a peak with a doping level 3 decades above 10^{16} atoms/cm^3. This corresponds to an increasing slope of 13 nm/dec. From SIMS, which is less accurate in this situation [2, 3], a value of 25 nm/dec is determined. The deposition rate depends on arsenic concentration but is approximately 95 Å/min, so a few minutes are needed to grow a peak. Since gas flows can be switched within seconds, the thickness of the peak can be controlled very accurately and the maximum peak doping level is determined by the growth time.

Fig. 5 illustrates the control of the peak doping. Peaks have been grown by applying both the high, and the low temperature desorption schemes used in Fig. 2 and 3, respectively. Only the deposition time and not the arsine flow has been varied to control the doping level. For the high temperature desorption scheme, the high temperature

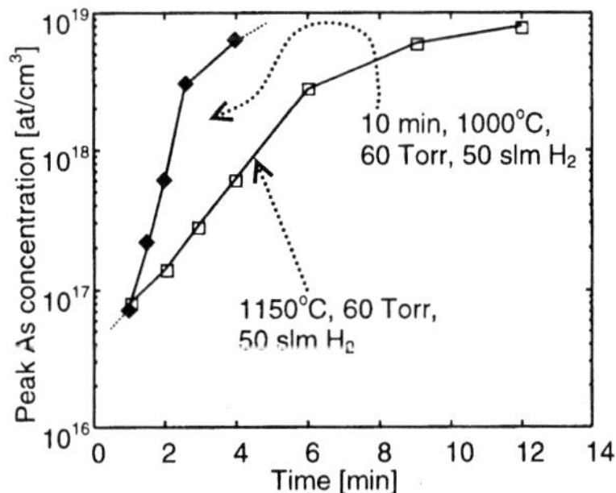

Figure 5: Time-stop control of peak doping. CV measurements for 1000°C desorption step (with 950°C top-layer) and with 1150°C capping layer are shown. Time indicates growth time of the peak.

during cap-layer growth widens the peak and reduces the maximum peak doping. Higher doping levels are measured for the 1000°C desorption scheme where the much lower thermal budget results in lower out diffusion of the peak.

THE ROLE OF SURFACE KINETICS

The incorporation of As is determined by the surface conditions during epitaxy. Fig. 6 shows all arsenic fluxes and state transitions that are involved. The rates of

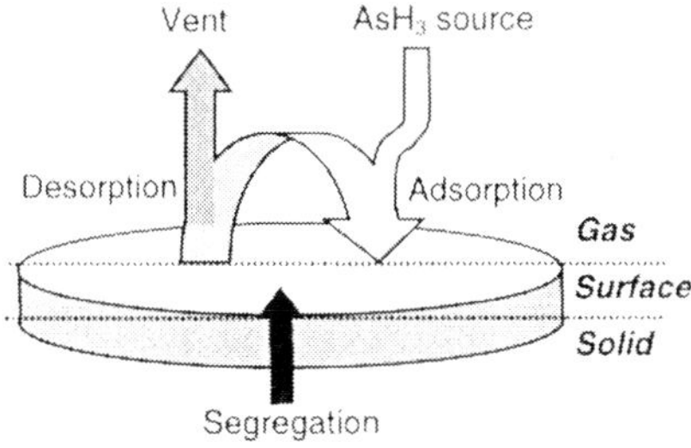

Figure 6: Schematic view of arsenic states (bold) and fluxes.

desorption and adsorption are very high at the temperatures used in the low temperature desorption schemes [8, 11]. Also diffusion in the gas phase should be rapid since it is possible to grow As doped layers at much lower temperatures (700°C). Despite the higher temperatures, the above experiments show that the transport of arsenic to the vent is still diffusion limited.

The stagnant gas layer on the wafer surface can be assumed to influence the diffusion of As away from the wafer. The situation can be described by the following analysis. The equilibrium of arsenic between the surface-bound and gas phase is given by

$$3H_2 + 2As\,(surface) \leftrightharpoons 2AsH_3\,(gas) \qquad (1)$$

Adsorption is kinetically favourable over desorption [8], so the left side of equation (1) dominates and the portion of arsenic in the gas phase is very low. When ad- and desorption reach equilibrium, the partial pressure of arsenic at the surface, P_{As}/P_{tot}, can be expressed as a function of surface concentration [12]:

$$\frac{P_{As}}{P_{tot}} = \frac{k_{de} C_{As}}{k_{ad}(C_t - C_{As})} \qquad (2)$$

where k_{ad} and k_{de} are rate constants for ad- and desorption, respectively. The concentration C_t represents the total number of sites where As can be adsorbed and C_{As} represents the number actually occupied by As. During the H_2 purge, diffusion through the stagnant gas layer is needed for arsenic transport towards the vent. The diffused flux ϕ_d, in which the flux ϕ is normalised to the density, N_t/V, can be calculated by using Ficks' law:

$$\phi_d = \frac{V}{N_t}\phi = D_{As}\frac{dn}{dx} = \frac{D_{As}}{W}\frac{P_{As}}{P_{tot}} \qquad (3)$$

where the concentration gradient in a stagnant gas layer of width W is assumed to be constant, and D_{As} is the diffusion coefficient of arsine in hydrogen gas. This diffusion coefficient is a function of both temperature and pressure.

Using equation (2) as a boundary condition for equation (3) gives:

$$\phi_d = \frac{D_{As}}{W}\frac{k_{de}}{k_{ad}}\frac{C_{As}}{(C_t - C_{As})} \approx \frac{D_{eff}}{W}\frac{C_{As}}{C_t} \qquad (4)$$

The approximation that has been made is valid when $C_{As} << C_t$. This approximation clearly illustrates that adsorption kinetics that have a low k_{de}/k_{ad} ratio, will strongly reduce the effective diffusion coefficient D_{eff}.

There are several alternative desorption schemes that may shift the balance in equation (1):

- a strained Ge wetting layer can reduce the free energy of adsorbed As,
- ex situ HNO_3 cleaning will remove the As surface coverage chemically,
- with a HCl purge, a few percent HCl will enhance desorption by etching.

Several experiments have been performed in order to evaluate the effectiveness of these strategies. The same As peak as in Fig. 3 is grown on all wafers and followed by different desorption schemes. After desorption a 300 nm top-layer was grown at 950°C. The auto-doping level in this layer is measured by CV-profiling at 150 nm. Table 1 shows that a Ge wetting layer deposited just after peak growth acts as a surfactant, greatly enhancing the net desorption. The strained Ge layer decreases the ratio between ad- and desorption, thus increasing the effective diffusion coefficient [13]. The total amount of Ge that is incorporated is very small, only a few monolayer, so its effect in a device would be

Scheme	P [Torr]	H_2 [slm]	Time [min]	T	Concentration [atoms/cm^3]
-	60	50	10	1000°C	$7x10^{16}$
Ge	60	50	10	1000°C	$3x10^{16}$
HNO_3	-	-	-	-	$1.2x10^{17}$
HCl	60	50	10	1000°C	$5x10^{16}$

Table 1 Autodoping for different desorption schemes.

negligible. The HNO_3 cleaning is less effective indicating that segregation takes place prior to top-layer growth. This problem is also encountered when growing epi on implanted As buried layers or As doped n^{++} substrates. The great importance of surface kinetics on the autodoping of arsenic is clearly demonstrated in these experiments.

CONCLUSIONS

A new method to grow very sharp As doped profiles has been presented. To allow a steep rise and high maximum doping levels, the profiles are grown at low temperature. Under these conditions the arsenic deposition time can be used to control the maximum doping level. To preserve the peak during subsequent Si epitaxy, a desorption step must be used to decrease the arsenic surface coverage and thus suppress autodoping. Relatively low temperature desorption schemes at 950°C and 1000°C have been sucessfully implemented and highly doped 500 Å wide As peak have been demonstrated.

The effective diffusion coefficient of As in the stagnant gas at the wafer surface is very important for the desorption process, as are the overall surface kinetics. New desorption schemes where a surfactant is used are promising candidates for reducing the minimum temperature needed for desorption.

REFERENCES

1. C.C.G. Visser, L. K. Nanver, A vd Bogaard in *Mat. Res. Soc. Proceedings* Vol. 533, pp. 105-110: 1998.
2. W.B.de Boer, in *Advances in Rapid Thermal and Integrated Processing* , Kluwer Academic, pp.443-463, Netherlands, 1996
3. T. I. Kamins and D. Lefforge in *J. Electrochem. Soc.*, **144**, 2, pp. 674-678: February 1997.
4. Carl O. Bozler in *J. Electrochem. Soc.*, **122**, 12, pp. 1705-1709: Dec 1975.
5. M. EL-Diwany, et al. in *IEDM tech. dig.,* pp. 245-248: Dec. 1989.
6. C.T. King, R.W. Johnson, T.Y. Chiu, J.M. Sung and M.D. Morris in *J. Electrochem. Soc.*, **142**, 7, pp. 2430-2434: July 1995.
7. P.D.Agnello, T.O. Sedgewick, M.S. Goorsky and J. Cotte, in *Applied Physics Letters* **60**, 4, pp.454-456: January 1992
8. A.K.Ott, S.M. Casey, S.R. Leone in *Surface Science*, **405**, pp. 228-237: 1998
9. Z.T. Zhong, D.W. Wang, Y. Fan, and C.F. Li. in *JVST-B* , **7**, 5 ,pp1084-1089: 1989
10. W M. -H. Xie, et al. in *Surface Science* , **397**, pp164-169: 1998
11. April L. Alstrin, Paul G. Strupp, Stephen R. Leone, in *Applied Physics Letters,* **63,** 6, pp.815-817: August 1993
12. H.Scott Foyler, *Elements of chemical reaction engineering*, Prentice Hall int. inc., New Jersey, 1986.
13. W.-X.Ni , et al. in *Surface Science*, **321**, pp. 131-135: 1998

Section VII

Equipment & Temperature Issues and Modeling

CRITICAL CONSIDERATIONS AND INTEGRATION ISSUES IN THE DESIGN OF AN RTP SYSTEM

Arnon Gat, Zion Koren, Paul J. Timans, and Randhir P.S. Thakur
AG Associates, 4425 Fortran Drive, San Jose, CA 95134, USA

Rapid thermal processing technology emanated from an earlier invention commonly known as laser annealing. The transition to a non-scanned source exposing a wafer in a thermally floating position took place in the early 1980s and was originated by AG Associates. The first product to utilize broad-band and wide-beam exposure was the Heatpulse 210T, which is briefly described in this paper. The methodology to design the first RTP system employed a simple but effective intensity simulation program that allowed uniform exposure of 4" wafers. After the introduction of this first RTP system, ten years of successive evolution gave birth to nine products, which covered the increase of wafer size, and the tighter process performance requirements. In 1995 we decided to re-visit our technology, as well as the design approach towards a new RTP system. An extensive effort was initiated to develop a 0.18μm-capable system, working with the newest available technologies in temperature measurement, simulation and temperature control. This paper reviews the design methodology for the AG Associates' Starfire system and describes the various trade-offs involved in optimization of such systems. We also describe the temperature measurement approach, which required multi-point sensing and a fast sampling rate. The trade-offs in the heater design between uniformity, controllability and power requirements are described with the performance data from two major iterations of the design. Highlights of the process performance from a completely integrated and tested Starfire system are included. The interactions between various subsystems within the Starfire are discussed together with some subtle phenomena observed during the integration and testing phase. Finally, general conclusions are shared concerning both the system and the process by which it was specified, designed and implemented.

INTRODUCTION

The pioneering work in rapid thermal processing (RTP) in the early eighties emanated from initial studies of laser in radiation of semiconductors at Stanford University (1). After realizing the deficiencies of laser heating we explored the use of a broadband non-scanning light source for rapid heating cycles which achieved novel results compared to those obtained from long heating cycles in conventional batch furnaces (2). The initial implementation of RTP was executed in a rudimentary fashion. A guess on a possible design was made, a breadboard was built and initial testing took place. A problem was discovered immediately. During the heating cycle the 3" wafers became severely distorted. Only at this stage the authors resorted to simplified modeling to estimate the light distribution inside the chamber. Measured illumination profiles in the system confirmed the simulation and showed a very non-uniform radiation pattern that caused a temperature gradient, stress and deformation. The same simulation program was used to examine various configurations for the number of heating lamps and the distances from lamp to lamp, lamps to the reflector and to the wafer. An optimum design was selected and implemented in a subsequent breadboard. Further testing verified a relatively uniform temperature distribution and that wafers remains flat even after high temperature cycles. This breadboard was later designed into the first bench top commercial RTP system called the Heatpulse 210T (3).

In the following ten years the Heatpulse systems went through many variations and improvements such as the introduction of a temperature measurement sensor to close the loop on the wafer temperature, addition of a controlled ambient to cause various desirable chemical reactions at the wafer surface, addition of automation to facilitate mass production and sophisticated control and human interface software (Table 1).

Table 1. AG Associates RTP product evolution. The numbers refer to products that grew in sophistication as wafer size increased and device dimensions decreased.

	Year	82	86	90	93
	Feature	> 1μ	> 0.6μ	> 0.4μ	> 0.25μ
Wafer Size	4" - 5"	210T 410			
	150 mm	610 2101	2106 2146	4100	
	200 mm			4108	8108 8800

Manual Module (rectangle): 210T, 410, 610
Automatic Module (oval): 2101, 2106, 2146, 4100, 4108, 8108, 8800

These RTP systems were used to develop many applications that replaced batch furnaces and enabled a range of new processes (4). However, the basic configuration of the heating chamber did not change drastically. In the mid-nineties new drivers created an urgent need to re-examine the basic configuration of the RTP chamber. The drivers were technological, application-related and commercial. AG Associates decided to embark on a fresh design for its next generation RTP system. Aggressive specifications were defined and are given in table 2.

Table 2. Specifications of the next generation RTP system.

Temperature range	400-1150°C
3σ-Total Component of Variance	3°C (ε range 0.4-0.85, 3mm edge exclusion)
Ramp Rate	5-125 °C/sec
Slip free	No slip after 1150 °C for 60 seconds
Ambient < 10 ppm O_2	Purge in < 20 seconds
Throughput	40-50 WPH, single module, 10 sec anneal
Module cost target	$158K
Cost-of-Ownership	$1.08/wafer, 2 chamber system

Four major development areas and teams were defined; Temperature measurement, heater, chamber subsystems, and integration and test (I&T). Modern techniques of product development were used with a very strong emphasis on simulation and testing of individual subsystems before integration.

A strategic philosophy was established very early in the development process. **The temperature measurement subsystem will determine all other configuration choices.** This led directly to the system configuration shown in figure 1, where the wafer is heated from above to permit multi-channel temperature measurement to occur from the backside.

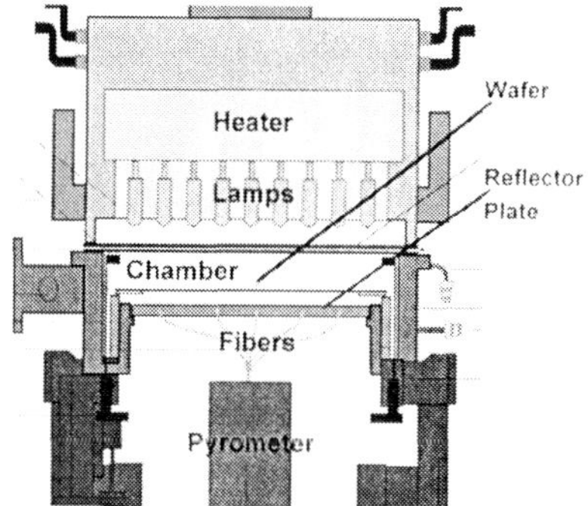

Figure 1. AG Associates next generation RTP system (Starfire) configuration.

TEMPERATURE MEASUREMENT

The specification required a sensitive, high-speed, multi-channel temperature measurement capability. This subsystem provides repeatable temperature measurements that are not affected by the wafer emissivity. The design had to provide low-noise measurements for temperatures between 400 and 1150°C on 7 channels at 50 Hz. When the initial system configuration was selected there were several possible temperature measurement technologies. These included conventional pyrometry, ripple pyrometry, direct contact thermocouples and methods based on measurement of wafer expansion or the velocity of ultrasonic waves in silicon. It was decided that a new type of emissivity-independent pyrometry would be used, at the wavelength of 2.7 μm. This wavelength was chosen because it can be filtered out by making the window between the wafer and the lamps from quartz glass which contains a high concentration of hydroxyl bonds (high-OH quartz) (5). This wavelength also permits low-noise measurements at low temperatures, even below 400°C (6).

Figure 2 shows the results from a test of the stray-light rejection. The signal is from the pyrometer channel which normally views the slip-free ring, for the case where a wafer was heated but the slip-free ring was not installed. As a result this pyrometer channel was directly exposed to the intense lamp radiation when the lamps were switched on at 70% of full power after 12 s. The result shows that the signal did not rise above the noise level indicating the complete elimination of stray light by the high-OH quartz window.

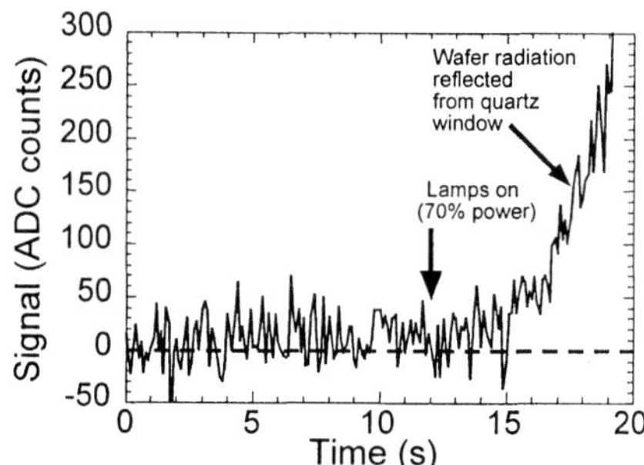

Figure 2. Stray light rejection test shows no stray light from the lamps.

A few seconds later the pyrometer signal starts to increase, because the wafer heats up and starts to emit radiation, some of which is reflected from the quartz window into the pyrometer channel. Stray light elimination greatly eases both the system hardware and software design and makes calibration simple, because the pyrometer signal is purely defined by Planck's radiation law. Figure 3 shows a graph relating pyrometer and thermocouple readings from an experiment where a wafer was heated up to 950°C. The combination of low-noise electronics and stray light elimination gives excellent linearity for temperatures above ~200°C.

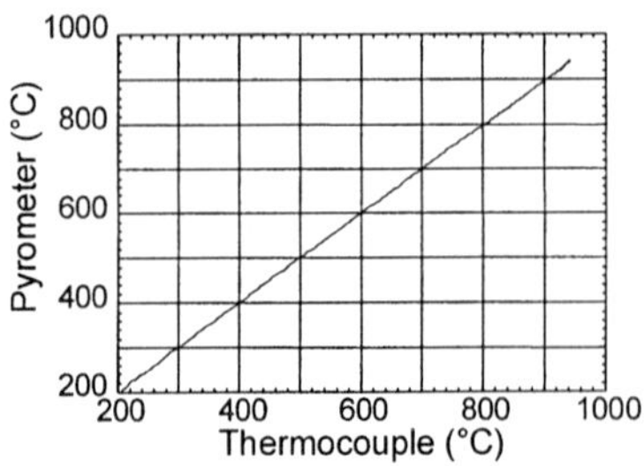

Figure 3. A plot of the pyrometer output against a thermocouple-wafer reading demonstrates the excellent linearity of the temperature measurement sub-system.

The approach taken for attaining emissivity independence was initially based on the use of pyrometer input fibers that observed the wafer backside through small apertures in a highly reflecting plate. This concept is based on the well-known idea of increasing the effective emissivity of a target by using an auxiliary reflector as shown in figure 4. The effective emissivity observed by the pyrometer increases towards one as the chamber wall reflectivity rises (7). This method resulted in very significant improvements in emissivity-independence as compared to earlier RTP systems.

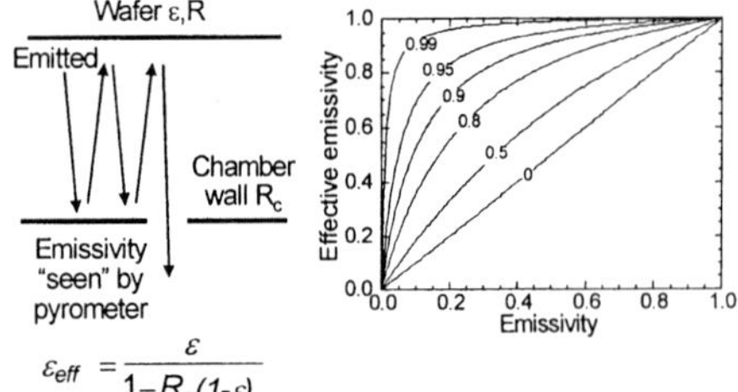

Figure 4. Multiple reflections in the chamber increase the effective emissivity, ε_{eff}, of the wafer as seen by the pyrometer. The graph shows that as the chamber wall reflectivity increases from 0 to 0.99, ε_{eff} tends towards unity for all wafer emissivities.

Figure 5 illustrates the oxide film thicknesses obtained in experiments where wafers with various backside coatings were subjected to an 1100°C, 30 s rapid thermal oxidation (RTO) process. In a conventional RTP system the range of emissivities from 0.22 to 0.94, as tested here, would result in a temperature spread of ~ 450°C. In figure 5, the use of the reflector plate alone reduces this range to ~13°C. This residual error is a result of practical limits on the plate reflectivity. A secondary subsystem was developed to improve the performance further. The results when the experiment was repeated using the secondary subsystem in addition to the basic reflective-enhancement show a temperature range of only 1.7°C. With the achievement of emissivity-independent operation, the two-year development effort of the temperature measurement subsystem had reached all its design objectives, including short and long-term repeatability.

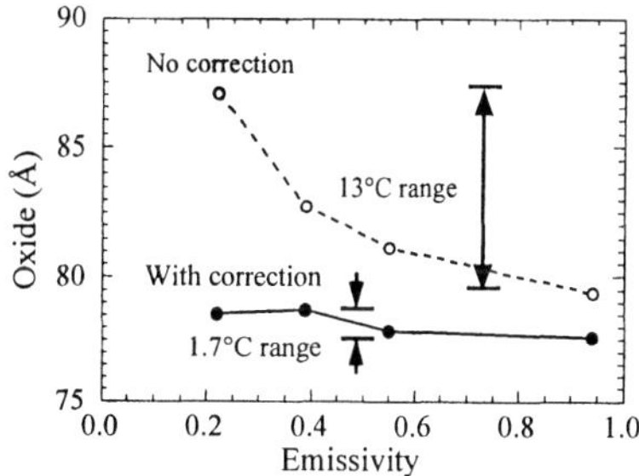

Figure 5. RTO results for wafers with emissivities between 0.22 and 0.94 show a 13°C range when there is no correction for emissivity apart from the reflective enhancement. An additional correction system reduces the error to a range of 1.7°C. RTO process sensitivity is 0.58Å/°C.

HEATER

The challenge of the new heater design was to meet the aggressive specifications of ±1°C uniformity, 125°C/sec ramp rate, and 1150°C maximum temperature. The goal of the design process was to minimize the number of iterations in heaters to save time and money. The approach taken is illustrated in figure 6. Ray-tracing calculations were used to assess design concepts and select lamp reflector test fixtures, whose illumination profiles were measured using a scanning fixture. These profiles were then used to predict wafer temperature uniformity and zone power levels by employing heat transfer simulation tools. The design constraints also included power, controllability, lamp selection, packaging and cooling.

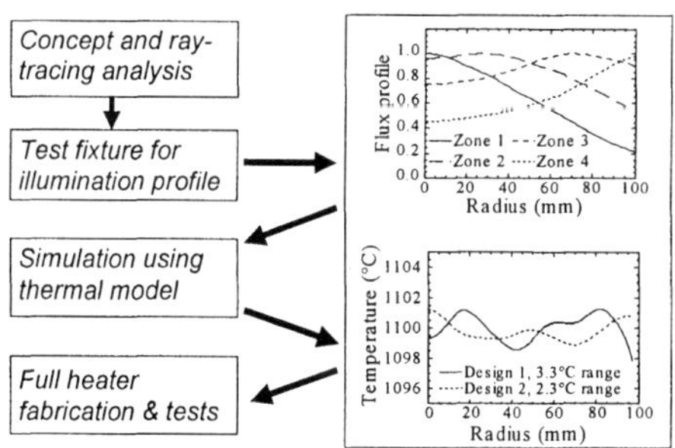

Figure 6. Heater development methodology

The initial design used individual cone reflectors around each lamp arranged in six concentric rings. A simple assessment of the power requirement indicates that it is necessary to have a power density of 50 W/cm^2 to ramp a wafer at 100°C/s at a temperature of 1150°C. This suggests that the total heater power should be ~62 kW if one assumed a 50% efficiency of power delivery to the wafer and slip-free ring. The first heater used 86 lamps rated at 800W.

Although the first process uniformity, which is shown in figure 7, was promising even without wafer rotation, the performance in power delivery shown in figure 8 did not meet the specifications. This graph shows the wafer temperature and maximum heating rate as a function of the highest lamp power, after tuning the zones for uniformity. Further analysis showed that the heater efficiency was much below the assumed 50%. The next design iterations employed different lamps and different reflector configurations.

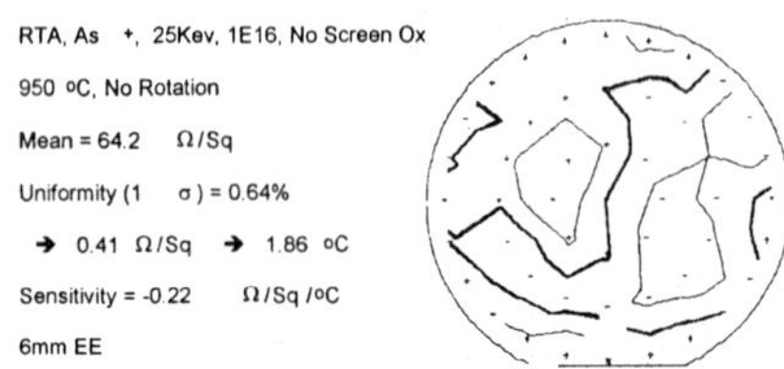

Figure 7. The first rapid thermal anneal (RTA) wafer processed in the prototype system showed good uniformity without wafer rotation.

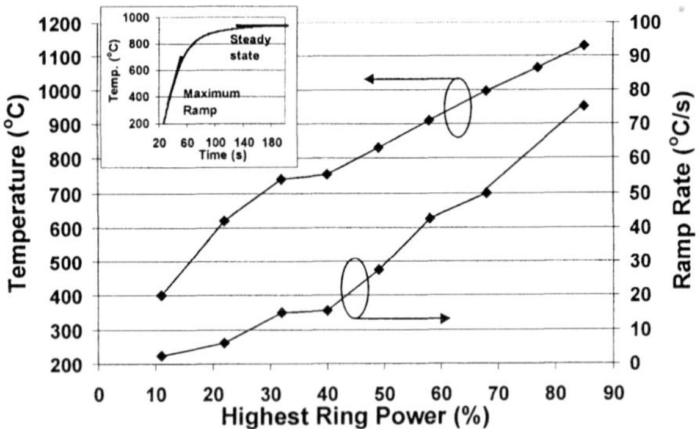

Figure 8. Open-loop heating tests of the first heater. The x-axis shows the power for the lamp zone with the highest zone ratio, after the zones had been tuned for wafer temperature uniformity. The heater required 85% of maximum power to achieve a steady state temperature of 1150°C and a peak ramp rate of 75°C/s.

The final configuration is shown in figure 9 and employs 65 lamps rated at 1.5 kW each, arranged in five concentric rings. The lamps are separated by reflective walls to achieve a balance between power delivery, wafer uniformity and controllability (8,9). The final heater performance is shown in figure 10 and 11 and in Table 3. Figure 10 shows a representative process uniformity wafer map from an RTO test performed at 1150°C for 30 s. The 3σ-uniformity is equivalent to a temperature of 1.5°C.

Figure 9. The final heater configuration uses 65 lamps rated at 1.5kW arranged in 5 rings.

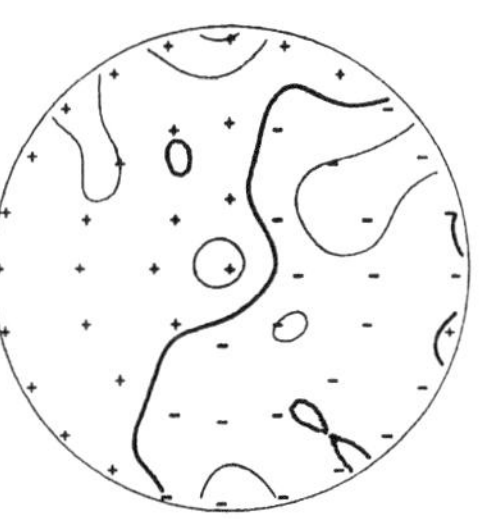

Figure 10. RTO process result from the final heater configuration demonstrated a 3σ uniformity of ~ 1.5°C.

Figure 11 shows that the heating power efficiency has improved significantly, since the highest lamp zone power required to reach 1100°C is only 62%, whereas it was ~80% with the first heater design. The ramp rate at 60% has also improved from 40°C/s to 100°C/s.

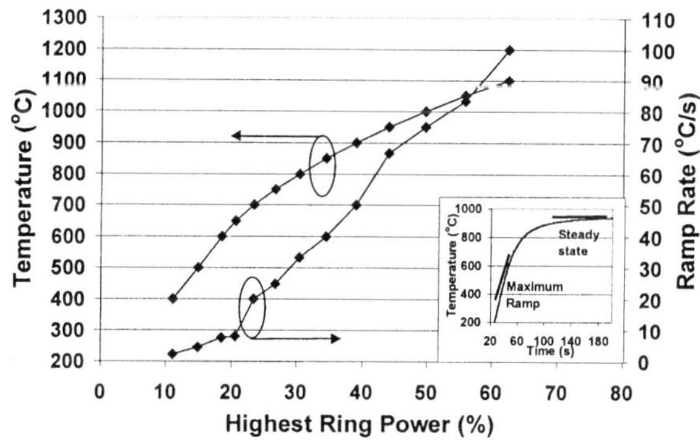

Figure 11. Open-loop heating tests of the final heater demonstrate a large improvement over the results in figure 8. The new heater only requires 62% of maximum power to achieve a steady state temperature of 1100°C and a peak ramp rate of 100°C/s.

With the completion of the tests on a range of processes, as shown in Table 3, the heating system had demonstrated that it met the basic capability defined in the requirements.

Table 3. Summary of process results from the 200mm Starfire system

Process	Temp. Time	Edge (mm)	Mean	Sensitivity	Result	
					1σ (%) Uniformity	°C
RTO - 125Å	1150°C 30s	3	125.0Å	0.75 Å/°C	0.31%	0.52
RTO - 80Å	1100°C 30s	5	81.8Å	0.58 Å/°C	0.29%	0.41
RTO - 50Å	1000°C 60s	3	51.2Å	0.4 Å/°C	0.69%	0.87
RTA As^+, 25Kev, 1E16, 100Å Ox	1000°C 10s	5	110.22 Ω/Sq	-1.06 Ω/Sq/°C	0.47%	0.49
RTA BF_2, 5E15, 25KeV	950°C 60s	3	128.7 Ω/Sq	-1.3 Ω/Sq/°C	Added non-unif.=0.88% Post Unif.=1.01% Implant non-unif.=0.5%	0.87
RTA B11, 3E15, 5KeV	1050°C 10s	5	59.37 Ω/Sq	-0.39 Ω/Sq/°C	0.95%	1.45
RTS, 600Å Ti	675°C 20s	5		-0.047 Ω/Sq/°C	Added non-Unif.=0.43% Pre Mean=10.6Ω/Sq Pre Unif.=0.77% Post Mean=7.1Ω/Sq Post Unif.= 0.88%	0.64

A further aspect of heater system design was the lamp reliability, which was investigated as part of the I&T activity described later.

CONTROL

Much of the Starfire design was based on the decision that excellent temperature control is the enabling factor to allow RTP to be used in the most advanced process applications. Temperature control has to be applied across the wafer in each heating cycle and across many wafers in a production environment, and the control approach plays a critical part in determining the total components of variance (TCV) obtained in process results. Traditional RTP systems typically used one temperature feedback point to control the heating cycle. These systems relied on lamp zones being slaved to the control signal by user-defined, recipe-dependent zone ratios, which were optimized to achieve process uniformity across the wafer. This meant that only one point on the wafer was under true closed-loop control, and the temperature at other positions on the wafer could change because of fluctuations in system facilities, lamp aging effects and other unexpected changes in the heat-transfer conditions. The new approach addressed this issue by incorporating multiple temperature measurement points and a multiple-input, multiple-output (MIMO) control algorithm. Figure 12 shows a schematic of the control scheme, which is based on multiple-loop proportional-integral-derivative (PID) controllers (10).

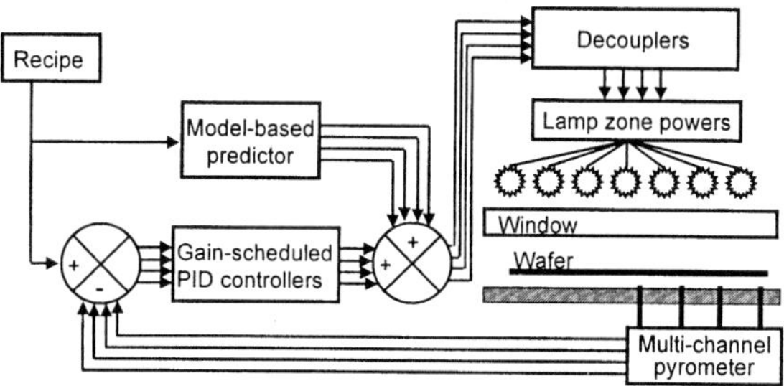

Figure 12. Schematic diagram of the multiple-input multiple-output (MIMO) controller.

The control scheme includes a feed-forward element called a predictor, which estimates the power needed at any given moment and hence reduces the magnitude of the control actions needed from the PID-feedback controller. The predictor and the PID controller use parameters that were obtained from a black-box model of the system response which was constructed through model-identification experiments. This allows the control scheme to handle the inherently non-linear nature of the RTP system through appropriate gain-scheduling and prediction methods. Although the controller has multiple loops where each pyrometer is controlled by one lamp zone, the control scheme also incorporates de-coupling elements that account for the interactions between lamp zones. Figure 13 shows MIMO control for a cycle where the wafer ramps up at 75°C/s to 900°C and then at 25°C/s to 1050°C . The four pyrometer readings are controlled so that they follow the set point closely and both the transient and steady-state spread in the pyrometer readings is ~2°C. The "bring-in" is the integral of the temperature difference between the recipe trajectory and the pyrometer readings as the control transitions from the ramp to the steady state. In figure 13 the bring-in is ~5.5°Cs, demonstrating a fast dynamic control. Once MIMO control had been successfully demonstrated on a number of sensitive processes, the control development had met its specified goals.

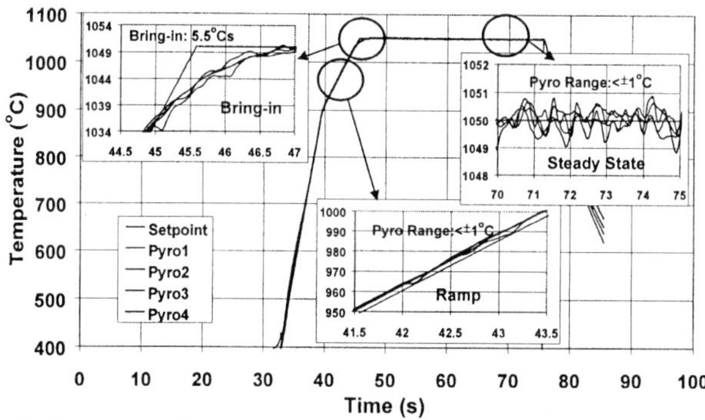

Figure 13. MIMO control performance for a heating cycle with a 75°C/s ramp to 900°C followed by a 25°C/s ramp to 1050°C. The figure shows the readings from the four pyrometers that view the wafer. The inset graphs illustrate the tight temperature uniformity control during the ramp, bring-in and steady state phases.

CHAMBER DEVELOPMENT

The cold-wall chamber was designed to include wafer rotation and a stationary slip-free ring which is in the plane of the wafer during processing. The volume was minimized to permit a fast atmospheric purge with a horizontal-flow configuration. The wafer support was designed so that the wafer was held near its edge at three points. The chamber interacts with a controlled-ambient atmospheric-pressure wafer handler which is purged to eliminate oxygen contamination. The combination allows the system to be purged to <10 ppm O_2 in under 20 s from when the chamber door closes. The construction is leak-tight and permits the use of toxic and corrosive gases. Tests showed that the system introduces less than 5 added particles/cm^2.

INTEGRATION AND TESTING METHODOLOGY

Figure 14 outlines the integration and testing methodology used in the development process. The product development was organized so that the basic building-blocks of the new technology could be tested in several phases prior to the final integration effort. The first phase tested the basic feasibility of the major innovations described above. This went from simple "Table Top" experiments to a manual test "Jig" and then to an automatic "Breadboard #1" system which was complete 6 months from beginning of the project. This stage successfully met the "Proof-of-Concept" requirements. In the second phase, more extensive process performance tests were performed on the "Breadboard#2" system, which included an improved chamber, multi-point temperature measurement, improved quartz cooling, MIMO control and an improved lamp power supply. The third phase was an alpha tool which included the new wafer handling platform and computer architecture. This system was where most of the "bugs" were worked out and integration was carried out on the platform with all modules in place. In all it took 3.5 years of extensive development and I&T, to complete the alpha phase of the new product. Much of the delay in the program arose from unexpected problems which were revealed when the subsystems were integrated. Some of the surprises are described in the next section.

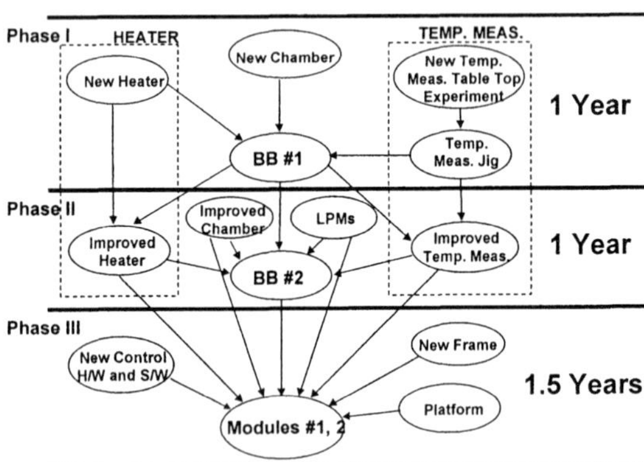

Figure 14. Schematic illustration of the 3 phases of the Starfire development program. BB#1 and BB#2 were the development breadboard systems.

SYSTEM DESIGN INTERACTIONS

One of the interesting aspects of RTP system design arises from the many potential interactions between system features, which require various trade-offs in the design. One aspect of RTP system design which has received much attention in theoretical analysis comes from the choices in illumination pattern, where there is a trade-off between the uniformity across the wafer and the ability to selectively affect the power delivered to any given region of the wafer (8,9). Radiation patterns which are tightly focused can selectively heat the wafer, but they may tend to lead to oscillations in the temperature distribution. On the other hand broad illumination patterns tend to create very smooth temperature distributions, but they can lead to problems in controlling and optimizing the temperature distribution.

Several "second-order" problems, which arose from interactions, were discovered during the I&T effort. One problem was a slow drift in process results which seemed to be related to fluctuations in the ambient temperature. In particular, as multiple cassettes of wafers were processed, it was also found that heating of the system cabinet and the components outside the chamber had an impact on the process results. Extensive troubleshooting revealed that the problem arose from a temperature-sensitive optical component inside the pyrometer. Figure 15 shows how the pyrometer temperature reading changed with the ambient temperature. The sensitivity was such that a 1°C change in the ambient temperature around the pyrometer caused its reading to change by ~1.5°C at a wafer temperature of 1000°C. The pyrometer had been designed with the capability for a sophisticated thermal stabilization, and this feature was used to mitigate the effect. Figure 16 shows that when the stabilization was employed, the pyrometer readings became immune to large fluctuations in the ambient air and water cooling temperatures.

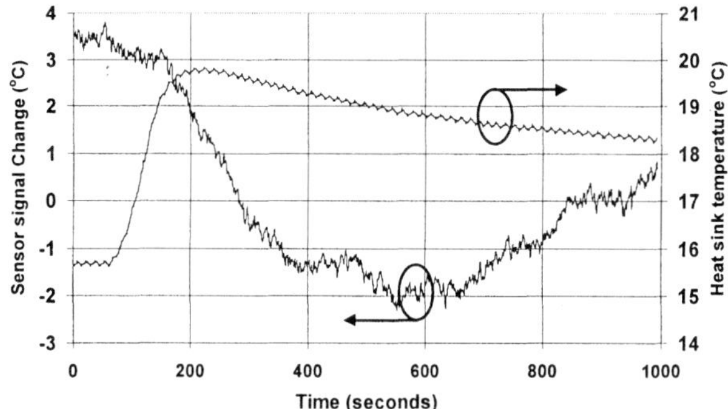

Figure 15. Pyrometer drift test results revealed that the temperature reading was sensitive to the pyrometer ambient temperature. The graph shows the variation in temperature when the pyrometer was observing a target at 1000°C.

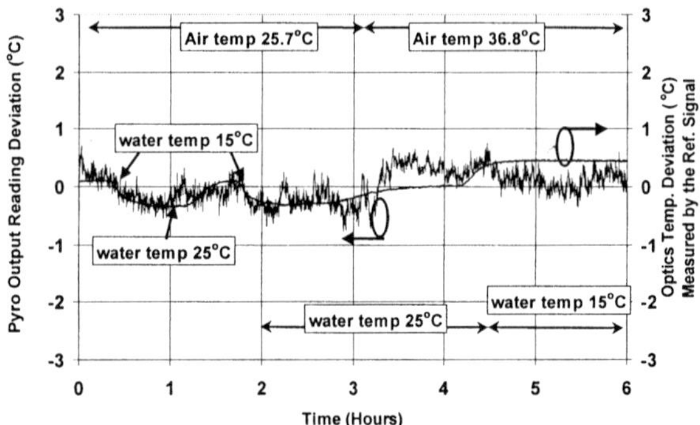

Figure 16. Pyrometer drift was eliminated by active thermal stabilization of the pyrometer optics. This drift test result shows very small changes in pyrometer reading as it observed a target at 1100°C even though the ambient air temperature and the water cooling temperature were deliberately varied by large amounts.

Another design interaction arises from the use of reflectors to gather lamp radiation and direct it towards the wafer. This has an unintentional impact on the temperature of the lamp filaments, tending to make them increase relative to the temperature they would reach in free space when run at the same power. This leads to a trade-off between reflector efficiency and lamp life. The reflecting environment also makes the seal at the base of the lamp top reach a much higher temperature than it would if the lamp was run at an equal power in the open air. In the early stage of the Starfire development failure of this seal was the main cause for poor lamp reliability. Figure 17 shows the temperature indicated by a thermocouple embedded in the lamp near the seal during multiple heating cycles. The peak temperature is over 500°C, well above the manufacturers recommended maximum. A mechanical redesign of the heater hardware had to be performed to permit the lamp base to be shielded from lamp radiation and to allow the use of a stiffer retaining spring to improve the thermal contact for more effective cooling of the seal area Figure 18 shows that these changes resulted in a peak seal temperature of 400°C, which greatly improved lamp reliability.

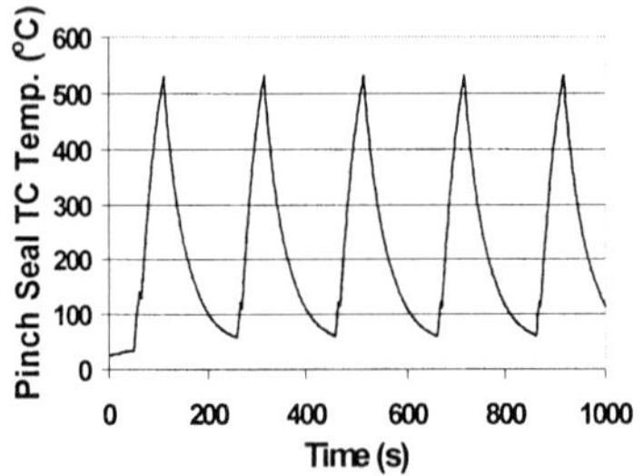

Figure 17. This graph shows that the lamp pinch seal temperature during multiple heating cycles exceeded 500°C, above the maximum permitted temperature for reliable operation.

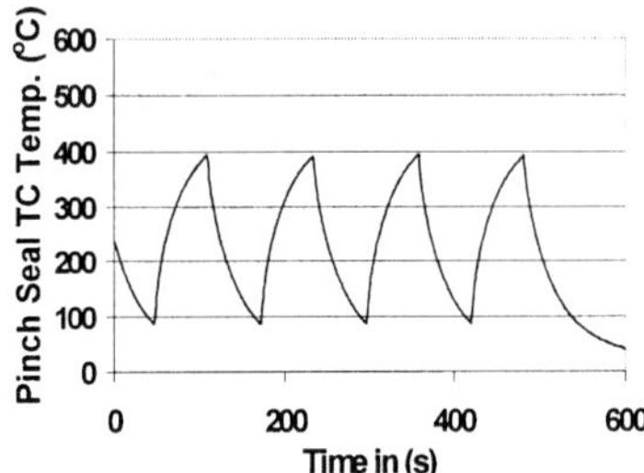

Figure 18. Lamp cooling improvements allowed the peak pinch seal temperature to be reduced to 400°C greatly improving the reliability.

The problems discovered during the integration phase significantly delayed the achievement of satisfactory results during alpha testing. One of the lessons learnt in this exercise is that the integration and testing of subsystems should be allowed a significant fraction of the time scheduled for product development.

CONCLUSIONS

AG Associates completed the development of a very complex and interactive system to meet the process requirements that were originally conceived and agreed upon. In parallel we also learned some very important lessons about the development process itself, which we share here. The heater iteration process taught us that no assumptions should be made without verifying their validity. The time taken to test the assumptions even in an approximate manner would have saved us considerable resources and time. A second lesson learnt during interaction of the teams developing various sub-systems was the benefit of using detailed written specifications. Substantial time is needed to establish specifications but, if properly used, they facilitate smooth interfaces between subsystems. Throughout the development process an enormous amount of data needed to be collected and digested. The team had invested in a comprehensive data collection and management network that enabled team members to access wafer maps, time-temperature profiles, intermediate and final reports throughout the development process. The sharing of such information among team members was critical for expedient decision making during the development process. Lastly, one conclusion of this program was the need for ample time for the integration and testing phase. As a rule of thumb, our experience showed that it is necessary to allocate 50% of the total development cycle to the integration and test phases.

It took 3 ½ years, $16M, a lot of hard work and learning the lessons above to produce the 200mm Starfire system. In contrast, once the radically new technology had been created, it was possible to produce a successful 300mm Starfire system in only 9 months. This difference came from the completion of the R & D groundwork and the greater experience of the team in both product development and advanced RTP technology.

ACKNOWLEDGEMENTS

The authors would like to acknowledge the whole development team at AG Associates who participated in this exciting and challenging program between the years of 1995 and 1999. The combined efforts, intelligence, and experience of this group enabled us to come up with process performance that exceeded all previous designs and established a new standard for RTP processing in the future.

REFERENCES

1. A. Gat, Ph.D. dissertation, "CW Laser annealing of ion implanted silicon", Stanford University, 1979.
2. R. Iscof, *Semiconductor International*, Nov 1981, pp. 73-82.

3. A. Gat, Heatpulse 210T, user manual, AG Associates, 1982
4. P.S. Burggraaf, *Semiconductor International,* Dec 1983.
5. H. Walk and T. Theiler, in *RTP'94*, R. B. Fair and B. Lojek, Editors, p. 194, RTP'94, Round Rock (1994).
6. R. P. S. Thakur, P. J. Timans and S. P. Tay, *Solid State Technology*, **41**, 171 (1998).
7. P. J. Timans, *Solid State Technology*, **40**, 63 (1997).
8. J. G. Li, P. J. Timans and R. P. S. Thakur, in *RTP'98*, T. Hori, B. Lojek, Y. Tanabe and R. P. S. Thakur, Editors, p. 37, RTP Conference, Round Rock (1998).
9. P. J. Timans, *Materials Science in Semiconductor Processing*, **1**, 169 (1998).
10. K. S. Balakrishnan, S. Shooshtarian, N. Acharya, P. J. Timans and R. P. S. Thakur, in these proceedings.

TEMPERATURE CALIBRATION IN MICROELECTRONIC MANUFACTURING

Peter Vandenabeele
SensArray Corporation
St.-Annastraat 85, 2500 Lier, BELGIUM
Email: PeterV@netvision.be

Wayne Renken
SensArray Corporation
3410 Garrett Drive, Santa Clara, CA 95054, USA
Email: Wayne_Renken@sensarray.com

ABSTRACT

Precise calibration of wafer temperature is critical to advanced microelectronic manufacturing. Direct measurement of silicon temperature, with sensors incorporated in an in-situ calibration wafer, provides a measurement standard for calibrating process temperature control systems. Several types of sensors can be incorporated in an instrumented calibration wafer such as thermocouples (TC), resistance temperature devices (RTD), and photoluminescence decay time-based fiber optic probes (FOT). The operating principles and the error mechanisms in these three types of embedded sensors in Si wafers are explained. To benefit from the measurement accuracy of an *in-situ* instrumented wafer, the instrumented wafer's temperature must be correlated to the process wafer's temperature. Examples of the use for temperature analysis or calibration of temperature are given in three applications: rapid thermal processing, photoresist hot plate baking, and high-density plasma etching.

INTRODUCTION

The control of wafer temperature and temperature uniformity is critical for many semiconductor processes. Calibration of the wafer-temperature control systems in semiconductor process tools is important for cross-tool and cross-fab process control and matching. Multi-point *in-situ* temperature measurement on silicon wafers is needed for evaluation of wafer heating and process optimization. On most process tools, only single point measurement and temperature control is available. The most accurate types of sensors that can be incorporated into a multi-point instrumented calibration wafer include thermocouples (TC), temperature-sensitive resistive devices (RTD), and photoluminescence fiber optic thermometry (FOT) probes (see Table I).

Table I: Comparison of Temperature Sensors

	Thermocouple	Resistive Temperature Device	Fiber Optic Thermometry
Temperature Range (°C)	0-1150	0-600	0-400
Accuracy (°C)	± 1.0	± 0.1	± 2.0
Precision (°C)	± 0.2	± 0.05	± 0.5
Operating Principle	Voltage generated by temperature gradient between two junctions	Resistance change of sensor	Temperature-dependent decay time of fluorescent signal
Materials	Metals or alloys– NiCr, NiAlSi, Pt, PtRh ...	Pure metal– Pt, Ni and others	Fluorescent material
Connections	Two lead– same material as sensor	Four leads– any conductor	Single fiber-optic cable
Instrumentation	μ Voltmeter and reference junction	Four-point resistance meter	Measurement of optical decay time
Noise Immunity	Sensitive at μ V levels	Sensitive at mV levels	Sensitive to certain optical interference
Environment	Oxygen-free (<10 ppm) ambient required at >600°C for Type-K	Not compatible with RF plasma	Compatible with RF plasma

THERMOCOUPLE MEASUREMENT

Thermocouple operation

Thermocouples are used for measurements up to the highest temperature (<1150 °C in a controlled atmosphere). Thermocouples generate a voltage by integration of a thermoelectric power over the entire length of the wire. The thermoelectric power is dependent on the local temperature derivative and the local material composition.

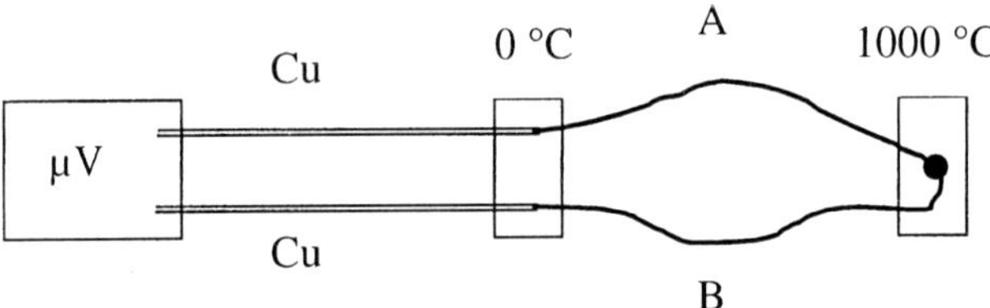

Fig. 1: Basic set-up of thermocouple measurement

Fig. 1 shows the basic set-up for a thermocouple with cold junction. The generated voltage can be described as:

$$U = \int a(T).\frac{dT}{dx}.dx - \int b(T).\frac{dT}{dx}.dx \qquad [1]$$

Where U [V] is the voltage on the microvoltmeter, $a(T)$ [$V.K^{-1}$] is the Seebeck coefficient of thermocouple material A, T [K] is the temperature, x [m] is the distance along lead A, $b(T)$ [$V.K^{-1}$] is the Seebeck coefficient of thermocouple material B, T [K] is the temperature, and x [m] is the distance along lead B. Important to notice is that the voltage is not generated at the hot junction, but is generated over the entire length of the wire, more specifically in those areas that are subject to a temperature gradient. For a correct measurement of the *hot junction* temperature, it is thus required that:

(1) In the area where materials A and B interact and the material composition of each lead is not pure (the hot junction area), a very uniform temperature is maintained. This is also required at the cold junction or reference junction, where A and B interact with the Cu connection leads, but is more easily achieved since this is a low gradient region and is close to ambient temperature.

(2) In the area where material A and B are exposed to a temperature gradient, the materials are perfectly uniform along their lengths and generate signals precisely to their standards for Seebeck coefficient.

Condition (1) requires that the area immediately adjacent to the junction (the "junction area") is in a uniform temperature field. The extension of this "junction area" is determined by the area immediately adjacent to the junction where the lead material could have a disturbed composition. A distortion of the material composition can be caused by cross contamination in the leads that was caused by the junction welding process and diffusion of thermocouple doping materials during exposure to high temperature. In addition, when the junction is mounted in immediate chemical contact with the wafer, diffusion of Si along the leads can cause a cross contamination with Si in the leads adjacent to the junction. For thermocouples with a lead diameter in the range of 0.003 to 0.005" (0.075 mm to 0.125 mm) it is assumed that an extension of maximum 0.5 mm away from the junction needs to be in a uniform temperature area. In this area, no contamination of the TC leads by any foreign element (such as e.g. Si or C) should be possible. This also indicates that it is not sufficient to have the junction alone in thermal contact with the Si wafer, but that the full junction area needs to be in an isothermal area.

Condition (2) requires that the material that is exposed to the temperature gradient (further away from the junction) needs to maintain a composition equal to the standard composition. As soon as a non-uniformity in composition is present (due to contamination), unreliable measurements will result. The measurement is unreliable because under certain conditions (when the contaminated lead section is in an area with uniform temperature) the modification of material will not create an offset in the measured signal. However, in a next measurement with different geometrical conditions (e.g. after unloading/loading the wafer) the modified lead section may be in a large temperature gradient (e.g. adjacent to the wafer edge) and a large effect may show up due to the material composition change.

For K-type thermocouple, maintaining a uniform material composition requires that the ambient contains a very low concentration of oxygen (< 10 PPM) during calibrations at temperatures beyond 600 °C. The combination of oxygen and higher temperatures will lead to the formation of Cr_2O_3 from the Cr in the NiCr lead and this will evaporate at high temperatures. This will cause the Cr concentration to drop and a deviation of indicated temperature could result eventually. This is a weakness of the application of type K thermocouple for RTP applications at high temperature. If oxidation of the lead system occurs after prolonged use of the wafer, errors in the order of a few degrees C may occur. This error is not immediately noticeable since the TC wafer still operates and gives a reasonable, but not fully accurate reading.

Thermocouple mounted in Si wafers

The thermocouple needs to be mounted in a Si wafer in such a matter that will create the conditions for a correct measurement. This requires the following conditions for the correct measurement of the Si wafer temperature:

(1) The junction and the lead section immediately adjacent to the junction needs to be in an isothermal area equal to the wafer temperature. For this reason, the TC junction needs to be embedded over enough distance so thermal conduction along the leads will not offset the junction temperature.

(2) The local wafer (Si) temperature should not be modified by the presence of the sensor.

To fulfill condition (1), the junction needs to be embedded in some sort of isothermal cavity with an electrical and chemical barrier between the Si and the thermocouple. If no barrier were present, the junction would include the Si and would extend up to the point where the metal/Si contact ends. However, from that point on, the thermocouple lead would enter a non-isothermal condition, while Si diffusion along the lead, could compromise the standard composition of that lead, adjacent to the junction (see fig.2). An additional requirement for fulfilling condition 1 is that enough length of lead material is embedded in the Si cavity, so that there is no relevant heat flux form the outside section of the leads onto the junction. From calculations, it was found that for a 0.075 mm diameter TC lead, approx. 1.5 mm length of TC lead needs to be inside the cavity. In practical implementations, due to limited wafer thickness, this is realized by mounting the TC leads in a circular shape inside the cavity.

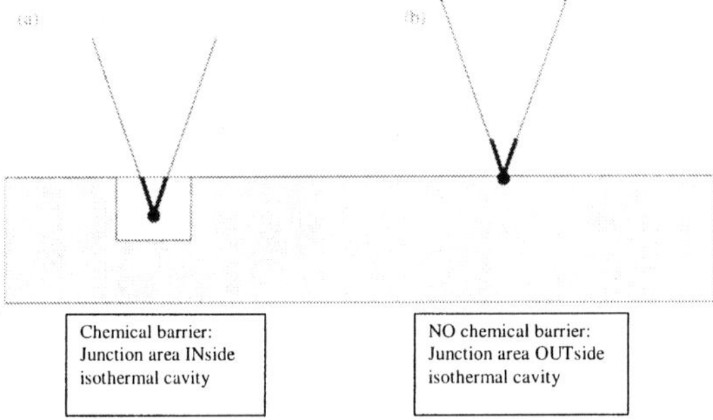

Fig. 2: When the junction area is in an isothermal area with a chemical barrier layer, the condition for accurate measurement is met. When there is no chemical barrier layer, the "junction area" (area with non uniform material composition due to Si diffusion in the leads) extends out of the isothermal area. The combination of non-uniform material composition with temperature gradient can cause a measurement error.

Condition 2 requires that the wafer temperature is not modified by the presence of the TC. Simulations and measurements conducted earlier have shown that the use of a typical cement showing an optically disturbed area of approximately 1mm (0.040") diameter, caused an offset up to 1°C of the local Si temperature. Since it is difficult to make the diameter of the cavity any smaller and provide for adequate lead length in an isothermal region, no immediate solution is available for this offset. Two solutions for this issue remain: (1) the thermocouple junction could be mounted away from the visible (absorption and emissivity altered) section of the chemical barrier in a buried structure, and (2) the TC material could be deposited in a very thin line on top of the Si wafer (see fig. 3). Option (1) is difficult to manufacture mechanically. One possibility to manufacture this is by attaching two wafers to each other and mounting the TC in between. A limit for having virtually no influence from the visible section of the chemical barrier onto the junction is that a distance of at least 4 mm is provided. It is possible to realize option (2) to a certain maximum temperature [6]. If the deposited lines are less than 0.05 mm wide, very little impact on the Si temperature (< 0.1 °C) is expected. The thermocouple materials used may show a different Seebeck coefficient when deposited as a thin film, compared to the material that was drawn from a bulk ingot and annealed. Thin film thermocouples of Pt and Pd have been successfully fabricated [6]. The Seebeck coefficient of the thin film material can be calibrated, but then again, the majority of the EMF is generated in the leads that connect the signal off the wafer. The same problem of an isothermal secondary junction between the TC lead and the thin film TC material shows up again (but to a lesser extent). One suggested solution is to use a specially engineered set of leads for picking off the signal from the thin film TC so that the Seebeck coefficient of each lead is identical to the Seebeck coefficient of the thin film material. In this case, the secondary junction between the thin film TC and the TC lead becomes irrelevant. This could be realized e.g. by using non standard compositions of a Pt/Rh or Pt/Pd mixture, where the Seebeck coefficient of thin film Pt and thin film Pd is matched by the discrete wire leads.

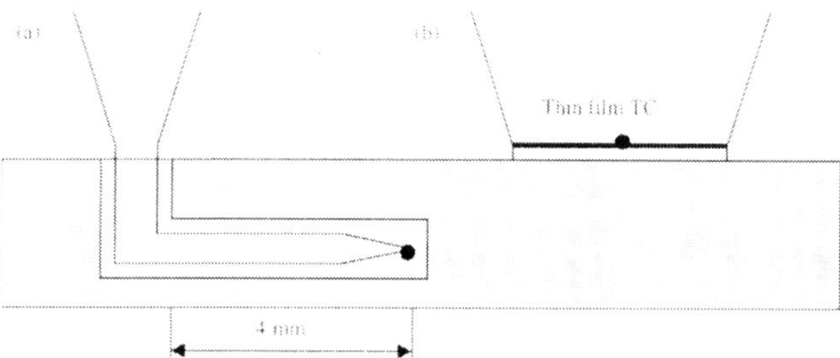

Fig. 3: Solution for the effect of the chemical barrier on the local Si are: (a) the junction is buried in the Si wafer and is moved a distance of 4 mm away from the visible area of the barrier and (b) a thin film TC is mounted on a thin chemical barrier layer on top of the Si.

Calibration of TC wafers

Form the above discussion, it is clear that there are two fundamental sources of offsets of the measurement that need to be calibrated: (1) general effects of the TC lead composition and (2) local effects in the immediate vicinity of the junction.

General lead effects. If the manufacturing of TC lead material is well characterized and known to be homogeneous, it may be calibrated by measuring a sample in an isothermal calibration furnace, and the calibration can be attributed to a longer length of the new unused lead material. This is the method used by standards laboratories world-wide.

If the lead material has been subjected to a change in composition, it will normally result in a non-uniform Seebeck coefficient along the length. As explained above, the exact location of the thermal gradient that will create the EMF is not known *a priori,* thus the calibration can vary over multiple applications of the same wafer due to changes in lead position and location of thermal gradients. It makes little sense to calibrate an instrumented wafer in an "isothermal" calibration system since it is only the composition of the leads at the furnace exit that is subjected to calibration. The full lead length that is inside the isothermal area is not tested since it does not contribute to the total EMF. The best *a priori* calibration of a wafer is to calibrate a number of lead sections and to find out what is the average response from these lead sections and what is the variability over the different sections. A simple calibration check to test for local change in material composition on a used wafer is to test its sensitivity to local thermal non-uniformity within a single lead by applying a local hot spot that is moved along the lead. No relevant signal should be caused by a local hot spot in a single lead of the TC pair (under the condition that the wafer and TC junction remain at a stable temperature, e.g by clamping the wafer on a cold chuck). If a lead has degraded, e.g. due to chemical composition change, this will certainly show in a hot spot test.

Local effects near the junction. The effects near the junction can only be tested on a wafer, in a thermal flux gradient that is equal to the typical usage condition of the wafer. For use in RTP, a similar type calibration system must be used for this test. When the same conditions are used for heating (lamp type, double side or single side, quartz/metal chamber, ...) time regime, distance to the quartz tube, cooling of the quartz tube etc. a precise correlation can be made between the actual Si temperature and the indicated TC temperature. A fundamental problem is that in most cases no other tools than a TC are available that yield a traceable value of the actual Si temperature. Some form of "absolute" pyrometer can be conceived, but is tricky due to potential lamp interference or chamber reflectivity effects). As another alternative, we have found one particular condition where, by means of a symmetrical set-up with two arrays of lamps and two identical wafers present in the chamber, the actual Si wafer temperature could be measured with great precision [3].

RTD MEASUREMENT

RTD Operation

RTD's (Resistive Temperature Devices) have a precise relationship between temperature and resistance. For certain metals, particularly Pt, this relationship is known very accurately. The resistance of the Pt RTD is also dependent on the stress in the material. For the highest accuracy, this requires a stress free mounting of a coiled Pt wire spiral on a substrate. However, it is not possible to fit such a structure inside an isothermal cavity in a Si wafer of only 0.027" (0.7 mm) deep. For this application, small thin film Pt RTD sensors are used where Pt is mounted on a alumina base. Due to the good adherence and the similar Coefficient of Thermal Expansion (CTE) between Pt and alumina, reliable operation with stable calibration is possible. The resistance/temperature relationship of the RTD is specific for each individual unit and should be calibrated per unit. With an RTD, a much better accuracy can be obtained, compared to a TC. This is due to a higher signal level with less electrical noise, and a well behaved resistance/temperature correlation and because the measurement is a true "point" measurement: the full resistance "signal" is generated at the sensor location and is not strongly dependent on the quality of the 4 leads connecting to the device. Figure 4 shows some details of RTD instrumented wafer structures.

RTD Calibration

Still, the small sensors need to be mounted in the isothermal cavity in the wafer and this may cause some minor amount of stress on the sensor. Also, the base resistance of the RTD is not known a priori. For an accurate temperature reading, a calibration of the completed wafer is required over the range of operation. This takes into account both the base resistance of the device, the temperature dependence and potential stress related effects. Due to the operation principle, an isothermal calibration is possible and preferred. In a temperature range of 10 to 220 °C, this can be achieved to an accuracy of < ± 0.1°C and a precision of ± 0.01°C in a stirred liquid bath. During the calibration, the wafer is protected from the liquid. The RTD calibration has been described in detail in [4].

Thin Film Sensor & Discrete Sensor RTD Instrumented Wafer

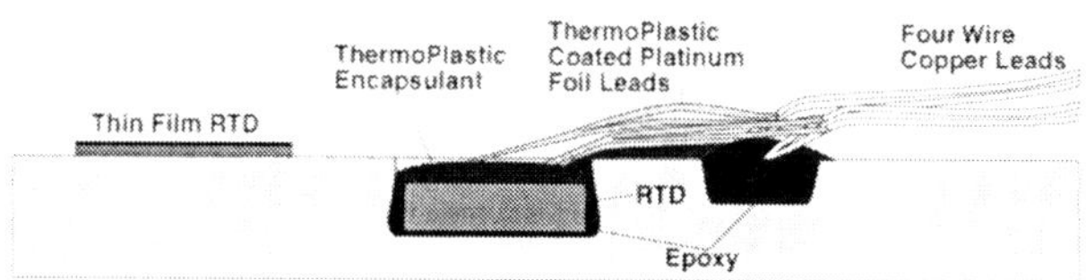

Figure 4: A typical cross section view of a discrete RTD chip sensor mounted into a cavity in a silicon wafer. Thin film RTD sensors directly deposited on silicon, which are under development, have potential for higher accuracy at high temperatures.

PHOTOLUMINESCENT FIBER OPTIC THERMOMETER

FOT Operation

Photoluminescent materials generate an amount of radiation after excitement with light of another wavelength. One typical behavior is that the decay time of the emitted light is in the order of milliseconds to microseconds, and is dependent on the material temperature.

For mounting an FOT sensor in a Si wafer, it is optimal to mount the photoluminescent material in an isothermal cavity in the Si and to read the signal with an optical probe in the immediate neighborhood, but not touching the sensing material. This avoids heat leaks through the probe. Figure 5 detail the construction of a FOT instrumented wafer.

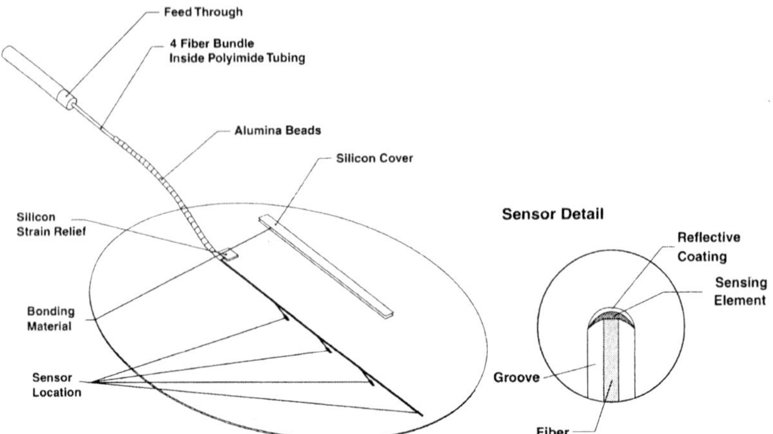

Figure 5: *FOT instrumented wafers, with a silicon cover protecting the sensors, survive power levels of >10 watts/cm^2 in active plasma without sensor degradation.*

FOT error mechanisms

The measurement errors can come from photo and thermal aging of the sensors, drift of the optics and electronics components in the instrument, and poor thermal coupling of the sensor material to the substrate. For best accuracy, the sensor material should be directly coated onto the substrate with the optical fiber interrogating the sensor material without touching the sensor layer. The thinner the coating layer, the closer the sensor temperature is to the substrate temperature. Intense fluctuating ambient light can interfere with the decay time measurement and cause an error in temperature reading. An obvious improvement is reached with a better optical shielding of the isothermal cavity from ambient radiation, especially in plasma chambers. Reconnect error, although small (~0.1-0.2C), also contributes to the overall accuracy of the measurement. Both systematic and random error can occur during the calibration process.

APPLICATION RESULTS

Rapid Thermal Processing (RTP)

The use of thermocouple instrumented wafers in rapid thermal processing systems has previously been described [1]. Instrumented wafers can be used to model, characterize and calibrate thermal control systems for RTP [2, 3]. Thermocouples (typically Type K for use in the 1000°C temperature range), are imbedded in cavities formed in silicon wafers. The preferred designed has a reentrant cavity. The TC is bonded using a high-temperature ceramic cement that fills the cavity. Each of the TC leads makes a 180-degree rotation around the perimeter of the cavity (see Figure 6); this reduces the heat loss from the junction along the TC wire since a longer amount of material is embedded in the cavity. To reduce the temperature difference between the welded thermocouple junction and the silicon, the junction is located near the edge of the cavity, close to the silicon-ceramic boundary. For this structure, it is believed that the junction temperature is within ±2°C of the silicon temperature. Instrumented wafers of this type exhibited an output drift of –0.5 to +1.0 ° C over 200 heating cycles at 1150°C. TC matching was within ±1.0°C.

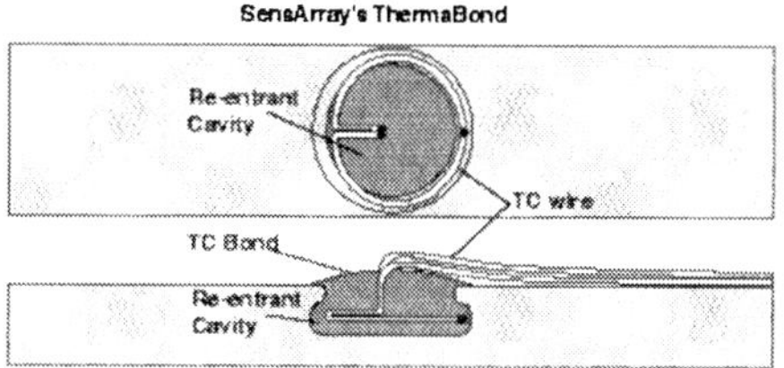

Figure 6: The sensor bond structure of a high accuracy TC instrumented wafer.

Variability in TC lead wire positioning can cause differences in the thermocouple lead temperature and this can influence junction temperature. A wafer tray or pallet on which the instrumented wafer and TC leads are permanently attached can mitigate this source of calibration error.

It was found that oxidation of K-type thermocouples can affect the calibration and oxygen impurities in the >1000°C nitrogen ambient or nitrogen of less than 99.999% (<10 PPM impurities) purity could cause TC drift. An oxygen- and moisture-free ambient and sufficient chamber purging during load-unload cycles are required to maintain TC stability

The optical properties (emissivity and absorption coefficient) of the process wafer are critical for both radiative heating and temperature measurement. In one RTP experiments it was determined that a systematic temperature drift upward of one wafer over 200 temperature cycles was related to the formation of a surface haze that changed the thermal properties of the instrumented wafer.

Photoresist Baking

For deep-UV photoresists, the post-exposure bake temperature and temperature uniformity are critical to the control of pattern dimensions [4]. The matching of hotplate from tool to tool and the calibration of hotplate temperature must have offset errors of less than 1 °C.

Figure 7 shows the time-temperature profile of a photoresist bake hotplate using an instrumented silicon wafer having an array of 17 embedded RTDs. Each trace represents the temperature of one sensor over a 3-minute time interval. The difference between the trace represents the temperature uniformity across the 17 sensor positions.

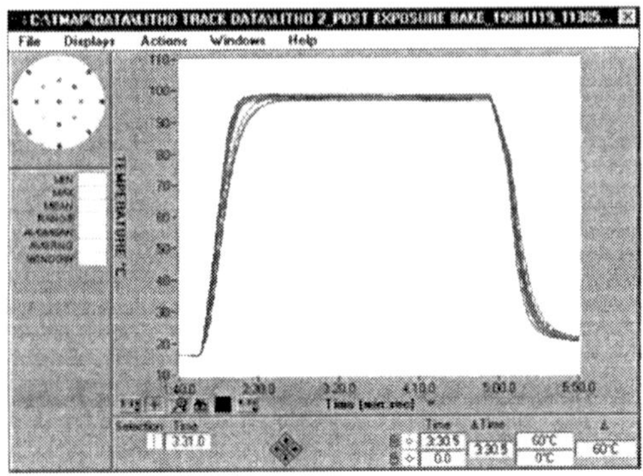

Figure 7: A graph of the heat up and cool down cycle of a 200 mm instrumented wafer, with 17 embedded RTD's, on a highly non-uniform photoresist hot plate.

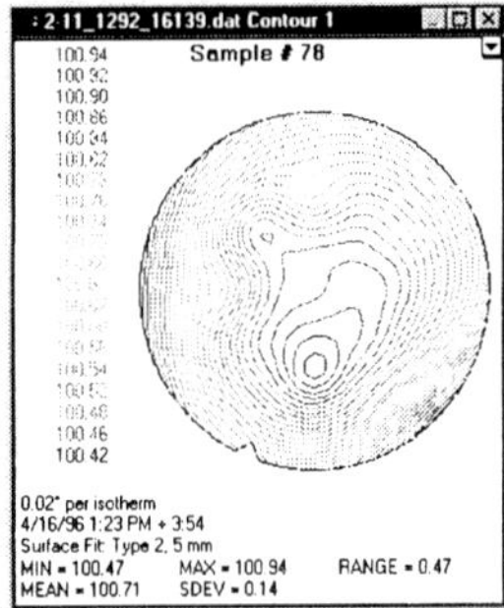

Fig. 8: Contour map of a RTD instrumented wafer on a typical photoresist post exposure bake hotplate.

Plasma Etching

In plasma etching, the etch rate, selectivity and CD control all depend on wafer temperature and uniformity, which depends on the thermal contact between the wafer and a temperature-controlled wafer chuck. The conditions that affect this heat transfer can be optimized using instrumented wafers having fiber optic photoluminescent sensors (5). These sensors, in contrast to TCs or RTDs can operate in a plasma environment and real time *in-situ* temperature measurements can be made under actual plasma processing conditions. The sensor is a fluorescent material that has a temperature-dependent decay time. The sensor material is coated directly onto the end of the grooves formed in the wafer. The optical fiber is then brought close to the coating to optically communicate with the sensor material. The temperature is measured by determining the decay time of the fluorescence initiated by a light pulse guided by the optical fiber.

A common means to control wafer temperature during plasma etching is by wafer backside helium flow in the wafer chuck at about 10 Torr. The two processing parameters that have the strongest influence on the wafer temperatures are the plasma RF power level and helium backside pressure. Studies of the effects of plasma etching parameters on wafer temperature using FOT-sensor instrumented wafers have revealed the expected effect of power as well as unexpected effects due to the nature of the process gases and the sensitivity to helium backside pressure above 10 Torr (5). Figure 9 shows the effect of power level on wafer temperature at four positions across the wafer diameter.

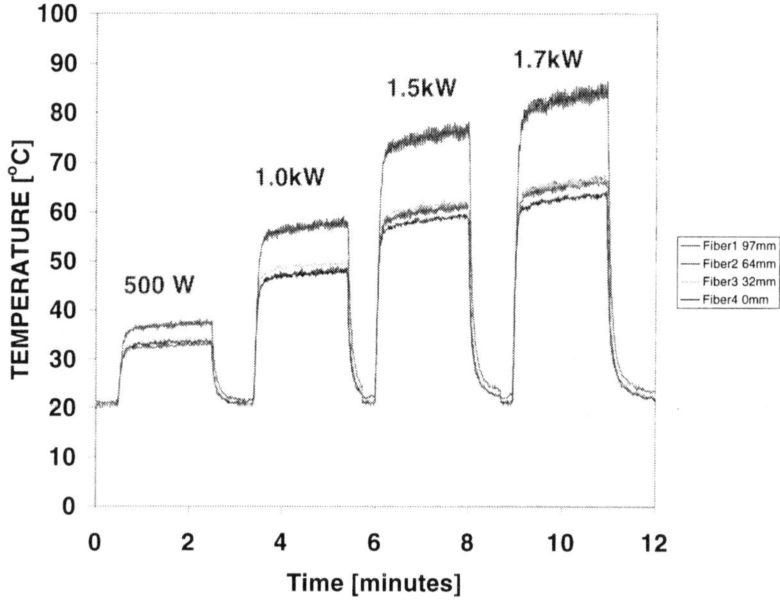

Fig. 9: Effect of power level on the wafer temperature profile in a high-density plasma oxide etch tool. Measurements were made with a 4-sensor fiber optic instrumented silicon wafer. Fiber 1 (near the edge) indicates a much higher temperature than the other probes (center and mid-radius).

CONCLUSIONS

Three different measurement techniques are available to measure and calibrate temperature in microelectronic manufacturing. TC measurement prevails for high temperature (> 400 – 600 °C) measurements. An accuracy of 1 °C can be obtained when taking care. When taking into account the sensitivity of K-type TC for oxygen above 600 °C, accurate measurements can be made over a long time. For medium temperatures (10 – 400 °C), RTD measurements can be far more accurate (0.1 °C accuracy, 0.05 °C precision). The calibration of RTD wafers can occur in a high accuracy isothermal bath, so NIST traceable measurements are possible. For application in a "live" plasma, the Fiber Optic Thermometer with temperature dependent photoluminescence can be used in the range up to 400 °C. This allows unique multi-point measurements of wafer temperature on cold chuck, while submitted to plasma heating from the front side of the wafer.

ACKNOWLEDGMENTS

The authors would like to thank Mei Sun for the review of the paper.

REFERENCES

1. P. Vandenabeele and W. Renken, "Study of Repeatability, Relative Accuracy, and Lifetime of Thermocouple Instrumented Wafers for RTP," *MRS Symposium Proceedings*, (Spring 1997).

2. P. Vandenabeele and W. Renken, "Model-Based Temperature Control in RTP Yielding ± 0.1°C Accuracy on a 1000°C, 2-second, 100°C/s Spike Anneal," *MRS Symposium Proceedings*, (Spring 1998).

3. P. Vandenabeele and W. Renken, "Calibration of Wafer Temperature to NIST Traceable Standards Using an Isothermal Cavity," *MRS Symposium Proceedings*, (Spring 1997).

4. J. Parker and W. Renken, "Temperature Metrology for CD Control in DUV Lithography," *Semiconductor International*, **20**(10), 111 (Sept. 1997).

5. C. Gabriel and E. Yeh, "*In-Situ* Wafer Temperature Measurement During Plasma Etching" *Solid State Technology* (to be published, 1999).

6. K. Kreider, *et al.*, " *MRS Symposium Proceedings*, (Spring 1997).

PASSIVE AND ACTIVE PYROMETRY IN RTP AND RTCVD SYSTEMS

E.D. Glazman*, A. E. Glazman*, Z. Atzmon**, H. Gilboa**, E. Iskevitch**
and A. Thon*
*3T – True Temperature Technologies, Theradion Industrial Park, Misgav 20179, Israel
** STEAG CVD Systems, PO Box 171, Migdal Ha'Emek 10051, Israel

Acquiring good accuracy in real time temperature measurement in semiconductor processing is difficult due to a-priori unknown and changing optical properties (emissivity) of the target wafer, and the need to eliminate direct and indirect stray light.
The potential accuracy and applicability to RTP and RTCVD processes of common passive and active pyrometry methods are discussed. A new passive – active combined technique is described and demonstrated. Results of temperature measurement during actual process conditions with a precision well within the range of ±1% are given for wafer emissivity ranging from 0.2 to 0.95 and temperature range of 425 to 1000°C.

INTRODUCTION

Non-contact temperature measurement in RTP / RTCVD processes is complicated not only by the unknown and changing emissivity of the wafer, but also due to the very high temperature ramp rate, the stray light from the heating lamps and the aggressive ambient conditions in the growth chamber. Any measurement system has to have the following characteristics in order to be efficient and useful:

1. Accuracy within the specific process window. A firm window of parameters characterizing every specific process should be maintained within a pre-set and usually small tolerance, as low as some fractions of a percent of the set point [1].
2. Fast response: while the operating temperature range is typically between 400°C to 1200°C, the ramp rate is very high and can exceed 100-150°C/sec
3. Robustness.
4. Non-intrusiveness towards the process.

Pyrometry has been widely recognized as the method of choice for temperature measurement in RTP and RTCVD. For the most part, enhanced emissivity [2], Ripple [3] and brightness [2] methods are being used. In this work we describe a novel technique, that has been demonstrated in real time, during rapid thermal chemical vapor deposition (RTCVD) in the *IntegraPro* cluster tool [4].

Various multi – spectral passive methods, Approximation, Ratio and Calibration are well known in the art [2]. Differences between them are found in the algorithms used to transform raw spectral data (brightness temperatures) to true temperature. However, the total accuracy of any method depends solely on how accurate the emissivity $\varepsilon(\lambda)$ is known or being approximated. None of the approximations for the spectral emissivity can express the required degree of accuracy in situations where the emissivity varies with process conditions and wavelength[5].

Active optical pyrometry is based on the use of an external light source to measure the surface reflectivity of the object. Given that the object is opaque, as usually found in Si wafers above ~420°C, the surface emissivity can be calculated by use of Kirchhoff's law:

$$abs(\lambda, T) + t(\lambda, T) + R(\lambda, T) = 1 \quad (1)$$

Where **abs(λ,T)**, τ(λ,T) and **R(λ,T)** are the absorption, transmission and reflection. Once the wafer emissivity is known, then the true temperature can be calculated based on the brightness temperature being measured. As simple as this technique seems, the actual implementation is much more complicated, since the apparent surface reflectivity is a strong function of the surface roughness. The roughness RMS, while on the order of the wavelength of measurement, usually differs both from that of pure diffuse (Lambertian) reflector and also from that of pure specular reflector. As a result, the measured intensity of the reflected signal can deviate from the intrinsic true reflectivity, which in turn can lead to errors in the calculated surface temperature.

The *integraPro*, shown in Figure (1), is a single wafer integrated RTCVD system consisting of a gas-phase-cleaning module and two RTCVD modules. It is being used for the deposition of dielectric materials such as silicon oxides and nitrides and polysilicon for gate stack and DRAM cell formation. The process chambers are clustered to a central transfer chamber with a base pressure of 10^{-8} Torr. The transfer chamber has a built in wafer aligning station and an active cooling station. Two load locks, each separately pumped to 10^{-7} Torr, are connected to the central transfer chamber. Wafers treated in the gas-phase-cleaning module are exposed to a gaseous mixture of HF, CH_3OH and N_2, which removes the native oxide layer from the wafer. After native oxide removal, a dielectric layer and a polysilicon layers can be successively grown in the two modules.

We report here results of experiments aiming to test and evaluate the emissivity independent operation of the new pyrometer under real process conditions: Poly Si deposition in high vacuum emulating HSG (Hemispherical Grained Silicon) formation. This process was chosen due to its high sensitivity to small changes in the surface temperature of the wafer.

EXPERIMENTAL

Poly silicon deposition was carried under pressure of 10^{-5} Torr and temperature of 615-635°C. Di-silane was used as a growth precursor, with flow rate of 30 SCCM. The deposition was on top of 1000Å oxide layer. To check for emissivity independence of the measurement system, the following wafer types with different emissivity, covering a wide emissivity range from 0.2 to 0.9 (at wavelength of 0.95 μm and temperature of 625°C) were used:

1. 1000A Oxide.
2. 1600 Poly/1000A Oxide.
3. 1950A Poly/1000A Oxide.
4. 1600 A Oxide.
5. 3650A Poly/5000A Oxide.

6. 8000A Oxide.
7. 2000A Poly/1000A Oxide.

The actual wafer temperature during growth was calculated by *ex-situ* measurment of the deposited layer thickness. This data was compared to a pre-calibrated layers that were deposited on top of standard wafers ($\varepsilon = 0.82$) at temperature of 615, 625 and 635°C.

Figure (2) shows the temperature and emissivity of nine wafer run as measured by the pyrometer. After conversion of the thickness data into the actual temperature prevailed in the reactor during growth, the data showed that the maximum temperature error, for the very low emissivity wafer was 6°C. This gives a variance in temperature measurement well within 1% of the setpoint temperature. Figure (2) also shows the fast response of the measurement system to the high ramp rate. It should be noted however, that the minimum detection temperature varies from wafer to wafer, according to the wafer emissivity. The transition from transparency to opaqueness depends on the doping level of the wafer. Hence, wafers with low doping level can be measured only from higher temperature than highly doped wafers showing high apparent emissivity.

Additional data (spectral brightness temperature and emissivity) can also be provided to the user in real time during growth. This information can be used for *in-situ* growth analysis, such as growth rate and quality and also to monitor the equipment cleaness and readiness.

SUMMARY

In the general case where the spectral emissivity changes as a function of wavelength and temperature, the maximum accuracy of all passive pyrometry methods does not depend on the specific method being chosen. Active pyrometry, on the other hand, is very sensitive to dynamic changes in surface roughness occurring during growth. A novel technique that combines the advantages of both multi-spectral passive and active methods was demonstrated during actual conditions of RTCVD process. Measurement accuracy and repeatability are within 1.0% of the set point temperature. The additional dynamical spectral data available can serve as *in-situ* indicators for changes in layer thickness, growth quality and as a monitor of the growth chamber status.

REFERENCES

1. The National Technology Roadmap for Semiconductors, SIA Semiconductor Industry Association, 1997.
2. DeWitt, D.P. and G.D. Nutter (editors), Theory and Practice of Radiation Thermometry, John Wiley, New York, 1988.
3. A.T. Fiory, at. al., *Mat. Res. Symp. Proc.*, **303**, 1993.
4. E. Glazman, A. Glazman and A. Thon, to be published in the proceedings of EUROPTO / SPIE symposium, May 1999.
5. E.D. Glazman, I.I. Novikov, *Proceedings of Optical Methods and Means in Temperature Measurement. Moscow, Science*, 1983 (in Russian).

ACKNOWLEDGEMENTS

Financial support for this work was partially provided by the Israeli Governmental MAGNET Consortium for 0.25 mm / 300 mm technologies.

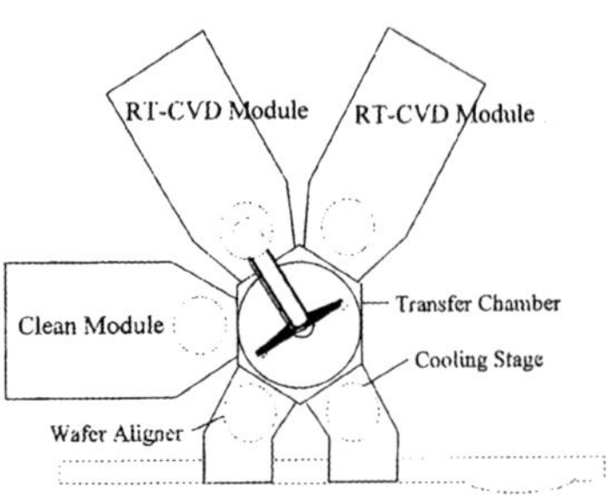

Figure 1: The *IntegraPro* cluster tool for HSG poly-silicon formation, consisting of two RT-CVD modules and one gas phase cleaning module, connected via a transfer chamber.

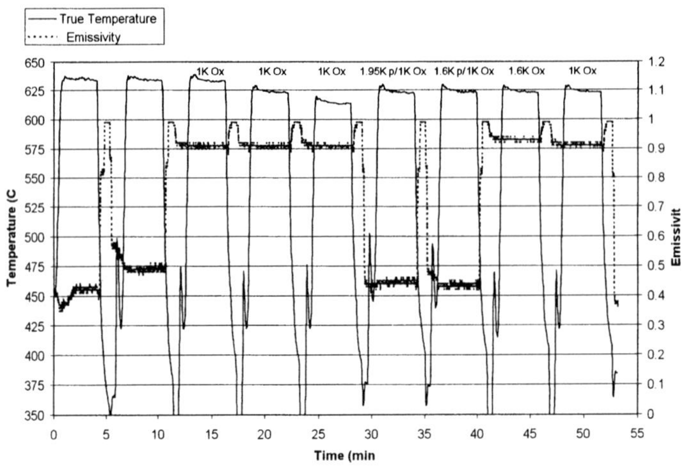

Figure 2: Closed loop pyrometer testing for repeatability and emissivity independent.

TEMPERATURE MEASUREMENT, UNIFORMITY, AND CONTROL IN A FURNACE-BASED RAPID THERMAL PROCESSING SYSTEM

Jeffrey Hebb and Ali Shajii
Eaton Corporation, Thermal Processing Systems, 2 Centennial Drive, Peabody, MA 01960, USA

In this paper we describe the operating principles of the Summit, a furnace-based rapid thermal processing system. Detailed performance results are presented with particular emphasis on wafer temperature measurement, uniformity, and control. Specifically, an emissivity compensation system combining *ex situ* and *in situ* reflectivity measurements is described. Implant anneal results for a large range of backside emissivities are presented which demonstrate the wafer-to-wafer temperature repeatability of +/- 2 °C. We also present implant anneal results which show within wafer temperature uniformity of better than +/-2.5 °C. Finally, a real time model based control scheme for the system is described, and shown to produce wafer temperature trajectories with linear ramp rates of 75 °C/s and overshoot less than 2 °C.

INTRODUCTION

In this paper we describe a furnace based rapid thermal processing (RTP) system (the Summit) which is inherently different from traditional lamp-based RTP systems. The paper focuses on the emissivity compensated pyrometry, within wafer temperature uniformity, and model based temperature control.

A schematic of the Summit system is shown in Fig. 1. The heating source is an axisymmetric silicon carbide bell jar that is heated by using a three-zone furnace. The base of the bell jar is water cooled, so that a vertical thermal gradient is established. Three spike thermocouples (TC's) are used to measure the bell jar temperature at all times. Before a process begins, the desired thermal gradient is established by adjusting the zone powers until the desired spike TC setpoints are reached. The zone powers continue to be actively controlled during wafer processing to maintain this stable thermal environment. The Summit is an atmospheric system, with ambient control achieved by flowing process gas through a quartz injector tube, as well as through the elevator tube.

To begin processing, the wafer is placed on a quartz wafer holder which sits on a quartz elevator tube. The wafer temperature is then changed by moving the wafer up or down in the reactor. The temperature trajectory of the wafer is determined by the position trajectory and the radiative properties of the wafer. Typically, the spike TC setpoints are set so that the top of the bell jar is approximately 200 °C higher than the desired processing temperature. Wafer temperatures up to 1200 °C and ramp rates of up to 150 °C /sec can be achieved in the system.

During processing, the temperature at the center of the wafer is measured using an emissivity compensated pyrometry system. The measured wafer temperature is then used in conjunction with real-time model based control of the elevator position to produce the desired time-temperature profile.

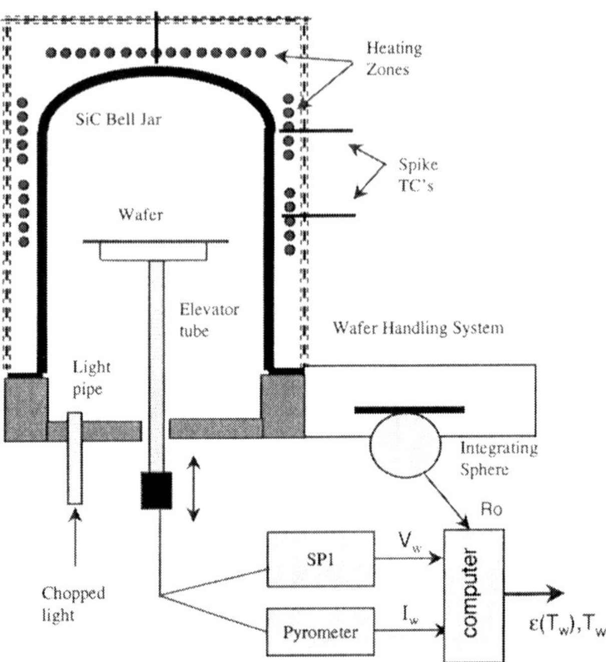

Fig 2: Schematic of the Summit, including the emissivity compensation system.

EMISSIVITY COMPENSATED TEMPERATURE MEASUREMENT

To make accurate wafer temperature measurements using pyrometry, the emissivity of the wafer backside at the pyrometer wavelength must be known. Typically, production silicon wafers have backside films which can drastically alter the spectral emissivity through interference effects, and potentially causing large temperature measurement errors for standard pyrometry. Furthermore, the emissivity is generally dependent on wafer temperature and backside surface roughness (1-4). All of these factors make real time *in situ* emissivity compensation a requirement to achieve acceptable temperature measurement accuracy.

The appropriate emissivity to use when performing pyrometry in an enclosure is referred to as the "effective" emissivity, which differs from the free space emissivity because of multiple reflections of the emitted light that find their way into the pyrometer. In our system, all of the surfaces that the wafer interacts with are generally far from the wafer, and have reflectivities less than 0.1 (SiC and quartz). Hence, the effect of multiple reflections is negligible, and the appropriate emissivity to measure is the "free space" emissivity.

The spectral and directional characteristics of the desired emissivity are defined by the pyrometer optics. In our system, it is required to measure wafer emissivity in the normal direction, averaged over a narrow spectral range ($\Delta\lambda$) and centered at approximately λ = 0.95 μm. Operating the pyrometer at this wavelength has the advantage that the wafer is opaque at all temperatures, thus eliminating transmitted light errors.

The objective of the emissivity compensation system is to measure the emissivity, ε, of the wafer backside during the processing of a wafer. This is achieved by determining the hemispherical directional reflectivity of the wafer backside, and applying the following equation:

$$\varepsilon(T_w) = 1 - R(T_w) \qquad [1]$$

where ε is defined as the normal emissivity averaged over the spectral range $\Delta\lambda$, and R is defined as the *hemispherical directional reflectivity* in the normal direction, also averaged over the spectral band $\Delta\lambda$. The hemispherical directional reflectivity in the normal direction is defined as the reflectivity one would measure by illuminating the target uniformly and hemispherically, and collecting the reflected light in the normal direction (5). If these conditions are not met for the reflectivity measurement, Eq.(1) is no longer valid. Our objective is to measure R at a spot on the backside of the wafer. In general, the conditions of uniform hemispherical illumination are difficult to achieve within the bell jar. Thus, a combination of *ex situ* and *in situ* measurements are used to achieve the high precision results presented in the paper.

The *ex situ* component of the measurement involves using an integrating sphere, as shown in Fig. 1. The hollow sphere interior is made of reflective-diffuse material. Three measurement ports are provided. Before the wafer enters the bell jar, it is placed on the integrating sphere, where a small spot at the center of the backside of the wafer covers the first port. Source light is introduced through the second port in such a way as to illuminate the spot on the backside of the wafer in a *hemispherically uniform* manner. The light reflected from the wafer in the near-normal direction is then collected by a fiber optic cable which is situated at the third port. The intensity of the reflected light is measured over the spectral range $\Delta\lambda$, and then normalized to the source intensity. The room temperature hemispherical directional reflectivity of the wafer, denoted by R_o, is determined from a calibration curve created using reflectivity reference standards.

Once R_o is obtained, the wafer enters the processing chamber, where the set-up presented in Fig. 1 is used to perform a relative reflectivity measurement. A chopped light source at frequency f is introduced into the chamber through a specially designed light pipe residing at a fixed position near the bottom of the chamber. The light pipe directs the chopped light upward where it illuminates the backside of the wafer. The total radiative flux from the wafer center (in the normal direction) is then collected by the pyrometer head, fixed at the bottom of the quartz elevator tube. The radiative flux includes light emitted from the wafer as well as the reflected portion of the chopped light. The optical signal is split, with one portion directed to a signal processing unit SP1 and the other directed to the pyrometer controller. The SP1 unit outputs a voltage proportional to the intensity of the reflected chopped light, denoted as V_w.

Recall that the goal here is to measure $R(T_w)$. Equation (1) can then be used to determine the emissivity. Since the *in situ* illumination of the wafer does not meet the requirements under which Eq. (1) was derived (i.e. uniform hemispherical illumination), we need to correlate R with V_w. Here, we assume that at any given time during the process, the hemispherical directional reflectivity is proportional to the reflected intensity V_w, i.e.,

$$R = KV_w \qquad [2]$$

where K is the proportionality constant, and is a strong function of the surface finish of the wafer backside, but not a function of wafer temperature. This is a well justified assumption, based on detailed studies of light scattering from the wafer backside at different wafer temperatures.

In order to obtain K, an elevator "sweep" is performed. As soon as the wafer is placed on the elevator tube, it is rapidly moved from the bottom to the top of the heating chamber, while doing a real time measurement of the z and V_w. Due to the finite thermal mass of the wafer, its temperature does not increase significantly during the sweep, which takes approximately 2 seconds (model predictions show that T < 100 °C at the end of the sweep). Therefore, we can assume that during the sweep, the wafer backside reflectivity (R) is equal to that measured using the integrating sphere (R_o). This yields the information necessary to determine the proportionality constant K:

$$K(z) = \frac{R_o}{V_{w,o}(z)} \qquad [3]$$

Where z is the wafer height in the reactor, and Vw,o is the measured reflected light during the elevator sweep.

From this point on, the reflected light intensity Vw is continuously measured, and the emissivity is determined by combining equations 1,2 and 3:

$$\varepsilon(T_w) = 1 - \frac{R_o}{V_{w,o}(z)} V_w \qquad [4]$$

The measured value of ε is then used in conjunction with the pyrometer reading to obtain the wafer temperature.

One of the main advantages of this approach is that no test wafers, thermocouple wafers, or implant monitor wafers are required to calibrate the *in situ* emissivity compensation system. The sweep is performed on every wafer, so that each actual product wafer has its own unique $K(z)$. The only portion of the system which has to be calibrated for absolute accuracy is the integrating sphere, which is in a fixed, low temperature, well controlled environment where calibration is relatively straightforward.

The effectiveness of the emissivity compensation system is evaluated by performing a series of implant anneal experiments. The wafers used had various backside films which yield an emissivity range of .20 to .76 at the pyrometer wavelength. The backside layers and emissivity of each wafer type are shown in Table 1. Four wafers of each type were annealed at 1040 °C for 10 seconds, using an average ramp rate of 65 °C/sec. All wafers have the same frontside implant (As, 1e16 cm^{-2}, 25 keV through at 100 A screen oxide) which is used to infer wafer temperature. The sheet resistance of each wafer was measured using a four point probe after stripping the screen oxide.. Figure 2 shows the average Rs for each wafer. The sensitivity for this implant is approximately 1 Ω/sq/ °C, so the results show a total range of 4 ohm/sq for all wafers, or +/- 2 °C. This corresponds

Table 1: Description of wafer backsides and emissivities at the pyrometer wavelength

Wafer Type	Emissivity	Backside layers
1	0.20	900 A poly / 1600 A SiO_2
2	0.22	900 A poly / 5300 A SiO_2
3	0.68	Bare Si
4	0.68	3800 A TEOS
5	0.76	1300 A Si_3N_4 / 3100 A SiO_2

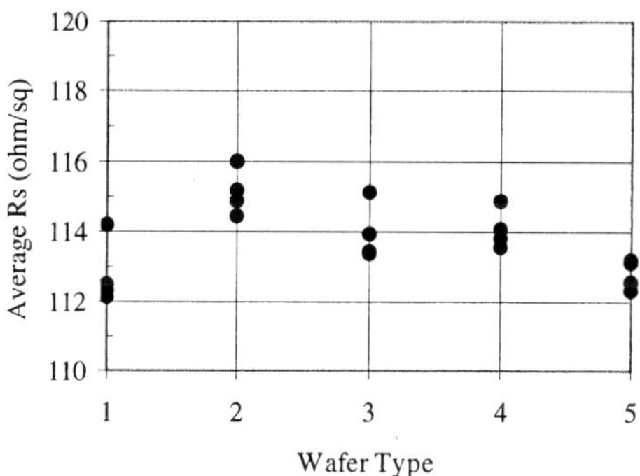

Fig. 2: Average sheet resistance for wafers with varying backsides, 1040 °C /10sec anneal.

to an accuracy of +/- .005 for the emissivity measurement.

WAFER TEMPERATURE UNIFORMITY

The within wafer (WIW) temperature uniformity is determined by the distribution of incident heat flux on the wafer during processing. The key aspects of the nature of the heating source are that it is spatially continuous, axisymmetric, and is in quasi-thermal equilibrium with the wafer. For a given bell jar temperature profile, there is a vertical position range (typically at the top half of the bell jar) over which a highly uniform temperature across the wafer can be maintained. To a first approximation, moving the wafer into this optimum processing zone of the bell jar is equivalent to placing the wafer in an isothermal cavity. It is well known that in such a cavity, a body will achieve a uniform temperature (5). The size and position of the optimum processing zone is determined by the bell jar vertical temperature gradient, which is controlled by the spike TC setpoints. We have developed detailed numerical models (not presented here) to guide and optimize the selection of the spike set points and the wafer processing height.

Figure 3 shows implant anneal data that demonstrates the across wafer uniformity and across cassette repeatability. The wafers are 200 mm bare Si wafers which have been implanted with As 25keV 1e16 cm^{-2}, with no screen oxide. They were annealed at

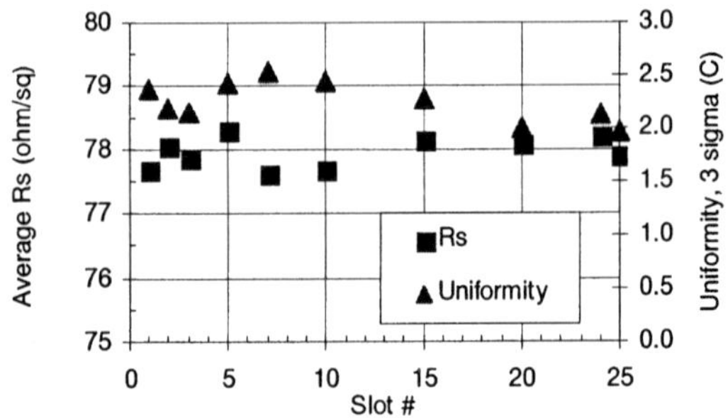

Fig. 3: Temperature uniformity and repeatability for 1040 °C/10sec anneal, across a cassette.

1040 °C for 10 seconds with a 65 °C/sec ramp rate in 600 ppm of O_2 to prevent outdiffusion. The sheet resistance was measured at 49 points with a 6 mm edge exclusion. The sensitivity of this implant for this anneal is 0.55 ohm/sq/ °C. Figure 3 shows the average sheet resistance of each wafer and the 3 σ temperature uniformity for each wafer. All uniformities are between 1.9 and 2.5 °C, with a 2.25 °C average uniformity. The across cassette repeatability is 1.29 °C (3σ). This data was part of a five day internal study which showed that the total temperature variation was 4.6 °C (3σ), including within wafer, within cassette, film-to-film, cassette-to-cassette, and day-to-day variations.

WAFER TEMPERATURE CONTROL

The temperature trajectory of the wafer is determined by the elevator position trajectory and the radiative properties of the wafer. Therefore, control of a specified time temperature profile is achieved by moving through the appropriate elevator-position trajectory. The Summit has several inherent advantages for temperature control:

1) The incident power on the wafer is determined by elevator position, which is a mechanical, repeatable and easily controlled parameter.
2) The quasi-thermal equilibrium nature of the system mitigates the effect of total radiative properties on thermal response of the wafer, so that different wafers will not have drastically different position trajectories to achieve the same temperature trajectory
3) Since Summit inherently delivers uniform temperature, only the temperature at the center of the wafer needs to be measured and controlled. This reduces the complexity of the control problem to a single input (T) – single output (z) system.

The goal of the temperature measurement and control system is to provide a repeatable temperature profile for any wafer, regardless of the radiative properties. This is achieved by using a real time model based temperature control approach, described below.

To develop a model for the wafer temperature, the general heat equation is written for the wafer in 2D cylindrical coordinates. This general model is then reduced to a simple ordinary differential equation (in time) that represents the average wafer temperature.

$$\rho C \frac{dT}{dt} = (\varepsilon_t + \varepsilon_b)\sigma T^4 + \alpha_t q_t(z) + \alpha_b q_b(z) \qquad [5]$$

where T is the average wafer temperature, ρ is the wafer density, $C(T)$ is the wafer heat capacity, ε_f and ε_b are total emissivities, α_f and α_b are total absorptivities, and q_f and q_b are total incident heat fluxes averaged over the wafer radius. Subscripts f and b denote wafer front and backside, respectively. As an approximation to the total radiative properties of a product wafer, we assume that the wafer is bare lightly doped silicon. The temperature dependent radiative properties are calculated using thin film optics theory and a model for the optical constants of silicon (6). To determine the averaged incident heat fluxes q_f and q_b, we assume that the presence of the wafer does not effect the bell jar temperature. This assumption allows us to separate the dynamics of the wafer and that of the bell jar, greatly simplifying the formulation. Thus, for a given set of furnace spike setpoints, the temperature distribution of the bell jar is calculated once (bell jar model not described here), and is then used as a known heat source to the wafer. For a given wafer position, the differential view factors from the bell jar to the wafer are calculated and integrated over the total areas to obtain q_f and q_b. Thus, for a given input $z(t)$, the output is $T(t)$.

The goal of model-based control is to use Eq. (5) in real time to guide the elevator trajectory during processing. In order to achieve this goal, we first invert Eq. (5) and solve for the wafer height as a function of wafer temperature and rate of temperature change:

$$z = z(T, dT/dt) \qquad [6]$$

If the radiative properties of the wafer where known exactly, and the bell jar temperature profile was highly accurate, Eq.(5) would be used to drive the elevator for a specified temperature trajectory. However, since the calculated bell jar temperature profile is only an approximation of the physical system at hand, and the radiative properties generally vary from wafer to wafer, we need to utilize Eq. (5) with a feedback servo in order to obtain acceptable performance. Extensive work is currently being conducted to determine the optimum servo type with appropriate parameters in the Summit system. One servo currently under investigation is given by:

$$z_{traj} = z(T_{traj} + D_p T_e + D_I \int_{t_i}^{t} T_e(t')dt', \frac{dT_{traj}}{dt} + D_d T_e) \qquad [7]$$

That is, for a given desired $T_{traj}(t)$, Eq. (6) would be solved to obtain $z_{traj}(t)$ to drive the elevator, where D_p, D_I, and D_d represent the proportional, integral, and derivative gain terms in the servo system, $T_e = T_{pyro} - T_{traj}(t)$, and T_{pyro} is the pyrometer temperature reading.

To evaluate the effectiveness of the servo described by Eq. (7), we ran the five types of wafers shown in Table 1 with the same recipe. Figure 4 shows the temperature and position trajectories for wafer type 5 (nitride/oxide backside). The instantaneous ramp rate calculated from the pyrometer data is also plotted. The recipe consists of the following: The wafer is introduced into the bell jar, the emissivity sweep is performed, and then the wafer is brought to its pre-soak position. When the wafer reaches 650°C, the desired temperature trajectory is to ramp from 650 to 900 °C at 75 °C /sec, decrease the

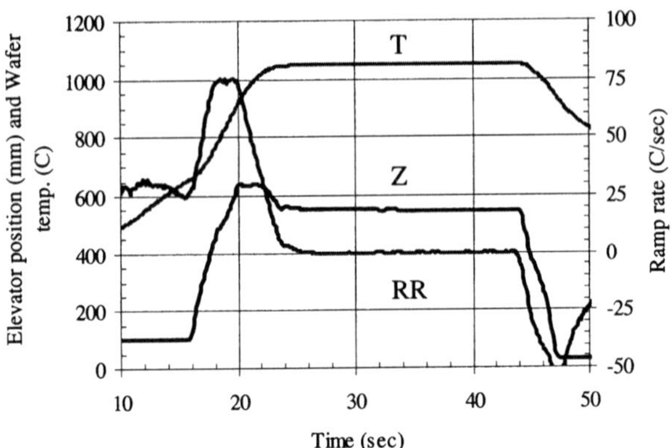

Fig. 4: Wafer temperature, position, and ramp rate using model based control

ramp rate linearly until the temperature reaches 1050 °C, and soak at 1050 °C for 20 seconds. The ramp rate trajectory shows that the control scheme is effective. This wafer demonstrates the success of the model based control scheme as the total radiative properties of this wafer are different from that of bare silicon due to the presence of backside films. The plots for the rest of the wafers are very similar, with acceptable ramp rate control, and overshoots less than 2 °C for all wafers. Work is currently underway to further decrease the transition time as the temperature trajectory goes from the ramp region to the soak region.

CONCLUSIONS

We have demonstrated effective thermal performance of the Summit. Implant anneal data shows that a total range of 4 °C, or +/- 2 °C, is achieved for wafers with a wide range of backside emissivities. Within wafer temperature uniformity of better than +/-2.5 °C (3σ) has also been demonstrated. Finally, the model-based control scheme shows effective performance with linear ramps and minimal overshoot for a wide range of wafer backsides.

REFERENCES

1. P.J. Timans in *Advances in Rapid Thermal and Integrated Processing*, ed. Fred Roozeboom, Kluwer Academic Publishers, The Netherlands, 35 (1996).
2. J.P. Hebb and K.F. Jensen, *IEEE Trans. on Semi. Man.*, **11** (1998).
3. P. Vandenabeele and K. Maex, *Proc. SPIE*, **1393**, 316 (1990).
4. C. Schietinger in *Advances in Rapid Thermal and Integrated Processing*, ed. Fred Roozeboom, Kluwer Academic Publishers, The Netherlands, 103(1996).
5. R. Siegel and J.R. Howell, *Thermal Radiation Heat Transfer*, Hemisphere Publishing Corporation, Washington, D.C. (1992).
6. J.P. Hebb, *Patterns Effects in Rapid Thermal Processing*, PhD Thesis, Massachusetts Institute of Technology (1997).

EMISSIVITY COMPENSATED WAFER TEMPERATURE MEASUREMENT USING INTENSITY-MODULATED LAMP LIGHT

M. Hauf, H. Balthasar, Ch. Merkl, S. Müller and Ch. Striebel
STEAG RTP SYSTEMS GMBH
Daimlerstrasse 10, 89160 Dornstadt, Germany

Abstract

An emissivity compensated wafer temperature measurement technique for use in double side heated RTP systems is described. This technique is based on an active compensation of the lamp radiation. In contrast to further attempts to realize a modulation based temperature measurement technique, the modulation is no longer a parasitic effect of the lamp power control, it is software driven and configurable. The data presented indicate a temperature error smaller than 3 °C within a wide emissivity ($\epsilon = 0.80 \ldots 0.25$) and temperature (600 °C...1100 °C) range. The modulation technique is also able to cope with high ramp rates (≥250 °C) and shows a very well short and longterm stability.

1 Introduction

The use of intensity-modulated lamp-radiation, is well-known as an emissivity-compensated temperature measurement technique (often called Ripple™-technique) for double-side heated RTP-Systems. Several works have been published [1, 2, 3, 4, 5, 6] showing the feasibility of emissivity compensated temperature measurement by the use of the Ripple™ method. Based on these previous attempts, a completely new and improved measurement technology was developed. The newly improved measurement technique is comprised of several modifications in mechanical and optical engineering as well as signal analysis to meet the demands of a state-of-the-art RTP tool. The measurement technique must be stable, must be emissivity independent over a wide range ($\epsilon = 0.2 \ldots 0.9$), cope with high ramp-rates ($> 250\ ^{\circ}\mathrm{C\,s^{-1}}$)[7, 9] and must be insensitive to mechanical bending and slight tilting of the wafer resulting from wafer rotation. Moreover, the measurement system must be able to deal with variations of the individual lamp power settings, should not be affected by the IR-absorption bands of ambient and process gases and should be insensitive to the quartz temperature. These properties are often hard to achieve or at least hard to combine. With this temperature measurement technique, being different a substantial performance improvement is achieved. Its successful implementation into all the AST RTP tools namely AST2800, AST3000 and AST EVOLUTION™ [1], is described in this contribution.

[1]Incorporates Ripple™ Technology from Luxtron Corporation

2 Measurement System

2.1 Modulation of lamp power

Essential to this measurement technique is the use of modulated lamp radiation. Most of the former attempts trying to realize this technique used a parasitic effect. The ripple of the lamp radiation was caused by the phase angle control (PAC) of the lamp power [8]. In contrast, we are working with pulse width modulation (PMW) and because of the high pulse rate there is nearly no detectable parasitic ripple of lamp radiation. However, the software controlled power electronics allows modulation of the effective electrical power to the lamps and hence the modulation parameters are well defined and can be freely chosen. This means the shape, the modulation depth, the frequency and the phase can be best adapted for measurement purposes and the modulation parameters are kept stable over the whole lamp power range. The modulation with a sine wave, frequency 20Hz and a constant modulation depth of 10% of the maximum power has proven to be very convenient for our measurement technique. The resulting modulation of the lamp radiation measured by the pyrometers can be seen in Fig. 1. The AC-coupled signals from the wafer pyrometer and the lamp pyrometer are shown during a ramp up of a lightly doped bare silicon wafer. The upper and the lower lamp field are modulated with the same frequency, but with a $\pi/2$ phase shift. Because of the partial transmissivity of the wafer the phase shifted modulated radiation from the upper lamp field is passing through the wafer and is also entering the wafer pyrometer causing a phase shift of the wafer pyrometer signal relatively to the lamp pyrometer signal. The relative modulation depth of the lamp radiation at the measurement wave length ($\lambda = 2.3\mu m$) is of the order of 5×10^{-3}.

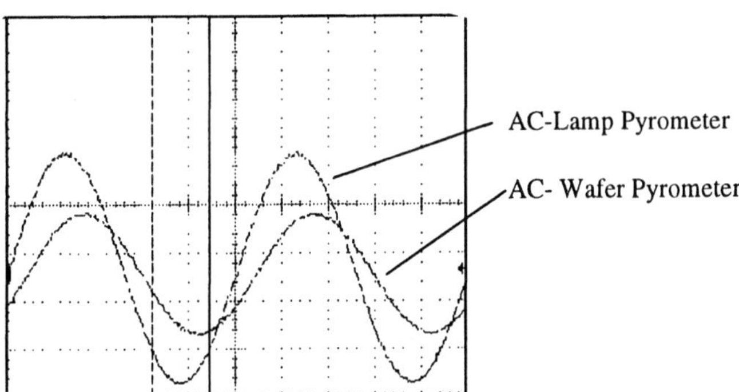

Figure 1: *Modulation of the lamp radiation. The signals are directly taken from the AC-outputs of wafer pyrometer and the lamp pyrometer. No smoothing of the signals was done while taking the oscilloscope screen shot. The relative modulation depth of the lamp radiation at the measurement wavelength ($\lambda = 2.3\mu m$)is of the order of* 5×10^{-3} .

2.2 Optics

The measurement system consists of two pyrometers: one pyrometer to monitor the lamp radiation (lp) and one pyrometer to monitor the wafer radiation (wp). The pyrometers were designed using classical lens optics for two principal reasons. First the lens optics allows to design a well defined field of view. Second, in contrast to the all fibre optic designs it is inherently robust because there is no bending of fibres and there is no trouble with input and output couplers. The special design of the wafer pyrometer allows the measurement of the hemispherical reflectivity, ρ, of the wafer whereas the design of the lamp pyrometer makes selective monitoring of the lamp radiation possible. The field of view of the wafer pyrometer is designed to be insensitive to mechanical bending and slight misalignment of the wafer's position and tilt caused, for example, from wafer rotation.

2.3 Pyrometer Heads

First of all, the pyrometer heads are expected to show an excellent short term and long term stability. In addition and in contrary to the fibre optic approaches, which are all dealing with remote detectors and amplifiers, we are dealing with classical lens optics and thus the pyrometer heads are always attached directly to the furnace. This causes some critical design constraints. The pyrometer heads must have a very compact design and must fit all AST furnaces. Because of high power consumption of the lamps it must also be robust against electromagnetic interference (EMI) and various ambient temperatures. Additionally the AC-coupling and the post-amplification of the modulated signal must be done by the pyrometer head electronics prior to feeding the analog signals to the data acquisition and processing unit. Signal to Background considerations lead us to choose to the measurement wavelength $\lambda = 2.3\mu m$. Also this wave length is free of IR absorption bands of process and ambient gases. The good suppression of EMI is demonstrated by the high quality of the AC-coupled readout captured in Fig. 1.
For the short term drift test a Micron M360 Black Body was used as a reference source. The nominal temperatures were 400 °C and 800 °C. Within 80h, the signal of the pyrometer differs around ± 0.5 °C. As can be seen from Fig. 2, this difference is similar to the stability of the calibration source. That means, that the drift and the noise visible in Fig. 2 are mainly due to the stability and the control oscillations of the calibration source. The longtime stability was checked on a AST2800 RTP tool in a production environment. An As-implant process was monitored over seven months. During this time a shift of the process result of smaller than 2.1 °C was found. These results indicates that our pyrometer technique is stable for short time as well as for long time measurements.

2.4 Signal Processing and Modelling

Each pyrometer has two output channels: a DC-coupled output and an AC-coupled pre-amplified output. From the two pyrometers, a four-dimensional measurement vector $\mathbf{X}$ is obtained:

$$\mathbf{X} = \left(X_{\text{lp}}^{\text{DC}}, X_{\text{wp}}^{\text{DC}}, X_{\text{lp}}^{\text{AC}}, X_{\text{wp}}^{\text{AC}}\right) \tag{1}$$

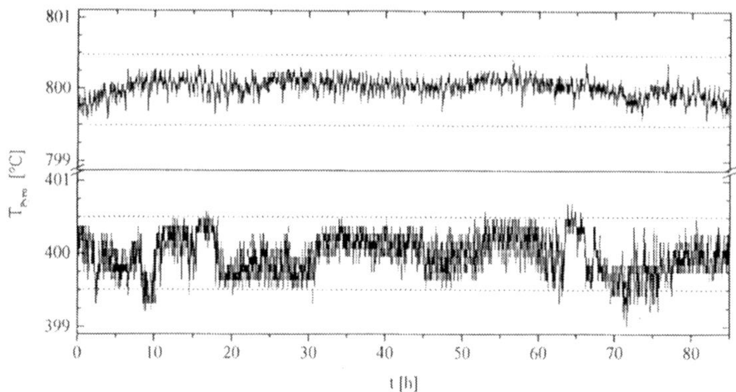

Figure 2: *Test of the drift stability of the pyrometer heads. The test temperatures were 400 °C and 800 °C. The dotted lines mark the stability specification of the reference source for a 8h period.*

From this vector, the wafer temperature can be calculated. The well known Ripple-formulas cannot be used, because they are based on a simplified system. To get a realistic model of the furnace, one has to take into account all the dominant radiation contributions including multiple reflections. So we developed an inverse model $f(\mathbf{X}, \alpha_1, \ldots, \alpha_N)$ of the dominant radiant interactions in the furnace. The model is matched to the real furnace by the parameter set $\{\alpha_1, \ldots, \alpha_N\}$. They are determinated by the start-up calibration of the measurement system (see below). Once the system is calibrated the temperature T, the reflectivity ρ and the transmission τ can be calculated according to

$$(T, \tau, \rho) = f(\mathbf{X}, \alpha_1 \ldots \alpha_N). \qquad (2)$$

3 Start-up Calibration Procedure

The calibration procedure can be looked on as a system identification procedure for the model $f(\mathbf{X}, \alpha_1 \ldots \alpha_N)$. The output of a calibration procedure is the parameter set $\{\alpha_1, \ldots, \alpha_N\}$. As usual in system identifications one needs to cover a wide range of the state space, where the system is expected to work later on. A subspace of the state space of the system is the $\{T, \epsilon\}$ space. This subspace will be covered uniformly by a set of wafers of different emissivities and a set of steady state temperatures. It is important to realize, that the calibration does not consist of creating look-up tables for these particular wafers, these wafers are only representatives of the states in the ε-axis of the state space. Hence the calibration of the measurement system is done using four S-Type thermocouple instrumented calibration wafers with different back-sides. The back side emissivity of these wafers ranges from 0.25 to 0.8. Calibration recipes containing a steady state of 120 s at temperatures between 600°C and 1100°C are used. To extend the life time of the calibration wafer set the upper temperature is limited to 1100 °C . Each calibration wafer is processed according to these recipes and the pyrometer data are recorded online. From the data, the parameter set $\{\alpha_1, \ldots, \alpha_N\}$ is calculated. Once the start up calibration is done, the temperature is calculated using the parameter set in Eqn. (2). Even though the temperature range used for calibration was limited, the temperature readout is also valid for temperatures up to 1250 °C .

4 Results

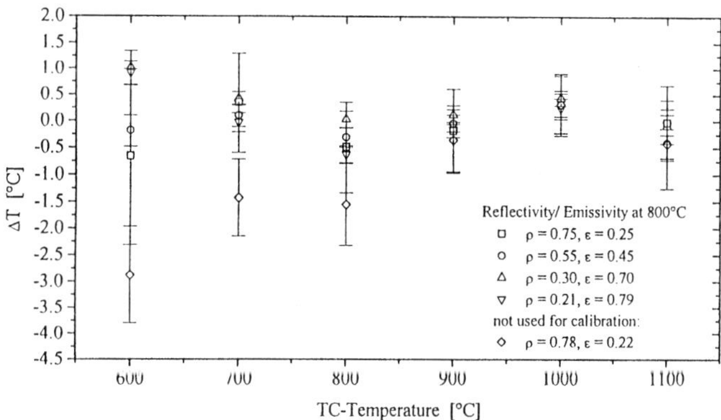

Figure 3: *Calibration Result on AST3000. The data points show the mean value of* $\Delta T = T_{\mathrm{TC}} - T\mathrm{Pyro}$ *for discrete temperatures from 600 °C to 1100 °C and emissivities from 0.25 to 0.79. The error bars are indicating the max-min errors for an 60s steady state.*

Fig. 3 shows a typical result of a calibration procedure on a AST3000. The temperature deviation $\Delta T = T_{\mathrm{TC}} - T_{\mathrm{Pyro}}$ is shown as a function of the wafer temperature for four different calibration wafers used in the start-up calibration. The data points are mean values over a 60s steady state and the error bars are indicating the upper and the lower limit of ΔT. The hemispherical reflectivity of the calibration wafers ranges from 0.21 to 0.75, which is equivalent to an emissivity range of $\epsilon = 0.25 \ldots 0.79$. Also data points from an additional wafer are shown, which exhibits an extremely low emissivity of 0.22 and which data set was not included in the calibration data set. Even though this wafer's low emissivity is outside the emissivity range covered by the calibration wafers, the data points are in close agreement with the data points of the calibration wafers. As can be seen from Fig. 3, in this range, the temperature error is smaller than 3 °C for all wafers and all temperatures from 600 °C to 1100 °C. This can only by obtained by using a physical model of the furnace.

4.1 High Ramp Rates

The active compensation of the lamp radiation is especially difficult for low emissivity, i.e. high reflectivity wafers, and for high lamp power. Thus a critical test

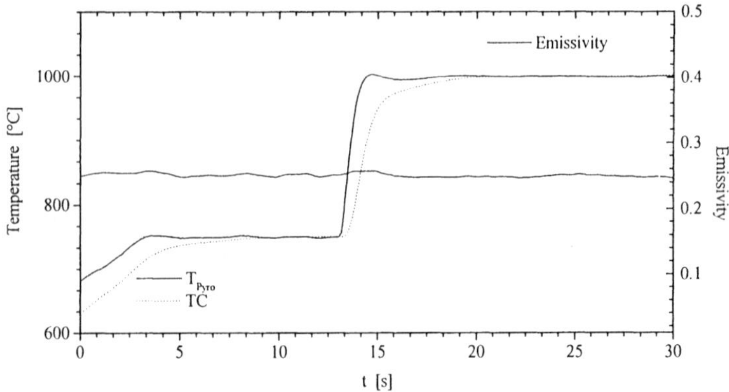

Figure 4: *Temperature readout of the pyrometer and TC and the online measured emissivity ϵ of the wafer along a recipe including a temperature ramp rate of* 250 °Cs^{-1}

to judge the quality of the active compensation of the lamp radiation is to run a process recipe containing a high ramp rate and using a wafer with a highly reflective backside. In Fig. 4 the temperature readout of the pyrometer along a temperature ramp from 750 °C to 1000 °C with a ramp rate of 250 °C s^{-1} is shown. Here a wafer with reflectivity of $\rho = 0.72$ was used. For comparison, the temperature readout of the TC is shown. As expected, the TC readout shows a longer response time than the pyrometer readout. Also in Fig. 4 the online measurement of the emissivity ϵ is shown. It can be clearly seen, that the online measurement of the emissivity ϵ of the wafer is independent of the extreme variations of the lamp power necessary to achieve these high ramp rates. This result also indicates, that the active compensation of the lamp radiation works very well, even at high ramp rates and for wafers with high reflective back-sides.

4.2 First experiments on low temperature measurement

Measurements in the low temperature regime are more difficult. The point is that measuring the intensity of the wafer radiation by pyrometric measurements would be very easy if there were no lamp light. Because of the presence of lamp light however, the wafer radiation is hidden in a huge background of lamp radiation. The signal to background ratio S/B is the most critical issue in expanding the measurement to lower temperatures.
To get an idea on how difficult low temperature measurement can be the S/B ratio decreases by the 10 to 20 fold when changing the temperature from 600 °C to 450 °C . This means that in the low temperature region the requirements are even harder to fulfil. Fig. 5 shows the result of our first experiments on low temperature measurements between 450 °C and 650 °C. It can be clearly seen that we are able to measure a reliable signal for $T \leq 450$ °C. Further developments will be done on temperature stability and temperature error in this low temperature range.

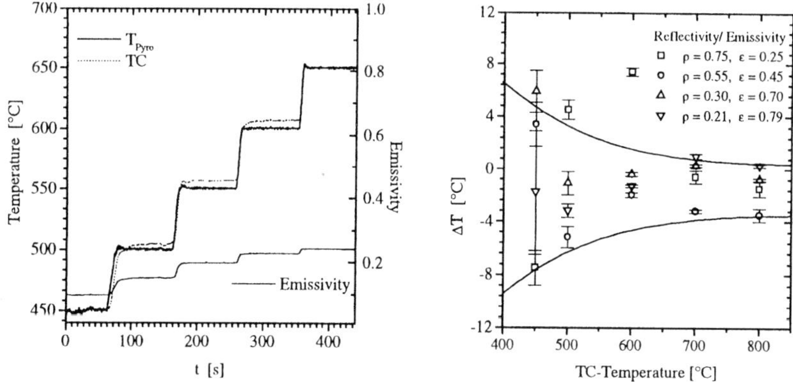

Figure 5: *Left: Temperature steps from 450 °C to 650 °C of a pyrometer controlled recipe. The test was done with the highly reflective wafer* ($\rho = 0.75$) *of the wafer set of the right figure. For comparison, TC datas are shown. Right:* $\Delta T = T_{TC} - T_{Pyro}$ *for wafers with different back side reflectivies (given for* $T \leq 800$ °C *).*

5 Conclusions

The current status of STEAG AST's emissivity compensated wafer temperature measurement system, which is characterised by a substantial improvement in system performance and measurement accuracy, was reported. This technique is based on a well defined and configurable pulse width modulation (PWM) of the lamp power, allowing the simultaneous determination of transmission, reflectivity, emissivity and wafer temperature.

In experiments we found that the measurement system is very stable for short time as well as for long time measurements. This technique was adapted successfully in the STEAG AST AST3000, AST2800 and AST EVOLUTION^{TM} RTP systems. The measurement system works at high ramp rates of more than 250 °C s^{-1}. It is shown for temperatures from 600 °C to 1100 °C that this temperature measurement technique enables a very precise and reliable control of the wafer temperature for any kind of RTP processes. However, the standard measurement range of the measurement system is 600 °C ... 1250 °C.

First measurements were also done in the low temperature regime. It could be shown that the measurement technique is able to get a reliable signal for $T \geq 450$ °C. Further developments of the measurement system are in progress and further efforts will be made on the electronic and the data processing units to improve the results, especially in the low temperature range.

Acknowledgements
The authors want acknowledge Helmut Sommer and Heinrich Walk for their great support and helpful and encouraging discussions. Thank you to Markus Glück for helpful discussions and support. We are thankful to Roland Mader, Frank Poxleitner and Jürgen Rau for doing a lot of work on development and testing of the measurement electronics and Thomas Knarr, Guido Bucher and Robert Kunz for their support in software development. Acknowledgements also to Ralf Reisdorf and Klaus Rührnschopf for doing a great deal of the process related tests of the measurement technique, Udo Jarisch and Frank Merkle for assistance in preparing all the experiments and Michael Grandy for his assistance in mechanical engineering.

References

[1] P. Vandenabeele, *Rapid Thermal Processing: Study Of Temperature Nonuniformity and Temperature Measurement*, Ph.D. Thesis, Leuven (Nov. 1994).

[2] Ch. Schietinger, B. Adams, Proc. of 5th. Int. Conf. on Advanced Thermal Processing of Semiconductors - RTP '97, p. 335 (1997)

[3] Ch. Schietinger, *Wafer Temperature Measurement in RTP*, Nato Summer School, Acquafredda di Matatea, Italy (1995)

[4] S.Abedrabbo, N.M.Ravindra, O.H.Gokce, F.M.Tong, M.Beggans, A.Pateel, V.Rajsekhar, D.Pattnaik, A.T.Fiory, B.Nguyenphu, A.Nanda, Proc. of 5th. Int. Conf. on Advanced Thermal Processing of Semiconductors - RTP '97, p. 163 (1997)

[5] M.Oh, S.M.Merchant, B.Nguyenphu, Proc. of 5th. Int. Conf. on Advanced Thermal Processing of Semiconductors - RTP '97, p. 199 (1997)

[6] B.Nguyenphu, M.Oh, Proc. of 5th. Int. Conf. on Advanced Thermal Processing of Semiconductors - RTP '97, p. 347 (1997)

[7] S. Shishgushi, A.Minji, T. Hayashi, S.Saito, Symp. VLSI Dig. Of Tech. Papers, p. 89 (1997)

[8] G.C.Xing, Z.H.Wang, Ch. Schietinger, Y.Wasserman, M.H.Sun, Mat. Res. Soc. Symp. Proc. 387, 119 (1995)

[9] W. Lerch, M. Glück, N.A.Stollwijk, H.Walk, M.Schäfer, S.D.Marcus, D.F.Downey, J.W.Chow, H.Marquard, accepted for publication in Mat. Res. Soc. Symp. Proc. 525 (1999)

FLOATING WAFER REACTOR: RTP BASED ON THERMAL CONDUCTIVE HEAT TRANSFER

V.I. Kuznetsov, S. Radelaar and E.H.A. Granneman
ASM International NV, Jan van Eycklaan 10, 3723 BC Bilthoven, The Netherlands

This article introduces an advanced RTP system based on a predominantly conductive heat transfer from the preheated reactor to the wafer. Heat-up rates of more than 250°C/sec, at peak power less than 20 kW, are achieved. The wafer is enclosed between two reactor plates at a very small distance (~0.1mm) from the plate surfaces. A high temperature uniformity across the wafer during heat-up is ensured by a high thermal conductivity of the reactor plates. The wafer comes in a thermal equilibrium with the reactor, therefore the wafer temperature control is simple and independent of the wafer emissivity. The system enables the wafer to cool down rapidly and has a throughput of up to 180 wafers per hour.

OBJECTIVES

Requirements for Rapid Thermal Processing have increased dramatically with a transition to technologies with feature sizes of 0.18 μ and less. The temperature uniformity has to be within ±1°C across the wafer. The dynamic temperature uniformity during a ramp-up becomes increasingly important, because the process time gets shorter: about one second for a "spike" anneal (1). It is also important for an anneal to heat-up wafers very quickly to minimize the effect of the transient enhanced diffusion (TED) (2). It is difficult to obtain a high temperature uniformity at high ramp rates. Conventional IR-lamp RTP systems use an array of lamps with differential power control. This control system has feedback from the temperature sensors measuring the wafer temperature at different points. Nevertheless, when the heat-up rate is as high as 250°C/sec, control becomes very difficult. The reason is that high power is required (~200 kW) at a high ramp-up rate. The wafer is transparent to the IR radiation at the start of heating. The parts of the wafer that reach the temperature at which Si absorbs IR radiation earlier, start to heat-up much faster than the rest of the wafer. Even a slight temperature non-uniformity across the wafer at high ramp-up rate (high power) spontaneously causes enhanced temperature non-uniformity when the wafer reaches the critical temperature for Si emissivity change. This is the reason why the heat-up process has to be stopped for a few seconds in the temperature range of 600-800°C: to give the wafer time to average the temperature. However, stopping the heat-up process can decrease process quality. It is known that B diffusion can be dominated by interstitial-defect diffusion (3), the junction formed at 750°C will be deeper than the junction formed at 1000°C (2). This means that stopping the heat-up process during ramp-up results in deeper junctions.

One of the ways to solve the problem is the "hot plate" anneal. When the wafer is placed on the preheated plate, heat transfer to the wafer occurs through thermal conductance. The wafer heat-up is still uniform in the temperature range where the wafer changes emissivity. The "hot plate" approach also has another advantage relative to lamp systems. It is difficult to repeat the exact target temperature from wafer to wafer for a one-second process when using a high power ramp-up. In a hot plate system the wafer reaches the temperature of the plate and overshoot is physically impossible. This makes temperature control easy for short processes like "spike" anneal.

However, in order to use a "hot plate" approach, one has to solve the following basic problem. When a wafer is placed on a hot plate it first touches the plate at a few points. Heat transfer at these points is much faster than in the areas where there is a gap between the wafer and the plate. This gap can be as wide as 75 μ referring to the wafer SEMI standard for wafer warp. The fast thermal wafer expansion in these places results in a greatly enhanced wafer warp and the gap between the wafer and the plate in the rest of the wafer becomes larger, worsening the uniformity of the heat transfer and reducing the overall heat-up rate. The resulting stresses in the wafer are sufficient to generate dislocations which propagate and form slip lines. Direct placement of the wafer on the hot plate is usually limited to a maximum temperature of 900°C. Such a temperature is too low for most processes. A higher temperature can be used when the wafer is placed on the hot plate more slowly. It gives the wafer time to average the temperature across its surface. However, it removes the potential advantage of the method: a high ramp-up rate.

In this work we introduce a new reactor for RTP, based on conductive heat transfer. This reactor enables a slip-free operation of up to 1100°C. The approach is similar to the "hot plate" method, but three principal innovations enable us to solve the problem:

- The wafer *does not touch* the plate, eliminating local overheating. The wafer is kept on a very small distance from the plate (about 0.1 mm) which ensures a high thermal conductance from the plate to the wafer.
- The wafer is placed *between two plates*. Wafer locations having a small distance from the bottom plate have a correspondingly larger distance from the top plate. Therefore the effective wafer-plates gap is uniform across the wafer. Note that the use of two plates doubles the heat-up rate.
- The wafer *is flattened* by the pressure of the gas supplied through the plates.

PRINCIPLE OF OPERATION

The principle of operation of the reactor is shown in Fig. 1(4). The wafer is loaded by the wafer transfer end effector when the reactor is open: the distance between the reactor plates is 15 mm. At a distance of 7 mm from each plate the wafer is predominantly subjected to radiative heating. The wafer is transparent to IR radiation at room temperature. This means that the wafer is heated-up slowly during insertion into the reactor. The bottom reactor plate moves up and the wafer is taken over from the wafer transfer end effector by the gas bearings in the bottom reactor plate. Then the reactor closes and the wafer floats between both reactor plates at a very small distance (0.1 mm) from the plate surfaces without actually touching them. If the wafer has an initial warp, it flattens during the time that the reactor is closing. The wafer comes so close to the plates, that conductive heat transfer dominates over radiative heat transfer. The simulated wafer

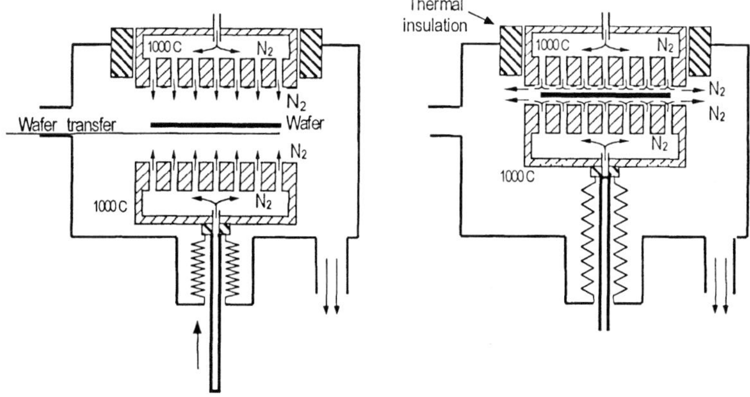

Fig. 1. The Floating Wafer Reactor principle of operation.

temperature dynamics and the ratio between conductive and radiative heat transfer during reactor operation are shown in Fig. 2. Note that during the time the reactor is closing, not only conductive heat transfer increases but the radiative heat transfer increases as well. The reason is that the wafer heats up and its emissivity changes. This happens rather abruptly at a temperature range of 250-500°C. As a result, the wafer absorbs more radiation. However, conductive heat transfer continues to prevail over radiative heat transfer considerably.

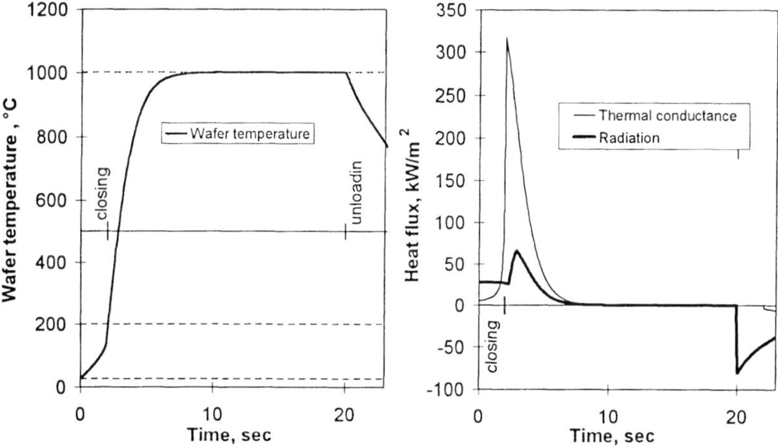

Fig. 2. Simulated wafer temperature dynamics and heat flux from each reactor plate to the wafer through thermal conductance and radiation.

GAS BEARINGS

The wafer is supported (floats) between two reactor plates by means of specially designed gas bearings. Gas is supplied through the reactor plates in a such way that the gas is evenly distributed across the wafer. Gas flows through the gaps between the wafer and the plates from the center of the reactor to the edge (Fig 3). To sustain the flow, the

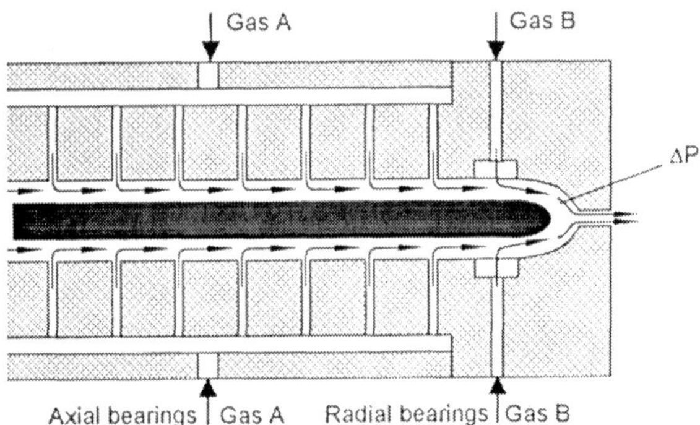

Fig. 3. Gas bearings.

pressure in the reactor center must be higher than the pressure in the edge. For a standard FWR gas flow of 3 slm on each wafer side, the pressure at the center of the reactor is about 1% higher than at the edge. The computer simulations and the experimental measurements of the pressure distribution are given in Fig. 4. The simulations have been done using two models. One model is for a conventional gas bearing where the gas flow restriction is formed by the area between the feed-hole edge and the wafer. In this case the pressure recovery beyond the restriction has to be taken into account. The prediction of this model is labeled as "model 1" in Fig. 4. The bearing gap in the FWR is larger than in a conventional bearing. The second model, where the restriction is formed by the cross-section of the feed holes, was used to simulate the FWR gas bearings. The data in Fig. 4 show that the predictions of this model (model 2) are in perfect agreement with an experiment. Experiments show that the bearings are stable at 1% pressure drop from the reactor center (1.01 Bar) to the wafer edge (1.00 Bar). As mentioned before, the wafer never touches the plates, even if it has an initial warp.

The gas bearings have to prevent the wafer edge from touching the reactor side wall. To this purpose, the gas outlet is narrowed at the wafer edge (see Fig. 3). This results in building up of a pressure difference ΔP between the outlet and the center. This pressure difference (ΔP) increases when the wafer approaches the gas outlet. Simultaneously, ΔP decreases at the opposite side of the wafer due to the larger distance from the wafer to the reactor edge on that side. This pressure difference forces the wafer to move back to the reactor center (5). Simulations and experiments show that the reactor can be tilted up to 10 degrees relative to the horizontal plane without the wafer touching the edge of the reactor.

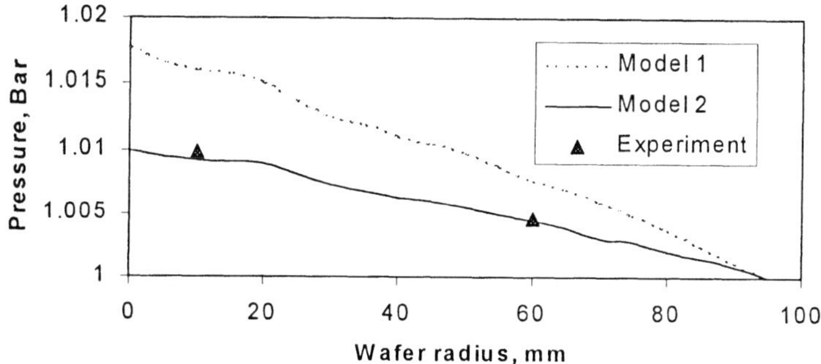

Fig. 4. Pressure distribution across the wafer

The reactor plates have a system of shallow grooves which gives the gas flow a certain direction and brings the wafer into rotation (6). The rotation speed depends on the gas flow. For typical gas flows (~3 slm) the rotation speed is 35 rpm.

EXPERIMENT: WAFER TEMPERATURE DYNAMIC AND UNIFORMITY

The temperature of the surface of the reactor plate facing the wafer has been measured by thermocouples. The wafer temperature is calculated based on these measurements. Results are shown in Fig. 5. The ramp rate is compared with a typical ramp rate which enables slip-free operation of a "hot plate" system. As can be seen in Fig. 5 the ramp rate in the FWR system is about five times higher than in the "hot plate" system.

The wafer reaches a target temperature (the reactor plates temperature) and stays at this temperature until it will be unloaded from the reactor. For very short processes with a soak time of less than one second, the reactor can be opened before the wafer reaches the plates temperature. The ramp-up rate slows down after opening the reactor due to decreasing conductive heat transfer. This slowing down can be controlled by the moment of opening of the reactor. The closer the wafer temperature is to the plate temperature, the slower the heat-up rate. At the slowed-down rate of 10-50°C/sec, the target temperature repeatability from wafer-to-wafer lies within ±1°C.

The wafer is unloaded from the reactor at process temperature. The wafer transport from the reactor to the cool-down station takes about two seconds. During this transport the wafer cools down at a rate of 90°C/sec and then much faster in the cool-down station, down to a temperature of 100°C. Fast cool-down is expected to improve the quality of junction formation due to a fast "freezing down" of the annealed wafer state.

The temperature uniformity across the wafer during the heat-up and the steady-state is determined by the temperature uniformity of both reactor plates. A high uniformity of the plate temperature is caused by a high thermal conductivity of the plate material (~50 W/m K at 1000°C). Nevertheless, when necessary it is possible to adjust the plate temperature profile, for example to correct 1% pressure difference shown in Fig. 4. Oxidation experiments have been carried out to estimate temperature uniformity. The oxide distribution uniformity is shown in Fig. 6. The temperature averages across the thick high-conductive plates so strongly, that the setpoint of the side heating element has to be changed by 25°C in order to change oxide thickness by 2% at the wafer edge.

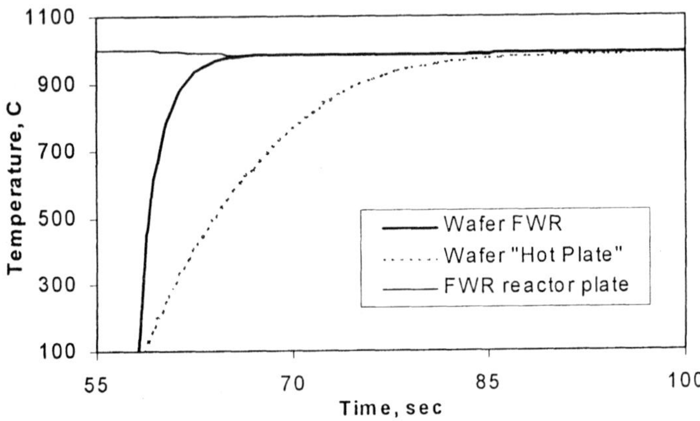

Fig. 5. Wafer heat up dynamics

Temperature control of the FWR system is simple and independent of the wafer emissivity variations. The wafer is placed inside a preheated "envelope" between reactor plates. It certainly reaches a thermal equilibrium with the reactor plates. It means that direct wafer temperature measurements are not necessary. Monitoring the temperature of the reactor plates using thermocouples, placed close to the surface facing the wafer, is sufficient.

The reactor plates are massive (25 kg) compared to the wafer mass (59 g). The temperature of the plates practically does not change when a wafer is loaded into the reactor (see Fig. 5). The energy losses to heat up a cold wafer are compensated by the heating element for each wafer. This compensation can take place during wafer loading, processing and unloading. This time is long enough to keep the peak power below 20 kW, contrary to lamp systems where power has to be supplied during wafer heat-up. When high ramp rates are used for a "spike" anneal, peak power is in the order of 200 kW. Such high power requires large power facilities which can be expensive and problematic for a plant using several RTP systems.

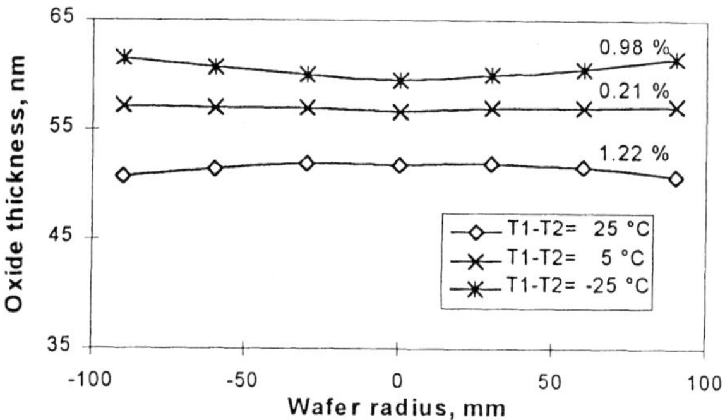

Fig. 6. Oxide thickness uniformity (standard deviation). Dry oxidation 1070°C (middle curve). Upper and bottom curves obtained when the setpoint of the side heating element was respectively increased and decreased by 25°C. Side heating element is located at the cylindrical surface of the reactor plates

THROUGHPUT

The kinematic throughput of the FWR system reaches 180 wafers per hour. Such high throughput is possible because the FWR system has a wafer carousel transport mechanism between three positions: reactor, cool-down station and I/O port. The carousel mechanism simultaneously transports a wafer from the I/O port to the reactor, from the reactor to the cool-down station and from cool-down station to I/O port. The throughput is determined by the process duration. Wafer cool-down and I/O port (un)loading occur in conjunction with reactor (un)loading and wafer processing.

CONCLUSIONS

A new concept of the reactor for rapid thermal processing is described. During a process the wafer is supported by gas bearings between two preheated reactor plates. A very small gap (0.1 mm) between wafer and the plates enables predominantly conductive heat transfer from the plates to the wafer. High ramp-up rates, exceeding 250°C/sec, are achieved. Such ramp rate is approximately five times higher than in conventional "hot plate" systems. No slip was observed in the wafers. Peak power is about 10 times lower than in lamp RTP systems. A high dynamic and steady-state temperature uniformity is ensured by high thermal conductivity of the reactor plates. Temperature overshoot is physically impossible. Simple temperature control based on plate temperature measurements by thermocouples is implemented. Fast wafer cool-down and a wafer carousel transport mechanism enable the FWR system to reach a throughput of up to 180 wafers per hour.

ACKNOWLEDGEMENT

The authors express their gratitude to engineering teams of ASM Europe and Philips CFT (Center for Manufacturing Technology). T. Ruijl and J. van der Sanden (Philips CFT) are gratefully acknowledged for their work on the wafer gas bearings design, H. Terhorst (ASM) for the computer simulations and R. Bast (ASM) for the experimental verification of the FWR system concept.

REFERENCES

1. M.A. Foad, D. Jennings, *Solid State Technology*, Vol. 41, No. 12, p. 43 (1998)
2. A.Agarwal, A.T.Fiory, H.-J. L.Gossmann, C.S. Rafferty and P.Frisella, *Materials Science in Semiconductor Processing*, Vol. 1 (3/4), p.237 (1998)
3. H.-J. Gossmann, in *Semiconductor Silicon*, ed. H.R. Huff, U. Goselle, and H. Tsuya, ECS Proc. Vol. 98-1, p. 884 (1998)
4. E.H.A. Granneman, F. Huussen, Patent pending
5. V.I. Kuznetsov, S. Radelaar, T. Ruijl, J. van der Sanden, Patent pending
6. V.I. Kuznetsov, S. Radelaar, T. Ruijl, J. van der Sanden, Patent pending

Dynamic Uniformity Control in a Rapid Thermal Processing System

K. S. Balakrishnan, S. Shooshtarian, N. Acharya and P. J. Timans
STEAG RTP Systems, Inc., 4425 Fortran Drive, San Jose, CA 95134
R. P. S. Thakur
STEAG Electronic Systems, 4425 Fortran Drive, San Jose, CA 95134

Temperature trajectory and uniformity control are key factors that affect process performance in rapid thermal processing (RTP) systems. In traditional RTP systems, single point control with slaved zone ratios has been used successfully, but the need for improved uniformity control and simpler recipe creation has stimulated the development of multiple input - multiple output (MIMO) control methods. The multiple inputs refer to temperature measurements at discrete points on the wafer using pyrometers and the multiple outputs refer to the power supplied to the different lamp zones in the heater. In this paper, we present a MIMO controller developed and implemented on an advanced RTP system. Uniformity and repeatability data for rapid thermal oxidation and ion-implantation damage annealing demonstrate temperature control with total components of variance (TCV) values less than 2.8°C, 3-sigma. It is shown that the use of MIMO control provides process performance that meets the needs of the next generation RTP technology.

INTRODUCTION

Temperature control in RTP is essential to meet the increasingly stringent requirements for process uniformity during semiconductor processing. At the high ramp rates and temperatures used in RTP, dynamic temperature uniformity control over the whole process cycle is necessary to avoid uneven processing and the formation of slip lines on the wafer. Closed-loop control for dynamic control of wafer temperature is the preferred mode of temperature control for most applications (1-8). In this paper, we describe a MIMO controller that has been developed to meet the specifications for the next generation of RTP technology. The temperature controller that is presented delivers a temperature range of less than ± 2 °C during the ramp, less than ± 1°C non-uniformity during steady state and no overshoot.

The Starfire RTP system described in this paper employs an axisymmetric heater and chamber configuration (9). The lamps in the heater are arranged in five concentric rings, and they are configured into four independently manipulated zones. Azimuthal temperature non-uniformity is minimized by wafer rotation (10). Temperature measurement across the wafer is accomplished using multiple pyrometers.

CONTROLLER DEVELOPMENT

A typical setpoint trajectory for RTP consists of a fast ramp region followed by a steady state region. The ramp regions in RTP usually employ heating rates between 25-125°C/s, and the steady state regions range from 0 seconds for a spike anneal to a few hundred seconds. The particular challenge in control of RTP processes is to ensure fidelity to the setpoint temperature over the whole wafer. This can, in principle, be achieved by either SISO or MIMO control techniques.

Single input – single output (SISO) control has traditionally been used for temperature control in semiconductor processing. In this scheme, the temperature at a single point on the wafer is controlled to the setpoint trajectory and temperature uniformity is obtained by tuning the power scaling ratios for different lamp zones in the system. However, single point control limits the ability to perform real-time adjustment of uniformity and requires significant optimization effort for different processes. As temperature uniformity specifications become tighter, the impact of small fluctuations in heat transfer conditions makes it difficult to maintain adequate process performance over extended periods of time. In MIMO control, temperature uniformity is obtained by dynamically adjusting the lamp zones independently during the run instead of fixing the scaling ratios for different recipe blocks by manual adjustments.

The controller described in this paper is a proportional-integral-derivative (PID) controller that is based on linear control theory. Linear models of the system are determined at equally spaced temperature intervals covering the processing range in RTP. These linear models are then used to determine the values for the controller parameters which are then gain-scheduled across the temperature range of processing (11).

Modeling

Models for the RTP system can be determined either from a physical model or through black-box modeling techniques (12, 13). In this work, black-box models were determined by making use the Matlab® System Identification toolbox (14). For the purpose of determining the models, small-signal temperature response data for open loop changes in zone powers were collected at a set of wafer temperatures.

A systematic procedure was used to identify the model parameters in order to minimize errors in the data collected. In the high temperature limit, the static system gains, K, approximately follow an inverse cubic profile with the absolute temperature, T, of the wafer (7). It can then be shown that the error in gain calculation, ΔK, scales linearly with deviation from target temperature, $\Delta T = T - T_{sp}$, where T_{sp} is the target temperature, and

$$\frac{\Delta K}{K} = -3\,\frac{\Delta T}{T}. \qquad [1]$$

In order to minimize this error, and errors arising from in-plane conduction, approximate temperature uniformity is first obtained in open-loop mode at the required temperature by manually tuning the scaling ratios for the different zones. Step changes in the power of each lamp zone are then made that yield approximately 10 °C changes in temperature at the pyrometer locations, and the corresponding temperature response data are recorded. The input (power) and output (temperature responses) data are used to determine the black-box models using system identification techniques. This procedure is repeated at several temperatures to obtain linear models across the temperature range of interest.

The RTP system can be characterized as having n inputs which correspond to the lamp zone powers and m outputs which correspond to the temperature measurements. Therefore, the inverse is true for the controller – it has m inputs and n outputs. For a 4X4 system, the model can be represented by Eq. (1).

$$\mathbf{T} = \mathbf{G}\,\mathbf{P} \qquad [2]$$

where

$\mathbf{T} \cong$ vector of temperatures $= [\, T_1\ T_2\ T_3\ T_4\,]^T$

$\mathbf{P} \cong$ vector of zone powers $= [\, P_1\ P_2\ P_3\ P_4\,]^T$

$$\mathbf{G} \cong \text{Transfer function matrix} = \begin{bmatrix} G_{11} & G_{12} & G_{13} & G_{14} \\ G_{21} & G_{22} & G_{23} & G_{24} \\ G_{31} & G_{32} & G_{33} & G_{34} \\ G_{41} & G_{42} & G_{43} & G_{44} \end{bmatrix}$$

In Eq. [2], each G_{ij} is the model for the interaction between the jth input, which is the power to lamp zone j, and the ith output, which is the temperature at pyrometer location i.

Controller structure

Figure 1 illustrates the structure of the controller. There are two main components, a predictor and a feedback controller. The predictor is a model that describes the nonlinear relationship between temperature and power. It provides an estimate of the power needed for wafer temperature to track the setpoint trajectory. The feedback controller dynamically corrects for mismatch between the predictor output and the real power demand. The use of the predictor reduces the magnitude of the error that the feedback controller has to eliminate, resulting in improved tracking characteristics.

The predictor and feedback controller combine to form the controller as shown in Fig. 1. The predictor accepts the setpoint as an input to provide the feed-forward open-loop power and the feedback controller acts on the difference between the setpoint and the temperature signal, the error, e, to provide the corrective action.

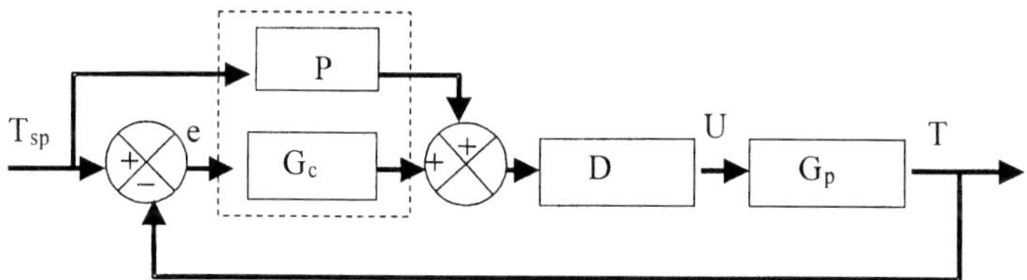

Figure 1. Block diagram of the feedback controller

In Fig. 1, T_{sp} is the setpoint temperature, T is the temperature feedback, P represents the predictor, G_c is the feedback controller, e is the difference between the setpoint and the feedback signal, U is the output to the RTP system, D is the decoupler and G_p represents the transfer function of the system.

The system described in this paper monitors and controls the temperature at several points on the wafer. The controller attempts to drive the temperature error to zero at all the sensor locations. The MIMO controller is described by a series of linear models defined by Eq. [2]. In the feedback control system, each input is paired with one unique output such that each temperature measurement is controlled by one zone. Therefore, the controller can be represented by the matrix shown below.

$$\mathbf{G_c} = \begin{bmatrix} G_{c1} & 0 & 0 & 0 \\ 0 & G_{c2} & 0 & 0 \\ 0 & 0 & G_{c3} & 0 \\ 0 & 0 & 0 & G_{c4} \end{bmatrix} \quad [3]$$

where each G_{ci} is the transfer function of a SISO PID controller (11). The controller combines proportional, integral and derivative actions through a set of gains and time constants determined from experimental data.

This design of such a PID MIMO controller will work well if the interactions denoted by the off-diagonal elements in the definition of **G** in Eq. [2] are minimal (11). The interaction terms are a function of the illumination profiles of the lamp zones and heat transfer characteristics of the system. The lamp zones are designed to provide flux profiles whose superposition produces a uniform temperature profile on the wafer. The need for a uniform temperature profile requires that the lamp zones are not focussed tightly, and as a result, there are inevitably significant interaction terms. This leads to a trade-off between the requirement for uniformity and the requirement for controllability of the system (9). Decouplers are controller elements that account for the interactions and incorporate them into the feedback control action. The feedback controller described in this paper consists of four SISO PID controllers combined with decouplers to handle the interactions.

A schematic of the decoupler applicable to a 2X2 system is shown in Fig. 2.

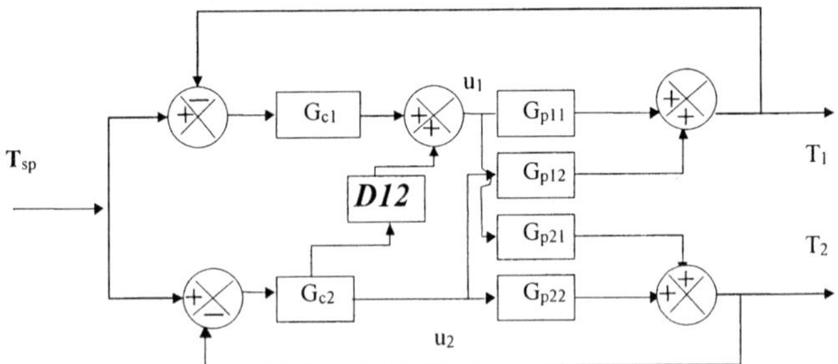

Figure 2. Block diagram of a decoupler for a 2X2 system

In Fig. 2, G_{pij} denotes the transfer function of the temperature at pyrometer location i, T_i, with respect to the lamp zone power, u_j. G_{ci} is the controller element that controls the temperature at location i, and D12 is the transfer function of the decoupler between u_2 and T_1. For the system shown in Fig. 2, the transfer function matrix between lamp powers u_1 and u_2, and temperatures T_1 and T_2 is given by

$$G_p = \begin{bmatrix} G_{p11} & G_{p12} \\ G_{p21} & G_{p22} \end{bmatrix} \quad [4]$$

The decoupler D12 is used to minimize the effect of the second input, u_2, on T_1. This is summarized by the equation

$$G_{p12} \cdot u_2 + G_{p11} \cdot D12 \cdot u_2 = 0 \quad [5]$$

which shows that the decoupler D12 is given by

$$\Rightarrow D12 = -\frac{G_{p12}}{G_{p11}} \qquad [6]$$

Decouplers, Dij, are defined for each combination of zone power, u_j, and controlled temperature, T_i.

RESULTS AND DISCUSSION

The MIMO controller described above was used for temperature control in the Starfire RTP system. Figure 3 shows temperature profiles at the four measured points on the wafer for a 1050°C, 60 s rapid thermal oxidation (RTO) recipe. The measurement points on the wafer are placed at equidistant intervals starting at the center of the wafer. The solid curve indicates the setpoint trajectory and the dotted curves are the temperatures recorded by the different pyrometers. The nominal ramp rate is 75°C/s to 900°C and 25°C/s from 900°C to 1050°C. The inset is an expanded image of the curves at high temperature. The controlled temperatures track the setpoint within ± 2°C throughout the ramp, and there is little overshoot. In the steady state, the temperatures are controlled to within ± 1°C. This figure shows the intrinsic ability of the controller to adhere tightly to the specified temperature setpoint trajectory.

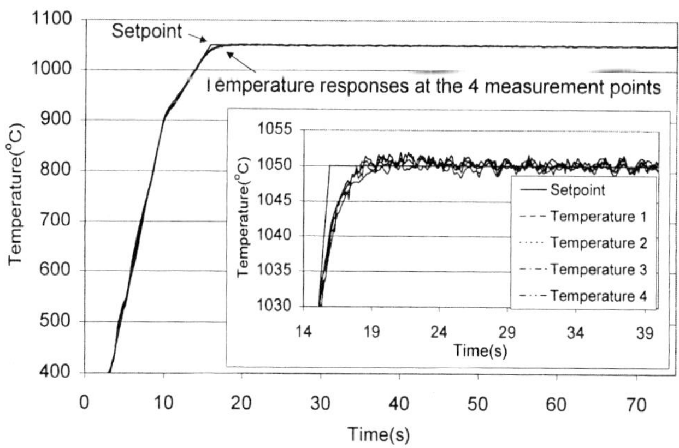

Figure 3. MIMO control of 4 temperature measurements across a 200 mm wafer for a 1050°C, 60 second recipe. The inset shows a close up of the corner between the ramp and steady state.

The controller was tested with an 1100°C RTO recipe with a steady state time of 30 s. Fig. 4 shows process results for a wafer measured with a 49 point polar scan on an ellipsometer system at 3 mm edge exclusion. The total range of oxide thickness over the wafer is 1.5 Å, which translates to a temperature range of <±1.4°C based on a sensitivity of 0.56 Å/°C. This shows that the successful application of MIMO control leads to good process uniformity on the wafer. The inherent uniformity obtained with MIMO control also enables the successful annealing of implanted wafers, a process that is highly sensitive to the temperature uniformity during the ramp-up. For the results shown in Fig. 5, the range in resistivity is ±1.2 ohms/sq. that translates into a temperature range of <±1.2°C based on a sensitivity of 1.0 ohms/sq./°C.

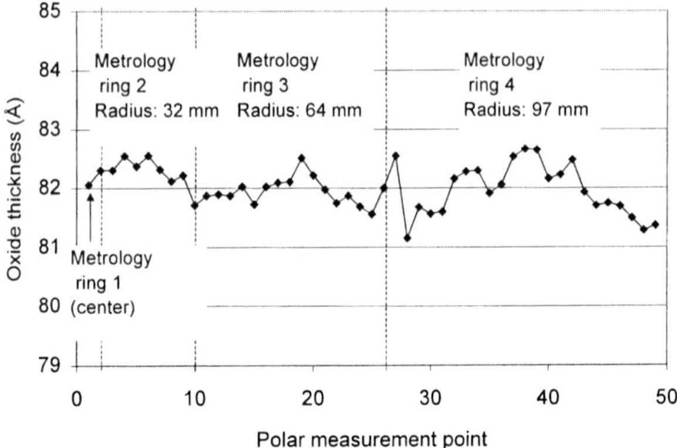

Figure 4. Process results for a 1100°C, 30 second RTO recipe. The recipe involves a ramp at 75°C/s to 900°C and then 25 °C/s to 1100 °C. The measurements are taken from a 49-point polar scan at 3 mm edge exclusion consisting of 4 rings as shown.

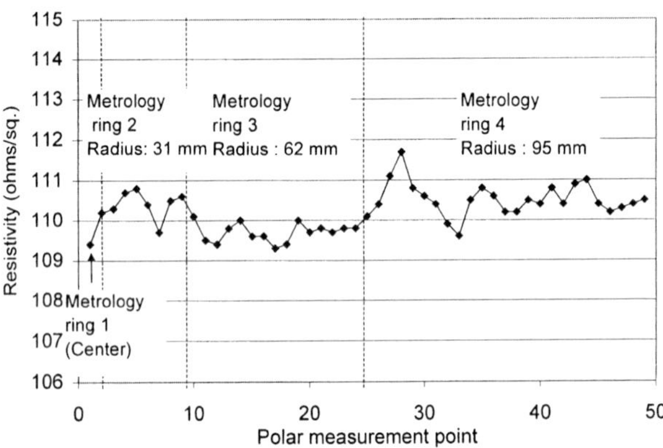

Figure 5. Process results for a 1025°C, 10 second RTA recipe. The recipe involves a ramp at 60°C/s to 900°C and then 25 °C/s to 1025 °C. The measurements are taken from a 49-point polar scan at 5 mm edge exclusion consisting of 4 rings as shown.

An advantage of the MIMO control methodology is its ability to compensate for long term drift in the thermal characteristics of system components. These variations can manifest themselves under SISO control as (a) the first-wafer effect where the process results for the first wafer in a cassette differ from those of later wafers because of gradual warm up of chamber components, or (b) long term drift arising from subtle changes in heat transfer conditions and aging of system components. Fig. 6 shows repeatability data for the RTO recipe of Fig. 4 over a 3-cassette run. For this run, RTO wafers were placed in slot numbers 5, 10, 15, 20 and 25 in the cassette with dummy wafers in other slots. All wafers were measured with an edge exclusion of 3 mm and the measurement points were in 4 rings with 1 point at the center, 8 points in the next ring, 16 points in the third ring and 24 points in the last and outermost ring. The data shows a 1-sigma repeatability of the mean oxide thickness of 0.2 %. Based on a process sensitivity of 0.56 Å/°C, the 3-sigma TCV over the three cassette run is 2.7 °C. A similar chart is shown in Fig. 7 for annealing of 25keV, $10^{16}/cm^2$ As^+ implant wafers over a 4-cassette run. These wafers were implanted through 100Å of screen oxide, and were measured after processing on a four point probe with a 49-point polar scan at an edge exclusion of 5 mm. The calculated 3-sigma TCV for this set of wafers is 1.9°C. These results demonstrate that MIMO control permitted the Starfire system to meet tight temperature control specifications required in advanced RTP processes. The control approach is also shown to be robust to small perturbations in the heat transfer characteristics of the system.

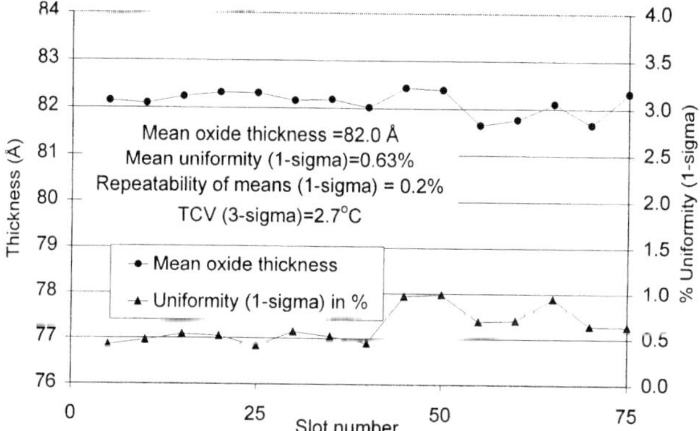

Figure 6. Uniformity and repeatability data for the 1100 °C, 30s RTO process over 3 cassettes. The process involves a ramp at 75°C/s to 900°C and then 25°C/s from 900 °C to 1100°C. The mean oxide thicknesses are for 49-point polar scans at edge exclusions of 3 mm.

CONCLUSIONS

This paper presents an implementation of MIMO control for an advanced RTP system. When implemented in conjunction with an accurate temperature measurement system, and stable operating conditions, this approach yields high quality process results. MIMO control strategies will be the preferred mode of operation for semiconductor equipment to meet the increasingly tight process specifications in the years ahead.

ACKNOWLEDGMENTS

The authors would like to acknowledge the help provided by Prof. Thomas F. Edgar at the University of Texas during this work.

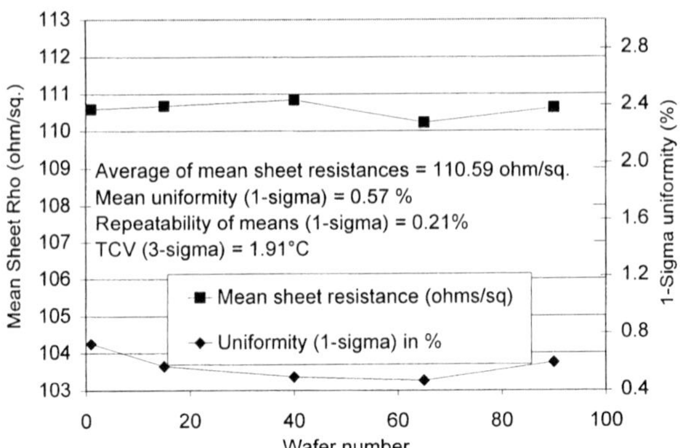

Figure 7. Repeatability results for a 1025 °C, 10s RTA recipe. The recipe used a ramp at 60°C/s to 900°C followed by a 25°C/s ramp to 1025°C. The mean sheet resistance measurements are for 49-point polar scans measured in 4 rings on the wafer at an edge exclusion of 5 mm.

REFERENCES

1. K. S. Balakrishnan, Ph.D. Dissertation, University of Texas, Austin (1998).
2. K. S. Balakrishnan and T. F. Edgar, *Special Edition of the Journal of Thin Solid Films,* accepted for publication (1998).
3. K. S. Balakrishnan, T. L. Cooper and T. F. Edgar, in *Proceedings of the 4th International Conference on Advanced Thermal Processing of Semiconductors, RTP'96,* R. B. Fair, M. L. Green, B. Lojek and R. P. S. Thakur, Editors, p. 279 (1996).
4. P. P. Apte and K. C. Saraswat, *IEEE Trans. Semiconductor Manuf.*, **5**, 180 (1992).
5. T. Breedijk, T. F. Edgar and I. Trachtenberg, in *Proc. Amer. Cont. Conf.*, 2980 (1993).
6. C. F. Elia, in *Proc. Amer. Cont. Conf.*, 907 (1994).
7. C. D. Schaper, M. M. Moslehi, K. C. Saraswat and T. Kailath, *Journal of the Electrochemical Society*, **141**, 3200 (1994).
8. J. D. Stuber, I. Trachtenberg, T. F. Edgar, J. K. Elliott and T. Breedijk, *Proc. IEEE Control and Decision Conference*, 1, 79 (1994).
9. J. G. Li, P. J. Timans and R. P. S. Thakur, *Proceedings of the 6th International Conference on Advanced Thermal Processing of Semiconductors, RTP'98,* T. Hori, B. Lojek, Y. Tanabe and R. P. S. Thakur, Editors, p. 37 (1998).
10. P. J. Timans, *Materials Science in Semiconductor Processing*, **1**, 169 (1998).
11. D. E.Seborg, T. F.Edgar and D. A. Mellichamp, *Process Dynamics and Control*, Wiley, New York (1989).
12. Y. M. Cho and T. Kailath, in *IEEE Trans. Semiconductor Manuf.*, **6**, 233 (1993).
13. L. Ljung, *System Identification – Theory for the User,* Prentice Hall, N. J.(1987).
14. L. Ljung, *The Matlab® System Identification Toolbox*, The MathWorks Inc., Mass. (1995).

A NOVEL FULL-QUARTZ OPEN CLUSTER PLATFORM FOR ADVANCED RAPID THERMAL PROCESSING

R. Bremensdorfer, H. Walk, E. Merz, S. Paul
STEAG RTP Systems

Daimlerstrasse 10, 89160 Dornstadt, Germany
E-Mail: r.bremensdorfer@steag-ast.de

As the requirements for Rapid Thermal Processing (RTP) are becoming more stringent and the application range of RTP is expanding, open and more flexible cluster tool concepts than those currently available are a key factor in addressing future needs in the market place. The trend in the industry is towards processing entire sequences on multi chamber, single wafer tools. The AST EVOLUTION system addresses these requirements. It runs both classic RTP applications in a parallel configuration, as well as advanced gate stack applications in the configuration of a sequential cluster. The system makes use of a novel wafer handling concept and a novel process chamber design.

Introduction

The trend towards smaller device geometries requires stringent control of transient temperature uniformity, emissivity independent pyrometry, and the ability to integrate different process sequences such as RTO, RTCVD in a clean and controllable ambient.

In recent years rapid thermal oxidation (RTO) has started to demonstrate competitive performance for dry, wet, and nitrided silicon oxide processes as used in gate- and capacitor stacks. For design rules requiring a layer thickness below 3 nm, silicon nitride based materials will replace silicon oxide. Compared to silicon oxide these materials have a higher dielectric constant and lower leakage current. Silicon nitride layers have to be grown by CVD. The dielectric materials road map for the effective oxide thickness as required for CMOS gates is shown in Fig. 1. The current standard process for DRAM capacitors is already based on silicon nitride, since there is a much higher need to reduce effective oxide thickness.

Integrating single wafer processing steps by means of cluster tools reduces the overall number of processing steps, avoids time linkage between steps, and allows contamination and particulate control. As an example Fig. 2 illustrates an EEPROM Gate stack flow using a conventional tool compared with clustered tools [1].

With the EVOLUTION System STEAG RTP Systems introduces a platform dedicated for the processing of 200 mm wafers for 0.18 μm and 0.15 μm, and beyond device generations. The EVOLUTION follows the SEMI standard for radial cluster tools. Its central wafer handler comprises of the Carousel™ rotating buffer stations to minimize wafer loading times.

In its basic configuration the EVOLUTION system is a dual chamber RTP system dedicated to run high volume production of 200 mm wafers. The advanced configuration will be integrating RTP, cleaning, and CVD modules. Its primary application is the formation of gate stacks.

The RTP module features dual sided heating, wafer rotation and independent lamp control to minimize pattern effects (e.g. pattern shift) and wafer stress. The RTP module comprises of an all-quartz process chamber. Due to a unique approach to venting the reaction chamber, the rotation of the wafer is readily incorporated into the chamber design. While the oven and quartz chamber show circular symmetry, both, top and bottom lamp arrays are using a linear lamp configuration. This is similar to the highly successful AST 2800 and 3000 RTP systems series.

Emissivity independent temperature control is provided by a modified and optimized version of Luxtron Corporation's Ripple[1] pyrometery. For calibration and dynamic uniformity verification purposes, a remote sensing system allows one to perform thermocouple readings on rotating wafers.

The AST EVOLUTION System

The AST EVOLUTION RTP Module

In order to integrate rotation with ambient flexibility and cleanliness, a unique process chamber design was devised. Fig. 3 shows a cross section of the rotational symmetric quartz chamber. The top plate as well as the cylindrical side wall are stationary whereas the bottom plate is mounted in a rotating ring (black portion in Fig. 3). The stationary and rotating parts are separated by a gap of 0.2 mm. Fig. 3 also shows the dovetail gas labyrinth which is an important part of the venting scheme.

Both, top and bottom lamp arrays are stationary. Inside the chamber, the wafer is supported by a quartz wafer tray which itself sits on top of the process chamber's bottom plate. Since the bottom plate is rotating the wafer is rotating as well. In contrast, the support of the silicon slip guard ring is mounted to the stationary parts of the chamber.

The quartz chamber is cooled from the top and from the bottom sides by air. This is done by two air blowers which are separately controlled. The typical flow for each blower is in the range of 150 to 200 m^3/h.

Process gas enters the chamber through a quartz tube and is decelerated by a gas distribution system. The resulting laminar gas flow is parallel the wafer surface. The gas then leaves the chamber through the 0.2 mm gap along the entire circumference of the reaction chamber. The dovetail gas labyrinth then deflects the gasflow into the chamber exhaust. The process chamber is entirely surrounded by purge gas which is vented together with the process gas. The chamber itself is sealed against the nitrogen-purged central handling unit by an aluminum VAT valve. This configuration ensures an optimized chamber purge and reduces the amount of impurities entering the reaction chamber during the loading cycle.

Following STEAG AST's proven dual-side heating method the AST EVOLUTION system is equipped with linear halogen lamps which are grouped into top and bottom heater banks to heat the wafers from both sides. The bottom lamp array includes 18 lamps, the top array 17 lamps, of different lengths. These 35 lamps are independently controlled during the course of a recipe. Besides this active method to control ramp and steady state uniformity the proven slip guard ring design passively reduces temperature gradients during temperature transients. The slip guard ring is mounted on the same level as the wafer. To allow wafer handling the whole bottom portion of the reaction chamber is lowered with the ring remaining in the top position.

---

[1] Ripple is a trademark of Luxtron Corporation

Together with top and bottom heating and wafer rotation this arrangement allows one to minimize the dynamic non-uniformity across the wafer and hence helps to prevent pattern shift effects [2],[3].

The wafer temperature is controlled by up to three fast-response optical pyrometers, which evaluate the wafer signal based on the Ripple-method. The pyrometer signals are compared and electronically translated into real wafer temperature [4]. This ensures a temperature measurement completely independent of wafer backside conditions.

In addition, a remote sensing system allows thermocouple readings to be performed on rotating wafers. By means of an HF-transponder working at 13.5 Mhz, the TC signals are transmitted to a pick-up antenna. After demodulation, these signals are fed into the analog inputs of the control computer (Fig. 4).

The architecture of the AST EVOLUTION RTP hardware and software follows the rules of an open, distributed, modular and standardized design. The communication protocol follows SEMI / MESC HSMS. A versatile VME system together with a soft PLC connects machine peripherals to the controlling system. At the top of the hierarchical control system is a UNIX based computer to communicate with the user, the module control computers, and factory control systems. The graphical interface on a TFT display with touchscreen allows easy menu driven machine control, recipe editing, data display and service access. The module is controlled by VME bus computers running under a realtime operating system.

The AST EVOLUTION Wafer Handler

The EVOLUTION Wafer Handler utilizes the wafer Carousel™ to increase throughput. Six integrated wafer buffer / cooling stations reduce the amount of lateral movement the robot arm has to perform. In addition, a special endeffector design allows the centering of the wafer during the handling sequence. The wafer alignment and hence the movement to an aligner station is omitted. Compared to a dual arm robot the main advantage of the Carousel is that no overhead time is consumed to load and unload cooling stations. After one process module has finished its recipe, the processed wafer is placed on an empty Carousel slot and the next wafer is put into the chamber. Fig. 5 illustrates an handling sequence for two process modules. Since the processed wafer can remain on the Carousel for cool down, there is no additional movement necessary to place the wafer into a cooling station and the next process module can be immediately loaded. I.e., in cases were the robot arm utilization to attend the process chambers is close to 100%, the Carousel significantly reduces handling time and therefore increases throughput. Fig. 6 shows simulated throughput numbers for a dual chamber system vs the time the wafer resides inside the chamber. The theoretical limit for zero process time approaches 145 wafers per hour. Since the EVOLUTION is a loadlock system, there is no need to purge the process chamber for each wafer. Hence, the residing time of a wafer inside the chamber is significantly reduced.

Experimental

TC Measurements On Rotating 200 mm Wafer

The remote sensing capabilities described above (see also Fig. 4) allow mapping the temperature homogeneity of the reactor by means of only a small number of thermocouples mounted into a silicon wafer. Since the wafer is rotating, the entire two-dimensional space of the wafer plane can be mapped. As will be illustrated below, this feature can be used to optimize the dynamic uniformity of a process without compromising the integral uniformity of e.g. a rapid oxidation or an implant anneal. This method allows significant reduction of the number of start-up monitor wafers needed.

The following experiment outlines the methodology. A TC-wafer with three embedded S-type thermocouples was loaded into the chamber. Thermocouple locations were along the wafer radius: in the center, 30 mm towards the edge, and 5 mm from the edge. A heat cycle with a 50°C/s ramp to 1020°C and a hold for 10 sec was performed. The temperature was controlled by the center thermocouple. The other two thermocouples were only used to record the temperature. By means of the three thermocouple readings, the relative lamp power values (lamp correction values) were set to give the best dynamic temperature uniformity while the TC-wafer was rotating. The set of lamp correction values for all 35 lamps comprise the lamp correction table. In the case described here, only two different lamp correction tables, one for the ramp up, the other one for the steady state were used. This approach is appropriate because the heating dynamics of a wafer inside a chamber with reflecting walls is dominated by two effects [5]. First, the photon box effect during ramp-up phases which primarily heats the wafer edge. Second, the edge effect which leads to a cooler edge and a hotter wafer center during steady state.

The resulting TC traces for the optimized lamp settings and for a rotational speed of 75 rpm are shown in Fig. 7. All three TC readings are within ±2.5°C in the ramp phase and ±1°C during the steady state. Since the thermocouples were located on the rotating wafer, the three TC readings represent a two-dimensional uniformity map of the entire wafer.

Without changing the lamp correction values a 200 mm implant anneal monitor wafer was run. The wafer was implanted with As^+ at 20 keV and a dose of $1.0 \cdot 10^{16}$ cm^{-2}. As can be seen in Fig. 8 the suggested TC-wafer uniformity is confirmed by the resulting sheet resistance mapping. Mean sheet resistance is 69.08 Ohm/sq with a 1σ standard deviation of 0.5%. Number of sites measured are 121 with 5 mm edge exclusion and 3σ sorting criteria. Based on the sensitivity of this implant, the range of 1.51 Ohm/sq translates to ±1.5°C total variation across the wafer.

Slowing down the rotational speed allows to trace a certain point of the wafer during its journey through the furnace. The information obtaind this way is analogous to the stationary case, when a TC-wafer with a dense mesh of embedded thermocouples is used to map the temperature distribution across a wafer. Fig. 9 shows the result of such an experiment. A heat cycle with a hold step at 1020°C was performed. Rotation was set to 4 rpm. The temperature was controlled by the center thermocouple. Again, the other two thermocouples were only used to record the temperature.

The effect of rotational speed on the uniformity is illustrated in Fig. 10. For rotational speeds of 40 rpm and higher the temperature variation of e.g. the edge TC drops below 2°C.

The significance of this kind of experiment becomes clear when looking at multipoint measurement and control schemes for rotating wafers. Here the across wafer uniformity is actively controlled by pyrometers located along the radius of the wafer.

Multipoint measuring systems using pyrometry are by design stationary to the process chamber, i.e. they can only provide the temperature of the part of the wafer that is in the field of view of the pyrometers at a given time. All pyrometers together may give the temperature distribution along a radius with a reasonable spatial resolution.

However, there is no information on the parts of the wafer that are not in the field of view, thus most of the wafer remains unobserved. Controlling such a system means to vary lamp power according to the temperature reading taken at a small section of the wafer. Which lamps are to be changed is not trivial and requires modeling. A change in the lamp power in one region of the furnace does not leave other regions unaffected. Controlling the lamp radiation according to this scheme may change the temperature on parts of the wafer that are not observed and are at that time in a different area of the furnace. Therefore, the temperature distribution across the wafer at a given time strongly dependends on the model used to control the unobserved parts of the wafer.

An example for a simple model is to asume that the same change that leads to a optimal temperature distribution at the location of the pyrometers will show the same effect at a different location in the furnace.

Conventional methodology to validate the model is based on indirect metods, like implant annealing, or rapid thermal oxidation. These indirect methods lack the time resolution of thermocouple measurements, since the temperature dependent parameter is integrated over time.

Another approach is a rather passive one. Here the intrinsic uniformity of the furnace wafer system is optimized by design, i.e. the wafer uniformity without rotation is already optimal. Adding rotation to such a system will always improve uniformity at all points of the wafer, at all times.

RTO Results

Bare 200 mm Si wafers have been oxidized in pure oxygen at 1100°C for 60 sec. There are various workable schemes for optimizing uniformity. The optimization sequence illustrated in Fig. 11 starts with optimization of the lamp correction values without rotation. Adding 75 rpm rotation to this configuration already improved the uniformity down to 1.1%, 1σ. Dedicated optimization for rotation improved uniformity to 0.4%, 1σ.

Conclusion

The AST EVOLUTION system presented here, is a high performance cluster system in full compliance to SEMI / MESC standards for both, classic RTP and gate stack applications. The system has a novel concept of handling wafers and a novel RTP process chamber design. However, the proven linear lamp array design and dual side heating concept of all STEAG RTP production systems is maintained. Emissivity independent temperature control is provided by the Ripple-method. All of which give rise to high troughput, stringent ambient control, improved dynamic uniformity, and high ramp rates.

Independent step by step lamp control is complemented by wafer rotation yielding in excellent temperature uniformity as shown in rapid thermal oxidation results. By utilizing the remote sensing capabilities for thermocouples the system has shown dynamic uniformities of ±2.5°C during ramp and ±1°C during steady state as measured by means of a rotating 200 mm wafer with three embedded TCs. Cross correlation with rapid thermal annealing of implanted monitor wafers validate these results. This technique allows a direct temperature measurement which is carried along with the rotating wafer; thus providing information that a conventional multipoint pyrometry is not able to reveal.

Acknowlegements

The authors would like to thank Werner Blersch, and Michael Maurer, STEAG AST Elektronik GmbH for their extensive contribution in the design of the system and for helpful discussions.

References

1. A.Gassl, A.Gschwandtner, A.Talg, G.Innertsberger, A.Mattheus, Proceedings, 5th Int. Conf. on Advanced Thermal Processing of Semiconductors (RTP'97), New Orleans, LA, Sep 1997, pp. 146
2. P.Vandenabeele, K.Maex, R.DeKeersmaecker, 1989 Spring Meeting of the Mat. Res. Soc., San Diego, Symp. B: Rapid Thermal Annealing / Chemical Vapor Deposition and Integrated Processing, April 25-28, 1989.
3. R.V.Nagabushnam, R.K.Singh, S.Sharan, and G.Sandhu, Mat. Res. Soc. Symp. Proc. **429**, 95, 1996
4. M.Hauf, H.Balthasar, Ch.Merkl, S.Müller, Ch.Striebel to be presented at the 19th Electrochemical Society, Seattle, May 1999
5. Z.Nenyei, H.Walk, T.Knarr, J. Electrochem. Soc. **140**, 1728 (1993)

Figures

Rule [nm]	250	180	150	130	100
EOT [nm]	4 - 5	3 - 4	2 - 3	2 - 3	1.5 - 2

Nitrided Silicon Oxide
1. SiO (N) Layer: RTO(N) 2. Polysilicon Layer: CVD

Si
1. SiO Interface: RTO 2. SiON Layer: CVD 3. Anneal: RTA 4. Polysilicon Layer: CVD

SiN
1. SiO Interface: RTO 2. SiN Layer: CVD 3. Anneal: RTA 4. Polysilicon Layer: CVD

Fig. 1 CMOS Gate Stack Roadmap for dielectric materials with corresponding process sequence [1]. EOT stands for Effective Oxide Thickness.

EEPROM Standard Sequence		EEPROM Clustered Sequence
Cleaning RTO / N (Tunnel Oxide) CVD amorphous Si (undoped)	time linked	Vapor Phase Cleaning RTO / N (Tunnel Oxide) RTCVD in-situ doped amorphous Si
Doping and Diffusion		
Etching		
Cleaning Bottom Oxide	time linked	
CVD Silicon Nitride		RTO Bottom Oxide + RTCVD Silicon Nitride in one chamber

Fig. 2 EEPROM Gate stack flow as an example for processing steps reduction due to tool clustering [1].

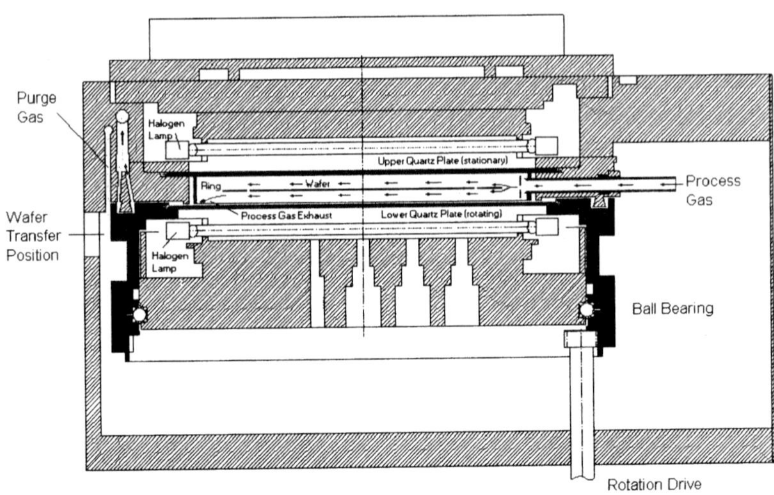

Fig. 3 Cross sectional drawing of the AST EVOLUTION process chamber. The hashed portion are stationary, whereas the solid black portions are rotating. There is a gap of 0.2 mm between the stationary top quartz plate and the rotating bottom quartz plate which accommodates the process gas vent.

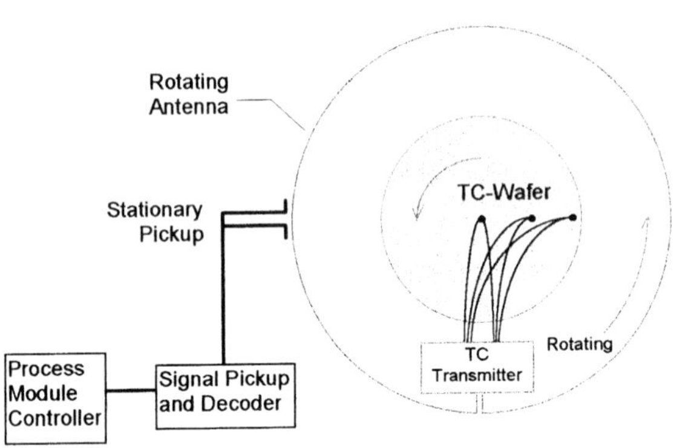

Fig. 4 Schematic of the remote sensing system for thermocouples. The thermocouple signals are transmitted via a rotating antenna mounted along the circumfence of the process chamber to a stationary pickup and finally to the control computer.

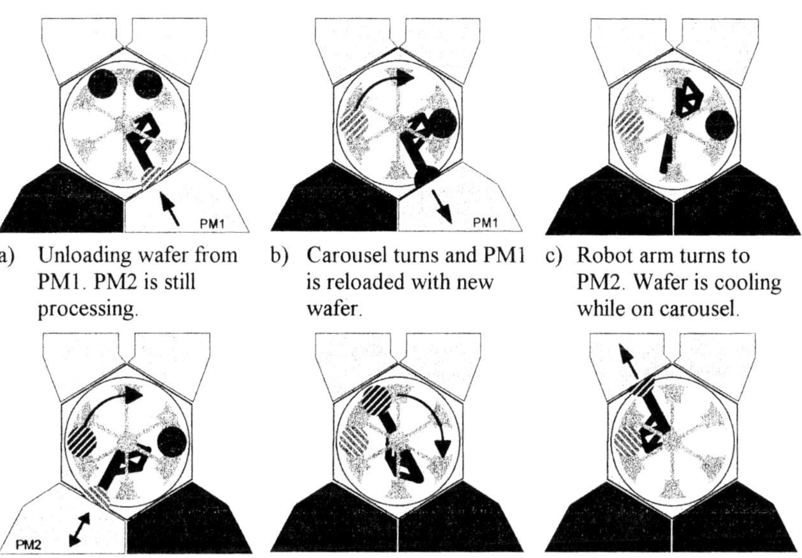

a) Unloading wafer from PM1. PM2 is still processing.

b) Carousel turns and PM1 is reloaded with new wafer.

c) Robot arm turns to PM2. Wafer is cooling while on carousel.

d) Unloading and re-loading PM2 with a new wafer.

e) While both modules are processing, there is time to return processed wafers to the loadlock.

Fig. 5 Illustration of a handling sequence using Carousel™. Since the carousel serves as cooling station no overhead time is needed to load or unload external cooling stations.

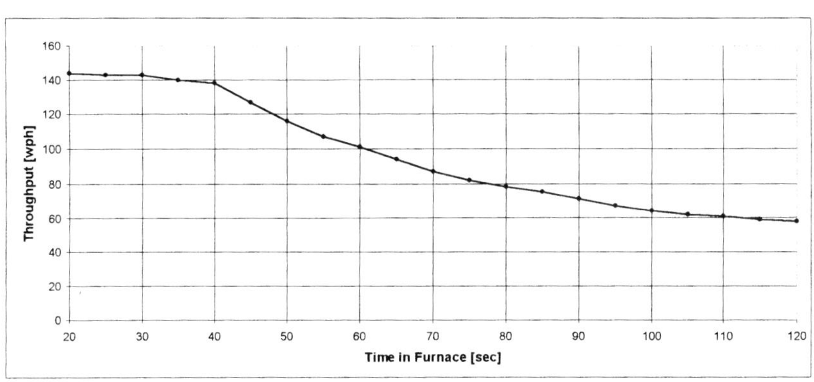

Fig. 6 Simulated throughput numbers vs time the wafer resides inside the furnace for processing. The assumed system configuration is the AST EVOLUTION wafer handler with Carousel™ and two process modules.

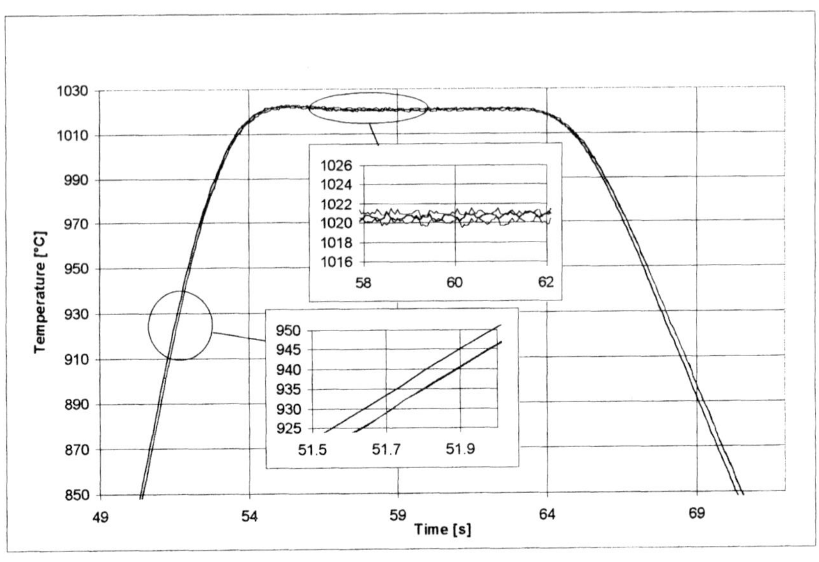

Fig. 7 Transient uniformity as measured with three thermocouples embedded along the radius of a 200 mm Si wafer while the wafer was rotating. TC positions were in the wafer center, 30 mm towards the edge, and 5 mm towards the edge. Deviations are within ±2.5°C in the ramp phase and ±1°C in the steady state.

Mean Rs: 69.08 Ohm/sq
Standard Dev. 0.5%
3σ sorting criteria

Fig. 8 Rapid thermal anneal (1020°C for 10 sec, 75 rpm) of a 200 mm Si wafer implanted with As^{+} at 20 keV and a dose of $1.0 \cdot 10^{16}$ cm^{-2}. The lamp correction settings are the same as in a).

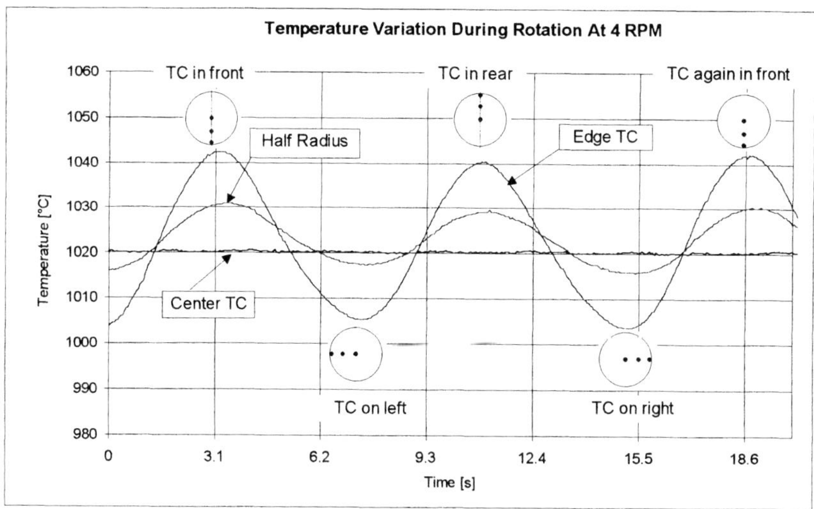

Fig. 9 Temperature variations along the radius during wafer rotation with 4 rpm. Since the embedded thermocouples are carried along with the rotating wafer the entire furnaces is being mapped. This method allows to unveil all variations a certain point on the wafer sees on its way through the furnace. A multipoint measurement that is at rest (and therefore only sees one section of the wafer at a time) is not able to provide this kind of information.

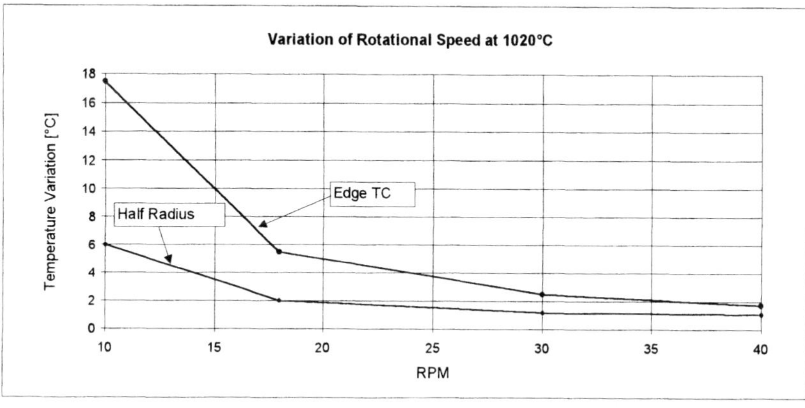

Fig. 10 TC temperature variation (max minus min) over rotational speed. Increasing the rotational speed results in improved uniformity. For a speed greater than 40 rpm the variation along the radius drops below 2°C.

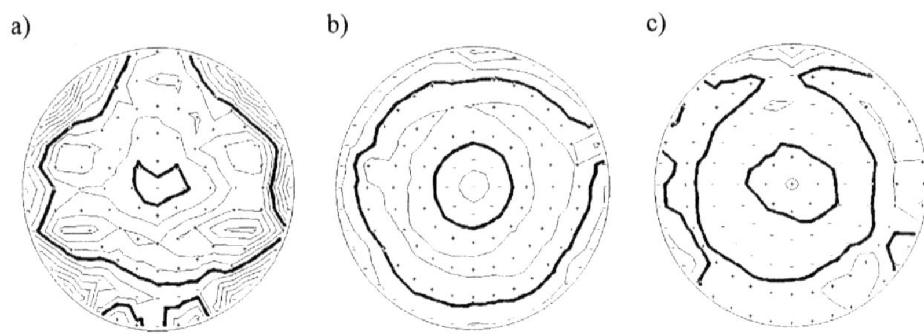

Without Rotation	75 rpm	75 rpm
Mean: 11.54 nm	Mean: 11.52 nm	Mean: 11.91 nm
Standard Dev. 2.6% 1σ	Standard Dev. 1.2% 1σ	Standard Dev. 0.4% 1σ
Temperature Range: ±9.6°C	Temperature Range: ±4.2°C	Temperature Range: ±2°C

Fig. 11 a) Oxidation thickness uniformity after 1100°C, 60 s RTO. Lamp correction optimization done without rotation.

b) Rotation added (75 rpm), same LCT as in a) .

c) Lamp correction re-optimized based on the result of b).

MODELING CHAMBER RADIATION EFFECTS ON RADIOMETRIC TEMPERATURE MEASUREMENT IN RAPID THERMAL PROCESSING

F. Rosa, Y. H. Zhou, and Z. M. Zhang
Department of Mechanical Engineering, University of Florida, Gainesville, FL 32611

D. P. DeWitt and B. K. Tsai
Optical Technology Division, National Institute of Standards and Technology, Gaithersburg, MD 20899

ABSTRACT

A detailed analysis of the radiation environment in the lower chamber of the RTP test bed at the National Institute of Standards and Technology (NIST) is performed to improve the accuracy of radiation thermometry. The models developed in this work consider the non-uniform temperature and/or heat flux distributions on the silicon wafer and the cold plate. The diffuse-gray enclosure model has been extended to predict the spectral effective emissivity as well as to include a specularly reflecting shield. A simplified Monte Carlo model is also developed to account for the directional dependence of the reflectance of the silicon wafer and the cold plate. Parametric studies are performed to evaluate the effect that the radiative properties, temperature, and geometric arrangement have on the effective emissivity. The importance of the radiometer opening size on the heat flux distribution is demonstrated. This work will contribute toward a better understanding of radiation heat transfer in RTP furnaces and improvement of radiometric temperature measurement in those systems.

INTRODUCTION

A serious challenge in the implementation of RTP for mainstream manufacturing applications has been accurate temperature measurement and control (1-4). Radiation thermometry (or pyrometry) is one of the most mature and widely used methods for non-contact temperature measurement (5). One way to improve the accuracy of temperature measurement by radiation thermometry is to construct a cavity that simulates a blackbody (6). This has been done to the NIST RTP test bed by adding reflective surfaces in the lower chamber (7). As shown in Fig. 1, the lower chamber of this RTP furnace consists of the area enclosed by the wafer, a guard ring, a guard tube, and a cold plate (covered with a reflective shield). All the surfaces in the chamber except for the wafer are highly reflective in order to simulate a blackbody enclosure.

Previously, a simple two infinite-surface model was used to predict the effective emissivity of such chambers (8). While this model is useful to demonstrate the feasibility of the reflective shield concept, it cannot give accurate results because it does not account for edge effects or the temperature and heat flux non-uniformity on the wafers inside RTP chambers. This paper presents enclosure models that provide a more realistic representation of the radiation environment in this chamber. The objectives are to estimate the influence of various chamber parameters on the effective emissivity of the wafer. Two approaches are considered: the classical net-radiation method and the Monte Carlo method.

The net-radiation method consists of performing energy balances on each surface (or zone) and obtaining their radiosities. The surfaces of known temperature are assumed to be either (a) diffuse and gray, (b) diffuse but non-gray, or (c) all diffuse; except for the specularly reflecting shield.

The net-radiation method cannot account for surfaces that have complicated directional radiative properties and cannot be treated as completely diffuse or specular. Because the back surface of the silicon wafers is usually not polished, their radiative properties can vary significantly with direction. The Monte Carlo method is a statistical method used to represent physical phenomena and can be used to study surfaces that emit and reflect non-diffusely.

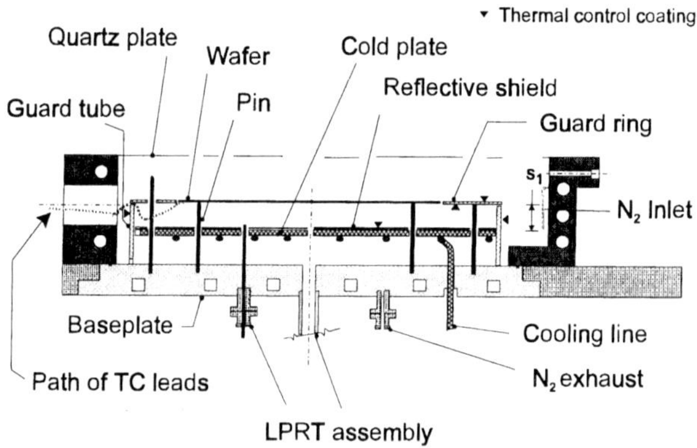

Figure 1 The lower radiation chamber of the NIST RTP test bed, where LPRT represents light pipe radiation thermometer

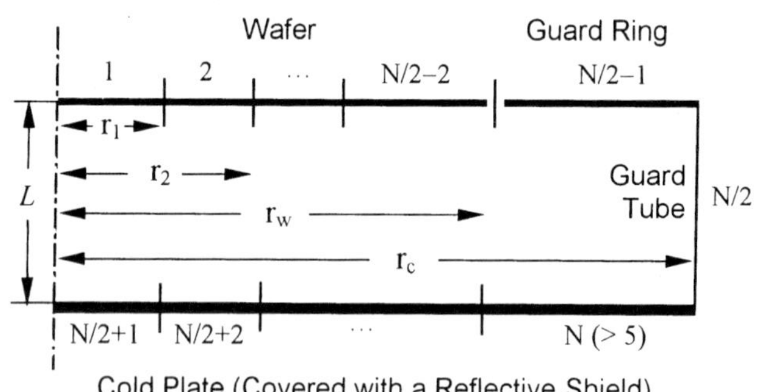

Figure 2 Axially symmetrical schematic of the enclosure model, where the *N*-zones represent the chamber surfaces

THE NET-RADIATION METHOD

Figure 2 depicts the model of the lower chamber enclosed by the wafer, guard ring, guard tube, and the reflective shield (cold plate). The wafer and the cold plate are further divided into concentric rings. In the net-radiation method, energy balances are performed for all surfaces using the concept of radiosity, i.e., the outgoing radiative heat flux from a surface by both emission and reflection. A set of algebraic equations can be obtained under the diffuse-gray assumption (9):

$$\sum_{j=1}^{N} [\delta_{ij} - (1-\varepsilon_i)F_{ij}]J_j = \varepsilon_i E_{bi}, \quad i = 1, 2, \ldots, N \quad [1]$$

where N is the total number of surfaces (zones), $\delta_{ij} = 1$ if $i = j$ and $\delta_{ij} = 0$ if $i \neq j$, ε_i and J_i are the emissivity and radiosity of the ith surface, E_{bi} $(= \sigma T_i^4$, where σ is the Stefan-Boltzmann constant) is the blackbody emissive power at T_i (i.e., the temperature of surface i), and F_{ij} is the view factor (or configuration factor) from surface i to surface j. The view factors can be calculated using the view factor algebra for given geometric arrangements (9). With the knowledge of the geometry, temperatures, and emissivities, Eq. [1] can be solved for the radiosities. The effective emissivity for each surface is defined as the ratio of the radiosity to the blackbody emissive power. In the following calculations, however, it is assumed that the radiometer views the first surface and hence $\varepsilon_{eff} = J_1 / E_{b1}$. The net radiative heat flux leaving the ith surface is (10)

$$q''_{net,i} = (E_{bi} - J_i)\varepsilon_i / (1-\varepsilon_i) \quad [2]$$

For non-gray surfaces, by introducing the spectral blackbody emissive power $E_{\lambda b,i}$ (which is given by Planck's law as a function of temperature and wavelength), the spectral emissivity $\varepsilon_{\lambda i}$, and the spectral radiosity $J_{\lambda i}$, Eq. [1] can be re-written as (9,10):

$$\sum_{j=1}^{N} [\delta_{ij} - (1-\varepsilon_{\lambda i})F_{ij}]J_{\lambda j} = \varepsilon_{\lambda i} E_{\lambda b,i}, \quad i = 1, 2, \ldots, N \quad [3]$$

which can be solved for the spectral radiosities and hence the spectral effective emissivity $\varepsilon_{\lambda,eff} = J_{\lambda 1} / E_{\lambda b,1}$.

The spectral effective emissivity is needed in the temperature measurement equation to determine the true wafer temperature from the radiance temperature obtained by the spectral radiometers (8). If all surfaces are gray, Eq. [3] can be integrated to obtain Eq. [1]. However, even though all surfaces are gray, i.e., $\varepsilon_{\lambda i} = \varepsilon_i$, the spectral effective emissivity $\varepsilon_{\lambda,eff}$ may be wavelength dependent.

Considering that the reflective shield (or cold plate) may be polished, another modification to the above-mentioned models is made that treats the cold plate as a specular surface using the virtual surface method (9). Virtual surfaces are constructed below the shield that create mirror images of the wafer, guard ring, and guard tube. The new radiosity equations are set up for all surfaces using new view factors that include the specular reflectance and virtual surfaces (10). Because the silicon wafer is assumed diffuse, the definition of the effective emissivity is unchanged.

THE MONTE CARLO METHOD

Monte Carlo is a name given to a family of modeling methods that uses the statistical characteristics of real processes to simulate physical events (11). The applications of the Monte Carlo method to radiative heat transfer problems can be found in the literature (12,13). In the study of radiative heat exchange between surfaces, the model consists of surfaces that emit bundles of energy; these bundles are then traced until they are either absorbed by one of the surfaces or missed.

In the simplified model presented here, only two surfaces, i.e., the silicon wafer and the cold plate, are considered (10). They are treated as equal-diameter concentric disks separated by a distance L. Any bundles that leaves one disk either strikes the other disk or is lost to the side. This is equivalent to modeling the vertical cylinder (guard tube) as a zero K blackbody. The silicon wafer and the cold plate are assumed to be diffusely emitting and their reflectance consists of a diffuse component and a specular component. The wafer and the cold plate are further divided into sub-surfaces (zones). The only geometric parameters needed are r_w and L, which define the enclosure. The input radiative properties are the emissivity and the percentage of specular reflectance of each surface. The effective emissivity of a surface is the number of emitted and reflected bundles from the surface divided by the number of bundles that a blackbody at the same temperature would emit. To calculate the effective emissivity at a given wavelength, the number of bundles emitted by surface i is proportional to $\varepsilon_{\lambda i} A_i E_{\lambda b,i}$, where A_i is the surface area.

RESULTS

A base case is defined to simulate the NIST RTP chamber, in which $r_w = 100\text{ mm}$, $r_c = 135\text{ mm}$, and $L = 12.5\text{ mm}$, see Figs. 1 and 2. The wafer is at a uniform temperature $T_W = 800\,°\text{C}$ and the cold plate is at $T_C = 25\,°\text{C}$. The emissivity of the wafer is $\varepsilon_w = 0.65$ and that of the cold plate is $\varepsilon_c = 0.2$. The temperature and the emissivity of the guard ring and guard tube are assumed the same as those of the cold plate. Although the wafer is at uniform temperature, due to the heat flux non-uniformity, it is important to divide it into smaller areas. When the number of zones is changed from N = 20 to 50, the change in ε_{eff} is less than 0.01%. Therefore, N is set to 24 in all calculations. As L decreases, the result approaches that for the two infinite parallel plates (10). However, the effective emissivity of 0.893 obtained for the base case is more than 1% lower than that obtained from the two infinite-surface model due to the edge effect. For larger L, the error of the two infinite-surface model will be even larger.

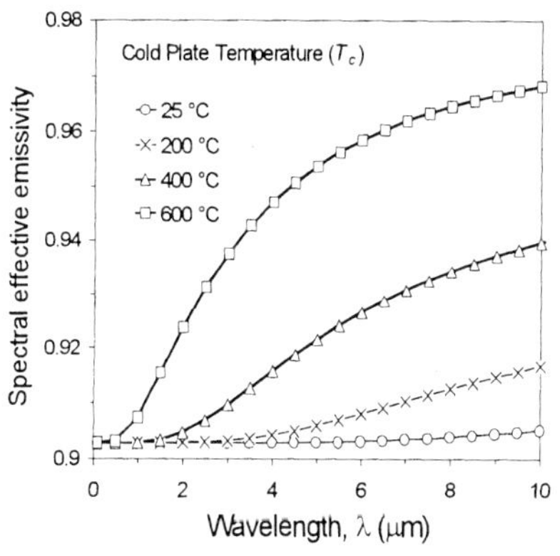

Figure 3 $\varepsilon_{\lambda,eff}$ versus T_c and λ for two infinite plates with T_w = 800 °C, ε_w = 0.65, and ε_c = 0.2

The influence of various parameters on the effective emissivity is now presented. The wavelength dependence of $\varepsilon_{\lambda,eff}$ is shown in Fig. 3 for different cold plate temperatures (T_c), based on the two infinite-surface model (10). Although both surfaces are gray, at high enough T_c or long enough wavelengths, $\varepsilon_{\lambda,eff}$ varies significantly with λ. This demonstrates the importance of spectral calculations.

Figure 4 shows variations in the emissivity and temperature of the wafer and cold plate on the effective emissivity. The effect that different parameters have on the effective emissivity is investigated by varying one or more parameters while keeping the rest the same as the base case. These results are obtained by assuming that the radiative properties do not change with the surface temperature. Because the spectral radiometer used at NIST is centered at λ_0 = 0.95 μm, only ε_{eff} and $\varepsilon_{\lambda_0,eff}$ are reported in these graphs.

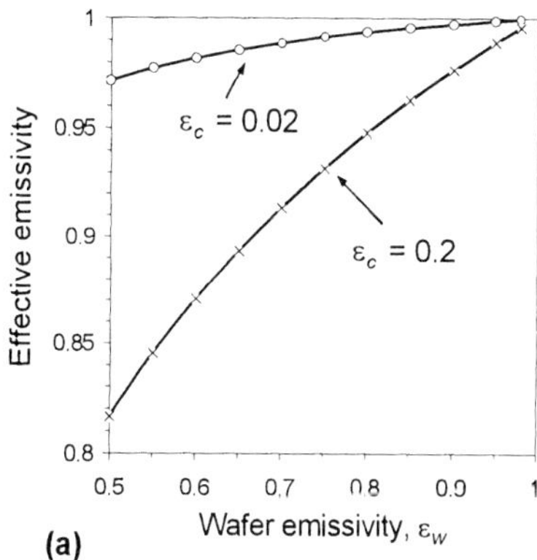

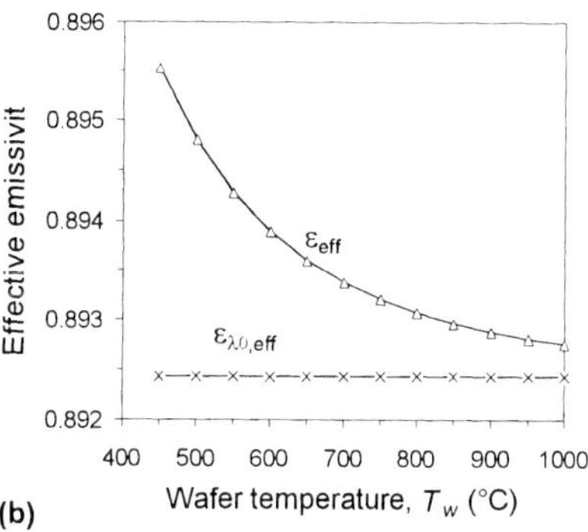

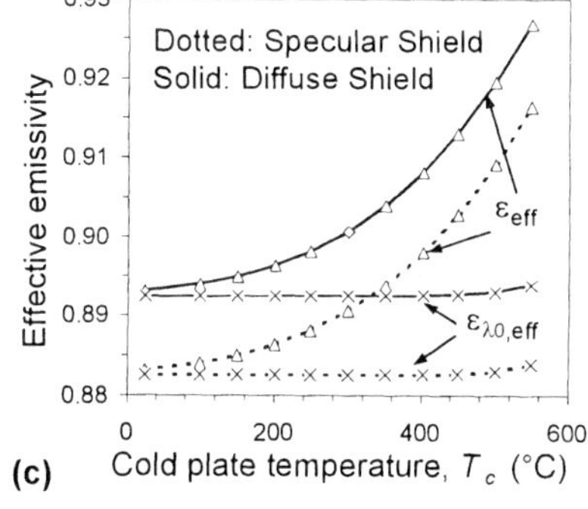

Figure 4 Variations of the effective emissivity with (a) ε_w and ε_c, (b) T_w, and (c) T_c

The variables that have the largest influence on the effective emissivity are the emissivities of the wafer and cold plate, as shown in Fig. 4a. Only ε_{eff} is shown in Fig. 4a, because $\varepsilon_{\lambda_0,eff}$ is too close ε_{eff} to be distinguishable in this graph. The wafer emissivity is not a variable that can be controlled in the day to day application of RTP, but emissivities of the cold plate, guard ring, and guard tube can be modified. By reducing the chamber (excluding the wafer) emissivity to 0.02, the effective emissivity for $\varepsilon_w > 0.60$ will be greater than 0.98. This will reduce the temperature measurement error greatly, because a 0.05 uncertainty in ε_w only causes a 0.005 uncertainty in the effective emissivity. As can be seen from Figs. 4b and 4c, for this particular wavelength, the spectral radiometer would be less sensitive to the temperature change of the wafer or the cold plate.

Figure 4c also shows the results when the reflective shield (cold plate) is treated as a specular surface. The effective emissivity is about 0.01 lower with a specularly reflecting shield than with a diffusely reflecting shield. Note that for two infinite parallel plates, the effective emissivity is the same with either a diffuse or a specular shield. With the guard ring and guard tube, however, the effective emissivity is different in the two cases, suggesting that directional properties may have non-negligible influence on the effective emissivity in the actual RTP chamber.

The foregoing calculations do not consider the radiometer opening on the cold plate. To study this effect, it is assumed that the opening is a circular disk at the center of the cold plate, which has a temperature equal to T_c and an emissivity of one (as a blackbody). The net heat flux of the wafer calculated from Eq. [2] is plotted in Fig. 5 for different radiometer

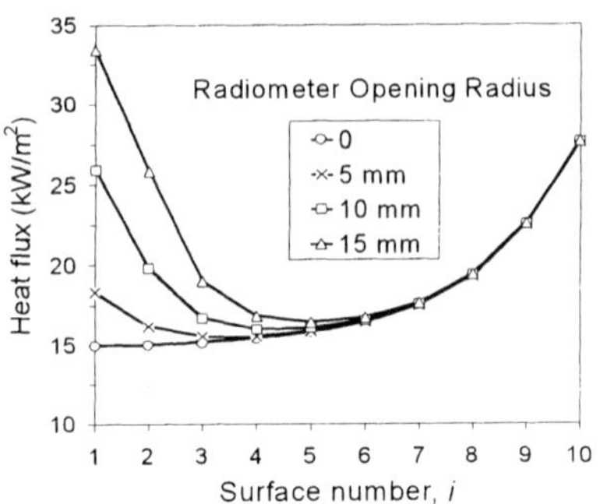

Figure 5 Effect of radiometer opening on the heat flux of the wafer

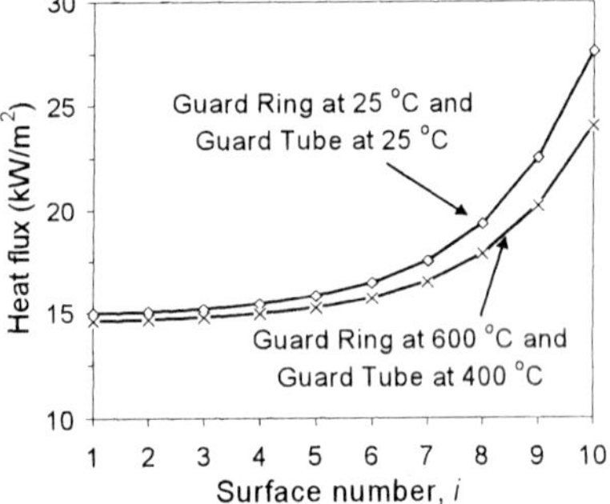

Figure 6 Effect of the temperatures of the guard ring and guard tube

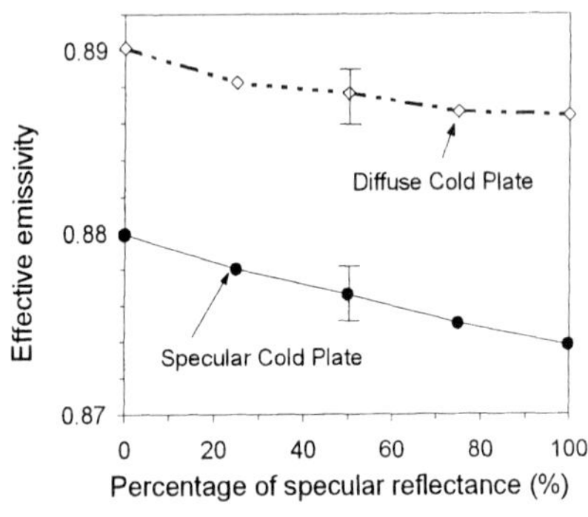

Figure 7 ε_{eff} versus the percentage of specular reflectance of the wafer

opening radii. The wafer is divided into ten surfaces of equal radial increments, with surface 1 being at the center and surface 10 being at the edge. Without the opening, the net heat flux increases with the radius because of the cold guard ring and guard tube. As the size of the opening is enlarged, the heat flux at the center increases significantly. The non-uniformity in heat flux at the bottom of the wafer may cause a temperature non-uniformity if the heat flux at the top of the wafer is not controlled accordingly. A decrease in temperature at the center of the wafer was observed in the NIST RTP test bed with a 25 mm diameter opening.

For the same wafer temperature, the radiosity will be smaller for larger net heat flux, see Eq. [2]. Therefore, the effective emissivity will decrease as the radiometer opening increases. The local effective emissivity will decrease towards the edge of the wafer due to the cold guard ring. For the base case, $\varepsilon_{eff} \approx 0.89$ if $q''_{net} = 15$ kW/m^2, $\varepsilon_{eff} \approx 0.82$ if $q''_{net} = 25$ kW/m^2, and $\varepsilon_{eff} \approx 0.75$ if $q''_{net} = 35$ kW/m^2.

Figure 6 shows the effect of the temperatures of the guard ring and guard tube on the heat flux distribution. By increasing the temperature of the guard ring to 600 °C and the guard tube to 400 °C, the heat flux at the edge can be greatly reduced.

Being a statistical method, the Monte Carlo model does not give exactly the same result from run to run. Therefore, the average effective emissivity for all runs is calculated and the standard deviation is used to test the accuracy of this method. It is found that when the total emitted number of bundles for each run is 15 million, the standard deviation in the effective emissivity for ten runs is approximately 0.0015. It takes about 7 min on a 400 MHz personal computer for each run with 15 million emitted bundles. The Monte Carlo model is verified by comparison with the net-radiation method, in which it is assumed that $r_c = r_w = 100$ mm (i.e., without the guard ring) and that the guard tube is a zero K blackbody. The results show that the Monte Carlo model agrees with the net-radiation method within the standard deviation (10). Figure 7 presents the effective emissivity obtained from the Monte Carlo model. The cold plate reflects either diffusely or specularly, whereas the emission from the cold plate is always diffuse. The emissivity of the silicon wafer is 0.65, while its reflectance consists of a diffuse component and a specular component. The effective emissivity decreases as the percentage of specular reflectance changes from 0 to 100 (i.e., its reflectance varies from completely diffuse to completely specular). The Monte Carlo method is capable of modeling properties that are wavelength, direction, and/or temperature dependent.

CONCLUDING REMARKS

A detailed radiative heat transfer analysis is performed for the lower chamber of the NIST RTP test bed to study the influence of various parameters on the effective emissivity of the chamber and the net heat flux distribution on the wafer. Even though all surfaces in an enclosure may be gray, the spectral effective emissivity is in general wavelength dependent. To simulate a blackbody cavity, the emissivity of the cold plate and the size of the radiometer opening must be made as small as possible. Furthermore, a diffusely reflecting shield (over the cold plate) is preferred to a specularly reflecting shield. The cold guard ring and guard tube may cause a large edge effect and reduce the effective emissivity. Increasing their temperatures can improve the wafer temperature uniformity. Further study on the effect of the emissivities of the guard ring and guard tube should be performed. The impact of the wafer temperature non-uniformity on the effective emissivity should also be investigated.

A simplified Monte Carlo model is built to consider the directional dependence of the radiative properties. The influence of the percentage of specular reflectance of the wafer is studied using this model. In the future, this model needs to be expanded to include the actual geometry, temperature, and material properties including the measured bidirectional reflectance distribution function (BRDF) of silicon wafers.

The models developed in this work include the temperature and/or heat flux non-uniformity, wavelength dependence, and diffuse or specular surfaces. They can be applied to other RTP systems as well. Studies such as this aid in understanding the causes of radiometric temperature measurement uncertainty and wafer temperature non-uniformity in RTP furnaces. They will also help researchers improve the design of future chambers.

ACKNOWLEDGMENTS

The support of the NIST Office of Microelectronics Program is much appreciated. Z.M.Z. is grateful for the support of the National Science Foundation through the CAREER Program.

REFERENCES

1. F. Roozeboom, in *MRS Symp. Proc.*, **224**, 9 (1991).
2. F. Roozeboom (ed.), *Advances in Thermal and Integrated Processing*, Kluwer Academic Publishers, The Netherlands (1996).
3. P.J. Timans, in *RTP '96*, 145 (1996).
4. P. Vandenabeele, *Rapid Thermal Processing: Study of Temperature Nonuniformity and Temperature Measurement*, Catholic University of Leuven Press, Belgium (1994).
5. Z.M. Zhang, in *Annual Review of Heat Transfer*, **XI** (1999) (in press).
6. D.P. DeWitt and G.D. Nutter (eds.), *Theory and Practice of Radiation Thermometry*, Ch. 10, John Wiley and Sons, New York (1988).
7. K.G. Kreider, D.P. DeWitt, B.K. Tsai, F.J. Lovas, and D.W. Allen, in *MRS Symp. Proc.*, **525**, 87 (1998).
8. D.P. DeWitt, F.Y. Sorrel, and J.K.Elliott, in *MRS Symp. Proc.*, **470**, 3 (1997).
9. R. Siegel and J.R. Howell, *Thermal Radiation Heat Transfer*, Chs. 6-9, Hemisphere Publishing Corporation, Washington DC (1992).
10. F. Rosa, M.S. Thesis, University of Florida, Florida (1999).
11. M.H. Kalos and P.A. Whitlock, *Monte Carlo Methods Volume I: Basics*, John Wiley and Sons, New York (1986).
12. W.J. Yang, H. Taniguchi, and K. Kudo, in *Adv. Heat Transfer*, Vol. 7 (1995).
13. J.R. Howell, *J. Heat Transfer*, **120**, 547 (1998).

EMISSIVITY OF BARE AND COATED Si WAFERS: THEORETICAL STUDIES

Bhushan Sopori, Wei Chen, Yi Zhang and Jamal Madjdpour, and N. M. Ravindra*
National Renewable Energy Laboratory, 1617 Cole Boulevard, Golden, CO 80401
* New Jersey Institute of Technology, Newark, NJ 07102

ABSTRACT

A software package, based on a generalized optical model, is developed to determine the optical parameters such as reflectance, transmittance, and emissivity of a semiconductor wafer. The wafer can have multilayer structures with polished, rough, or textured interfaces. The software takes into account the dopant and temperature dependencies of n and k. The results of the spectral emissivity for bare and coated Si for different surface morphologies are presented

INTRODUCTION

The emissivity, ε, of a Si wafer is required for a number of applications, including pyrometric measurements and RTP modeling. Because ε depends on a variety of wafer parameters such as thickness, surface characteristics, temperature, and the doping type and concentration, the experimental data for ε can be quite tedious to generate and is often not available. Hence, it is desirable to have a theoretical model that can accurately predict the behavior of ε. Some earlier work has successfully dealt with the calculation of the spectral emissivity (ε) of a bare, double-sided polished (DSP) Si wafer (1-2). However, in many applications, the wafer may not be DSP or bare, and the surface characteristics may change during the process. For example, during RTP oxidation or nitridation, the ε can vary with the thickness of the coating. In other cases, the wafer may have rough or textured surfaces. Indeed, the latter is true in the fabrication of photovoltaic devices where the wafers can have rough surfaces and antireflection coatings. Thus, for practical applications, the ε-modeling requires a detailed optical approach that can handle polished, rough and textured surfaces, and the surface with dielectrics, epitaxial, and other semiconductor layers. In newer applications of RTP, such as used in metal silicidation, it is also necessary to include heavily absorbing layers.

NREL EMISSIVITY MODEL

We have developed an optical model and software for calculating the ε for Si wafers with any surface characteristics—polished, rough, textured, including dielectric and metal coatings (3). Because modeling ε requires the values of refractive index (n) and the extinction coefficient (k) to be known as a function of temperature and dopant concentration, the emissivity model must include a methodology for calculating these parameters. Hence, our software involves optical and electronic modules. The optical module calculates the total absorption in the wafer while the electronic module supplies values of n and k needed for the calculations.

Our optical model uses a combination of ray and wave optics to address thick and thin regions of the wafer. A description of the optical modeling approach is given elsewhere (4). However, it should be emphasized that our model identifies the optical path of light propagating through the wafer, including multi-reflections within the wafer, and determines the net absorption. Thus, the model calculates absorption (and ε) of the sample independently, and not simply as [1-(T+R)]. The values of optical constants, n and k, are calculated for the desired wavelength range for the known dopant concentration and temperature of the wafer through a series of empirical and exact formulas(2, 5-10). Particularly the calculation of k involves a host of contributions due to direct and indirect band-to-band transitions, and free carrier absorption through a plasma model. This methodology will be discussed in detail in a forth-coming publication (11).

RESULTS AND DISCUSSION

Because the calculations for DSP-bare wafers are well established, the main objective of this paper is to address the emissivity of commercial wafers that may have a dielectric coating and may have rough surfaces. We will, however, use DSP wafers as an example to compare our calculated results with experimental results, because their **ε**-behavior is well established. This comparison is aimed at establishing the accuracy of n and k calculations. Furthermore, we will include results on the transmittance (T) and the reflectance (R) because a valuable insight into the mechanisms leading to changes in the emissivity can be gained through their behavior. The first example we consider is the emissivity of DSP samples of different resistivity. Because it is difficult to find the results of well-characterized samples, we have attempted to acquire this information through measurements at room temperature. We used commercial SSP, boron-doped samples of different dopant concentrations. Their resistivity was measured with a scanning four-point probe system. The rough surfaces of the samples were then chemically-mechanically polished using standard procedures to ensure a low-damaged surface. The transmittance and the reflectance spectra were measured at room temperature with two spectrometers — Cary 5G, from 0.4 μm to 2 μm, and in a Nicholet 800 FTIR, in the wavelength range of 2 μm to 20 μm. Figures 1a, 1b, and 1c compare theoretical and experimental values of the reflectance, transmittance, and emissivity, as a function of wavelength, for two wafers with dopant concentrations of 10^{17} and 8×10^{14} cm^{-3}. The solid and dotted lines correspond to measured and calculated data, respectively. There is a step in the measured reflectance plot at 2-μm wavelength, which appears to be due to the differences in the reflectance standards used in the two spectrometers.

These figures show an excellent agreement between the measured and calculated data. One may note the presence of the absorption bands due to oxygen appearing around 9 μm in the experimental data. These are absent in our calculations because the impurity absorption is not considered here. We also see that, for higher resistivity, the reflectance is nearly independent of wavelength. As the resistivity is reduced, the reflectivity decreases with λ due to free carrier effects— suppression of n and increased k. However, the reflectance due to front surface (i.e. at short λ's) does not change significantly; the major change in the reflectivity is due to reduced contribution from the backside. Because the free carrier absorption increases with a decrease in resistivity, the transmittance of a lower resistivity material decreases more rapidly with an increase in λ. Thus, the emissivity dependence seen in Figure 1c is as expected, and the calculated values are in a good agreement with the experiment. We also note that the ε of the low-doped wafer drops from about 0.7, at a wavelength of 1 μm, to nearly zero for longer wavelengths (except in the impurity absorption bands) because of nearly negligible free carrier absorption. The ε is higher for higher-doped wafer and exhibits a typical free-carrier absorption behavior.

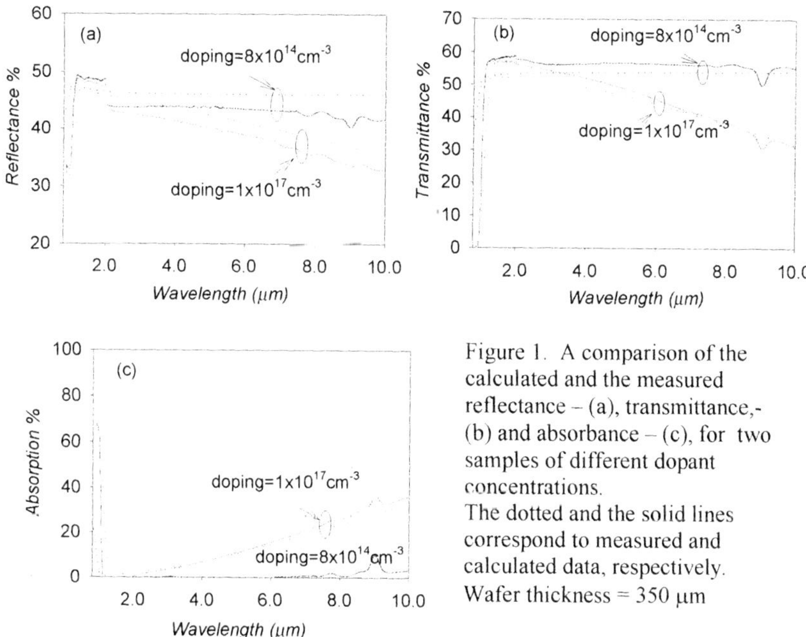

Figure 1. A comparison of the calculated and the measured reflectance – (a), transmittance,- (b) and absorbance – (c), for two samples of different dopant concentrations.
The dotted and the solid lines correspond to measured and calculated data, respectively.
Wafer thickness = 350 μm

In order to maintain the brevity of this paper, a detailed discussion of the temperature dependence is deferred to later papers. However, we have previously shown that the calculated results for the emissivity of a low-doped, DSP-Si wafer are in excellent agreement with Sato's results(1) at higher temperature(4). The excellent agreement between experimental and calculated data indicates an accurate modeling of n and k.

The second example, we apply our theory to examine the effect of a dielectric coating on a DSP wafer. Figure 2 shows the calculated values of R, T, and ε for a 300-μm thick wafer, coated on one side with a 1-μm thick SiO_2 layer. These calculations are performed for 1100 K looking from the SiO_2 side. As expected, this figure shows the presence of interference fringes in the reflectance spectrum. The transmission is limited to a narrow wavelength region near the band-edge of Si because of the temperature-related high absorption for the longer wavelengths. Thus, the emissivity spectrum also exhibits interference effect, but the maxima and minima in the R- and ε-spectrum are inverse of each other. Figure 3 shows results of similar calculations looking from the Si side. It can be noted that the modulation of ε due to SiO_2 coating is almost unobservable in Figure 3. This is because very little light is reflected from the back surface because of high absorption in the wafer. Concomitantly, the interference contribution from the back SiO_2 is negligible, and the reflectance is primarily due to the front surface. Hence, the ε is nearly constant as a function of wavelength, with a value of about 0.7. This example also shows that ε depends strongly on the characteristics of the surface of a given wafer and can be significantly different by changing the surface treatment.

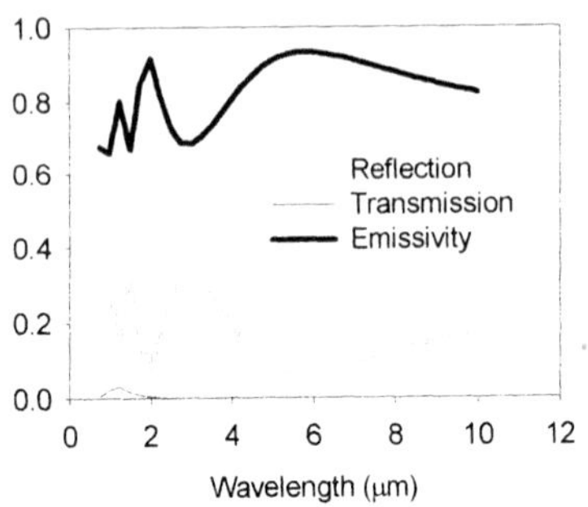

Figure 2. Calculated R, T and ε of a 300 -μm DSP-wafer with a 1-μm SiO_2 layer, at 1100 K. Dopant concentration = $3\times10^{15}cm^{-3}$. The incident side is the coated side

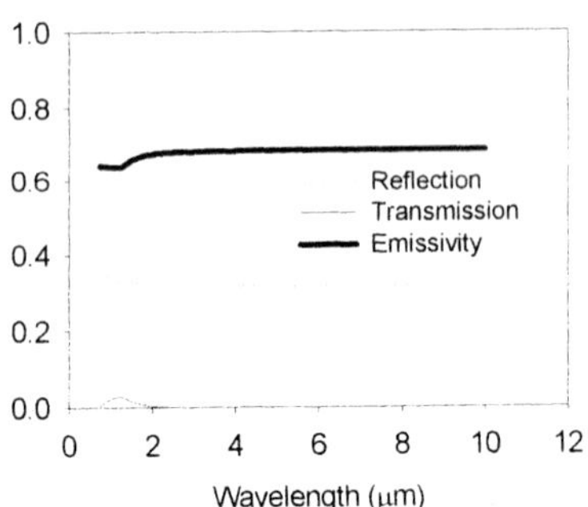

Figure 3. Calculated R, T and ε of a 300 -μm DSP-wafer with a 1-μm SiO_2 layer, at 1100 K. Dopant concentration = $3\times10^{15}cm^{-3}$. The incident side is the bare side

The third example considers the effect of texturing on emissivity. Texturing is extensively used in solar cells to produce surfaces of low optical reflectance in the useable solar spectrum. An interesting case is a wafer that is textured on the front and polished on the back-side. Figure 4 shows the calculated R, T, and ε for such a wafer. The thickness of the wafer is 300 μm, its dopant concentration is 10^{16}, and the temperature is 300 K. It is seen that R is high, near 0.8, while T is low in spite of the low reflectance (R ~0.1) at the front surface. This behavior can be attributed to the fact that most of the light that reaches the back surface on its first pass is totally reflected. Some of the light is transmitted from the backside as a result of subsequent passes due to multi-reflection. The back surface has an effective R of about 0.9. These results can be

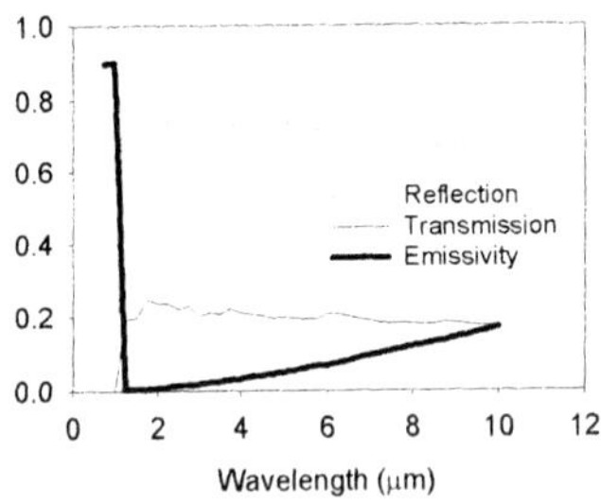

Figure 4. Calculated R, T and ε of a front textured/back polished, 300-μm wafer. Dopant concentration = 10^{16} cm^{-3}. Temperature = 300K

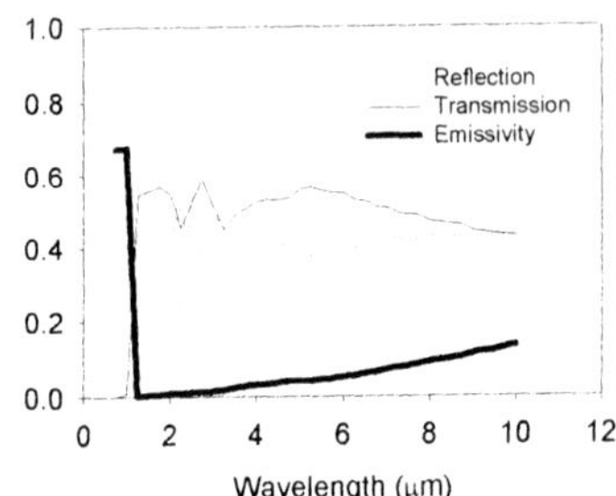

Figure 5. Calculated R, T and ε of a front polished/back textured, 300-μm wafer. Dopant concentration = 10^{16} cm^{-3} Temperature = 300K

compared with those in Figure 5, which shows calculated results for a front polished and back textured wafer. It is seen that the reflectance is lower and T is higher than in Figure 4—a result of R=0.3 at the front and 0.1 at the back surface. However, because of the low dopant concentration in the wafer, the absorption is low. The emissivity is about the same for both cases.

If the temperature of the wafer is raised, the front and the backside start to become optically decoupled, leading to higher values of ε. The calculated results are shown in Figures 6 and 7, for a front textured/back polished wafer of Figure 4, using wafer temperatures of 800 K and 1100K, respectively. These figures also show the characteristic low reflectance of a textured surface that yields ε–values close to 0.9. It should be emphasized that texturing should not be confused with the roughness observed on commercial semiconductor Si wafers. For example, texturing of (100) wafers yields a pyramidal structure, whereas typical rough wafers have flat tops and valleys. These differences are discussed in a previous paper (4). This is why such high ε–values are not typically observed when measurements are made for the rough side of such a SSP wafer. To elucidate this, we have performed calculations for a commercial wafer, with front-rough and back-polished surfaces. The roughness profile of a typical SSP wafer was measured and used in our calculations. These results are shown in Figures 8 and 9 for two different temperatures – 300 K and 1100 K, respectively. The dopant concentration of the wafer is 10^{16} cm^{-3}. In these cases, the ε reaches about 0.75, which is in agreement with the experimental results.

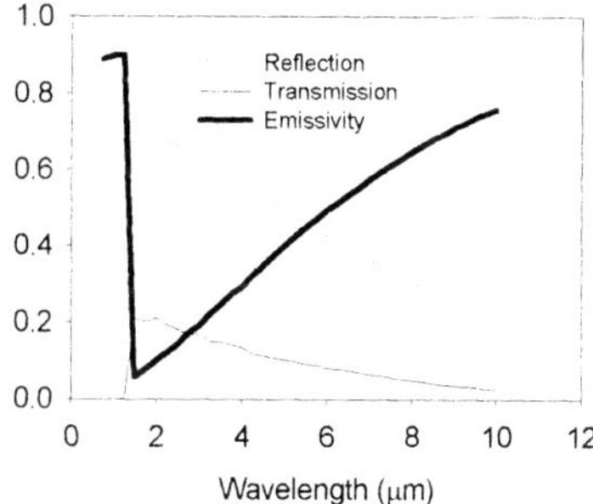

Figure 6. Calculated R, T and ε of a front textured/back polished, 300-μm wafer. Dopant concentration = 10^{16} cm^{-3}. Temperature = 800 K

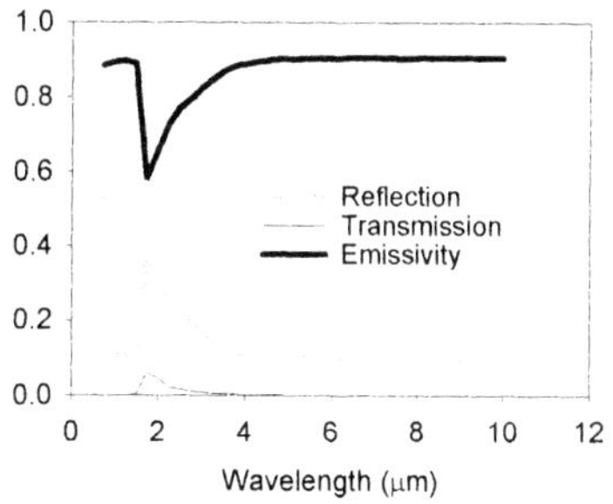

Figure 7. Calculated R, T and ε of a front textured/back polished, 300-μm wafer. Dopant concentration = 10^{16} cm^{-3}. Temperature = 1100 K

In the above discussions, we have seen that the back reflection plays an important role in determining the emissivity of a thin wafer. The effect of the backside reflectance can be calculated in a straightforward manner for a DSP wafer, but sophisticated software is needed when the morphologies of the wafer surfaces deviate from the polished case. However, in many situations it suffices to have a qualitative understanding of how the emissivity can change with wafer parameters. In such cases, one can acquire a good

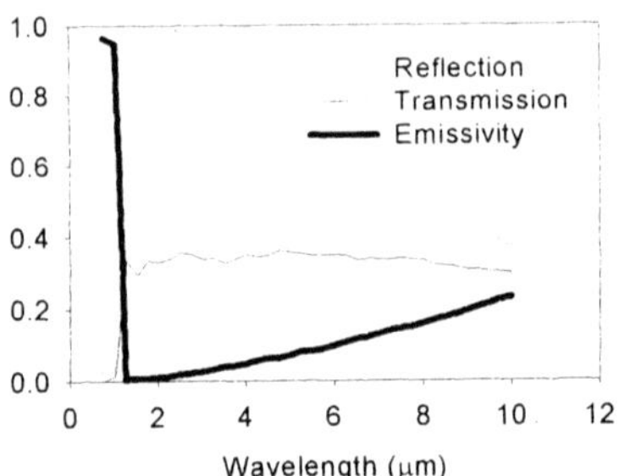

Figure 8. Calculated R, T and ε of a front commercial textured/back polished, 300-μm wafer.
Dopant concentration = 10^{16} cm^{-3}.
Temperature = 300 K

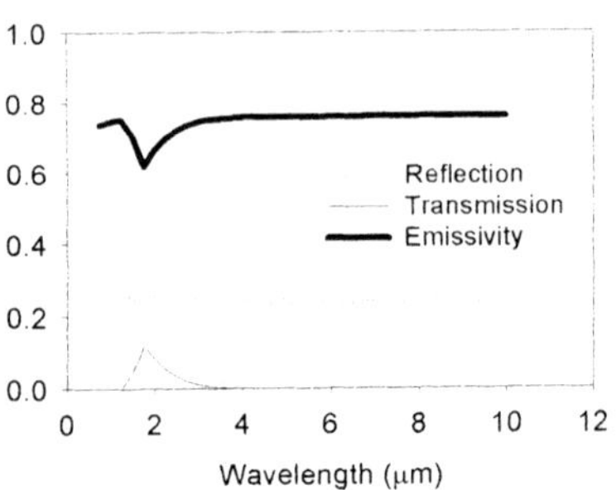

Figure 9. Calculated R, T and ε of a front polished/back commercial textured, 300-μm wafer.
Dopant concentration = 10^{16} cm^{-3}.
Temperature = 1100 K

understanding by assigning effective reflection coefficients to the front and back-surfaces, and simply examining the net result of front and the back reflections (keeping in mind the absorption within the wafer). As an example, let us examine the effect of temperature on the ε of the front-textured and back-polished wafer of Figure 4. Here, the front reflection coefficient is about 0.1, and the back reflection coefficient is about 0.9. If the wafer is low-doped, the bulk absorption is low at the room temperature, resulting in a net reflectance of about 0.81. However, as the temperature is increased the back reflectance effect will diminish because of increased bulk absorption. At temperatures above 800K, the back reflectance will be quite small, yielding an ε-value of about 0.9. Likewise, if the front surface has commercial roughness, the front R-value increases to about 0.25, yielding an ε-value of about 0.75 at high temperatures.

CONCLUSIONS

We have briefly described our *emissivity* software that includes electronic and optical modules to calculate (n and k), and (R, T, and ε), respectively. The results for the DSP wafers agree very well with the experimental data, giving confidence in the n and k calculations. A number of examples were discussed to show the applications of the software to rough and textured wafers, for different doping and for different operating temperatures.

ACKNOWLEDGEMENT

This work was supported by US Department of Energy under Contract #DE-AC36-98-G010337. The authors would like to thank John Webb of NREL for his help with FTIR measurements.

REFERENCES

1. T. Sato, Jap. J. Appl. Phys., **6**, 339 (1967)
2. P. Vandenabeele and K. Maex, J. Appl. Phys., **72**, 5867 (1992).
3. B. L. Sopori, Laser Focus, **34**, 159 (Feb. 1998)
4. B. L. Sopori, W. Chen, S. Abedrabbo and N. M. Ravindra, J. Elec. Mat., **97**, 1341 (1998)
5. Handbook of Optical Constants of Solids, Edited by E. D. Palic, Academic Press, 547 (1985).
6. N.M. Ravindra, W. Chen, F. M. Tong, and A. Nanda, Transient Thermal Processing in Electronic materials, Edited by N. M. Ravindra and R. K. Singh, 159)(TMS, ,PA, 1996)
7. H. H. Li, J. Phys. Chem. Ref. Data 9, 561, (1980)
8. G. E. Jellison and F. A. Modine, J. Appl. Phys., vol. 76, No. 6, 3758 (1994)
9. G. E. Jellison and D. H. Lowndes, Appl. Phys. Lett. , vol 41, no.7, 594 (1982)
10. T. S. Moss, Optical Properties of Semiconductors, Butterworths Scientific, 1959):
11. B. L. Sopori, W. Chen, S. Abedrabbo and N. M. Ravindra, J. Elec. Mat., in press.

An Advanced Radiation Model for Thermal Processing of Wafers

Sandip Mazumder[1]
CFD Research Corporation, Huntsville, AL 35805, USA
and
Alfred Kersch
Infineon AG, Munich, D-81730, Germany

Thermal Radiation is the most effective heating process for Rapid Thermal Processing (RTP) and Rapid Thermal Chemical Vapor Deposition (RTCVD) of wafers. The Monte-Carlo (MC) method for radiation is the most accurate method for modeling radiative transport in RTP and RTCVD reactors. This article presents a fast Monte-Carlo scheme for thermal radiative transport, which was developed within the framework of CFD Research Corporation's commercial block-structured flow code, CFD-ACE. The importance of high spectral resolution and temperature dependent radiative properties have been highlighted here in the context of RTP simulations.

INTRODUCTION

Radiative heating is widely used in the semiconductor industry for thermal processing of wafers. In such processes, it is necessary to heat the wafer to a desired temperature. Typically, high power lamps provide radiative energy. The relationship between lamp power and wafer temperature is extremely complex. It is strongly dependent on the lamp orientations, the optical behavior (spectral and reflective characteristics, both of which are often temperature dependent) of the various surfaces and objects within the RTP or RTCVD reactor and finally, on the pressure and flow rates of the reactants [1, 2]. Since the number of factors affecting this relationship is so large, it is expensive to tune lamp powers by trial and error methods. It is now well-known in the semiconductor processing industry that the design cost can be reduced drastically if effective modeling techniques are used to numerically predict the temperature field. Since radiation is the dominant mode of heat transfer in such systems, it so happens that in order to predict meaningful temperature distributions, a high level of accuracy of the radiation model is necessary [3]. First, a very accurate nongray description is necessary. Secondly, the model should be able to treat specular and partially specular surfaces. Thirdly, it should be able to treat transport through stacks of semitransparent solids. Finally, during processing of wafers, thin films are often deposited on a substrate, and these films drastically change the optical behavior of the surface, often causing interference and/or diffraction. The radiation model should be able to model the effects of these thin films.

The one method that can successfully account for all of the above physical effects is, undoubtedly, the Monte-Carlo method. This article discusses a Monte-Carlo radiation module, which was developed as an add-on module to a commercial CFD code (CFD-ACE). The unified code can be used to produce extremely accurate results

[1]Author for correspondence; E-mail: sm@cfdrc.com

for RTP and RTCVD applications, as has been demonstrated here for rapid thermal processing of a silicon wafer for various doping levels.

TECHNICAL APPROACH

In the absence of a participative medium, the net radiative heat flux on any patch i, is a net sum of the incident radiation from all other patches and self-emission. Mathematically it is written as:

$$Q_i = q_i A_i = \sum_{j=1}^{N_s} (R_{ij} - \delta_{ij} \epsilon_j) \sigma T_j^{\,4} A_j \,, \tag{1}$$

where N_s are the total number of radiatively active patches within the domain. ϵ_j, T_j and A_j are the spectrally integrated emissivity (henceforth, referred to as the 'effective emissivity'), temperature and area, respectively, of patch j. Q_i is the radiative flux and q_i is the radiative flux density on patch i. δ_{ij} is the Kronecker delta. R_{ij} is the normalized radiative exchange matrix, and represents the fraction of radiation that is emitted by patch j and absorbed by patch i.

A packet of energy (commonly referred to as a 'ray'), when emitted from a patch j undergoes several events before it is actually absorbed by another surface. These events may include reflection (diffuse, specular, or partially specular), refraction and total internal reflection. Each of these events will depend on the direction of propagation of the ray and its frequency, and the optical characteristics and geometric location and orientation of all the intermediate patches. Mathematically, this reduces to the solution of a multidimensional integral. In general, it is difficult to perform this integral analytically if specular and nongray effects are present. An alternative is to perform the integral stochastically, by drawing samples from independent stochastic spaces, and finally performing a discrete summation. With large number of samples, the discrete sum tends towards the continuous integral.

Determination of Radiative Properties

The fundamental property that governs the optical behavior of a material is its complex index of refraction, expressed as $r = n - ik$, where n and k are the real and imaginary parts of the refractive index, respectively. In general, these are temperature dependent, and this dependence can be described either in tabular form or by curve-fits. For most materials prevalent in semiconductor applications, curve fits are available, and therefore, this approach was adopted here. The most comprehensive source is the handbook by Palik [4]. The curve-fits are expressed as follows:

$$\begin{aligned} n_i &= n_i^{(0)} + n_i^{(1)}\,\Theta + n_i^{(2)}\,\Theta^2 + n_i^{(3)}\,\Theta^3 \,, \\ k_i &= \exp(k_i^{(0)}\,\Theta)\,[k_i^{(1)} + k_i^{(2)}\,\Theta + k_i^{(3)}\,\Theta^2] \,, \end{aligned} \tag{2}$$

where Θ, the non-dimensional temperature is expressed as $\Theta = (T - 300)/1000$, where T is the absolute temperature. The spectrum was subdivided into sixty nominal spectral intervals with twenty spectral intervals in each decade of wavenumber (this was not a matter of choice but, rather, prompted by the desire to model radiative phenomena between 300 K and 6000 K). The refractive index data available in Palik [4] and elsewhere were interpolated carefully to provide values for these sixty spectral intervals. The subscript 'i' indicates the spectral interval in question. The eight curve-fit coefficients in equation (3) were stored in a database for various materials.

Once the complex index of refraction has been calculated, the Fresnel complex amplitude reflection and transmission coefficients for the P and S polarized components can be calculated using the Fresnel formulae and Snell's law. For details the reader is referred to the text by Azzam and Bashara [5]. The emittance, reflectance and transmittance were finally computed by invoking the McMahon approximation [3] and using a matrix multiplication approach, discussed in Section 5.6 of Azzam and Bashara. This method was preferred over other methods on account of its generality, and the fact that multi-layer coatings can be treated elegantly by introducing layer matrices in the final matrix multiplication.

The presence of thin films often lead to interference. To account for this effect, a phase averaging technique was implemented. This is necessary in two cases: the first case is one where the coherence length is smaller than the layer thickness, in which case the Fresnel formulae are invalid. The reflectivity and transmissivity (which are the result of the Fresnel formula) have to be substituted by values averaged over the phase related to the layer in question. Here, the averaging was performed using a six-point Gaussian quadrature. The second case is when the phase difference of the radiation between two interfaces becomes very large. From a numerical point of view, phase averaging is then desirable because of very large exponents. The choice of the maximum allowed phase difference is, of course, somewhat arbitrary.

RESULTS

The starting point of development of this code was CFD Research Corporation's commercial flow solver, CFD-ACE. CFD-ACE is a code which solves for the conservation of mass, momentum, energy and species in multi-domain block-structured grids for any arbitrary geometry. As a first step, the Monte Carlo method, described above, was developed as a stand-alone module, and validated extensively. It was then coupled with the overall energy equation for all possible boundary condition types to obtain temperature fields. The fundamental conservation equations and their numerical solution technique may be obtained from the CFD-ACE Theory Manual [6].

Simulations were performed, in which a Silicon wafer was heated rapidly by radiation to 1000K. The geometry of the projection plane of the axisymmetric reactor chosen for the simulations (the Jipelec Reactor) is shown in Fig.1. All walls are diffuse reflectors. The side walls and the inlet have gray emissivities of 0.15, while the two outlets were considered black. The power supplied by the lamp heating zone (made of tungsten) was simulated by prescribing a constant temperature boundary condition on the surface. The water layer was removed to better simulate RTP conditions. All side walls have isothermal boundary conditions with a temperature of 300K. The temperature of the lamp heating zone needed to be adjusted by trial and error so as to produce an average ramp-rate of 100 K/s for all simulations. The radiative properties of all semitransparent solids (quartz, and silicon) were calculated from first principles, as discussed earlier. The reactor pressure was 10 Torr (1333 Pa). The simulation was

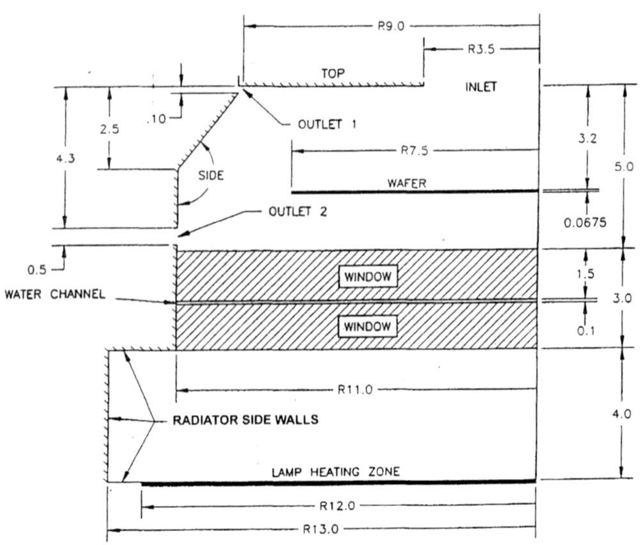

Figure 1: Reactor geometry

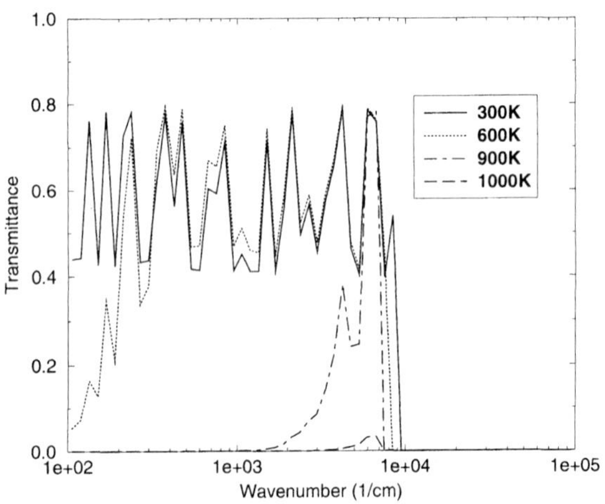

Figure 2: Temperature dependence of transmittance of a 0.675 mm undoped silicon wafer

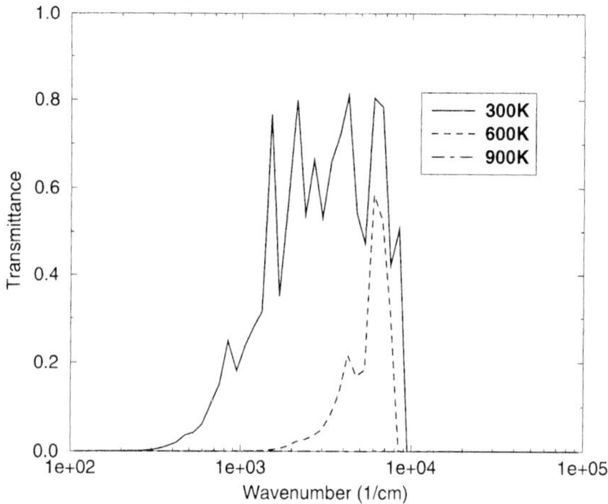

Figure 3: Temperature dependence of transmittance of a 0.675 mm medium doped silicon wafer

run with 1 million rays and the Quasi Monte Carlo Method [7,3]. Three RTP simulations were run: in the first simulation, the wafer was undoped silicon, in the second it was medium doped silicon, and in the third, it was heavily doped silicon. The transmittances of the undoped and medium doped silicon wafers are shown in Figs. 2 and 3, respectively. Heavily doped silicon is opaque at all temperatures and wavelengths and has an emittance of 0.7. The silicon data used here are modified forms of the data provided by Sturm [8]. Sturm's data was validated only for wavelengths close to the bandgap and no proof of its accuracy is available for the infrared wavelengths. Therefore, this data was adjusted in the infrared to match the model predictions of Sato [9].

The transient temperature evolution of the wafer, in each case, has been illustrated in Fig.4. The edge to center nonuniformity of temperature during the RTP process is of great interest in such processes, and this has been shown in Fig.5. It is clear that with increase in doping, the temperature uniformity of the wafer is enhanced, as has been observed experimentally at the beginning of the prestabilization phase in a RTP process. Such a behavior stems from the temperature dependent spectral nature of silicon's radiative properties, and it was possible to simulate this behavior with the current 'first principles' approach for the calculation of radiative properties and the high spectral accuracy of the computation.

CONCLUSIONS

Results were presented for Rapid Thermal Processing of a Silicon wafer, with and without doping. The validity and feasibility of the use of a Monte Carlo radiation model for commercial semiconductor processing applications has been demonstrated. The self-consistent calculation of radiative properties from fundamental optics is imperative to accurately describe strong nongray and, often, temperature dependent

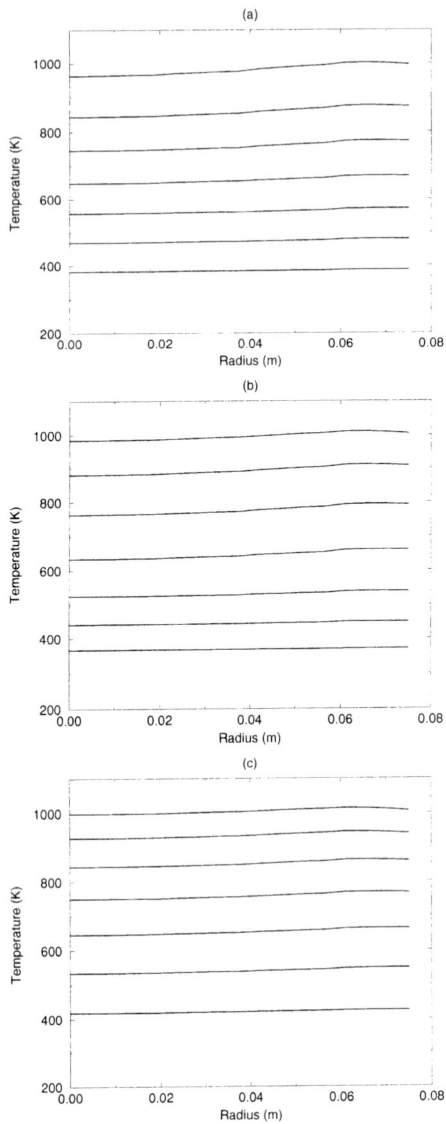

Figure 4: Temperature profiles during RTP of a Silicon Wafer: (a) Undoped wafer, (b) Medium-doped wafer, (c) Heavily-doped wafer. The curves are at intervals of 1 second from bottom to top.

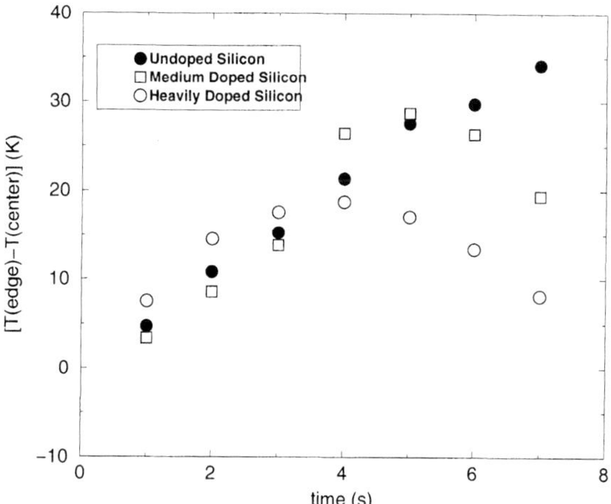

Figure 5: Center to Edge Temperature Nonuniformity of a silicon wafer during a RTP process with various dopant levels

behavior (consistent with reality) of most materials occurring in semiconductor processing applications. It was demonstrated that these effects play a big role in rapid thermal processing of wafers.

Acknowledgments

This work was supported in part by Applied Materials and in part by the NASA Langley Research Center. Their contributions are gratefully acknowledged.

References

1. Chatterjee, S., Trachtenberg, I., and Edgar, T., "Modeling of a Single Wafer Rapid Thermal Reactor," *J. Electrochem. Soc.*, Vol. 139, 1992, p. 3682.
2. Merchant, T., Lie, K., Cole, J., and Jensen, K., "Strategies for Modeling of Rapid Thermal Processing Systems," *RTP93*, 1993.
3. Kersch, A. and Morokoff, W., *Transport Simulation in Microelectronics*, Birkhauser, Basel, 1995.
4. Palik, E., *Handbook of Optical Constants of Solids*, Academic Press, New York, 1985.
5. Azzam, R. and Bashara, N., *Ellipsometry and Polarized Light*, Elsevier North-Holland Inc., New York, 1977.
6. CFD Research Corporation, 215 Wynn Drive, Huntsville, AL35805, *CFD-ACE: Theory Manual*, 4 ed., 1998.

7. Kersch, A. and Morokoff, W., "Radiative Heat Transfer with Quasi Monte Carlo Methods," *Proceedings SISDEP93, Wien*, 1993.

8. Sturm, J. and Reaves, C., "Silicon Temperature Measurement by Infrared Absorption: Fundamental Processes and Doping Effects," *IEEE Transactions on Electronic Devices*, Vol. 39, 1992, p. 81.

9. Sato, T., "Spectral Emissivity of Silicon," *Japan Journal of Applied Physics*, Vol. 6, 1967, p. 339.

Section VIII

Author Index and Key Word index

AUTHOR INDEX

A
N. Acharya, 399
A. Agarwal, 133
P.D. Agnello, 217
K.Z. Ahmed, 81
Z. Atzmon, 371

B
G. Bai, 39
G. Bailey, 229
A. Balakrishna, 3
K.S. Balakrishnan, 399
H. Balthasar, 383
I. Ban, 179
M. van de Berg, 171
J.G.M. van Berkum, 125, 187
W.B. de Boer, 309
A. van de Boogaard, 335
K.K. Bourdelle, 133
R. Bremensdorfer, 407

C
D.M. Camm, 133
M. Carroll, 319
C. Chen, 3, 249
L.J. Chen, 237, 257
W. Chen, 427
L.W. Cheng, 237
S.L. Cheng, 237
M.Y. Chern, 299
J.Y. Choi, 15
A. Claverie, 125
N.E.B. Cowern, 125
F. Cristiano, 125
F.N. Cubaynes, 187

D
J.H. Das, 15
D.P. DeWitt, 419
D.F. Downey, 141, 151

E
S. Everist, 89

F
S.W. Falk, 151
H. Fang, 207
S.B. Felch, 151
A.T. Fiory, 133
B. Froeschle, 31
J. Fulford, 45

G
M. Gardner, 45
A. Gat, 345
J. Gelpey, 45
M. van Gestel, 187
H. Gilboa, 371
A.E. Glazman, 371
E.D. Glazman, 371
F. Glowacki, 31
E.J.G. Goudena, 171
E.A.H. Granneman, 391

H
M. Hauf, 383
J.R. Hauser, 81, 95
J. Hebb, 375
R. Hegde, 23
H.H. Heinisch, 81
S. Hong, 249
S.T. Hsu, 291
Y.Z. Hu, 229
H.Y. Huang, 257
H.G.A. Huizing, 125

I
E. Iskevitch, 371

J
Y.B. Jia, 15
H.S. Joo, 3
S.-K. Joo, 285

K
A. Kersch, 435
B.Y. Kim, 45

C.-H. Kim, 285
G.-B. Kim, 285
T.-K. Kim, 285
T.-J. King, 277
H. Kitajima, 105
Z. Koren, 345
V.I. Kuznetsov, 391
D.-L. Kwong, 45

L
H. Lazar, 95
B.-I. Lee, 285
M.H. Lee, 299
M.E. Lefrancois, 133
W. Lerch, 141
W. Li, 95
C.W. Liu, 299
D. Lopes, 3, 249
H.F. Luan, 45
G. Lucovsky, 69

M
T.P. Ma, 57
J. Madjdpour, 427
Y. Maeda, 249
J.W.H. Maes, 327
M. Mahler, 95
R. Mallee, 171
G. Mannino, 125
M. Mansoori, 327
S.D. Marcus, 45, 141
T.Y. Matsuda, 105
S. Mazumder, 435
C. Merkl, 383
E. Merz, 407
K. Min, 81
A. Mineji, 105
G. Miner, 3, 23, 249
V. Misra, 95
M. Mulkarni, 95
S. Müller, 383

N
L.K. Nanver, 171, 335
J. Nelson, 89
W.D. van Noort, 335

O
P.A. O'Neil, 207
C.M. Osburn, 81, 197
M.C. Öztürk, 179, 207

P
V. Parihar, 163
C. Parker, 81
S. Paul, 407
C. Pomarede, 327
Y.V. Ponomarev, 187
K.F. Poole, 163

R
I.J. Raaijmakers, 327
S. Radelaar, 391
N.M. Ravindra, 427
K.G. Reid, 23
Q.W. Ren, 171
W. Renken, 359
A. Rohatgi, 265
F. Rosa, 419
F. Roozeboom, 125, 187

S
N. Sacher, 31
E. Sanchez, 3
M. Schäfer, 14
J. Schmitz, 187
J. Schuur, 15
E.G. Seebauer, 117, 207
A. Shajii, 375
R. Sharangpani, 15, 89
S. Shishiguchi, 105
S. Shooshtarian, 399
R. Singh, 163
A. Singhal, 249
J. Slabbekoorn, 171
J.W. Slotboom, 335
P.M. Smith, 89, 229
S.C. Song, 45
B. Sopori, 427
A. Srivastava, 81
P.A. Stolk, 125, 187
N.A. Stolwijk, 141
C. Striebel, 383
J.C. Sturm, 319

T
S.P. Tay, 89, 229
H. Takeuchi, 277
D. Terpstra, 309
B. Timberlake, 81
R.P.S. Thakur, 15, 89, 163, 229, 345, 399
A. Thon, 371
P.J. Timans, 345, 399
N.N. Toan, 125
B.K. Tsai, 419
H. Tseng, 23
Y.D. Tseng, 299
D. Tweet, 291

V
P. Vandenabeele, 359
S. Venkataraman, 163
C.C.G. Visser, 335
E.M. Vogel, 81

W
H. Walk, 407
Z. Wang, 81, 95
C.W. Werkhoven, 327
W.M. van de Wijgert, 187
P.H. Woerlee, 125, 187
J.J. Wortman, 81
D. Wristers, 45
W.F. Wu, 257

X
G. Xing, 3, 23

Y
M. Yamazaki, 249
J. Yang, 229
M. Yang, 319
R.P. Yang, 257
K.F. Yee, 81
Y. Yokota, 3
H.A. Yoon, 249
Y.-G. Yoon, 285

Z
P.C. Zalm, 187
Y. Zhang, 427
Z.M. Zhang, 419
Y.H. Zhou, 419

KEY WORD INDEX

A

arsenic
- / - implantation, 151
- / - surface passivation, 207

^{75}As, 151

B

^{11}B, 151
B^+ and BF_2^+ implanted oxide, 187
$^{49}BF_2$, 151
bond ionicity, 70
boron, 81, 133
- / - implants, 133
- / - penetration, 81

C

calibration, 359, 386
charge balance, 69
charge-to-breakdown, 31
chemical bonding/ constraints, 69
Cl_2, 179
cleaning, 31
cluster, 407
cluster tool, 407 / radial -, 407
cold-wall reactor, 309
conductive heat transfer, 391
contact, 171, 237
cool-down rates, 151
$CoSi_2$, 217, 237, 257
crystallinity, 327

D

damascene W, 217
DCS, 249
Deal-Grove equations, 89
defect etching, 299
defects, 125, 163
- / {113} -, 125
- / ion-implantation generated -, 125

dichlorosilane, 249
dielectric passivation, 265
diffusion, 179
direct tunneling, 69
doping, 151, 179
dynamic logic, 217

E

ELA, 171
elevated junctions, 171
elevated source/drain, 197
emissivity, 310, 383, 427
- / - compensation, 376, 383
- / effective -, 348, 376, 419

enhanced diffusion, 105, 151
excimer laser annealing, 171

F

floating wafer reactor, 391
Fourier transform infrared spectroscopy, 299
Frenkel-Poole conduction, 57
FTIR, 299

G

gate dielectrics, 23
gate leakage current, 81
gate oxide, 15, 23
gate stack, 39, 45, 57, 277
gate stack dielectrics, 81
grain growth, 257
growth kinetics, 89

H

heat transfer, 391, 419
- / conductive -, 391
- / radiative -,

heater development, 349
Heterojunction Bipolar Transistor, 309
high-k dielectrics, 39, 69

High-Temperature Sputtering Deposition, 257
HTSD, 257
hydrogen content in Si-nitride, 95

I

implanted oxide, 187
incandescent-lamp annealing, 133
interface Si/SiO_2, 23
interface traps, 95
ion implantation, 151, 237
- / ^{11}B, $^{49}BF_2$, ^{75}As, and ^{31}P, 151

J
Jet Vapor Deposition, 57
junction depth, 133, 187

K
kinetic analysis, 117

L
laser annealing, 171
loading effect, 310
local interconnect, 217
localized leakage, 217
low energy implantation, 105, 133, 151
low temperature measurement, 388
low-resistivity, 237, 249
low-temperature epitaxy, 309
low-temperature RTCVD, 319
LPCVD, 319, 327

M
metal contacts, 237
MIMO, 399
mobility, 291
modeling, 89, 125, 400, 419, 427
modulation, 383
MOSFET, 105
multiple input/output, 399
multiprocessing, 45

N
Ni-germanosilicides, 277
nitric oxide/NO, 31, 45, 89
nitridation, 31, 141
nitride, 45
nitrous oxide/N_2O, 3, 31, 45, 81
NMOSFETs, 95
NO-O_2 gas, 89

O
Ostwald ripening, 125
outdiffusion from implanted oxide, 187
oxidation ambients, 3, 15, 23, 31
oxidation enhanced diffusion, 105
oxidation, 391
oxidation, 407
oxynitride, 89

P
^{31}P, 151
phosphorus doping profiles, 319
PLAD, 151
plasma doping, 151
plasma etching, 359
PMOSFET, 291
PMOSFETs / 0.1 μm, 81, 95
polycide, 249
poly-metal structure, 277
poly-$Si_{1-x}Ge_x$ /polycide stack gate, 277
polysilicon, 285
/ in-situ doped -, 327
pyrometer, 385
pyrometry, 371, 375

R
radiation, 435
radiative heat transfer, 391, 419
ramp rate effects, 125, 151, 229
ramp rate, 229
ramp rate, 387
rapid thermal annealing, 141
Rapid Thermal Chemical Vapor Deposition 45,
-----81,179, 207, 435
rapid thermal nitration, 249
rapid thermal outdiffusion, 187
rapid thermal oxidation, 407
Rapid thermal processing, 105, 291, 345, 359,
-----371, 391, 407, 419, 435
rapid thermal silicidation, 229
rate selectivity, 117
reflective shield, 419
reliability, 60
remote plasma, 95
remote sensing, 410
ripple, 383
RTCVD, 45, 81, 179, 207, 371, 435
RTP, 345, 359, 371, 375, 391, 399

S
SALICIDE, 229
scaled CMOS, 69
scanning Rapid Thermal Annealing, 285
Secondary Ion Mass Spectroscopy, 31
segregation, 319
selective silicon, 179
selective, 207
self-interstitial injection, 179
semiconductor wafer, 391
sensors, 359
sequential cluster, 407
shallow junctions, 171, 187, 197

sheet resistance, 187
Si_2H_6, 179
Si-Ge, 291, 309
Si-Ge-C quantum wells, 299
SIMS, 31
SILC, 59
silicidation, 229
silicide, 229
/ - as a diffusion source, 197
silicided junctions, 197
silicon nitride deposition, 57
silicon-compatible oxides, 39
simulation, 345
$Si-Si_3N_4$, 95
software package, 427
solar cells, 265
source/drain engineering, 105
source/drain extension, 151
spike anneals, 117, 133, 151
steam oxygen, 3, 15
steam oxidation, 15, 23
strain relaxation, 299
sub-0.1 μm CMOS, 39
submicron devices, 197
surface morphologies, 427
surface passivation, 207
surface roughness, 327
system design, 355

T
Ta_2O_5, 69
TED, 125, 171
temperature
/ - control, 345, 380, 399
/ - measurement, 310, 345,347, 376, 383, 419
/ - uniformity, 379, 399

thermal stability, 257
thermal stress, 163
thin film transistor, 285
Ti-germanosilicides, 277
TiO_2, 69
TiO_2/ silicon nitride gate stack, 57
$TiSi_2$, 207, 237
transformation temperature, 229
transient enhanced diffusion, 125, 171
transistor characteristics, 60
transition metal oxides, 69
tungsten silicide, 249
tunneling, 58
tunneling current, 45

U
UHV-RTCVD, 179
ULSI technology, 229
Ultra-High Vacuum RTCVD, 179
ultra-low energy implantation, 105, 133
ultrashallow junctions, 105, 133, 141,
-----151, 179, 187, 237
ultra-thin dielectrics, 81
ultrathin gate oxide, 31

V
Vapor Phase Cleaning, 31

W
wafer temperature, 359, 383
WSi_x, 249

X
X-ray diffraction, 299

Z
zero-depth junctions, 179
$Zr(Hf)O_2-SiO_2$ alloys, 69
ZrO_2, 39